PROGRESS IN COLLOID & POLYMER SCIENCE
Editors: H.-G. Kilian (Ulm) and G. Lagaly (Kiel)

Volume 75 (1987)

Permanent and Transient Networks

Guest Editors: M. Pietralla and H.-G. Kilian (Ulm)

Springer-Verlag Berlin Heidelberg GmbH

ISBN 978-3-662-15680-3 ISBN 978-3-7985-1696-0 (eBook)
DOI 10.1007/978-3-7985-1696-0
ISSN 0340-255 X

Originally published by Dr. Dietrich Steinkopff Verlag GmbH & Co. KG, Darmstadt in 1987
Softcover reprint of the hardcover 1st edition 1987

Chemistry editor: Heidrun Sauer; Copy editing: Deborah Marston; Production: Holger Frey.

Preface to Gomadingen

In October 1986, a conference was held in the Swabian Alps and attended by some 60 distinguished scientists – the "Gomadingen Discussion Meeting on Networks".

The intention behind this assembly was to discuss, by means of invited lectures, the current problems still standing in the way of a consistent understanding of permanent and transient networks.

Scientists were deliberately invited as representatives of very different schools of thought. This gave rise to the special structure of this Progress volume. The inclusion in the publication not only of prepared abstracts and contributions, but also the majority of the free discussions of them, is derived from the underlying aim of the meeting: to bring to light the contrasts and consensuses in our understanding.

H.-G. Kilian
M. Pietralla

Contents

Preface V

Permanent Networks

Rigbi Z: On the lack of experimental evidence for certain structures in polymers 1

Vilgis TA: Entanglements in rubbers 4

Ilavský M, Dušek K: Formation, structure and elasticity of model imperfect networks prepared by endlinking 11

Havránek A: The dependence of the retardation spectrum width on the network chain length and on its distribution 21

Pechhold W, Böhm M, v. Soden W: Meander model of polymer melts and networks 23

Dubault A, Deloche B, Herz J: Deuterium magnetic resonance on elongated networks: orientational order versus network structure 45

Oppermann W, Rennar N: Stress-strain behaviour of model networks in uniaxial tension and compression 49

Enderle HF, Kilian H-G: General deformation modes of a van der Waals network 55

Böhm M, Grassl O, Pechhold W, v. Soden W: Dynamic shear compliance of swollen networks and its dependence on crosslink density 62

Godovsky YuK: Thermomechanics of polymer networks 70

Picot C: Polymer network structure as revealed by small angle neutron scattering 83

Transient Networks

Winter HH: Evolution of rheology during chemical gelation 104

Laun HM: Orientation of macromolecules and elastic deformations in polymer melts. Influence of molecular structure on the reptation of molecules 111

Laun HM: Nonlinear rheologcial properties of polymer melts and the prediction based on the relaxation time spectrum 136

Stadler R: Transient networks by hydrogen bond interactions in polybutadiene melts 140

Demarmels A, Meissner J: Multiaxial elongations of polyisobutylene and the predictions of several networks theories 146

Rigbi Z: Viscoelasticity and thermal equilibrium 149

Boué F, Bastide J, Buzier M, Collette C, Lapp A, Herz J: Dynamics of permanent and temporary networks: Small angle neutron scattering measurements and related remarks on the classical models of rubber deformation 152

Pakula T: A Model of cooperative motions in dense polymer systems by means of closed dynamic loops 171

Keller A, Müller AJ, Odell JA: Entanglements in semi-dilute solutions as revealed by elongational flow studies 179

Filled Networks

Vidal A, Donnet JB: Surface properties of fillers and interactions with elastomers 201

Kilian H-G: Filled van der Waals networks 213

Weymans G, Eisele U: Influence of the filler on the reinforcing mechanism of styrene-butadiene rubbers 231

Mergenthaler D, Pietralla M, Kilian H-G: Filler-matrix coupling in rubbers as revealed by heat conduction experiments 234

Zentel R: Interrelation between the orientation of the polymer chains and the mesogenic groups in crosslinked liquid crystalline polymers 239

Comment: Networks and Theory

Vilgis TA: Some principal problems in statistical mechanics of networks and their relationship to other topics in physics and materials science 243

Author Index 248

Subject Index 249

Progress in Colloid & Polymer Science Progr Colloid & Polymer Sci 75:1–3 (1987)

Permanent Networks

On the lack of experimental evidence for certain structures in polymers

Z. Rigbi

Technion – Israel Institute of Technology, Haifa, Israel

This paper is intended to be a brief survey of some of those structures which are postulated to exist in certain polymers and a discussion of the experimental evidence which supports these postulates. In this respect, I take the liberty of selecting the structures for discussion according to my own preferences; this means that possibly few in this honoured audience will agree that the list is exhaustive, or that I have chosen the most important structures, or that you in this audience will agree with my evaluation of what consists of acceptable experimental evidence. The survey should, I hope, be a proper basis for a valuable discussion on the subject, here or elsewhere.

I begin by excluding certain specific structures for which, in my opinion, there is incontrovertible experimental evidence. For example, the evidence for crosslinking derived from the chemistry of unsaturation, and by the use of peroxides and other such reactants on model compounds, has completely determined and absolutely justified the existence of crosslinked structures. The X-ray studies by Bunn [1] and by others on polyethylene have shown, once and for all time, that crystallite formations do exist in this polymer and that its properties are to a large extent the results of their size and distribution. However, some may state that the fine structure of the lamellae formed, in particular the chain-folding claimed on the basis of small-angle X-ray scattering (SAX) remains to be confirmed using more reliable techniques.

Other structural postulates are to a greater or lesser extent more doubtful. At one end of the scale, for example, we have the interpretation of the effects of alternating hard and soft sectors in block polymers, almost certainly explained as a result of studies on simple diblock copolymers of very definite chemistry. The only doubt remaining in this case is in the effect of the formation of crystallites of hard sections along and between chains. That these crystallites are formed is certain, but I do not know of any experimental work which has established a model for their effect on the mechanical behaviour of the block copolymers.

At the other end of the scale we find structures which have been postulated on the basis of observed mechanical behaviour but for which there is little if any direct evidence from basic chemical or physical studies, as opposed to circumstantial evidence. I refer to such structures as entanglements, catenates, reptating units and saltating bonds. Although the concept of free volume is very well-established, I do not know of any hard experimental evidence for its existence. I will not discuss the latter, except to state here that I am of the opinion that all evidence for the existence of free volume is circumstantial, but indeed very strong. In any court of law, it would certainly result in a unanimous verdict of "guilty" or "innocent" as the case may be, depending on what you would call its existence. Nevertheless, the evidence is never more than circumstantial.

Entanglements are a well accepted concept in polymer science. They were proposed in the first instance to explain away otherwise unexplainable observations, or better said, deviations from the statistical theory of rubber elasticity based on the recognition by Meyer, von Susich and Valko [2] that flexible rubber molecules are statistically distributed in random configurations under thermal motion. The subsequent work by Guth and James [3], Wall [4], and Flory and Rehner [5] established its value.

The earlier theory ascribed the elasticity of a polymeric mass to the additive effect of single polymeric molecules anchored at certain points, these moving in an affine manner. While this was capable of giving a gross explanation of behaviour, it very quickly transpired that the theory was too primitive to be able to explain the finer details of polymeric behaviour.

The network described by the primitive theory mentioned has more recently been incorporated in the latest reformulation in which it has been renamed as "phantom network". The elasticity of the real network is said to be that of the phantom network modified by the very substantial spacefilling characteristics of other chains and junctions, and the inability of real network chains to transect each other. These are viewed as "entanglements", and I find it proper to quote Flory and Erman [6] in this connection.

"... it should be noted that we employ the term »entanglement« to denote diffuse interspersion of chains and junctions in ways that render them inseperable. This view contrasts with the more conventional one of discrete entanglements consisting of well-defined loops of one chain about another. These latter are supposed to act as crosslinkages between specific chains thus intertwined." The conventional view referred to has thus been eloquently refuted.

Thus we have at our disposal a model which will explain and interpret the most accurate observations carried out on the elasticity of swollen networks. Nevertheless, the necessity for swelling before "reliable" observations are made only removes or minimizes the effects of these entanglements, without telling us what they really are or why they act as they do. An attempt to do so was first formulated by de Gennes [7], and elaborated by Doi and Edwards [8], and by Greassley [9] among others. A process called "reptation", closely allied to diffusion, is invoked which allows macromolecules to travel along tubes formed in the mass around them. This is a mechanical concept of a probabilistic, not deterministic, nature. This, however involves molecules not linked to the rest of the network and does not therefore play a part in the asymptotic behaviour of the mass.

I wish to diverge for a few moments from this survey to address a question of philosophy: We are faced with theories which, under varying circumstances and provided proper conditions are imposed, can explain two different but closely related sets of phenomena in the same material. Can they be put together into one coherent theory capable of being experimentally tested? By this phrase, I wish to insist on experiments which measure behaviour unrelated to stress strain and swelling on the one hand or diffusion on the other.

Now, it should be noted that while the phantom network is a mathematical and thermodynamic concept, it is used together with the concept of entanglements in defining the elastic free energy. Mechanically speaking, the two cannot coexist because, while the real network chains cannot mutually transect, those of the phantom network do. We are therefore in the difficult situation that we are dealing with a model with incompatible parts, and though it seems to work, in my view it is simply not good enough, although certainly better than nothing at all.

In 1927, Einstein enunciated a principle stating that "... it is the theory which decides what we can observe." Apart from the suspicion of experiment which characterized some of Einstein's thinking, there is something very important for us in this statement. We must develop our theories so that we can use experimental techniques available, or else we must devote our energies to devising (?) such techniques as may help to explain phenomena: it works both ways. In the next lecture we hope to hear how SANS of marked chains in unmarked polymers has helped to determine which model is most appropriate to describe the model for the deformation of a network.

The motion that a reptating unit is actually performing is not quite clear. Measurements of diffusion and of mechanical dynamic properties results in concepts, in one paragraph of a recent paper [10] such as "tube formation", "step length of reptative motion", "dynamics of tube length fluctuations", and "chain slippage through entanglement links". These are set in opposition to "an assumption of enhanced chain-end mobility" of another model. You will, I believe, agree with me that this view want considerable tidying up, but requires much more experimental evidence than heretofore presented before it can be accepted as a "way of life" of polymer behaviour. I am neither a chemist nor a physicist, but I am certain that techniques are available today which would make it possible to investigate reptation. I am thinking, perhaps mistakenly, of a sufficiently long linear polymer capped with labelled compounds moving through the same unlabelled polymer, their motion followed by a suitable grating. Perhaps, though, this suggestion is rather naive. Apart from SANS, we also have techniques such as positron emission tomography and magnetic resonance imaging, and we are all waiting to know how these can resolve some of the problems.

Catenates already referred to another one of these nice concepts. Napoleon is said to have admonished his general never to form a picture, yet the picture of a catenate or an "olympic polymer" is to me, at least quite appealing. We are aware of the fact that cyclization can occur during chain extension, and there seems to be no good reason that this should not occur simultaneously or consecutively in two or more chains to form interlinking loops. The possibility has been investigated by Chen [11] and by Jacobsen [11] and on thermodynamic grounds, it is found that there exists a statistical probability that catenate structures can be formed on chain extension. In my own work on the subject, I have demonstrated that under certain circumstances insoluble polymers are formed without any crosslinks and for the moment I believe that they are catenates [12]. Again, direct experimental evidence on these is not available.

Finally, I should like to refer to my model for the reinforcing effect of carbon black in a noncrystallizing crosslinked polymer [13]. This I introduced under the name of saltation. The model describes many – as far as I am aware all – the phenomena associated with the action of carbon black, including swelling restraint, stress softening or the Mullins effect, strain-rate effects etc. but I have not been able to carry out a single experiment to show that saltation, or the movement of polymer chains over the surface of carbon black particles actually exists.

I hope that I have planted some doubts in your minds. It would be delightful if all our theories were simple, but Nature is certainly not simple. Yet in order to understand the workings of Nature we should direct more of our efforts to devising experiments to observe as directly as possible the mechanisms involved, rather than to develop theories concerning the phenomena.

Received October 6, 1986;
accepted April 14, 1987

Authors' address:

Zvi Rigbi
Technion-Israel Institute of Technology
Haifa, Israel

Discussion

KRÖNER:

You made a short comment on reptation: did you imply reptation is a possible process in melts but not in networks, permanent networks at least?

RIGBI:

I meant to imply that there is no experimental evidence for reptation and that if reptation does exist then it will exist in melts as well as in solid polymer.

D. RICHTER:

May I comment also on reptation. Last year we did some careful studies with neutron-spin echo in order to see reptation on a microscopic scale, that means on the time scale of down to 10^{-6} s. Unfortunately, we were not able to see reptation there. Even if you put trapped chains into networks with a relatively small size of chains between the crosslinks, you cannot see any signs of the microscopic model of reptation, which DeGennes put forward. So, in my opinion, the case is still open, because one might argue one should measure for longer times and over larger space ranges, but at the moment it doesn't look very good for reptations on a microscopic scale.

WINTER:

I could assume that a very long linear molecule coiled would also be insoluble, so maybe somebody could be asked, who knows about statistics of this chemical reaction, if it might be more probable to get a very large molecule and then get an insoluble polymer this way.

RIGBI:

Even samples of polyisobutylene of molecular length of some 3 million were perfectly soluble. There is no evidence that I have seen that infinitely long molecules are insoluble if they have no crosslinks or loops of another sort.

Progress in Colloid & Polymer Science Progr Colloid & Polymer Sci 75:4–10 (1987)

Entanglements in rubbers

T. A. Vilgis

Max-Planck-Institut für Polymerforschung, Mainz, F.R.G.

Abstract: How topological constraints change deformation behaviour in rubbers is reported. The theory predicts short and high strain behaviour in fair agreement with the real world. Precise expressions for the trapping factor for the modulus, entanglement sliding, and finite extensibility in dense systems are given.

Key words: Entanglements, networks, topological constraints, slip link, finite extensibility, neutron scattering, thermoelasticity.

Introduction

The classical theories of rubber elasticity [1] contain an unrealistic assumption: firstly that polymer chains can pass through each other like "phantoms" held together only by crosslinks. Secondly, that the whole system is prevented from collapsing by imposing a density constraint (incompressibility condition) generating the bulk modulus. In this paper we want to disprove one of the above assumptions – the first one – but still keep the second one.

The basic idea can be considered as follows: Take a polymer melt with its given topology at some time. Crosslink the melt ideally (and instantanously) and quench the topology at the time of crosslinking. Then deform the system and average over all possible topologies given at the quench time [2]. Obviously at the crosslinkage the topology of the melt is fixed forever, as long as the crosslinks are chemical linkages between two chains and not able to open. Since, in the melt, the elasticity is dominated by entanglements, the number of them must be large. The assumption then is that in crosslinking the melt, most of the entanglements become trapped between two crosslinks. This trapping acts most severely because the reptation has been stopped by the crosslinking. The conclusion that we now want to stress is that the number of trapped entanglements becomes a conserved quantity. There is no way for them to escape during any thermodynamic change of the systems (except by burning of the sample).

Looking at the problem in this way, rubber is a classical example of a system with quenched disorder [3]. The quenched disorder are the crosslinks: once the crosslink is formed it links two segments for ever. The chain segments between two crosslinks of one chain can move more or less freely except near the crosslinks, where different things can happen [4]. This connects the statistical mechanics of rubbers to those of spin glasses [5]; both systems can be treated by the same formulation.

Modelling entanglements

The crosslinks are very different from the entanglement. As described above, the crosslinks are representing some kind of quenched disorder with no internal degrees of freedom. It is meant by this that the crosslink can move as a whole by some Brownian motion but the crosslinked chain segments are always tight together. In the case of the entanglement the constraint is much weaker. An entanglement (any topological constraint) can be represented by a crude picture such as Figure 1.

Of course, there are much more complicated situations, such as in Figure 2.

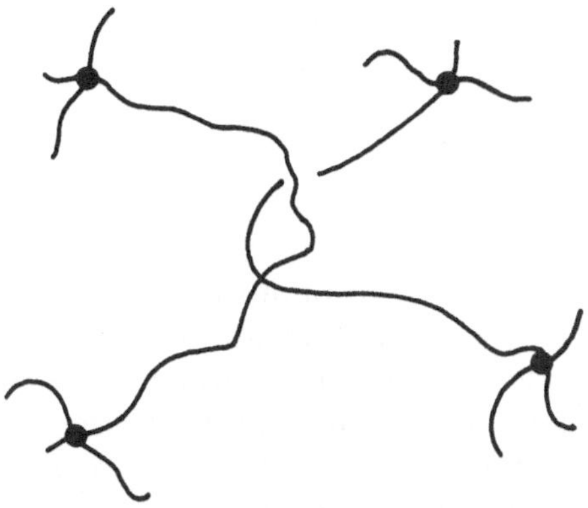

Fig. 1. The simplest topological constraint forming an entanglement

Fig. 2. Many entanglements in a crosslinked rubber

But we assume that in some way Figure 1 is a typical one. Its basic property is that there is *no way* to loose one entanglement in any time scale or by any other process, saying that the process shown in Figure 3 is *not* allowed.

In order to apply some mathematics we have to invent a model which makes this problem feasible. A possible model is the slip link [6] in which the situation in Figure 1 is replaced by something like that in Figure 4.

The ring can slide along the chains to some extent. This is the simplest and crudest way to replace a many-chain problem by an effective one-chain problem, but it seems nevertheless adequate for our purpose. This figure now shows the difference of the entanglement in comparison to the crosslink. The entanglement given in Figure 4 still has some internal degree of freedom given that the ring can slide. A crosslink could be defined by a ring with diameter zero, if one wishes. The picture can also be extended to larger length scales if there are more entanglements along the chain. The Edwards primitive path is used in this case [7] (Fig. 5).

The entanglements are shown by the dots. Normally, in the case of a polymer melt, the chain "replates„ through the rings (which may be replaced by some sort of tube [7]). The crosslinks which fix one chain to the rest of the (well-linked) network is not able to "replate" at all, but slippage still remains possible. To conclude this section we should keep in mind that the entanglements can be thought of as weak crosslinks with some degree of freedom.

The free energy of deformation

The first aim is to calculate the free energy of deformation for the fully entangled (or crosslinked) melt system. This has been done in a recent paper [8] so that we do not need to give the mathematical details here. Firstly, we must comment on the slippage process. Hence we are not interested in the dynamics of a rubber containing entanglements [9] but of the statistical properties of the additional degree of freedom given by

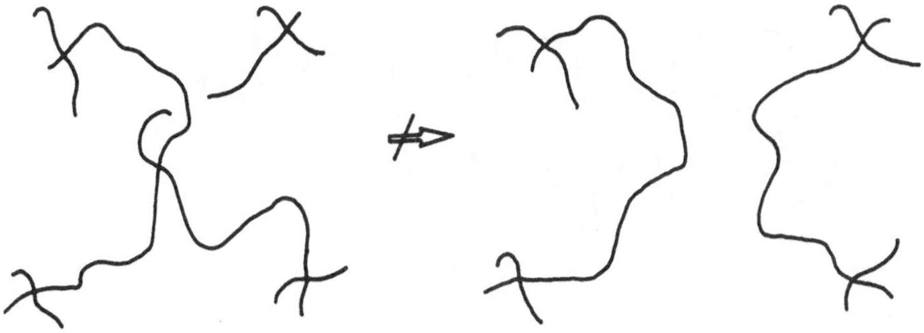

Fig. 3. Real chains cannot pass through each other, and the entanglement forms a topological constraint

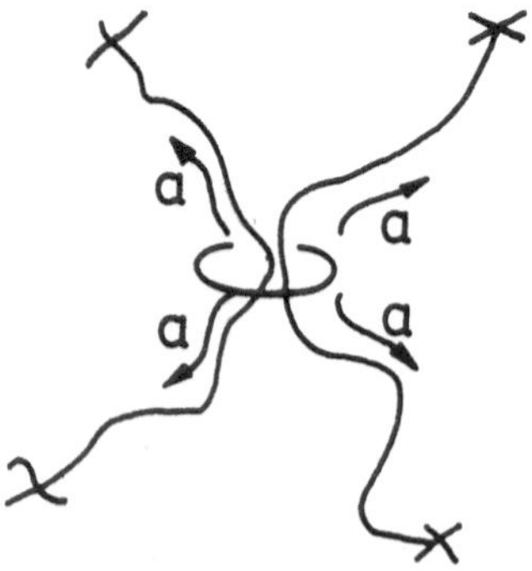

Fig. 4. The entanglement is replaced by a ring, sliding along both chains within a distance a

the slip link (Fig. 4). If we neglect entanglements, the crosslinks can be treated by a constraint

$$\int_0^L ds_\beta^{(j)} \int_0^L ds_\alpha^{(i)}\, \delta\,(R_\alpha\,(s_\alpha^{(i)}) - R_\beta\,(s_\beta^{(j)})) \tag{1}$$

joining the i-th segment of the α-th chain to the j-th segment of the β-th chain together. The integration of s_α, s_β means that the crosslinks can be every where along the contour length L. The entanglement can be modelled by a weaker constraint like

$$\int_{-a}^{+a} \frac{d\tau}{2a} \int_{-a}^{+a} \frac{d\tau'}{2a} \int_0^L ds_\beta^{(j)} \int_0^L ds_\alpha^{(i)}\, \delta\,(R_\alpha(s_\alpha^{(i)} + \tau) - R_\beta(s_\beta^{(j)} + \tau')) \tag{2}$$

where $\{\tau\}$ are the slip variables and a the extent of how far they can slip [6], a is a unknown quantity. The number of entanglements N_s is also assumed to be given. The free energy may be calculated by means of the replica approach (often used in systems with quenched disorder) [6] and one obtains

$$\beta F = \frac{1}{2} N_x \sum_{i=1}^{3} \lambda_i^2 + \frac{1}{2} N_s \sum_{i=1}^{3} \left\{ \frac{(1+\eta)\,\lambda_i^2}{1+\eta\,\lambda_i^2} + \log\,(1+\eta\,\lambda_i^2) \right\} \tag{3}$$

where $\beta = (KT)^{-1}$ and N_x the number of crosslinks, N_s the number of entanglements and η the new slip variable. If $\eta = 0$, the ring in Figure 4 has diameter zero and the entanglement acts as a crosslink and the modulus is given by $G = KT\,(N_x + N_s)$, while for $\eta \to \infty$ the entanglement is a very weak constraint and one has an ideal gas term $\eta \log(\lambda_1 \lambda_2 \lambda_3)$ in addition to the crosslink term; the modulus is given by $G = N_x KT$, as it must be. The assumption that the entanglement can slip between two crosslinks and if there is only one entanglement, as in Figure 7, suggests $\eta = 0.2$.

For some arbitrary value of η, the modulus is

$$G = N_x KT + N_s\, KT \frac{1}{(1+\eta)^2} \tag{4}$$

which is already of the Greassley-Langley form $G = G_{crosslink} + G_{entangle}\, T_e$, where T_e, the "trapping factor", can be read from Equation (4) [10] Equation (3) can be derived by a much simpler model and technique. In Reference [8] it has been derived without the replica trick. In that case it is easier to generalize the theory to large extensions where Equation (3) fails (see Fig. 6).

The physics of the high deformation regime are served by Figure 5. Obviously in all the classical theories the Cauchy length dominates the whole picture since the only length that matters is the distance between the crosslinks. With the entanglements, a new length scale is introduced which is far below the distance between the crosslinks. We expect that the entanglement spacing dominates the free energy. The entan-

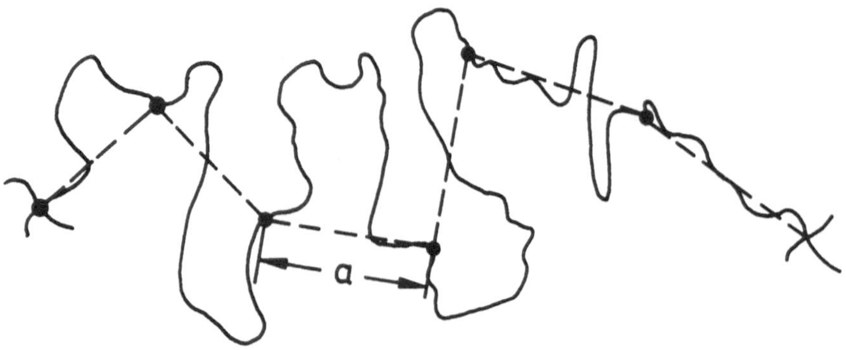

Fig. 5. The Edwards primitive path (dashed line) connects entanglements (●) with a random walk of step length a. The real chain is given by a random walk of step length $l \ll a$ (∼)

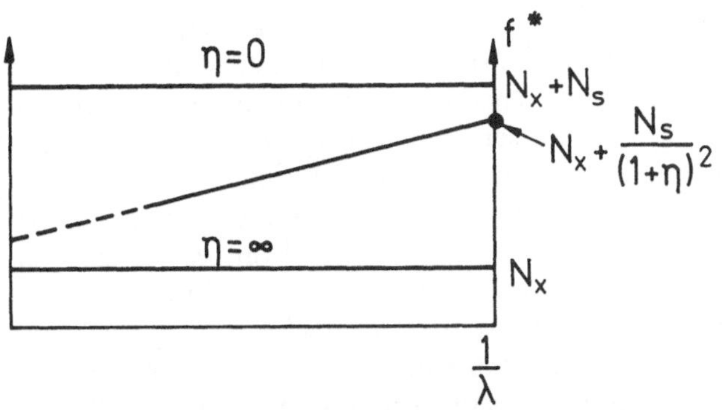

Fig. 6. Mooney plot of the force corresponding to Equation (4) for the two marginal cases of zero slip ($\eta = 0$) and infinite slip ($\eta = \infty$). Real experiments are around $\eta \sim 0.2$

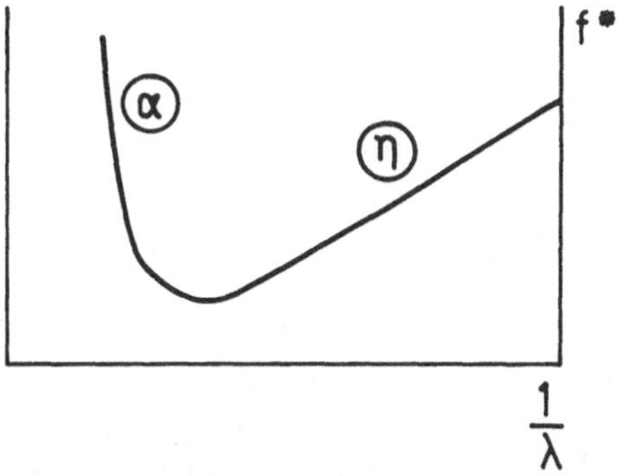

Fig. 7. Mooney plot of the force corresponding to Equation (7). The small deformation regime is dominated by the slippage η, while the high deformation range is ruled by α

glement spacing is an average quantity and not a real fixed length because of the slippage of the entanglements due to Brownian motion. The line in Figure 5 connecting all entanglements is called the primitive path and it is obvious that it is a random walk with a much larger step length than the Kuhn length of the real polymer [7]. Most of the real polymer is slack outside the primitive path. If we assume that both the primitive path (pp) and the polymer are random walks with the same end-to-end distance, we find the important relationship

$$N_{pp}\, a^2 = N l^2 \tag{5}$$

where N_{pp} is the number of pp steps, a the entanglement distance, N the number of Kuhn segments and l the length of the Kuhn segment. The most important consequence of the pp model for rubbers is the effect on the finite extensibility. It is easy to show that for deforming the material the pp deforms affinely and the length of the pp is given by

$$L_{pp}(\lambda) = L_{pp}(\lambda = 1)\,(\textstyle\sum \lambda_i^2)^{1/2}\,.$$

The polymer itself can only be deformed until all the slack is used up giving the maximum extension (here in uniaxial deformation)

$$\lambda_{\max} \simeq \left(\frac{a}{l}\right) \equiv \alpha^{-1} \tag{6}$$

which is much smaller than the maximum deformation of a free chain $\sqrt{N}$. This is consistent with experiments since $\sqrt{N}$ has values larger than $\lambda_{\max}$ given by experiments. The occurrence of a maximum deformation gives rise to a singularity in the free energy (and in the force, of course). Again without technical details, the free energy can be calculated and the result is [8]:

$$\begin{aligned}\beta F = {} & \frac{1}{2} N_x \left\{ \frac{(1-\alpha^2) \sum_i \lambda_i^2}{1-\alpha^2 \sum_i \lambda_i^2} + \log\left(1 - \alpha^2 \sum_i \lambda_i^2\right) \right\} \\ & + \frac{1}{2} N_s \left\{ \sum_i \frac{\lambda_i^2 (1-\alpha^2)(1+\eta)}{(1-\alpha^2 \sum_i \lambda_i^2)(1+\eta \lambda_i^2)} \right. \\ & \left. + \log\left(1 - \alpha^2 \sum_i \lambda_i^2\right) + \log\left(1 + \eta \lambda_i^2\right) \right\}. \end{aligned} \tag{7}$$

By calculating the uniaxial force f we have shown that experimental data can be fitted reasonably [8]. The uniaxial force has the typical shape in the Mooney plot as shown in Figure 7.

The free energy Equation (7) returns to the classical expression if $\eta \to \infty$ and $\alpha \to 0$ which means that the entanglement spacing becomes very large.

To conclude this section we can make the following important remarks. Firstly, there are four quantities in the theory (N_x, N_s, η, α) which are in principle not known, but they all can be related [9], by saying, for example, that η measures the slip between two entanglements, then η is determined by the entanglement spacing (a) as well as α, so both α and η are of the same order.

On the other hand, N_s is given by the number of entanglements between two crosslinks, but knowing the length of a strand (i.e. by endlinking) and the entanglement distance a, N_s and N_x are related and so are η and α, too. Secondly, since the "softening" in the small deformation regime (positive Mooney slope) and the hardening in the high deformation regime are of the same origin (entanglements) this theory always suggests a positive Mooney slope. This is because the free energy depends only on α^2 and only on η rather than $\alpha\,\eta^2$ or other combinations [8].

By adding good solvent to the already established rubber, the slippage becomes larger because the good solvent keeps the chain segments as far apart as possible (η becomes larger [9]). This means that the Mooney slope decreases with increasing swelling ratio as knowns from many experiments [1].

Neutron scattering

We have seen that stress-strain data can be fitted by the above theory. A further crucial test is the deformation dependence of the neutron scattering form factor of a labeled chain in a network [11]. This might be of importance in the question of whether entanglements contribute to the elasticity or not. The question of whether there is a contribution of entanglements to the modulus is still disputed [12] but there seems to be evidence that they do [15]. According to our theory, there is a strong contribution of entanglements to the modulus

$$\frac{G}{kT} = N_x \frac{1 - 2\,\alpha^2 + O(\alpha^4)}{(1 - 3\alpha^2)^2} + N_s \frac{1 - 2\alpha^2 + 3\alpha^2\eta + 4\alpha^2\eta^2 + O(\alpha^4)}{(1 - 3\alpha^2)^2\,(1+\eta)^2} \qquad (8)$$

which is again of the Langley-Greassley type. This dispute comes mainly from the restricted junction fluctuation theory of Flory [16], in which it is assumed that the only effect of entanglements is to restrict the fluctuations of the crosslinks. Despite this unrealistic assumption, the theory of Flory seems to fit a great deal of experimental results, too. The most important fit parameter ($\varkappa$ in Flory's theory) describes the severity of the entanglement constraint (see the original reference for details). This parameter is more or less an unmeasurable quantity. Flory's theory extrapolates between the two classical models of James and Guth [13] and Kuhn [14] by giving $\varkappa$ a range from o to infinity. But it is clear that the stress-strain experiment never can decide whether entanglements are important or not – it is not sensitive to molecular details. Support must come from neutron scattering. In Reference [17] we showed that Flory's theory is not able to explain the neutron data (see Fig. 8) based on results discussed more extensively by Bastide and Boué to more extent [21]. A further test, then, would be to compare the prediction for the neutron scattering for the model presented above. Calculation of $S(k) = \sum_{i,j} \langle e^{ik(r_i - r_j)} \rangle$ for a deformed chain yields, for the slip link [17]

$$S(k,\lambda) - \sum_{i,j} \exp\left[-\frac{l}{6}\,k^2 |s_i - s_j| - \frac{Q(\lambda^2 - 1)}{1 + \eta\left\{\left[\frac{1}{2}Q\lambda^4 + \frac{1}{2}(2\lambda^2 - 1)\,Q - (3\lambda^2 - 1)\right] \Big/ (\lambda^2 - 1)\right\}} \right] \qquad (9)$$

with

$$Q = -\frac{l}{6}\,k^2 \frac{|s_i - s_j|}{L}.$$

First estimates to compare this with experiments are proceeding well and a complete analysis of the data is in progress [18]. We have also provided $S(k)$ for the high deformation range [17], which should be included as $\lambda \gtrsim 2.5$.

Thermoelasticity

During the conference, Godovsky raised the question of thermoelasticity in this model. For our model this turns out to be very simple. The reason is as fol-

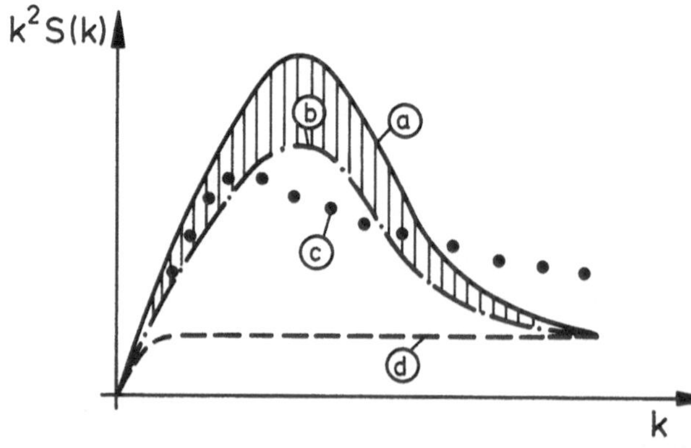

Fig. 8. Schematic Kratky plot of the form factor $S(k)$, for some deformation λ. (a) Kuhn model; (b) James & Guth; (c) range of typical experiments; (d) isotropic case ($\lambda = 1$). The shaded area covers Flory's constraint model by $0 \le \varkappa \le \infty$

lows: The elasticity in this model is mainly due to entanglements and the main parameter determining all the others is the number of trapped entanglements between two crosslinks. This number is given by the formation conditions and is fixed forever. By changing the temperature after fabrication of the rubber, one can expect only the "normal" thermal changes due to energy elasticity [1] (i. e. transgauche energies, etc.). These quantities are not taken into account in this model, but on the ground of a modified version of the Flory level, one can argue that there is a "memory term" in the modulus $\langle r^2\rangle/\langle r_o^2\rangle$ which relates the radius of gyration of a typical chain in the network to the radius of gyration of a free chain. Due to classical elasticity [1], we expect two contributions to the elastic force $f = f_h + f_s$ where f_h is the enthalpic part and f_s the entropic part. Fortunately the two parts are related by thermodynamics

$$f_h = f - T\left(\frac{\partial f}{\partial T}\right)_{L,p} \tag{10a}$$

$$f_s = T\left(\frac{\partial f}{\partial T}\right)_{L,p} . \tag{10b}$$

If we write for the uniaxial elastic force

$$f = kT\frac{\langle r^2\rangle}{\langle r_o^2\rangle}\left(\lambda - \frac{1}{\lambda_2}\right) g(N_x, N_s, \eta, \alpha) \tag{11}$$

with

$$g = N_s\left[\frac{(1-\alpha^2)(1+\eta)}{1-\alpha^2\phi}\left\{\frac{(1-\eta^2\lambda)\lambda^2(1-\alpha^2\phi)}{(1+\eta\lambda^2)^2(\lambda+\eta)^2} + \alpha^2\left(\frac{\lambda^2}{1+\eta\lambda^2} + \frac{2}{\lambda+\eta}\right)\right\} + \frac{\eta\lambda}{(1+\eta\lambda^2)(\lambda+\eta)} - \frac{\alpha^2}{1-\alpha^2\phi}\right] + N_x\left[\frac{1-\alpha^2}{1-\alpha^2\phi} - \frac{\alpha^2}{1-\alpha^2\phi}\right] \tag{11a}$$

and $\phi = \lambda^2 + 2/\lambda$ the thermoelastic equations can be derived very simply, since α, η, N_s are *not* temperature dependent. The temperature dependence occurs only in λ, $\langle r^2\rangle$, $\langle r_o^2\rangle$ in the usual fashion, and all the interpretations for the model hold in the way explained in References [1, 19].

Discussion

This paper discusses the influence of the entanglements on the deformation behaviour in a crude, but nevertheless adequate, way. It appears that in dense systems, the entanglements may dominate the whole deformation regime. The modulus depends, in this model, directly on the number and the slippage. Moreover, the presence of entanglements changes the shape of the stress-strain curve in a way that enables us to draw the following conclusion. The small deformation regime is dominated by the slippage, giving rise to the softening (in the Mooney plot) due to the increase of phase space during deformation. For large deformations, the entanglements are responsible for the hardening due to less phase space and using up the slack.

Clearly this is an extreme model of a fully entangled system. The question is how this model should be checked by experiments. This will be most difficult from the experimental point of view and involves a whole series of preparations and measurements.

Firstly, a polymer melt should be characterized by viscoelastic means. The plateau modulus and the relaxation times give, according to theories of viscoelasticity [20], the number of entanglements and the ratio $(a/1)$ (from the relaxation times). Then the melt should be crosslinked in a definite way so that the number of crosslinks N_x is known. Supporting measurements of the swelling properties of the rubber can be taken to get a further estimate of N_x. After that careful measurements, of the stress-strain behaviour and the neutron scattering should follow. The only (maybe naive) way to control the number of entanglements is then to do the crosslinking in solution and vary the solvent concentration in order to get different values of N_s (and hence a, η, α). Mechanical and neutron measurements should then be done in the same way, after characterizing the entangled solution by viscoelasticity. It is clear that this experimental problem is feasible (probably) only in the naive theorist's mind and I can imagine there bing a lot of additional problems from chemistry that I have not thought of. But there is still, nevertheless, a big challenge to both expermentalists and theorists in all problems in networks that we have discussed during this meeting.

Acknowledgements

First of all I would like to thank the organizers of the meeting for a pleasant atmosphere during the meeting and for giving me the opportunity to participate.

It is a pleasure to thank Prof. H. G. Kilian, Prof. Yu. K. Godovsky, Dr. B. Ewen, Dr. D. Richter, Dr. C. Picot for discussions of the subject. Some of the work has been done in collaboration with Prof. S. F. Edwards and Dr. F. Boué and the author is indebted to them for scientific and personal friendship.

References

1. Treloar LRG (ed) (1975) The Physics of Rubber Elasticity, Clarendon Press, Oxford
2. Deam RT, Edwards SF (1976) Phil Trans R Soc 11:317
3. Vilgis TA (1987) Progr Coll & Polym Sci 74:4
4. Richter D (1987) Progr Coll & Polym Sci 74: ; Warner M (1981) J Phys C 14:4985; Boué F, Vilgis TA (1987) to be published
5. v. Hemmen L, Morgenstern I (1983) Heidelberg colloquium on Spin Glasses, Lecture Notes in Physics 192, Springer, Heidelberg
6. Ball R, Doi M, Edwards SF, Warner M (1981) Polymer 22:1010
7. Doi M, Edwards SF (1978) Trans Farad II 74:1798
8. Edwards SF, Vilgis TA (1986) Polymer 27:483
9. Vilgis TA (1987) to be published
10. Langley NR (1968) Macromolecules 1:348
11. Boué F (1987) Progr Coll & Polym Sci 74:
12. Mark JE (1986) ACS Polym Preprints 27:1
13. James HM, Guth E (1943) J Chem Phys 11:455
14. Kuhn W, Grün F (1946) Kolloidzeitschr 101:248
15. Oppermann (1987) Progr Coll & Polym Sci 74: ; Ilavski I (1987) Progr Coll & Polym Sci 74:
16. Flory PJ, Erman B (1982) Macromolecules 15:800
17. Vilgis TA, Boué F (1986) Polymer 27:1154
18. Boué F (1987) private communication
19. Kilian HG (1983) Kautschuk + Gummi 36:959; Godorsky YuK (1987) Progr Coll & Polym Sci 74:
20. Greassley WW (1982) Adv Pol Sci 47:67
21. Bastide J. Boué F (986) Physica 140A:251

Received December 11, 1986;
accepted May 15, 1987

Author's address:

T. A. Vilgis
Max-Planck-Institut für Polymerforschung
Postfach 31 48
D-6500 Mainz, F.R.G.

Discussion

KILIAN:

Model networks formed by silica-to-filler bonds have only been prepared by Wolff [1]. The average particle diameter is about 10 nm. Hence, each of the filler particles represents a very heavy multifunctional crosslinkage. There are reasons for the assumption that fluctuations of the crosslinks are behind the global interaction in networks: the van der Waals interaction parameter should now disappear in filler-networks constituted by multifunctional heavy crosslinks. The stress-strain pattern of the networks mentioned above could indeed only be described under this assumption: The mechanical behaviour corresponds to that of a Langevin-network comprising phantom-chains (flexible mass-fibres) of finite length.

My question put forward is whether you can understand this in terms of your model.

VILGIS:

Well, it is difficult to speak about functionality. This model is done with the functionality 4, because the calculation is performed with a huge self-crosslinked chain and this always makes the functionality 4. The generalization to higher functionality is not so trivial in this approach and one can employ some methods of field theory to do it. Concerning the other point you mentioned with the Langevin approach, can you say something about the distance between the fillers and do they agree with your fit to the data?

KILIAN:

The mean length of the network chains can be evaluated from the mean distance of the filler particles [1]. One parameter is then left for fitting the stress-strain curves. This parameter is related to the surface density of the filler-to-matrix bonds. This parameter assigned to a fixed value, the stress-strain curves of the whole set of rubbers loaded to different degrees with the silica, can be nicely computed. The second Mooney-coefficient is negative – in principal accord with the Langevin approach – so I have problems in going along with your approach. Why is slipping not present in that case?

VILGIS:

Well, I have no answer at the moment for this more complicated system. I think it should be there.

KILIAN:

What about the calculation of the Mooney-plot of networks simply extended in the swollen state [2]? These patterns can be represented under the assumption of constant chain length and a nearly invariant interaction parameter. The system "knows" the prehistory of swelling, since the maximum extensibility is reduced due to the swelling process (which corresponds to an equitriaxial deformation of the network). The results can be interpreted under the assumption of having an invariant number of permanent crosslinks. In the quasi-static limit there are no hints that slip-links become operative.

D. RICHTER:

As I will show you later, it is not possible to measure directly the range of fluctuations of network points by this neutron scattering and I would like to ask you: What is the constraint you impose on these fluctuations in your model, compared to more simple models?

VILGIS:

Well, I have cheated a little bit, because I have introduced only the Kuhn model. I have treated my data only with the affine model. I didn't allow for fluctuation of the crosslinks. But if you allow the crosslinks to fluctuate freely, you will recover 1/2 in the modulus again. But if I want to treat the fluctuations of the crosslinks in a Flory manner, I have to make some model, which tells me how the fluctuations are changing with deformation. Flory simply puts the constraint of the fluctuation, let's say the potential. And if you deform it uniaxially, the potential becomes deformed as well and he makes an ellipsoid of it, so it loses some constraints in the direction of uniaxial stretching. But this is a model and no-one knows how to prove it. I have introduced the model, which might or might not be realistic in treating the fluctuations of the crosslinks as strain-dependent. I very simply assumed that there is no fluctuation. If there is free fluctuation, I will have the James Guth limit again. This is the factor of 1/2 in the modulus and if I fix the crosslink, I will have the Kuhn-modulus again, so there is something inbetween, but the dependence is not easy to find.

Progress in Colloid & Polymer Science Progr Colloid & Polymer Sci 75:11–20 (1987)

Formation, structure and elasticity of model imperfect networks prepared by endlinking

M. Ilavský and K. Dušek

Institute of Macromolecular Chemistry, Czechoslovak Academy of Sciences, Prague, Czechoslovakia

Abstract: The extraction, viscoelastic, equilibrium mechanical and optical behaviour of polyurethane (prepared from poly(oxypropylene) triols-diisocyanate-monofunctional alcohol or monoisocyanate in the dry or dilute states) and epoxy (prepared from diamine-diepoxide and monoepoxide) networks has been analysed. In all cases, the theory of branching processes adequately describes the network formation and the weight fraction of sol w_s. The networks are homogeneous and their viscoelastic behaviour has a model character. The equilibrium deformational behaviour of polyurethane networks indicates the presence of an entanglement contribution in the experimental modulus G_r. The G_r values can be adequately described by the theory of branching processes with the chemical, G_c and the topological, G_{ent}, contribution to both systems. The deformational behaviour of epoxy networks, however, can also be adequately described by the theory with the front factor $A = 1$, without the entanglement contribution.

Key words: Formation of networks, equilibrium modulus, sol fraction, polyurethane networks, epoxy networks, concentration of elastically active network chains, retardation spectrum.

Introduction

The structure and elasticity of polymer networks is described by theories of network formation (branching theories) and theories of rubber elasticity. The network formation theories can be grouped into two major categories [1, 2]: (a) Theories not associated with dimensionality of the space (statistical and branching theories in which spatial correlation can be approximated by mean field approach), and (b) Simulation of network formation in d-dimensional space (e. g. percolation).

The applicability of these theories can be tested using measurements of the gel point, evolution of the degree of polymerization changes, the sol fraction, and equilibrium elasticity. The latter measurements involve rubber elasticity theories, the most prominent of which are as follows: (a) restricted junction fluctuation theory of Flory, in which all topological constraints are reflected only in the magnitude of fluctuations of crosslinks [3, 4], (b) the van der Waals type model of Kilian in which the effect of constraints is modelled semi-empirically using parameters of a van der Waals gas [5, 6], (c) trapped entanglements model [7, 8] (Langley-Graessley) which specifically considers the contribution to the stress by permanent (unrelaxable) constraints, but does not aim to describe the shape of the stress-strain dependence, (d) the tube model [9] which simulates the local topological constraints using a mean field approach and can describe the whole equilibrium stress-strain dependence.

Verification of the applicability of both network formation and rubber elasticity models requires a study of networks with a broad variation in structure. Such networks can be prepared by varying not only the crosslinking density (concentration of elastically active network chains), but also the network topology (dangling chains, effective functionality of crosslinks). Well-defined polymer networks, prepared from telechelic prepolymers by endlinking, have been widely used in this respect in recent years [10-18].

In this paper, two sets of model networks obtained by stepwise polyaddition have been examined: (a) polyurethane networks prepared by endlinking of poly(oxypropylene) triols (PPT) with 4,4'-diphenylmethane diisocyanate (MDI) and (b) epoxy networks prepared from the diglycidyl ether of bisphenol A (DGEBA), phenylglycidylether (PGE) and 4,4'-diaminodiphenylmethane (DDM). In some cases a diluent was also used in the preparation of networks.

The viscoelastic and equilibrium mechanical and optical behaviour of both kinds of networks was studied. Attention was mainly devoted to an examination of values of the equilibrium modulus G_e and of the gel fraction w_g, in terms of existing theories of network formation and rubber elasticity. The experimental details and main results have been published elsewhere [10, 11, 19, 20, 22, 23].

Structural information obtained by applying the branching theory

The branching theories represent theoretical tools for deriving structural characteristics as a function of the initial composition of the system, reaction mechanism and possibly reaction kinetics or conversion of reactive groups into bonds. The classification and applicability of branching theories, categorized according to theoretical approaches and the presence of long-range correlations, have been recently analyzed in Reference [1] (cf. also a review on network formation in the curing of epoxy resins [2]).

The model polyurethane and epoxide networks discussed below fall into the category of networks formed by alternating stepwise reactions, possibly with the first shell substitution effect and with weak or negligible cyclization. The statistical methods of network build-up are applicable as a good approximation for the weak spatial correlations resulting in cyclization. Furthermore, it has been found that, even in the presence of weak cyclization, the network parameters can be calculated to a good approximation, using a ring-free model and by considering the conversion of reactive groups only as a conversion into intermolecular bonds; a small percentage of groups have been wasted in intramolecular bonds. The intramolecular bonds are not effective in structural growth. Although a direct experimental procedure allowing us to distinguish between inter- and intramolecularly reacted groups is impossible as a rule, the extent of intermolecular reaction can be calculated from the sol fraction [10, 11, 20]. In this way, one obtains a correlation between post-gel structural characteristics which have no adjustable parameters. The extent of cyclization and its effect on structure-sensitive properties can be estimated using appropriate models (cf. e.g. Ref. [27]).

In statistical theories, the network build-up occurs from structural units usually corresponding to the initial components having a different number of reacted functional groups. Examples of structural information in the post-gel stage which can be obtained by applying the theory are summarized below, together with an experimental method sensitive to the change of the given parameter:

Table 1.

Structural parameter	Experimental method or response
Weight fraction of the sol	Extraction of soluble material
Molecular weights and composition of the sol	Methods for molecular weight averages and LC separations
Concentration of elastically active network chains (EANC)	Equilibrium elasticity
Trapping factor in trapped entanglement contribution	Equilibrium elasticity
Distribution of the degrees of polymerization of EANC's and of dangling chains	Viscoelasticity

The only necessary information for application of the branching theory is the distribution of constituent units having a different number of (intermolecularly) reacted groups mentioned above. This distribution is obtained from chemical kinetics. In the theory of branching processes, the distribution is expressed through a probability generating function (PGF) F_0 for the number of bonds issuing from a monomer unit. Here we give the forms for the PGF F_0, corresponding to the polyurethane and epoxy-amine networks discussed below.

In the case of polyurethane networks, the OH-containing polyols were composed of tri-, bi- and monofunctional components with OH groups of independent and equal reactivity; in the diisocyanate and monoisocyanate, the NCO groups were also of independent and equal reactivity. Thus, the PGF F_0 for the OH- and NCO-containing components can be written as [10]

$$F_{0H}(z_I) = n_{H1}(1 - \alpha_H + \alpha_H z_I) + n_{H2}(1 - \alpha_H + \alpha_H z_I)^2 + n_{H3}(1 - \alpha_H + \alpha_H z_I)^3 \quad (1)$$

$$F_{0I}(z_H) = n_{I1}(1 - \alpha_I + \alpha_I z_H) + n_{I2}(1 - \alpha_I + \alpha_I z_H)^2 \quad (2)$$

where n_{H1}, n_{H2}, n_{H3} are molar fractions of mono-, bi- and trifunctional components with OH groups, respectively; n_{I1} and n_{I2} are molar fractions of molecules with 1 and 2 isocyanate groups, respectively; α_H and α_I are molar conversions of OH and NCO groups, respectively; and z_I and z_H are auxiliary variables of the PGF, where the subscript indicates the direction of the bond from a unit to a covalently-bound neighbouring unit (a bond from a unit in the root to a unit in generation 1). The distribution n_{Hi} corresponds to functionality distribution in the macrotriol; in some cases a monofunctional alcohol (cyclohexanol) was added.

For crosslinking of a diepoxide and monoepoxide with epoxide groups of independent and equal reactivity, and a diamine with amino groups of independent reactivity, but with a difference in the reactivity of hydrogens of primary and secondary formed amine (substitution effect), the components of the PGF F_0 are [19]

$$F_{0A}(z_E) = (a_p + a_s z_E + a_t z_E^2)^2 \quad (3)$$

$$F_{0E}(z_A) = n_{E1}(1 - \alpha_E + \alpha_E z_A) + n_{E2}(1 - \alpha_E + \alpha_E z_A)^2 \quad (4)$$

where a_p, a_s and a_t are, respectively, the fractions of primary, secondary and tertiary amino groups; n_{E1} and n_{E2} are molar fractions of mono- and diepoxide, respectively; α_E is the molar conversion of epoxy groups, and z_E and z_A are again the PGF auxiliary variables.

All other functions and relations are derived from these basic distributions (Eqs. (1)–(4)). The PGF F_0 for units occuring on generation $g > 0$ are obtained by the differentiation of PGF F_0 with respect to its auxiliary variables. The threshold conversion for infinite continuation of the structure (gel point) is obtained by the differentiation of F_0. The postgel stage is determined by the extinction probability v which is a conditional probability that, given that a bond exists, it has a finite continuation in the structure with probability v. Thus, in the sol, all bonds must have only finite continuations. The EANCs are chains between units that must have at least three bonds with infinite continuation. The classification of the units according to the number of bonds with finite and infinite continuation, and application of the cascade substitution, also enables one to generate the degree of polymerization distribution of EANCs and dangling chains [21]. Also, the trapping factor, T_{eg}, in the trapped entanglement contribution to the equilibrium elasticity can be calculated from this distribution [10, 11, 19] because it is equal to the probability of contact of two segments of EANCs. The derivation of the network parameters for the urethane and epoxy-amine systems mentioned above can be found in References [2, 10, 19]. The branching theories are able to offer a number of other structural parameters which may be needed in the future for a correlation between models of physical behaviour and experiments.

Experimental results and discussion

Polyurethane networks

Sample preparation

Two poly(oxypropylene) triols (PPT) [Union Carbide NIAX, LHT-240, $M_n = 710$, number average functionality $f_n = 2.89$ and LG-56, $M_n = 2630$, $f_n = 2.78$ and 4,4′-diphenylmethane diisocyanate (MDI)] were used in the preparation of networks. Three series of networks were prepared: samples of series A were prepared in the bulk state from PPT and MDI with various initial ratios of reactive groups $r_{HT} = [OH]_{PPT}/[NCO]_{DI}$ in the range $0.6 < r_{HT} < 1.7$. In the second series, B, cyclohexanol (CL) or phenyl isocyanate (PI) were also used in the preparation, in addition to PPT and MDI. Networks of the latter series, B, were prepared in the same range of PPT-MDI ratios, r_{HT}, but in the range of excess of OH and NCO groups, PI and CL, respectively, were dosed so as to make the total ratio $r = ([OH]_{PPT} + [OH]_{CL})/([NCO]_{DI} + [NCO]_{PI})$ equal unity. The series C was prepared from short PPT (LHT-240) and MDI in an excess of OH groups ($r_{HT} = 1.0$–1.6) with various contents of xylene as the diluent (the volume fraction of the polymer at network formation $v^0 = 1 - 0.4$).

Viscoelastic behaviour

The time dependences of super-imposed tensile creep compliance curves $D_p(t)$ of samples of series A given in Figure 1 show that the decisive effect on the

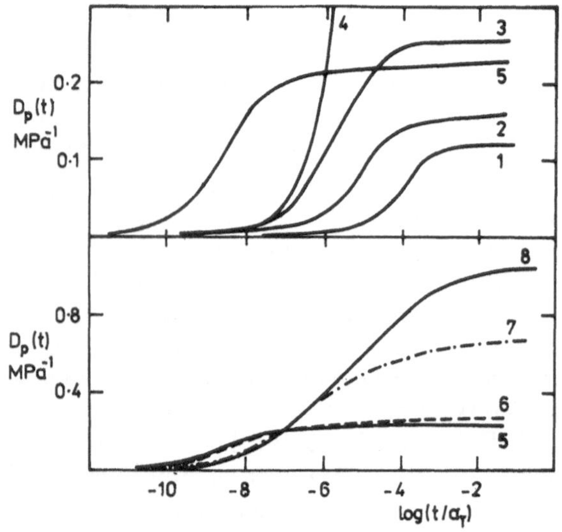

Fig. 1. Dependence of superimposed compliance $D_p(t)$ (MPa^{-1}) on time t/a_T (s) for reference temperature $T_0 = 340$ K. Networks $M_n = 710$: (1) $r_{HT} = 0.96$; (2) 1.07; (3) 1.28; (4) 1.74. $M_n = 2630$: (5) 0.95; (6) 10; (7) 1.10; (8) 1.20

time (or temperature) position of the main transition region of networks is exerted by the MDI content. This is indicated by the quick shift in the main transition towards shorter times with an increasing r_{HT} value of samples of both PPT, and also by the finding that networks obtained with the short PPT have their main transition at temperatures higher by ~ 50 K than those prepared from long PPT. With increasing r_{HT} the equilibrium value $D_p(t)$ in the rubberlike region also increases.

The dependences of reduced retardation spectra $L(\tau)/L_m$ (L_m being the $L(\tau)$ value in the maximum; the $L(\tau)$ values were calculated [23] from the compliance $D_p(t)$, using the Schwarzl-Staverman approximation method) on the reduced retardation time τ/τ_m (τ_m corresponds to L_m) suggest a model character of the networks, (Fig. 2). The samples of both PPT with $r_{HT} = 1$ were found to have narrow, virtually symmetrical spectra $L(\tau)$; as expected, the $L(\tau)$ spectrum of the network made from longer PPT is broader than that of the network prepared from short PPT. With increasing r_{HT}, the widths of the spectra also increase in all samples; in particular, the fraction of longer retardation times increases (the spectra become distinctly asymmetrical). The increase in the width of the spectra is determined by the increasing length of EANCs, and also by the increasing number and distribution of dangling chains with increasing deviation from the stoichiometric ratio. Detailed dynamic mechanical measurements led to the conclusion [24] that two retardation processes exist in these networks, one of which (at shorter times) corresponds to the glass-rubber transition, while the other (at longer times) is probably related to the motion of dangling chains.

Sol fraction and stress-optical coefficient

The weight fraction of the sol, w_s, as a function of r_{HT} of all three series of networks is given in Figure 3. As expected, with networks prepared from long PPT, the w_s values in the whole range of r_{HT} are higher than those of networks from short PPT; this finding is mainly due to the different functionalities f_n of both PPT. The experimental data of series A and B are compared with the theory of branching processes reported in the preceding part, in which only the formation of urethanes without cyclization was considered. Quite evidently, this theory adequately describes the experiment in the case of networks of series A in the range of the excess of OH groups ($r_{HT} > 1$), assuming that the conversion of minority (NCO) groups to intermolecular bonds is $\xi_{NCO} = 0.98$ (for LHT-240) and $\xi_{NCO} = 0.95$ (LG-56). For samples of series B the theory adequately describes the experiment within the whole r_{HT} range, at the same constant values of conversion of minority groups (ξ_{OH} for $r_{HT} < 1$ and ξ_{NCO} for r_{HT}

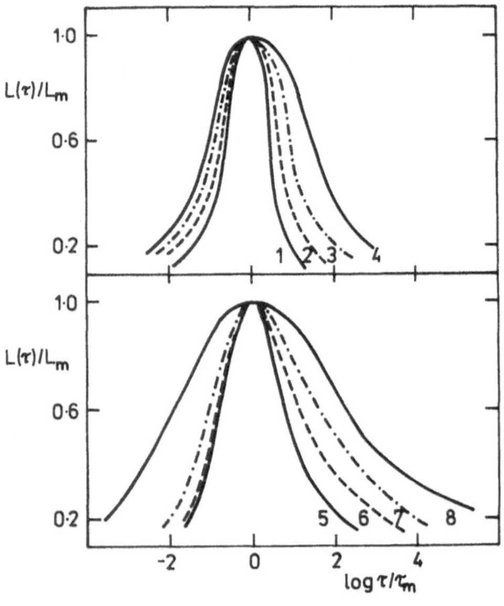

Fig. 2. Dependence of reduced retardation spectra $L(\tau)/L_m$ on reduced retardation time τ/τ_m. Samples denoted as in Figure 1

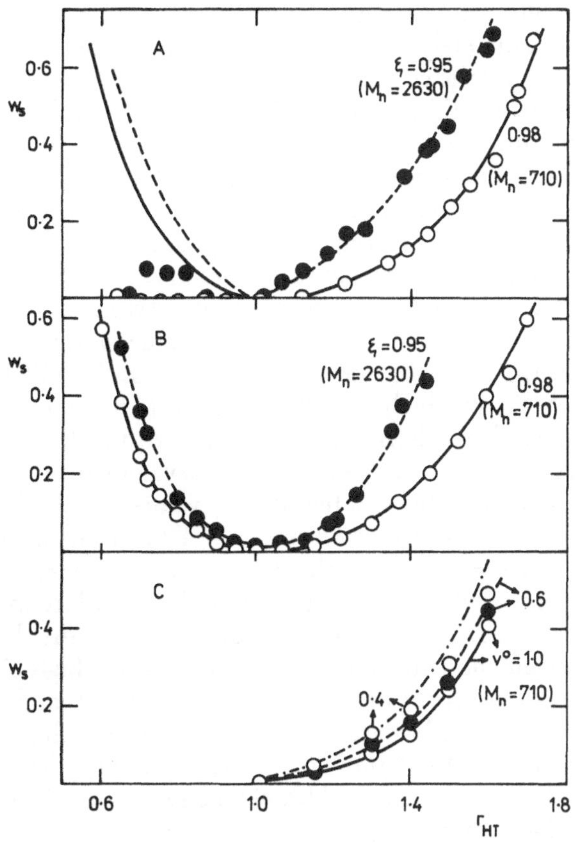

Fig. 3. Dependence of weight fraction of sol w_s on ratio of reactive groups r_{HT} for three series of polyurethane networks. M_n corresponds to PPT, ξ is conversion of minority groups; (——, – – –, —·—·—) are theoretical dependences

> 1). The results make it clear that in the networks of series A, side reactions take place in the excess of NCO groups, which are not considered in the theory (probably, formation of allophanates, by a reaction between NCO and the urethane group, leading to the formation of a trifunctional crosslink [25, 26]) and which reduce the w_s value. The fact that the networks of series B (prepared in the presence of monofunctional components) agree with theory within the whole r_{HT} range, suggests that when the total concentration of OH groups equals that of the NCO groups, virtually no allophanates are formed at all.

As expected, the increasing xylene concentration at network formation (series C) raises the sol value w_s (Fig. 3), as a consequence of the increasing ring formation with increasing dilution. Experimental data of series C are compared with the theory of branching processes which considers the ring formation [11, 27]. It can be seen in Figure 3 that the theory predicts a stronger effect of cyclization on the sol values than would correspond to the experiment. The theoretical prediction is very sensitive to the conversion of NCO groups, however. The theoretical dependences in Figure 3 have been calculated, assuming that the conversions of ξ_{NCO} do not depend on dilution v^0 and are the same as for the undiluted samples. A complete agreement between theory and experiment can be achieved by assuming that with increasing dilution, ξ_{NCO} increases by 0.008–0.014. This small rise lies within the limits of accuracy of determination of the concentration of NCO groups.

It can be seen in Figure 4 that the stress-optical coefficients C_e are sensitive to the formation of allophanates. For networks in which no allophanates are formed, the C_e values are independent of r_{HT}; this finding is a consequence of the same structure of EANCs (the chain contains multiples of units 2/3 PPT + MDI). The lower C_e values of networks made from long PPT are given by the lower MDI content. In the range of formation of allophanates (series A and $r_{HT} < 1$) the C_e values increase due to the increasing concentration of polar groups in the network.

Equilibrium elasticity

The experimental small-strain equilibrium moduli G_d, measured with dry sol-containing samples (in the

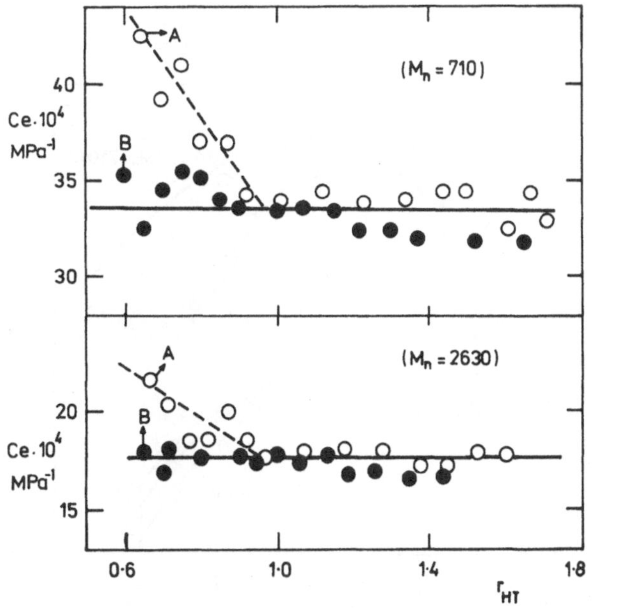

Fig. 4. Dependence of stress-optical coefficient C_e (MPa^{-1}) on ratio r_{HT} of networks of series A and B

series C after evaporation of xylene), have been analysed with respect to the effect of magnitude of the front factor, the possible contribution of permanent interchain interactions (entanglements), and the formation of elastically inactive cycles (EIC) on the equilibrium mechanical behaviour. In agreement with the theory of rubberlike elasticity, for the experimentally reduced equilibrium modulus G_r we may write

$$G_r = G_d/(v^0)^{2/3}\ RTw_g \tag{5}$$

where $w_g = 1 - w_s$, R is the gas constant and T is the temperature of measurement. If we assume that G_r is composed of both the chemical (G_c) and entanglement (G_{ent}) contribution, then

$$G_r = G_c + G_{ent} = A\upsilon_{eg} + \varepsilon T_{eg} \tag{6}$$

where A is the front factor (for the phantom network $A_{ph} = (f_e - 2)/f_e$, f_e is the effective functionality of the elastically active crosslink and $A = 1$ for a network with completely suppressed fluctuations of crosslinks), υ_{eg} is the concentration of elastically active network chains (EANC) in the dry gel, T_{eg} is the trapping factor, and ε is a constant.

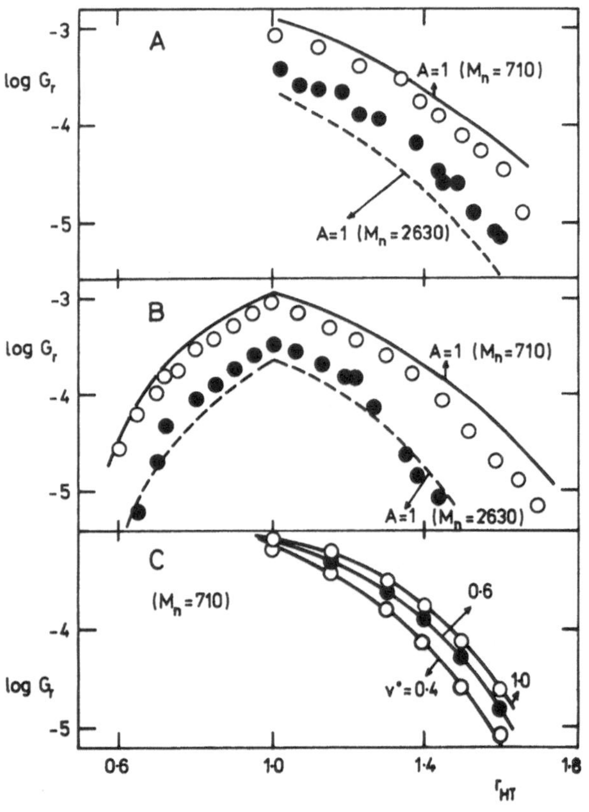

Fig. 5. Dependence of reduced modulus G_r (mol cm^{-3}) on r_{HT}. For series A and B (——, – – –) are theoretical dependences with $A = 1$

The dependence of moduli G_r on r_{HT} is given in Figure 5; the networks of series A, B are compared with the theoretical prediction $A\upsilon_{eg}$. As expected, with increasing deviation from stoichiometry, the values of modulus G_r quickly decrease in all networks; those made from long PPT have lower G_r values within the whole r_{HT} range than networks made from short PPT. The theory in which only the formation of urethanes is considered, predicts a decrease in both the EANC concentration, υ_{eg}, and in T_{eg} with a deviation from the stoichiometric ratio. While the G_r values of samples from the series made from short PPT lie in the range between $A = 1$ and the phantom value $A_{ph} = 1/3$, the G_r value of networks of long PPT lie above the limiting value $A = 1$. It is obvious that one must assume the entanglement contribution εT_{eg} to explain the high G_r values of networks of long PPT. It was found that the theory adequately describes the experiment with $A_{ph} = 1/3$, $\varepsilon = 5 \cdot 10^{-4}$ mol cm^{-3} and T_{eg} and υ_{eg} determined from the theory of branching processes within the whole range $r_{HT} \geq 1$.

Networks of series B behave similarly to samples of series A. While the experimental G_r values of networks of short PPT lie below $A = 1$ (between $A_{ph} = 1/3$ and $A = 1$) in the whole range of r_{HT}, those of networks of long PPT in the range $0.7 < r_{HT} < 1.3$ are higher than predicted by the theory with $A = 1$ without G_{ent}. In this case too, the experimental data for all networks of series B can be adequately described by the theory of branching processes, with $A_{ph} = 1/3$, $\varepsilon = 5 \cdot 10^{-4}$ mol cm^{-3} and υ_{eg} and T_{eg} determined from the theory. The agreement between theory and experiment can be seen in Figure 6, which shows a generalized log G_r vs. log w_g plot (it has been shown earlier [10, 20] that this plot reduces the experimental error in the determination of conversion ξ and of the ratio r_{HT}). It can be seen in Figure 6 that the theory predicts a somewhat different dependence of υ_{eg} and T_{eg} on w_g for networks in the range $r_{HT} < 1$, compared with samples in the range $r_{HT} > 1$, which is given by the different molecular weights of PPT and MDI. It can be seen in Figure 6 that the experimental entanglement contribution $\Delta = G_r - \upsilon_{eg}/3$ is adequately described by the theoretical prediction εT_{eg} with $\varepsilon = 5 \cdot 10^{-4}$ mol cm^{-3}.

In networks of series C the expected decrease in modulus G_r takes place with increasing dilution, and

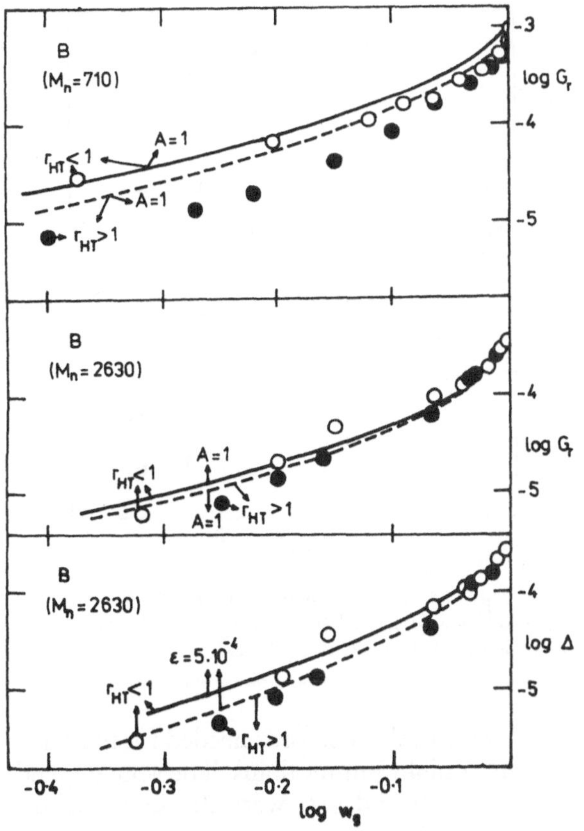

Fig. 6. Dependence of reduced modulus G_r (mol cm^{-3}) and topological contribution Δ (mol cm^{-3}) on weight fraction of gel w_g of networks of series B. (○, ●) experimental data; (——, – – –) theoretical dependences with given values A (for modulus G_r) and ε (for contribution Δ)

the magnitude of the decrease in G_r increases with increasing r_{HT} (Fig. 5). The theory also predicts a decrease in the concentration of υ_{eg} with dilution, due to the increasing number of elastically inactive cycles (EIC). It is expected, however, that the trapping factor T_{eg}, which is determined by the probability of contacts of EANC segments and decreases with dilution, due to both the increasing cyclization and the direct proportionality of $(v^0)^2$, will be affected much more. Hence, dilution at network formation has a decisive effect on the magnitude of T_{eg}, and thus also on the entanglement contribution G_{ent} to modulus G_r. A comparison between theory and experiment is shown in Figure 7 in the generalized log G_r vs. log w_g plot (the plot also reduces the effect of cyclization on the chemical concentration υ_{eg}, i. e. the dependence of υ_{eg} on w_g is universal, irrespective of dilution v^0). The quick decrease of experimental G_r values with dilution compared with G_r vs. w_g dependence observed for networks prepared in the bulk state, unambiguously favours the contribution of G_{ent} to the modulus G_r. Figure 7 also allows us to state that the experimental contribution $\Delta = G_r - \upsilon_{eg}/3$ is adequately described in the first approximation by the theory εT_{eg} with the same $\varepsilon = 5 \cdot 10^{-4}$ mol cm^{-3} value as in the networks of series A and B. The quick decrease in G_r with dilution is mainly due to the decrease in the T_{eg} value ($T_{eg} \sim (v^0)^2\ F(v^0)$).

Epoxy networks

Sample preparation

The networks were prepared from 4,4′-dihydroxydiphenyl-2,2-propane diglycidyl ether (DGEBA), phenylglycidyl ether (PGE) and 4,4′-diaminodiphenylmethane (DDM). Networks were prepared with a varying fraction of amino and epoxy groups r_A = 4[DDM]/([PGE] + 2 [DGEBA]) (in the range $1 \leqq r_A \leqq 2.3$) with four constant PGE contents characterized

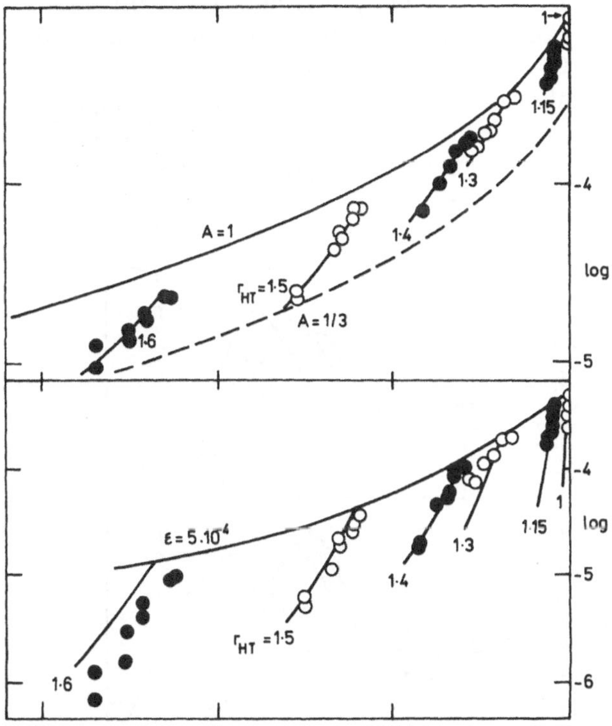

Fig. 7. Dependence of reduced modulus G_r (mol cm^{-3}) and topological contribution Δ (mol cm^{-3}) on weight fraction of gel w_g of dilute networks of series C. (○, ●) experimental data; (——, – – –) theoretical dependences with given A (for G_r) and ε (for Δ) values

by the fraction $s = [PGE]/([PGE] + 2\,[DEBA])$ ($s = 0$, 0.2, 0.33 and 0.5). The crosslinking [20] took place at 65 °C for 16 h and then at 150 °C for 5 h.

Viscoelastic behaviour

Figure 8 shows the dependence of the loss dynamic Young modulus E'' on the temperature of selected samples. With increasing excess of diamine at a constant monoepoxide content ($s = 0.33$) the main transition is displaced towards lower temperatures, and the height of the maximum in the E'' vs. T dependence increases (Fig. 8 a). Both these facts are due to the decrease in the network density with increasing r_A at constant s. The increasing PGE content (at constant modulus) displaces the main transition towards lower temperatures, and the height of the maximum increases (Fig. 8 b) as a consequence of the decreasing content of rigid DDM units in the networks. In all cases, however, the maxima in the E'' vs. T dependences have a roughly symmetrical shape, typical of amorphous systems, which also indicates completion of the crosslinking reaction.

For the most loosely crosslinked networks, for all PGE contents (prepared at the highest r_A), also at $T = 150$ °C, the time dependences of force, $f(t)$, and of optical retardation, $\delta(t)$ could be observed (Fig. 9). In all cases, these dependences could be described by Chasset-Thirion relation [20], thus yielding extrapolated equilibrium values of force needed for the calculation of the equilibrium modulus. The time dependences of both $f(t)$ and $\delta(t)$ were the same, so that $\delta(t)/f(t)$ is virtually time-independent (Fig. 9). This finding suggests homogeneity of the networks.

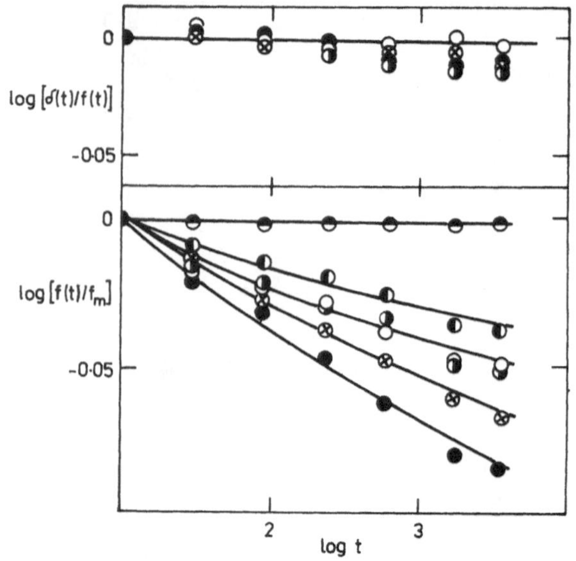

Fig. 9. Time dependence of reduced force $f(t)/f_m$ and ratio of optical retardation and force $\delta(t)/f(t)$. Samples denoted: (◓) $s = 0$, $r_A = 1.9$; (◐) 0, 2.1; (◑) 0, 2.3; (●) 0.2, 1.9; (⊗) 0.33, 1.6; (○) 0.5, 1.2

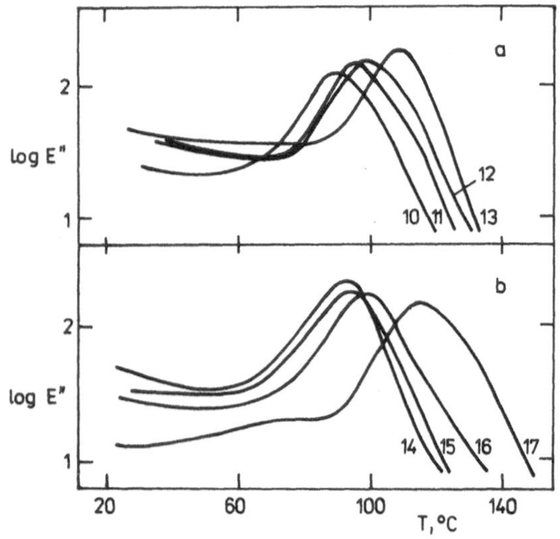

Fig. 8. Temperature dependence of dynamic loss modulus E'' (MPa). Networks denoted: a) $s = 0.33$; (10) $r_A = 1.6$; (11) 1.4; (12) 1.3; (13) 1.2; b) (14) $s = 0.5$, $r_A = 1.15$; (15) 0.33, 1.5; (16) 0.2, 1.8; (17) 0, 2.1

Sol fraction and critical ratios r_{Ac} at the gel point

Both the increasing excess of diamine (at constant s) and the increasing content of monoepoxide (at constant r_A) raise the weight fraction of the sol w_s (Fig. 10). The critical ratio values r_{Ac} rapidly increase with increasing s due to the fact that PGE produces free network ends. Using the r_{Ac} values and theoretical relations, the reactivity ratio [20], $\varrho = 0.19$, of the primary and secondary hydrogen atoms in the amino group DDM was calculated (experimentally determined conversions of epoxy groups ξ_E were considered in the calculation). Since it was found, moreover, that for $s = 0$, r_{Ac} is independent of the dilution of the system with chloroform ($r_{Ac} = 2.53$, 2.52, 2.53 and 2.52 for $v^0 = 1$, 0.8, 0.6 and 0.4), it can be said that in this system cyclization is not operative; this fact is determined by the high rigidity of the structural units.

A comparison between experimental w_s values and theory is given in Figure 10. It is obvious that the theory which considers the substitution effect in DDM and the final conversion of epoxy groups ξ_E adequately describes the experimental data.

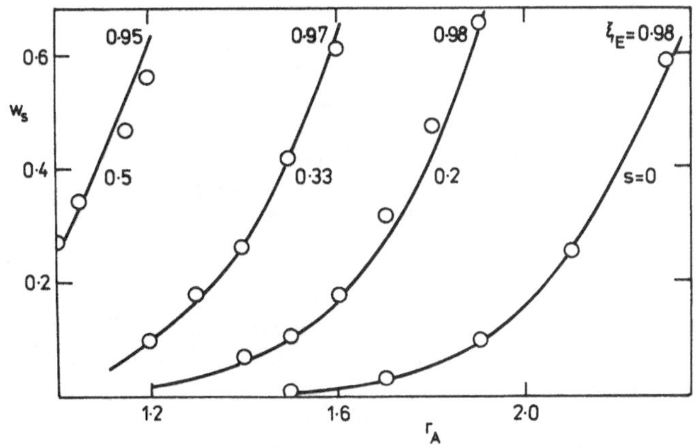

Fig. 10. Dependence of weight fraction of sol w_s on ratio of reactive groups r_A of epoxide networks. (○) experimental data; (——) theoretical dependences calculated for given conversions ξ_E

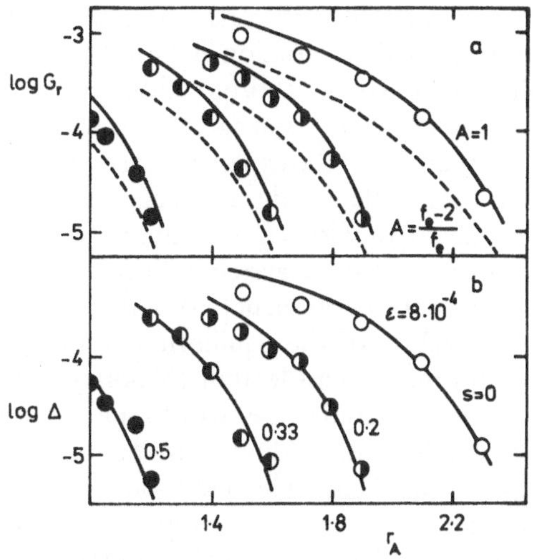

Fig. 11. Dependence of reduced modulus G_r (mol cm^{-3}) and topological contribution Δ (mol cm^{-3}) on ratio r_A. (○) experimental, (——, – – –) theoretical dependences with given A (for G_r) and ε (for Δ) values

Equilibrium elasticity

The values of the equilibrium moduli G_r (measured using dry unextracted networks at $T = 150\,°C$) quickly decrease with both increasing excess of diamine and increasing content of monoepoxides (Fig. 11). Figure 11 a also shows plots of the theoretical dependences $A\upsilon_{eg}$ both for $A = 1$ and for $A_{ph} = (f_e - 2)/f_e$ (the conversion ξ_E and the substitution effect in the amino group are taken into account in the calculation of υ_{eg} and f_e). It can be clearly seen that the experimental G_r data are adequately described by theoretical dependences with $A = 1$. It is also assumed (similarly to polyurethane networks) that the topological contribution $\Delta = G_r - (f_e - 2)\,\upsilon_{eg}/f_e$ (the dependence of which on r_A is demonstrated in Fig. 11 b) also contributes to the modulus G_r. The experimental Δ values as a function of r_A and s can adequately be described by the theory (εT_{eg}), using the constant $\varepsilon = 8 \cdot 10^{-4}$ mol cm^{-3}, independently of the monoepoxide and diamine content.

The generalized log G_r vs. log w_g dependence also leads to the same interpretation of the experimental equilibrium deformational behaviour of epoxy networks (Fig. 12). The superimposed G_r vs. w_g data can be adequately described in terms of the theory with the front factor $A = 1$ value without the entanglement contribution, or using $A_{ph} = (f_e - 2)/f_e$ with the contribution of trapped entanglements Δ (which can be described by theory, εT_{eg}, with $\varepsilon = 8 \cdot 10^{-4}$ mol cm^{-3}). The fact that it is not possible to decide which alternative more adequately describes the experimental behaviour is connected with the same form of theoretical dependences of the log υ_{eg} and log w_s on r_A. In this respect, the epoxy networks differ from polyurethane networks (cf. Fig. 7).

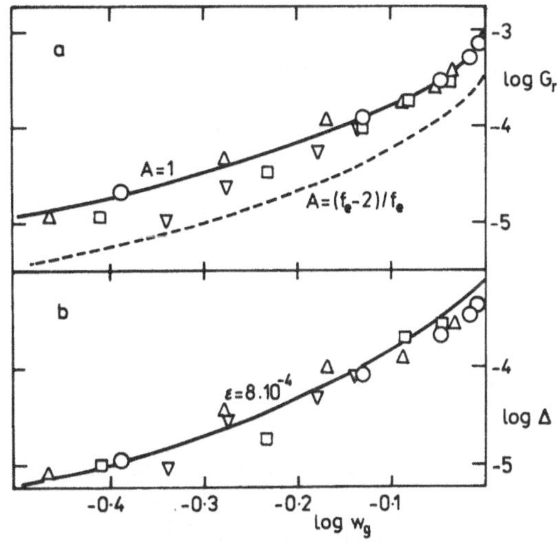

Fig. 12. Dependence of reduced modulus G_r (mol cm^{-3}) and topological contribution Δ (mol cm^{-3}) on weight fraction of gel w_g. Experimental data: (○) $s = 0$; (△) $s = 0.2$; (□) $s = 0.33$; (▽) $s = 0.5$; (——, – – –) theoretical dependences with given A (for G_r) and ε (for Δ) values

Conclusions

The relatively good agreement between the theoretical and experimental sol fraction for various model polyurethane systems, demonstrates the applicability of the branching theory. Also, the variation in the equilibrium modulus caused by variation in the initial composition (off-stoichiometry and/or addition of a monofunctional component) is in agreement with the theory. The comparison of the experimental equilibrium modulus with the calculated values suggests that a contribution, due to topological constraints on the modulus, is operative in these systems. In contrast to Flory's assumption, the permanent part of this contribution, i. e., the values of the equilibrium modulus extrapolated to the infinite tensile strain (Mooney-Rivlin constant C_1) or measured in the swollen state, is not negligible. The equilibrium modulus can be correlated with the simple trapped entanglement model of Langley and Graessley, based on the assumption of additivity of contributions, due to the covalent and trapped entanglement crosslinks.

References

1. Dušek K (1985) Brit Polym J 17:185
2. Dušek K (1986) Adv Polym Sci 78:1
3. Flory PJ (1979) Polymer 20:1317
4. Erman B, Wagner W, Flory PJ (1980) Macromolecules 13:1554
5. Kilian HG (1981) Polymer 22:209
6. Kilian HG (1985) Coll & Polym Sci 263:30
7. Langley NR, Polmanteer KE (1974) J Polym Sci Polym Phys Ed 12:1023
8. Pearson DS, Graessley WW (1980) Macromolecules 13:1001
9. Heinrich G, Straube E (1983) Acta Polymerica 34:589
10. Ilavský M, Dušek K (1983) Polymer 24:981
11. Ilavský M, Dušek K (1986) Macromolecules 19:2093
12. Allen G, Egerton P, Walsh DJ (1976) Polymer 17:65
13. Wals DJ, Higgins JS, Hall RH (1979) Polymer 20:951
14. Stanford JL, Stepto RFT (1977) Br Polym J 124:121
15. Stepto RFT (1979) Polymer 20:1324
16. Sung PH, Mark JE (1981) J Polym Sci Polym Phys Ed 19:507
17. Mark JE, Sung PH (1980) Eur Polym J 16:1223
18. Macosko CW, Benjamin GS (1981) Pure Appl Chem 53:1505
19. Dušek K, Ilavský M (1983) J Polym Sci Polym Phys Ed 21:1323
20. Ilavský M, Bogdanova LM, Dušek K (1984) J Polym Sci Polym Phys Ed 22:265
21. Dušek K (1984) Macromolecules 17:716
22. Ilavský M, Dušek K (1982) Polym Bull 8:359
23. Havránek A, Nedbal J, Berčík Č, Ilavský M, Dušek K (1980) Polym Bull 3:497
24. Havránek A, Ilavský M, Nedbal J, Böhm M, Soden WV, Stoll B (1987) Coll & Polym Sci 265:8
25. Dušek K, Ilavský M, Matějka L (1984) Polym Bull 12:33
26. Ilavský M, Bouchal K, Dušek K (1985) Polym Bull 14:295
27. Dušek K, Vojta V (1977) Br Polym J 9:164

Received January 29, 1987;
accepted March 27, 1987

Authors' address:

M. Ilavský
Institute of Macromolecular Chemistry
Czechoslovak Academy of Sciences
162 06 Prague 6, Czecheslovakia

Discussion

WINTER:

We did some experiments on PDMS which may be a little more tricky than the networks you are looking at. We measured the critical extent of reaction at the gel point and found that very different from the theoretical prediction, so I am impressed how close your theroretical predictions are with your experimental findings.

ILAVSKY:

I cannot judge your calculations. You do not believe you have any substitution effect in your network?

WINTER:

Could you please define the "substitution effect"?

ILAVSKY:

On the amine groups you have NH_2 groups. If the first hydrogen atom has already reacted, for the second hydrogen atom the reactivity *is slowed down or speeded up*. This can vary the critical conversion at the gel point very much. We, in our institute, are trying to solve this problem in a complex way. We are doing some experiments on low molecular weight substances to get more information about the reactivity, and to find the conversion really achieved. But I cannot judge how it is in your case.

WINTER:

Maybe I should specify a little further. We measured the functionality of our molecules, but then we assumed equal reactivity of the N's, so we did not have this additional adjustable parameter.

ILAVSKY:

This is, of course, not adjustable. You can determine it and so you can use it in theory as the value which is measured.

PAKULA:

One of the essential points of your theory is the calculation of the cycles. Could you explain in more detail how you do it?

ILAVSKY:

Well, this is only a first approximation to the problem. This is based on the Gaussian approximation, calculating the probability that two unreacted groups meet on the same place.

Progress in Colloid & Polymer Science Progr Colloid & Polymer Sci 75:21–22 (1987)

The dependence of the retardation spectrum width on the network chain length and on its distribution

A. Havránek

Department of Polymer Physics, Charles University V Holešovičkách 2, Prague, Czechoslovakia

Abstract: The relative portions and length distributions of the elastically active network chains and dangling chains of the idealised networks composed from difunctional and trifunctional units – the ratio of the units has been varied – have been calculated by the theory of branching processes. The results have been compared with the viscoelastic parameters obtianed by the measurement of model polyurethane networks. The main results of the comparison are given.

Key words: Polymer, viscoelasticity, model network, branching process, dangling chain.

Introduction

The theory of branching processes (TBP) (e. g. [1]) has been widely used in evaluation of equilibrium characteristics of polymer networks. Here I shall present how some structural insight to the viscoelastic characteristic of a network may be given on the basis of the TBP calculations. The viscoelastic characteristic under consideration is the retardation spectrum width.

The mean length of the elastically active network chains (EANC) and their distributions (polydispersities) as well as the relative portions and lengths of dangling chains (DC) have been calculated for idealised networks composed from difunctional and trifunctional units. The varying parameter in the calculations has been the ratio of the two types of units. The calculated results were compared with the experimentally obtained main transition zone viscoelastic spectrum of the polyoxypropylene triol (POPT) - diphenylmethane diisocyanate (MDI) model networks prepared from POPT with two different molecular masses [2]. The molar ratio r of the POPT functional units (OH terminal groups) to the MDI isocyanate groups has been varied in the samples. The width h of the transition zone is measured as the width of the retardation spectrum peak at its half-height.

From the comparison of the theoretical and experimental results we can make an estimate of the relative contributions of the EANC length, EANC polydispersity, and DC number and length to the spectrum width h. The widths h of the samples prepared from the POPT of larger molecular mass (long triol) are at the same r higher than the widths of the samples prepared from the POPT of smaller molecular mass (short triol). Therefore the width h increases with increasing length of EANC. The networks are nearly perfect if $r = 1$, polydispersity is 1 and nearly no DC are present. In this case the spectrum width Δh_e difference is only due to the difference of the EANC length of the samples prepared from different triols. From the TBP calculations we may find at which r the short triol sample has the same mean EANC length as is the EANC length of the perfect ($r = 1$) long triol sample. The width h of such a short triol sample has been found to be wider by Δh_d than is the width of the perfect long triol sample; Δh_d is approximately twice as large as Δh_e. The Δh_e represents the EANC length contribution and Δh_d the network disorder (polydispersity and DC) contribution to the width difference of the compared networks.

From this example and also from some more refined comparisons and evaluations we have concluded that the spectrum width increases with increasing EANC length and also with increasing disorder in the network. The disorder contribution to the spectrum width seems to be more important. Within the disorder contribution the role of DC is more decisive than the role of polydispersity of EANC.

The problem will be explained in more detail in the article which is now prepared with Dr. Dušek, Dr. Ilavský and Dr. Krakovský from the Institute of Macromolecular Chemistry, Prague. These men have done most of the work I referred here about.

References

1. Dušek K, Prins W (1969) Adv Polym Sci 6:1
2. Havránek A, Nedbal J, Berčik Č, Ilavský M, Dušek K (1980) Polym Bull 3:497
3. Havránek A, Dušek K, Ilavský M, Krakovský (1987) contributed to Coll & Polym Sci

Received February 2, 1987,
accepted March 5, 1987

Author's address:

A. Havránek
KFPY-MFF UK
V Holešovičkách 2
180 00 Prague 8, Czecheslovakia

Progress in Colloid & Polymer Science Progr Colloid & Polymer Sci 75:23–44 (1987)

Meander model of polymer melts and networks

W. Pechhold, M. Böhm, and W. v. Soden

Abteilung Angewandte Physik, Universität Ulm, Ulm, F.R.G.

Abstract: After reviewing the basic ideas of the meander model together with a short comparison with the coil model, the mechanical relaxation processes in polymer melts – fractions and mixtures – are discussed. This is done by decomposition of master curves of the dynamic shear-compliance into (a) glass relaxation with its plateau compliance J_{eN}, (b) the shearband process ΔJ_B^∞ and the rearranged shearband process ΔJ_B^0, and finally (c) flow relaxation and viscous flow. It will be shown how crosslinking decreases the relaxation strengths J_{eN} and ΔJ_B^0.

Key words: Shear compliance, mechanical relaxations, polymer melts, polymer networks, meander model.

Introduction

The meander model has been developed to describe molecular and supramolecular order in amorphous polymers, and to correlate all their macroscopic properties with the microscopic structure and its molecular parameters. First attempts have been made to understand consistently important features of polymer melts, e. g. viscous flow, rubber elasticity [1, 2], glass relaxation [3, 4], or crystallization [5] in the frame of this model. Moreover it is suggested for describing thin spread polymer films [2], thermoreversible gelation and (even dilute) polymer solutions [6]. In this paper, the relaxation processes controlling the dynamic shear-compliance of polymer melts and networks will be thoroughly discussed and compared with experimental results in dependence on molecular weight and crosslinking density. The effect of swelling on the shear compliance will be tackled in subsequent papers [6, 7].

The radius of gyration under unperturbed condition – comparison between the coil and the meander model

Average end-to-end distance R_e and radius of gyration R_g of a macromolecule of n segments of length l are defined by

$$R_e^2 = \left\langle \left(\sum_1^n \vec{l}_i\right) \cdot \left(\sum_1^n \vec{l}_j\right) \right\rangle \tag{1}$$

and

$$R_g^2 = \frac{1}{n} \left\langle \sum_1^n s_i^2 \right\rangle = \frac{1}{n} \left\langle \sum_1^n (\vec{r}_i - \vec{A})^2 \right\rangle$$

$$= \frac{1}{n} \left\langle \sum_1^n r_i^2 \right\rangle - \frac{1}{n^2} \left\langle \left(\sum_1^n \vec{r}_i\right) \cdot \left(\sum_1^n r_j\right) \right\rangle . \tag{2}$$

In Equation (2) the definition of the centre of mass, $\sum_1^n \vec{s}_i = 0$, $\vec{A} = \frac{1}{n} \sum_1^n \vec{r}_i$, was taken into account.

For the coil model of a freely orienting chain (Fig. 1) it is assumed that

$$\vec{l}_i \cdot \vec{l}_j = \delta_{ij} l^2 . \tag{3a}$$

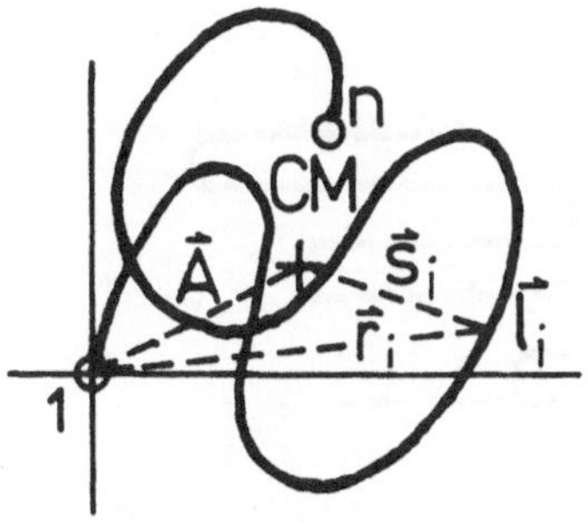

Fig. 1. Random coil model with its centre of mass (CM)

Therefore Equations (1) and (2) become, using $\vec{r}_i = \sum_1^i \vec{l}_i$,

$$R_e^2 = nl^2 \tag{1a}$$

$$R_g^2 = nl^2/6\,. \tag{2a}$$

For the coil model of a real chain with n denoting the number of backbone atoms, and l_{cc} its next neighbour distance, it was shown [8–10], that

$$R_e^2 = C_\infty n l_{cc}^2 \tag{1b}$$

$$R_g^2 = C_\infty n l_{cc}^2/6\,. \tag{2b}$$

C_∞ is called the characteristic ratio and can be calculated from the basic structure of the chain or experimentally determined via Equation (2b) e. g. by scattering methods or viscosimetry at θ-condition.

In the meander model, a two step procedure must be carried out to obtain R_e and R_g:

(i) Statistical tight folding and reptation of each molecule in the bundle (composed of 3 to 10 molecules) results in a one-dimensional (quasiparallel) coil appearance of its overall geometry. The correlation function, i. e. the scalar product $\vec{l}_i \cdot \vec{l}_j$ between the project ions of the main chain bonds i and j, is given by

$$\vec{l}_i \cdot \vec{l}_j = \sum_{f_k=0}^{k} (-1)^{f_k} \binom{k}{f_k} p_f^{f_k} (1-p_f)^{(k-f_k)} = (1-2p_f)^k \tag{3c}$$

in which p_f is the probability of tight folding per flexible main chain bond, $k = |i-j|$, the distance between bonds i and j, and $f_k < k$, a possible number of folds between these bonds. Applying Equation (3c) to definitions (1) and (2) and carrying out the various (up to four) summations, one obtains

$$R_e^{*2} = \frac{nl^2}{p_f}\left[1 - \frac{1-P^n}{2np_f}\right] \tag{1c}$$

$$R_g^{*2} = \frac{nl^2}{6p_f}\left[1 - \frac{3}{2np_f} + \frac{3}{2n^2p_f^2} - \frac{3(1-P^n)}{4n^3p_f^3}\right] \tag{2c}$$

with the abbreviation $P \equiv (1-2p_f)$. The asterisk indicates that these values refer to the straight bundle (without superfolding). Figure 3 gives a plot of the expressions in the brackets and a simple approximation to the second term which will be of practical use later on. It seems worth specifying the limiting cases:

$$R_e^{*2} = nl^2/p_f\,,\quad R_g^{*2} = nl^2/6p_f\,,\quad \text{for } np_f \gg 1\,,$$

i. e. for long chains, and

$$R_e^{*2} = n^2l^2\,,\quad R_g^{*2} = n^2l^2/12\,,\quad \text{for } np_f \ll 1\,,$$

i. e. for oligomers, which appear rodlike in the straight bundle.

(ii) Superfolding of the topological bundle or tube of (still unknown) diameter r – in which the statistically folded chain is arranged lengthwise – leads to a geometrical contraction of R_g^*: assuming 9-fold superfolding of the tube [11, 12, and below], and accepting space diagonal linking of the (cubic) meander blocks being formed (Fig. 4), that part of the tube $\Lambda^* = R_g^*\sqrt{12}$ which is labeled by the molecule under consideration gets shortened in length to $\Lambda \approx (\Lambda^*/9)\sqrt{3}$. Laterally, an averaged diameter of $3r$ must be taken into account. Taking the R_g-formula for a cylindrical block, the

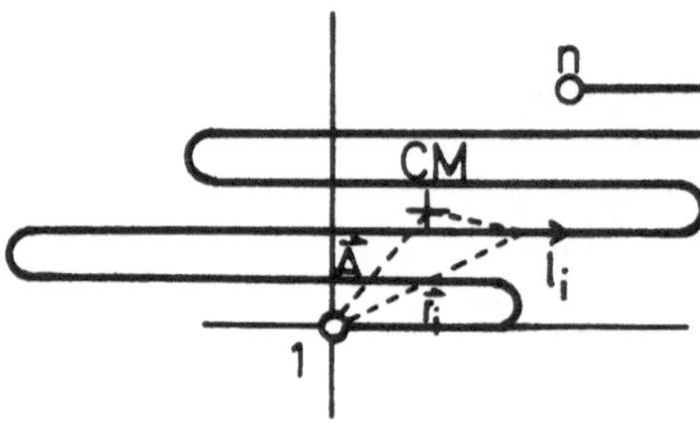

Fig. 2. Quasiparallel coil by statistically tight folding of the chain

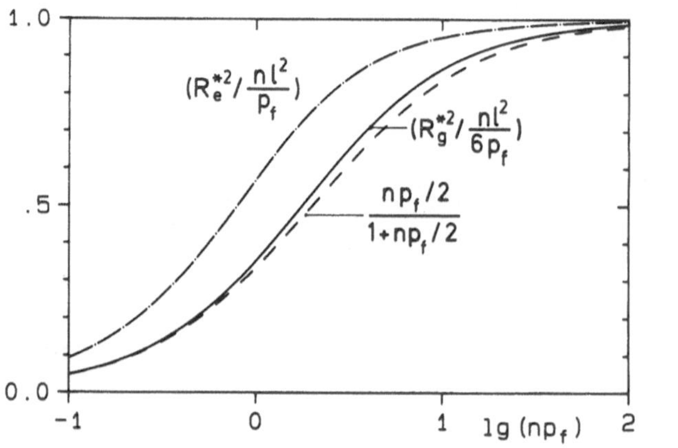

Fig. 3. Reduced molecular weight dependence of R_e^{*2} and R_g^{*2}

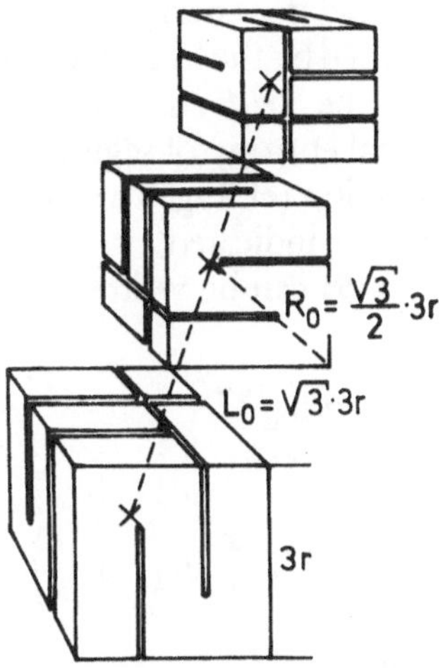

Fig. 4. Superfolding of one bundle into cubic meander blocks, linked across the space diagonal

squared radius of gyration of a labeled chain within the meander superstructure approximately reads [2, 11]:

$$R_g^2 \approx \frac{(3r)^2}{8} + \frac{R_g^{*2}}{12} = \frac{(3r)^2}{8} + \frac{nl^2\alpha}{162p_f} \tag{4}$$

α being given in the brackets in Equation (2c), plotted in Figure 3.

Postponing the dependence of r on molecular weight [2], which is discussed later, one may compare the expressions (2b) and (4) for higher M, when the first term in Equation (4) can be neglected, and $\alpha = 1$. From this comparison one obtains the relation

$$p_f^{-1} = 27\, C_\infty (l_{cc}/l)^2 \tag{5}$$

i. e. the reciprocal fold probability per main chain bond is proportional to the characteristic ratio. In Table 1, reciprocal fold probabilities (main chain bonds per fold) have been calculated using experimental C_∞-data gathered in [13]. $l_{cc} = 1.54$ Å was used throughout, whereas the bond length projection l (in the quasiparallel chain direction) is given approximtely in Table 1. Also inclueded are Δg_f values calculated via

$$p_f^{-1} = \exp(\Delta g_f / kT_m) \tag{6}$$

at the melting temperatures T_m indicated[1]), and (for some polymers) fold energies ΔU_f (actually ranges) determined by semiempirical potential calculation [2, 14]. For comparison of Δg_f with theoretical predictions one should add 2 to 4 kJ/mol to ΔU_f to account for an increase of vibrational free energy at the fold.

In the following, $n = M/M_o$, l_o and p_f^o will refer to the monomer: for vinyl-polymers n gets half the value as before, l and p_f are to be doubled; for BR and IR a factor of 4 must be applied. Equation (4) stays unchanged. But it can be rewritten using q, the average number of parallel stems (per cross-section of the

[1]) The temperature-dependence of p_f may be small because of immobilisation of 4–6 flexible bonds in a tight fold [11].

Table 1. Inverse tight fold probabilities p_f^{-1} per main chain bond and free energies of formation, derived from the characteristic ratio for various flexible polymers

Polymer	C_∞	l/Å	p_f^{-1}	$\Delta g_f/kT_m$	T_m/K	$\Delta g_f \Big/ \frac{\text{kJ}}{\text{mol}}$	$\Delta U_f \Big/ \frac{\text{kJ}}{\text{mol}}$
Polystyrene	9.5	1.1	500	6.2	(500)	26	16–22
Polyisobutene	6.3	1.1	335	5.8	(320)	16	12–16
Polyethylene	6.7	1.2	300	5.7	420	20	14–17
Isot polypropylene	5.0	1.1	265	5.6	460	22	17–25
Polybutene at.	6.6	1.1	350	5.9	(415)	20	
Polybutene isot.	18	1.1	950	6.9	415	24	
Polybutadiene	4.5	1.1	240	5.5	280	13	
Polyisoprene	4.7	1.0	300	5.7	300	14	
PVAc	8.8	1.1	465	6.1	(450)	23	
PMA	8.0	1.1	425	6.0	(360)	18	
PMMA	6.9	1.1	365	5.9	(450)	22	
PDMS ($l_{sio} = 1.65$)	5.2	1.2	265	5.6	233	11	

tight-folded molecule), instead of the fold probability: it is plausible to put

$$\Lambda^{*2} = 12\, R_g^{*2} = (nl_o/q)^2 \tag{7}$$

which yields, according to Equation (2c) and the approximation (cf. Fig. 3) $\alpha \approx \frac{np_f^o/2}{1+np_f^o/2}$

$$q = (1 + np_f^o/2)^{1/2} \tag{8}$$

and herewith

$$R_g^2 = \frac{(3r)^2}{8} + \frac{1}{324}\left(\frac{nl_o}{q}\right)^2. \tag{4a}$$

Meander model of polymers in the bulk

The meander model of amorphous (bulk) polymers relies on:

(i) The cluster-entropy-hypothesis (CEH) and

(ii) The assumption of chain parallelism of longer distances (i. e. the bundle concept) [2,11,12].

The excess free energy Δg_s^{sf} of superfolding per segment (cf. Eq. (2) in [2]) must include contributions form

(i) Superfolding of the bundles (segment line of r/d segments of length s)

(ii) Meander cube rotation (segment line of $(r+x)/d$ segments across the cube)

(iii) Shear fluctuation (molecule layer of $(r+x)^2/s \cdot d$ segments).

The statistical elements used for deriving the respective contributions Δg_{fold}, Δg_{rot} and Δg_{def} are given in the brackets. The latter two contributions have been discussed in [11, 15]: $\ln 3 > (-\Delta g_{\text{rot}}/kT) > 0.13$ (i/n-transition), depending on temperature and on the respective polymer; $(-\Delta g_{\text{def}}/kT) \approx 3 \ln 2 + \ln (3r/d)$.

The total excess free energy per chain segment in the meander reads

$$\frac{\Delta g_s^{sf}}{kT} = \frac{d}{r}\,\frac{s}{r+x}\left[\frac{\Delta g_f^*}{kT} + \frac{r}{d}\,\frac{\Delta g_k}{kT} - \ln\frac{[(r+x)/s]!}{(x/s)!(r/s)!}\right] + \frac{d}{r+x}\,\frac{\Delta g_{\text{rot}}}{kT} + \frac{d\cdot s}{(r+x)^2}\,\frac{\Delta g_{\text{def}}}{kT}. \tag{9}$$

The term in brackets, the superfold free energy per half meander layer Δg_{fold}, is determined by the free energies of one superfold Δg_f^* and of the $2r/d$ chain bends $(\Delta g_k/2)$, and by the orientational entropy of segment-lines which differ in chain direction (cf. Fig. 5 and the representative half meander layer indicated).

After rearranging, Equation (9) can be written as

$$\frac{\Delta g_s^{sf}}{kT} = \frac{s/d}{1+x/r}\left(\frac{d}{r}\right)^2\left[\frac{\Delta g_f^*}{kT} + \frac{\Delta g_{\text{def}}/kT}{1+x/r}\right] - \frac{1}{1+x/r}\,\frac{d}{r}\left\{\ln\left(1+\frac{x}{r}\right) + \frac{x}{r}\ln\left(1+\frac{r}{x}\right) - \left[\frac{\Delta g_{\text{rot}}}{kT} + \frac{s}{d}\,\frac{\Delta g_k}{kT}\right]\right\}. \tag{9a}$$

Minimization for d/r (at fixed x/r) yields the equilibrium bundle diameter r_o

$$\frac{r_o}{d} = 2\,\frac{s}{d}\,\frac{\Delta g_f^*/kT + \Delta g_{\text{def}}/kT\,(1+x/r)}{\ln\left(1+\frac{x}{r}\right) + \frac{x}{r}\ln\left(1+\frac{r}{x}\right) - \left[\frac{\Delta g_{\text{rot}}}{kT} + \frac{s}{d}\,\frac{\Delta g_k}{kT}\right]} \tag{10}$$

denoting the curved bracket $= B$, and introducing d/r_o, Equation (9a) gives

$$\frac{\Delta g_s^{sf}}{kT} = \frac{B/2}{1+x/r}\,\frac{d}{r}\left[\frac{d}{r}\,\frac{r_o}{d} - 2\right]. \tag{9b}$$

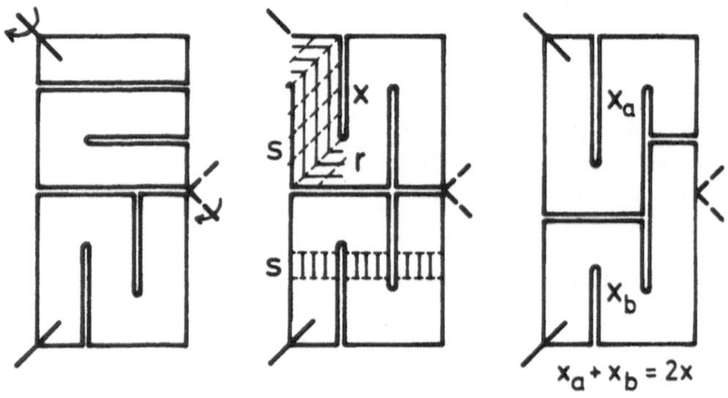

Fig. 5. Definition of the statistical elements for superfolding and cube rotation and the respective fluctuations

In equilibrium (relative to d/r) this becomes, using Equation (10)

$$\left.\frac{\Delta g_s^{sf}}{kT}\right|_{eq} = -\frac{B/2}{1+x/r}\frac{d}{r_o} = -\frac{d/4s}{\Delta g_f^*/kT}\,\frac{\left\{\ln\left(1+\frac{x}{r}\right)+\frac{x}{r}\ln\left(1+\frac{r}{x}\right)-\left[\frac{\Delta g_{\rm rot}}{kT}+\frac{s}{d}\frac{\Delta g_k}{kT}\right]\right\}^2}{1+x/r+\Delta g_{\rm def}/\Delta g_f^*}\,. \quad (11)$$

To discuss the most probable x/r (out of 2, 4, 6 and so on corresponding to meander cubes with 9, 25, 49 etc, times folded bundles), Equation (11) is plotted versus x/r in Figure 6, assuming $\Delta g_{\rm def}/\Delta g_f^* = -1$, with $\left[\frac{\Delta g_{\rm rot}}{kT}+\frac{s}{d}\frac{\Delta g_k}{kT}\right]$ as a parameter.

For $\left[\frac{\Delta g_{\rm rot}}{kT}+\frac{s}{d}\frac{\Delta g_k}{kT}\right] < 0.5$, $x/r = 2$ is favoured, i. e. the smallest meander cubes are the most probable ones. For larger bend energies ($\Delta g_k/2$), i. e. stiffer molecules, bigger meander cubes might be favoured, and at the limit, extended chains might be probable in the melt.

Fluctuations in x/r and r/d have to be allowed for (otherwise they must not be used as statistical variables) but must not voilate topology. Relying on CEH, these fluctuations can take place by reversibly coupling any two adjacent cubes in a suitable relative orientation (x/r-fluctuation, Fig. 5) and by changing the contour of the bundle cross-section without varying its area (r/d-fluctuation).

Because $\Delta g_{\rm def}/kT$ varies only slightly with r/d, the molecular weight dependence of r_o/d in Equation (10) must be due to $\Delta g_f^*/kT$: at lower M, pairs of chain ends are assumed [2] to be substituted for the tight chain fold within the superfold, thereby reducing $\Delta g_{f\infty}^*$ (the surperfold free energy for $M \to \infty$) to

$$\frac{\Delta g_f^*}{kT} \approx \frac{\Delta g_{f\infty}}{kT}(1-fw)\,, \quad (12)$$

$w = 3(r+x)/nl_o$ being the geometrical probability of finding a pair of chain ends within the contour length $3(r+x)$ per tight fold (within the superfold)[2]), and f a factor $\lesssim 1$[3]). Introducing Equation (12) into Equation (9a), after minimization for d/r, one obtains the equilibrium bundle diameter r_o, which now depends on molecular weight (r_o/d from Eq. (10) is now written as r_∞/d)

$$\frac{r_o}{d} = \frac{r_\infty}{d}\left[1+\frac{\Delta g_{f\infty}^*}{kT}\,f\,\frac{3s(1+x/r)}{Bnl_o}\right]^{-1}$$

$$= \frac{r_\infty}{d}\left[1+\frac{f}{2nl_o}\,\frac{3r_\infty(1+x/r)}{1+\Delta g_{\rm def}/\Delta g_{f\infty}^*(1+x/r)}\right]^{-1}. \quad (10a)$$

This expression can be shown to describe well the molecular weight dependence of several properties (e. g. of the plateau compliance or of the Cotton-Mouton effect) related to the meander cube size. Equation (10a), introduced in Equation (4a), should explain the M-dependence of the radius of gyration in the unperturbed case also for lower molecular weights. Figure 7 shows a plot of $\log R_g$ versus $\log M$ calculated for polystyrene, using the parameters $(1/p_f^o) = 250$, $x/r = 2$,

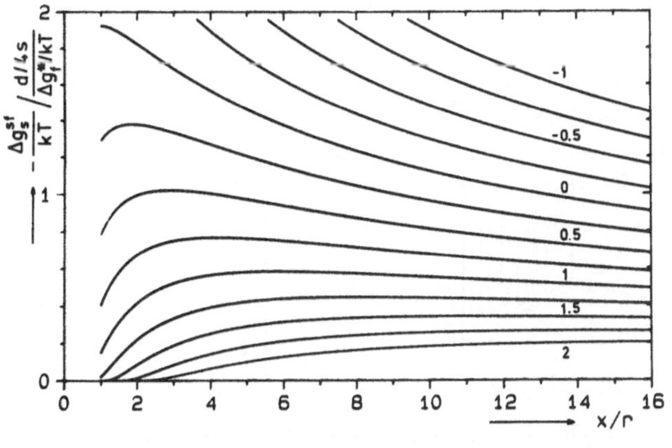

Fig. 6. Reduced free energy of superfolding in dependence on x/r

[2]) In the preceding papers [1, 2, 3, 11, 15] it had been assumed that $w = r(r+x)/nl_o d$, i. e. that a pair of chain ends within a *half meander layer* would reduce its superfold energy. This is in contradiction with the strict bundle concept. Moreover, the actual superfold free energy Equation (12) was introduced into Equation (10) after minimization, a procedure also inadmissable. By chance, both mistakes almost cancel each other out, and the result (for $f = 0.5$) is not far from the above Equation (10a) with $f = 1$.

[3]) $f < 1$ must be applied, either if not all of the chain ends, but only a fraction, f, of those for idealized geometry are substituted into the superfolds, or if all chain ends do enter the superfolds but reduce $\Delta g_{f\infty}^*$ to $\Delta g_{f\infty}^*(1-f)$ rather than to zero.

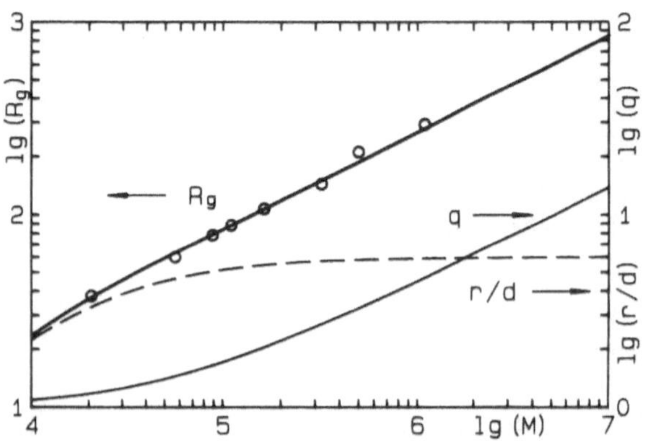

Fig. 7. Molecular weight dependence of the radius of gyration R_g, reduced bundle diameter r/d, and average number of parallel stems in a molecule as calculated for aPS

$r_\infty/d = 6, f = 1, M_o = 104$ g/Mol, $d = 2s = 4\, l_o = 8.8$ Å. The thick drawn curve is a nearly straight line with slope 1/2 and nicely fits the data points from SANS on bulk PS [16]. The other two curves give the M-dependence of r/d Equation (10a) and q Equation (8).

Plateau-compliance J_{eN} of polymer melts and networks as explained by intra-meander shear deformation

The most remarkable property of polymer melts or networks is rubber elasticity, which can only be explained by assuming paraelastic segmental movement yielding entropic restoring forces. This can be nicely studied by analyzing dynamic shear compliances versus frequency. For melts and amorphous networks, the frequency/temperature superposition is valid, as a rule, yielding mastercurves (over more than 20 decades in frequency, Fig. 8) and the proper shift factors on both axes.

In the coil model, the plateau-compliance (also well defined in higher molecular weight polymer melts between the glass-relaxation and the flow region) is attributed to the number ν of network strands per unit volume between entanglements or crosslinks and reads in the simple case

$$J_{eN} = (R_o^2/R^2)/\nu kT\,. \tag{13}$$

In the meander model, the plateau compliance can be explained by the relaxation-strengths of intra-meander shear deformation modes (Fig. 9) applying Reuss-averaging. In the statistical derivation of the bulk anisotropic compliances, layers of molecules are allowed to be displaced by one chain distance d against each other. This theoretical model has been worked

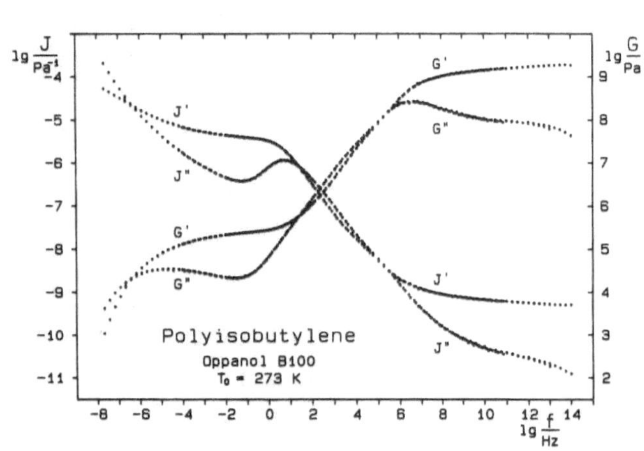

Fig. 8. Shear compliance — and shear modulus — master curves as deduced from measurements over 6 decades (10^{-4} — 200 Hz) at 6 temperatures

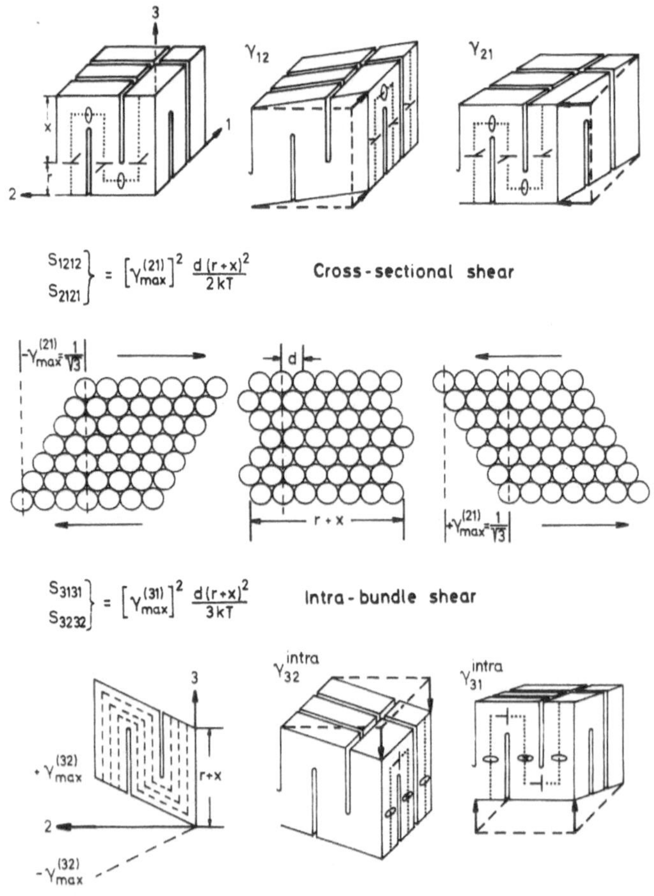

Fig. 9. The two main intra-meander shear modes demonstrated on single meander cubes with dislocations indicated on the surfaces

through in detail in preceding papers [2,15] and yields, for the plateau compliance J_{eN}^{o} of an uncrosslinked polymer melt

$$J_{eN}^{o} = \frac{1}{5}\begin{Bmatrix} S_{1212} + S_{2121} \\ + S_{3131} + S_{3232} \end{Bmatrix} = \frac{d(r+x)^2}{9kT}. \qquad (14)$$

Taking J_{eN}^{o} from the decomposition of master curves of various polymer melts, Equation (14) represents probably the best method to determine the cube side length r or the average number of quasiparallel stems $(r + x)/d$ in a layer of molecules [3].

If chemical crosslinks are present, say p_c per polymer segment, they will bridge adjacent layers of molecules in the meander cube and possibly hinder their relative displacements. But only a fraction p_c^* of the crosslinks present will effectively block the one d-displacement in the intrameander shear, as is illustrated in Figure 10 (for the cross-sectional shear): in the first two lines, uneffective crosslinks (open symbols) are shown, which can always be formed and which do not block (21)- and (12)-shear deformation. The third type of crosslink (filled symbol) can only be formed in one position and will block subsequent shear deformation (otherwise it must be broken). From this scheme, one concludes a geometrical probability $w^* = 1/7$ for a given crosslink to be effective. But the realistic w^* will be smaller, because of different crosslinking reaction probabilities. These should be larger for the uneffective crosslinks (between chains remaining in contact during shear deformation) than for the effective crosslinks (between chains being in contact only for short times). We can therefore write

$$p_c^* = w^* p_c \; (w^* < 1/7) \qquad (15)$$

and will get from experiment $w^* \approx 1/28$ for radiation crosslinked IR as below.

To consider how effective chemical crosslinks influence plateau compliance (Eq. (14)) – which until now has been generated by topology – we multiply the maximum intrameander shear $\gamma_{max} = 1/\sqrt{3}$ by $P(p_c^*)$ and, after [2,15] get

$$J_{eN} = P(p_c^*)\, d \cdot (r + x)^2/9kT. \qquad (14a)$$

$P(p_c^*)$ is the probability of finding no single effective crosslink between adjacent meander layers (consisting of $m = (r + x)^2/d \cdot s$ segments). Using Poisson distribution, one may write

$$P(p_c^*) = \exp(- m p_c^*). \qquad (16)$$

Figure 11 gives preliminary results from an analysis of the master curves of radiation crosslinked IR (cf. Figs. 19, 20). The scatter at low doses is probably due to an uncertainty of specimen area and thickness (within the double sandwich holder). Evaluating Figure 11, Equation (16) yields $mp_c^* \approx 4.6 \cdot 10^{-4} \cdot$ Dose/kGy. From equilibrium swelling, $p_c \approx 2 \cdot 10^{-5} \cdot$ Dose/kGy, approximately. Combining these two results with Equation (15), $m \cdot w^* \approx 23$. Assuming for IR $3r/d \approx 21$,

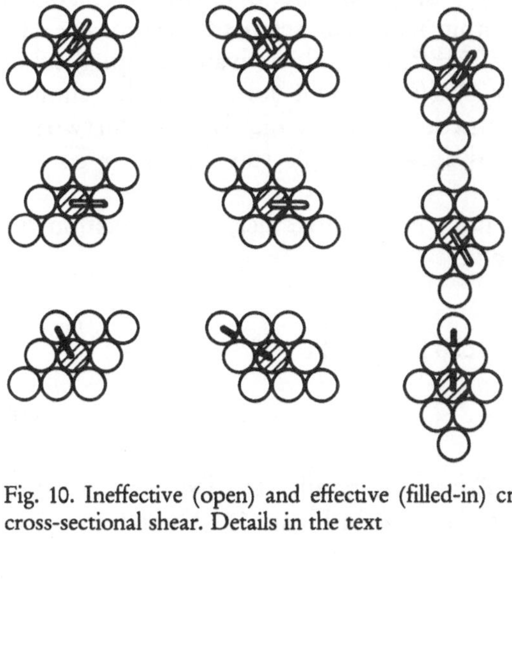

Fig. 10. Ineffective (open) and effective (filled-in) crosslinks for cross-sectional shear. Details in the text

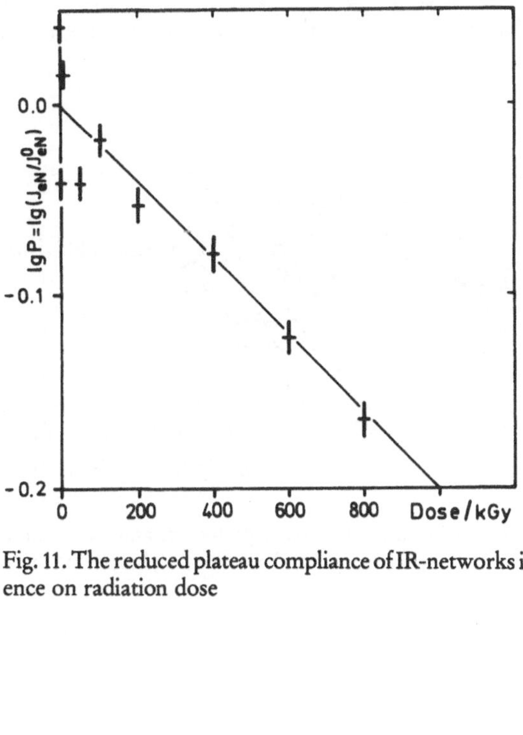

Fig. 11. The reduced plateau compliance of IR-networks in dependence on radiation dose

$d \approx 6$ Å, $s = 4.05$ Å, one obtains $m = (3r)^2/d \cdot s \approx 650$ and $w^* \approx 1/28$. From Equation (14) $J^0_{eN} \approx 2.8 \cdot 10^{-6}$ m²/N follows in agreement with experiment.

Glass relaxation

The concept of chain parallelism, realized by tight-folding of molecules into bundles which are superfolded to meander cubes, offers the possibility of introducing (quasi-)dislocations in the equilibrium melt. Whereas the viscous flow in an entangled system can only take place by chain reptation, paraelastic shear fluctuations (i.e. rubber elasticity) will be mediated by dislocation movement, as in plastic flow and the preceding anelasticity in crystalline solids.

To put such a dislocation concept into a quantitative form [3, 4], one has to start with the idea of dislocation walls (or small angle boundaries), because the stress fields of the edge-dislocations tend to cancel other in a wall. Figure 12 shows a cross-section of parallel chains with two edge-dislocation walls of opposite sign, which may have been formed by elastic shear fluctuations generating dislocation rings (neighbouring dislocations of opposite sign are assumed to be closed by parts of the screw-type within the superfolds). Possible dislocation arrays can be seen (following the dotted lines) on the surfaces of the meander cubes in Figure 9. They might have been created at the superfold and edges on one side and might leave that area on the other side, thereby producing one b-displacement across the surface.

Theres is, of course, no chance for such a dislocation network to be thermally generated – not even for a single dislocation ring – if its total energy has to be compared with kT. But if one takes the CEH into account, one has to ask for the representative statistical element (and its free energy) which can from such a network by appropriate clustering. The proper element is the "dislocation-segment line" or "s-dislocation", the number of chain segments in one line connecting the two walls. Its energy, ε_s, follows from the wall energy per unit length of a dislocation and becomes approximately [3, 4, 17]

$$\varepsilon_s \approx 0.3 \frac{Gb^2 d}{4\pi(1-\nu)} \tag{17}$$

with G being the shear modulus of the glassy state (extrapolated to the temperature of interest), b the Burgers vector of the dislocations, d the average chain distance and $\nu(\approx 1/3)$ the Poisson ration of the glass.

To determine ε_s – and from this the Burgers vector b which is the only unknown quantity in Equation (17) – the activation curves $f_{max}(10^3/T)$ obtained from G''- or ε''-maxima for example must be analyzed in the frame of the meander model. The relaxation frequency of a chain segment is determined by two factors:

(i) By an Arrhenius type factor $(\nu_o/\pi)\exp(-Q_y/kT)$ accounting for the intermolecular part of the activation and

(ii) By an intermolecularly caused factor, representing the probability that an intramolecularly activated segment will really jump. For the intrameander shear fluctuations to be fully activated, it is assumed that every segment line across coupled cubes contains *at least one* segment which is part of a dislocation wall and therefore has increased its energy by ε_s. If $\exp(-\varepsilon_s/kT)$ is the probability of finding such a segment at a fixed place [3], the probability of having at least one of this kind within $3r/d$ chain segments becomes $1-[1-\exp(-\varepsilon_s/kT)]^{3r/d}$. For fully activated coupled cubes, some $3(3r/d)^2 d/s$ segment lines must be simultaneously in this defect state (which is realized if two perpendicular dislocation walls run through a coupled cube).

Putting both factors together, the relaxation frequency reads:

$$f_{max} = \frac{f_o}{\pi} \exp\left(-\frac{Q_y}{kT}\right)\left[1-\left(1-e^{-\frac{\varepsilon_s}{kT}}\right)^{\frac{3r}{d}}\right]^{3\left(\frac{3r}{d}\right)^2 \cdot \frac{d}{s}}. \tag{18}$$

Formula (18) has been fitted to the dielectric relaxation frequencies (from ε''_g) in Figure 13. These frequencies

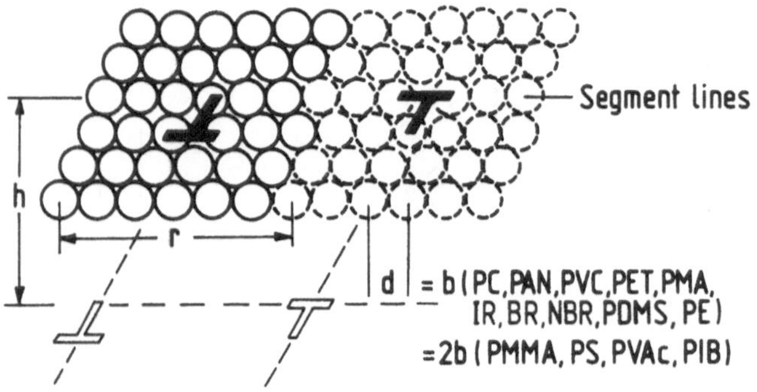

Fig. 12. Edge-dislocation wall concept for intrameander shear deformation. b, Burgers-vector; d, chain distance; h, spacing of the dislocation array

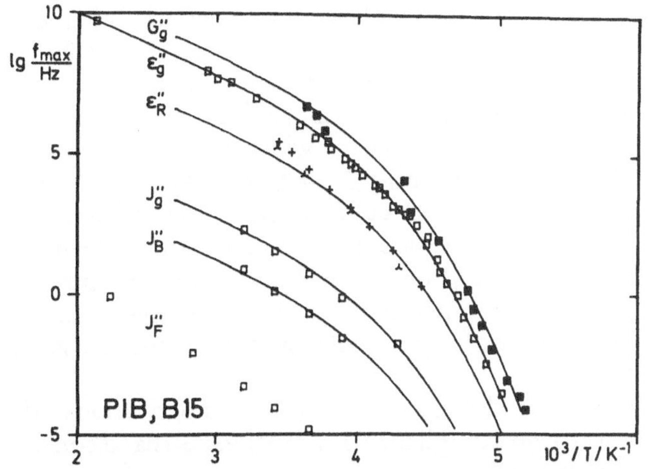

Fig. 13. Activation diagram for poly(isobutylene) from dielectric and mechanical measurments. Glass relaxation frequencies taken from G''-, ε''-, J''-maxima. Cube rotation from label relaxation ε''-maxima. Flow relaxation and shearband relaxation from J-master-curves

are approximately equal to those determined from volume and enthalpy relaxation, as has been shown for PVAc [18]. A good fit (in Fig. 13) wa achieved with the parameters[4]):

$f_o = 5 \cdot 10^{14}$ Hz, $Q_y = 40.5$ kJ/Mol, $\varepsilon_s = 2.55$ kJ/Mol, $3r/d = 18$ (and using $d = 6.5$ Å, $s = 4.6$ Å). The dielectric relaxation frequency spectrum, with its high frequency tail, corresponds to the equilibration of the dislocation distribution which finally gives the total relaxation strength. During equilibration all segments were touched upon by a moving dislocation that provided the free volume necessary for the rearrangement of the individual dipoles.

Deformation modes of the meander superstructure

Interestingly, the glass relaxation frequency of the shear modulus G is higher by a factor of about 6 than the dielectric one (Fig. 13). If one assumes that single steps of a dislocation[5]) are the paraelastic events for the onset of the G-relaxation, it is of no importance in which of the r/d layers of stems in a bundle the dislocation falls in, whereas in the dielectric case it takes $r/d \approx 6$ times longer for it to touch nearly all layers.

In Figure 13, the dielectric relaxation frequencies of another process (with subscript R), taken from [19], have been added. This process is due to label molecules mixed into PIB with 1 to 3 weight percent, and is absent in pure PIB. The crosses refer to tetraisobutenyl-succinic anhydride (1.4 nm in length) and the stars to poly-isobutenyl-succinic anhydride (PIBBSA, $M_n = 1100$, i. e. 4.6 nm in length). The anhydride dipole probably has a component in chain direction and can only relax either by rotation around the molecule's short axis or by rotation of the whole meander cube (around its space diagonal). Because the relaxation frequencies are about equal for both labels – though differing in length by a factor of more than 3 – the meander cube rotation seems to be the most probable explanation, at least for the longer label PIBBSA. Further investigations with other dipolar labels in the chain direction and in other matrices have been started [20] and are in progress. This rotational relaxation is important also in LC-polymers with mesogenic side groups carrying an aligned dipole, and probably explains the δ-process dielectrically observed in these polymers [21, 22].

Whereas the meander cube rotation may contribute to dielectric relaxation and to orientation correlation phenomena (e. g. in magnetic birefringence or dynamic light scattering) it does not directly[6]) manifest itself in mechanical relaxation. But there are three other mechanically active relaxation processes in a commercial polymer melt which can be analyzed, e. g. by decomposing its shear-compliance master-curve into a few (symmetric) components. This has been done in Reference [1] – using Fuoss-Kirkwood functions – for PIB B15 ($M_v = 82800$) yielding relaxation strengths, widths and frequencies for (a) the glass process in the compliance, (b) the shearband process and (c) the flow relaxation additional to the viscous flow. The activation diagram (Fig. 13) gives the relaxation frequencies for these processes (determined from J''_g-, J''_B-, J''_F-maxima in [1], Fig. 1) which are not far from parallel to the more rapid processes already mentioned.

a) The glass process

The strength of the glass relaxation is the plateau compliance, discussed above. Its relaxation frequency

---

[4]) In this example, the frequency factor is extremely high and can only be understood by assuming $Q_y = Q_y^{\text{intra}} + Q_{y_0}^{\text{inter}} (1 + T\partial \ln G/\partial T)$, i. e. with the temperature dependence of the shear modulus G of the glass.

[5]) From the average concentration of the (somewhat abstract) s-dislocations, $x = \exp(-\varepsilon_s/kT)$, which is about 0.2 near the glass-relaxation, one may conclude that on average there will be one dislocation per bundle cross-section. This conclusion relies on the assumption of similar distances between dislocations in a wall and between walls (cf. Fig. 12).

[6]) Nevertheless, meander cube rotation is a presupposition for the formation and annihilation of shearbands and guarantees orientational disorder within the polymer melt or network at a scale of several cube diameters.

deviates from that of the shear modulus by 5.9 decades in PIB (Fig. 13). In other polymers, this factor is found to be much lower: e. g. PVAc, 3.7 ([18], Figs. 5, 6, 16) or PMMA 4.0. Its lower limit, for one single relaxation frequency, should be $\lg(f_G/f_J) = \lg(J_{eN}/J_\infty)$ which is 3.8 for PIB, 3.2 for PVAc and about 3.0 for PMMA (J_∞ being the compliance of the glass). The larger experimental ratios are certainly a consequence of a spectrum of relaxation frequencies which still has to be discussed in the meander model. It is an open question whether one could explain the different amounts of broadening (2.1 decades for PIB, 0.5 for PVAc and 1.0 for PMMA).

Moreover, it should be kept in mind that there is a broad high frequency tail of the G_g-relaxation (in Fig. 8) probably due to the differences in activation energy of single dislocation steps. In the dielectric spectrum, the asymmetry of the ε_g-relaxation is less pronounced (i. e. averaged over several steps). In the frequency range below the G_g''- and ε_g''-maxima the relaxation frequency spectrum narrows (for some polymers near the glass temperature, nearly to the Debye case). This may be due to the existence of a lattice of dislocation walls with well defined creation and annihilation probabilities for the dislocation rings.

b) The shearband process

So far, the intrameander shear, i. e. shear fluctuations — restricted to one chain distance per layer of molecules — of coupled meander cubes have been discussed. In equilibrium, they explain the plateau compliance Equations (14), (14a), but the can account only for a maximum shear $\gamma_{max}^{intra} \approx 1/\sqrt{3}$ (Fig. 14a). Beyond these fluctuations, large scale deformations become possible by interbundle displacement, i. e. by unfolding of suitably arranged meander cubes (Fig. 14b). To this purpose, whole files of meander cubes or even layers of them must cooperate in a shearband process, the stress-strain behaviour of which was first derived in Reference [15] with $\gamma_{max}^{inter} = 9$ (because of the 9 half meander pieces per cube). From this characteristic of an already developed shearband, the stress-dependent concentration β of shearbands and the superposition of intra- and interbundle shear, the nonlinear stress-strain curve of any type of deformation can be evaluated.

In the following, we will present a derivation of the shearband-compliance ΔJ_B and the shearband concentration β, taking into account that (i) only a fraction, ϕ, of segments (e. g. the polymer content in a swollen sample) has lost its free energy of rotation Δg_{rot} within the shearband, and (ii) only a fraction, $\phi\xi$, of it has kept the orientation correlation in the deformed shearband. In Figure 15, a 3-dimensional view of a short piece of a

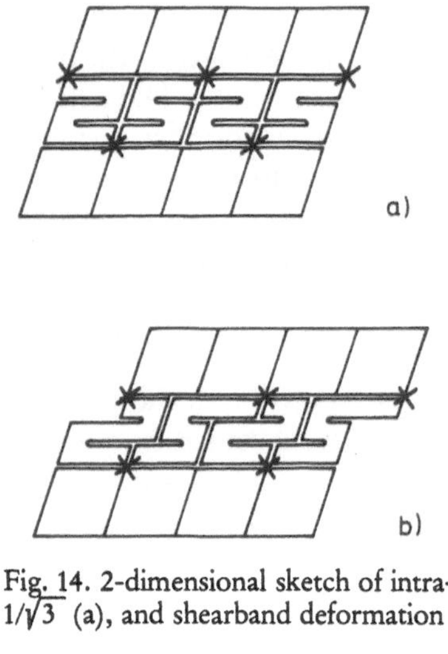

Fig. 14. 2-dimensional sketch of intra-meander shear with $\gamma_{max}^{intra} = 1/\sqrt{3}$ (a), and shearband deformation with $\gamma_m \approx 9$ (b)

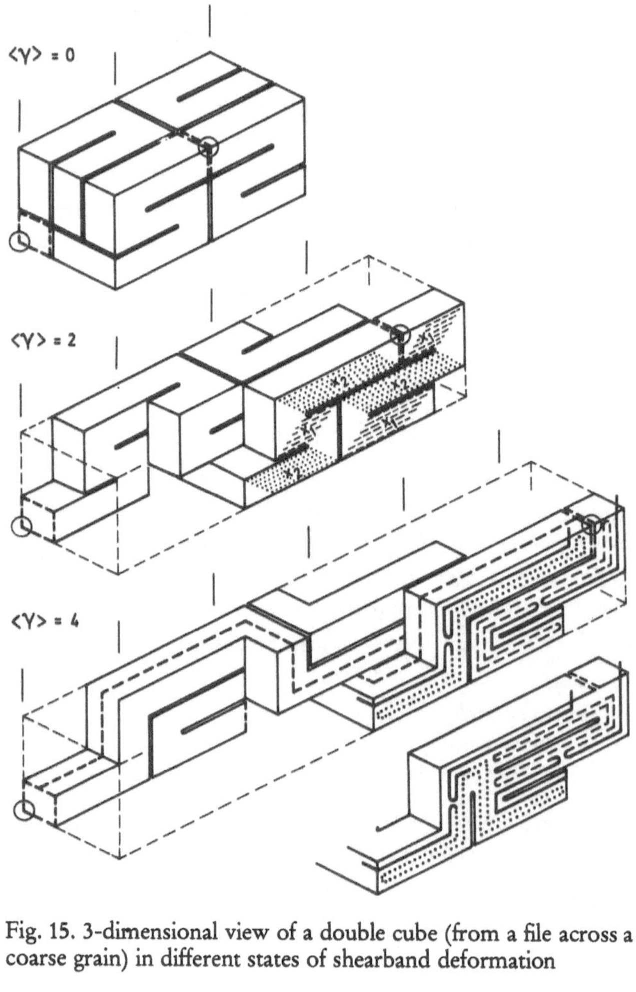

Fig. 15. 3-dimensional view of a double cube (from a file across a coarse grain) in different states of shearband deformation

shearband (two cubes from a file across the coarse grain) is shown in different states of shear deformation γ_{inter}. It is obvious that this deformation subdivides the set of fold heights, x, into two classes with averages x_2 and x_1, and can be most simply defined by

$$\gamma_{inter} = \gamma_{max} \frac{x_2 - x_1}{2x} = \gamma_m \cdot \gamma_R \,. \tag{19}$$

To describe the change in free energy Δg_{inter} during formation and deformation of shearbands, a statistical element, representative of the geometrical changes but as small as possible, must be chosen. We accept one strand of $3(r + x)/s$ segments from a cube layer and obtain, using Equation (9a)

$$\frac{\Delta g_{inter}}{kT} = -\phi \cdot \xi \cdot 3\frac{d}{s} \cdot \frac{1}{2} \left\{ \begin{aligned} &\ln\left(1 + \frac{x_1}{r}\right) + \frac{x_1}{r}\ln\left(1 + \frac{r}{x_1}\right) - 2\ln\left(1 + \frac{x}{r}\right) \\ &+ \ln\left(1 + \frac{x_2}{r}\right) + \frac{x_2}{r}\ln\left(1 + \frac{r}{x_2}\right) - 2\frac{x}{r}\ln\left(1 + \frac{r}{x}\right) \end{aligned} \right\}$$

$$-\phi \cdot 3\frac{d}{s} \cdot \frac{\Delta g_{rot}}{kT} + 3d^2(r + x)\frac{\sigma}{kT}\left[\gamma_{inter} - \gamma_m \cdot \gamma_R\right] + \lambda(2x - x_1 - x_2) \,. \tag{20}$$

The first two terms describe the deviation from the normal superfold free energy per strand (keeping the number of superfolds, the chain bends and the intrameander shear fluctuations constant), the last two terms comprise both geometrical conditions with Lagrangian multipliers as front factors. $3d^2(r + x)$ is the volume of a representative strand; $\sigma = \partial(g/V)/\partial\gamma$.

After minimization of Equation (20) for x_1/r and x_2/r one obtains

$$\frac{\gamma_R}{1 + \frac{x}{r}(1 - \gamma_R^2)} = \tanh\left\{\frac{s \cdot d \cdot r(r + x)}{x \cdot \phi \cdot \xi}\frac{\gamma_m \sigma}{kT}\right\} \tag{21}$$

and the equilibrium free energy

$$\left.\frac{\Delta g_{inter}}{kT}\right|_{eq} = -\phi \cdot \xi \cdot \frac{3}{2} \cdot \frac{d}{s}\left\{\left(1 + \frac{x}{r}\right) \times \ln\left[1 - \left(\frac{\gamma_R}{1 + r/x}\right)^2\right] - \frac{x}{r}\ln\left(1 - \gamma_R^2\right)\right\}$$

$$-\phi \cdot 3\frac{d}{s}\frac{\Delta g_{rot}}{kT} + 3d^2(r + x)\frac{\sigma}{kT} \cdot \gamma_{inter} \,. \tag{22}$$

From Equations (21) and (19) the anisotropic shear compliance of a given shearband can be deduced and reads, in the limit of $\sigma \to 0$,

$$2S_{3131}^{inter} \equiv \frac{\partial\gamma}{\partial\sigma} = \frac{s \cdot d \cdot (r + x)^2}{x \cdot \phi \cdot \xi}\frac{\gamma_m^2}{kT} \,. \tag{23}$$

To get the macroscopic relaxation strength due to shearband fluctuation one has to know the average concentration β of shearbands. To this aim, the free energy Δg_{mix} per representative strand – a fraction β of it in shearbands and $(1-\beta)$ in normal meander cubes – is written down (applying CEH)

$$\frac{\Delta g_{mix}}{kT} = \beta\frac{\Delta g_{inter}}{kT} + \beta\ln\beta + (1 - \beta)\ln(1 - \beta)$$

$$+ 3d^2(r + x)\frac{\sigma_{31}}{kT}\left[\gamma_{31} - \gamma_{intra} - \beta\gamma_{inter}\right] \tag{24}$$

with the last term expressing the superposition of a homogeneous intrameander shear and a contribution from active shearbands. The concentration β follows by minimization of Equation (24) and substituting $\Delta g_{inter}/kT$ from Equation (22)

$$\beta = \left\{1 + \exp\left[-\phi\, 3\frac{d}{s}\frac{\Delta g_{rot}}{kT}\right] \cdot A(\gamma_R)\right\}^{-1} \tag{25}$$

where

$$A(\gamma_R) = \{(1 - \gamma_R^2)^{x/r}/[1 - \gamma_R^2/(1 + r/x)^2]^{(1 + x/r)}\}^{\frac{3}{2}\frac{d}{s}\phi \cdot \xi}$$

tends to unity with $\sigma \to 0$.

Combining Equations (23) and (25) and applying Reuss-averaging, the macroscopic shearband relaxation strength in the linear case can be written

$$\Delta J_B = \beta(\phi)(S_{3131} + S_{3232})/5$$

$$= \beta(\phi)\frac{s \cdot d(r + x)^2\gamma_m^2}{5x\,kT\,\phi\xi} = \frac{\beta(\phi)}{\beta(1)}\frac{\Delta J_B^\infty}{\phi\xi} \tag{26}$$

where $\beta^{-1}(\phi) = 1 + \exp\left[-\phi\, 3\frac{d}{s}\Delta g_{rot}/kT\right]$, and ΔJ_B^∞ denotes the relaxation strength for the bulk polymer melt without any loss of orientation correlation within the shearbands

$$\Delta J_B^\infty = \beta(1)\; s \cdot d(r+x)^2 \gamma_m^2 / 5x\, kT\,. \tag{26a}$$

In conclusion, it should be kept in mind that Equations (21) and (25) should also be valid for larger stresses and allow one to derive the main features of the stress-strain behaviour, provided the orientation of the shearbands can be properly described for any macroscopic deformation mode (cf. the model for uniaxial deformation in Ref. [15]) to replace the isotropic Reuss-averaging.

c) *The flow relaxation process*

Besides viscous flow, polymer melts show a recoverable strain if the stress is released. In dynamic mechanical investigations, this corresponds to a paraelastic relaxation process taking place in the same frequency range as the onset of the $J'' = 1/\omega\eta$ behaviour, i. e. when chain reptation becomes effective. In the meander model, this paraelastic deformation is attributed to an extra shear γ_r (Fig. 16a) caused by the reptating chains, when leaving one junction and entering a neighbouring one. To minimize interfacial energy of these spread junctions, two assumptions will be made: (i) Only the upper and lower interface of a shearband is appropriate to adapt extra chain sections with a minimum of chain crossing. (ii) The extra sections (indicated in Fig. 16a by fully drawn, dashed and dotted lines) are the shortest bridges between the respective junctions, i. e. no back and forth chain folding takes place therein.

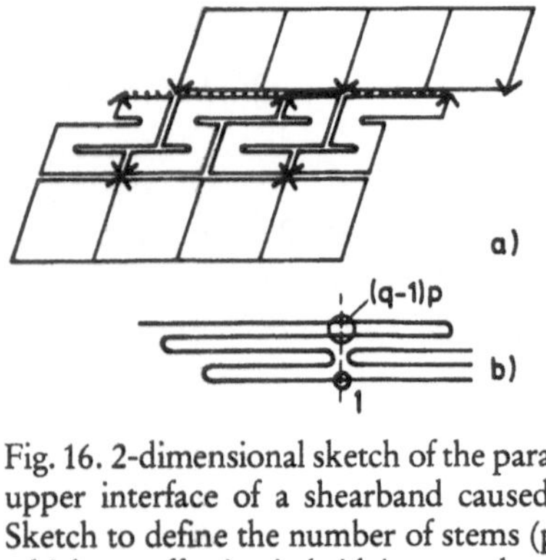

Fig. 16. 2-dimensional sketch of the paraelastic deformation on the upper interface of a shearband caused by spread junctions (a). Sketch to define the number of stems (per junction cross-section) which are effective in bridging to other junctions (b)

For a quantitative treatment, we assume the number of stems per junction effective in bridging to be

$$\nu = \frac{2}{q}\left(\frac{r}{d}\right)^2 [1 + (q-1)\,p]\,. \tag{27}$$

(Two bundles per junction of one shearband, $(r/d)^2/q$ number of molecules per bundle cross-section, p the probability of a tight fold lying across the junction instead of stopping at it, to increase its short range shear fluctuation; cf. Fig. 16b).

Starting at an ideal position ($\gamma_r = 0$), possible bridges can be formed by the effective stems to the neighbouring junctions, causing partial deformations $\gamma_i = 0, \pm 2, \pm 4, \pm$, and so on (if each of them were the only one), but increasing their free energies by $\delta g_i/kT = |i|\, p_f^0 6r/l_o$[7]). Introducing probabilities z_i for bridging the 0- to i-junction, one may define the recoverable shear per interface by

$$\gamma_r = \sum_{i=-m}^{+m} z_i \gamma_i \tag{28}$$

m denotes the maximum index for a given length of the molecules and therefore becomes $m = (l_o/6r)\, M/M_o$. Now the free energy per spread junction can be written as

$$\frac{\Delta g_F}{kT} = \nu\left[\sum_i z_i \ln z_i + \sum_i z_i \frac{\delta g_i}{kT} + \lambda\left(\sum_i z_i - 1\right)\right] + 2(3r)^3 \frac{\sigma}{kT}\left[\gamma_r - \sum_i z_i \gamma_i\right] \tag{29}$$

taking into account both conditions to be fulfilled and added with Lagrangian multipliers $[2(3r)^3$ being the shearband volume per junction$]$.

Minimization of Equation (29) for the probabilities z_i yields the equilibrium recoverable shear per interface

$$\gamma_r^{eq} = \frac{\sum_{i=1}^{m} 4i \cdot \exp[-ip_f^0 6r/l_o] \cdot \sinh[4i(3r)^3\sigma/\nu kT]}{1 + 2\sum_{i=1}^{m} \exp[-ip_f^0 6r/l_o] \cdot \cosh[4i(3r)^3\sigma/\nu kT]} \tag{28a}$$

---

[7]) The free energy of back and forth folding per monomer of a molecule, can be shown to be $\Delta g_m^f/kT = p_f^0\,\{1 - p_f/p_f^0\,[1 - \ln(p_f/p_f^0)]\}$ which becomes p_f^0 if it is stretched out. For simplicity the smallest cubes ($x/r = 2$) have been assumed and the junctions are therefore spaced by $6r$.

and after Reuss-averaging and including both interfaces per shearband, the flow relaxation strength for $\sigma \to 0$ becomes

$$\Delta J_F = \frac{2}{5}\,\beta(\partial\gamma_r^{eq}/\partial\sigma) = \frac{2}{5}\,\beta \cdot 216\,\frac{q\,d^2 r/kT}{1+(q-1)\,p} \cdot \frac{\sum_{i=1}^{m} i^2 \exp\left[-ip_f^o 6r/l_o\right]}{1+2\sum_{i=1}^{m} \exp\left[-ip_f^o 6r/l_o\right]}\,{}^{8)}\,. \qquad (30)$$

The viscoelasticity of polymer fractions in the melt and its interpretation in the meander model

In Figure 17a, shear-compliance master curves for 10 aPS fractions of $M_w/M_n < 1.1$ and with molecular weights between 10^4 up to $2.4 \cdot 10^6$ g/mol are presented[9]). The reference temperature is $T_o = 405.5$ K in all cases and each pair of J', $J''(f)$-curves is shifted by one decade in the ordinate against the preceding one, starting at the bottom.

If one tries to analyze these curves by decomposing each pair into a few symmetric relaxation processes – characterized, e.g. by Cole-Cole distributions[10]) – it

[8]) The last term can be summed up and becomes, with $Q \equiv \exp\left(-p_f^o 6r/l_o\right)$

$$\frac{Q(1+Q)-Q^{m+1}\left[(m+1)^2-(2m^2+2m-1)\,Q+m^2Q^2\right]}{(1-Q)^2\,(1+Q-2Q^{m+1})}\,.$$

[9]) The measurements with the mechanical spectrometer (10^{-4} to 200 Hz) including the double sandwich sample holder and other details of the device are described in Reference [23].

[10]) The details of the decomposition procedure and the quality of the fit are also given in Reference [23].

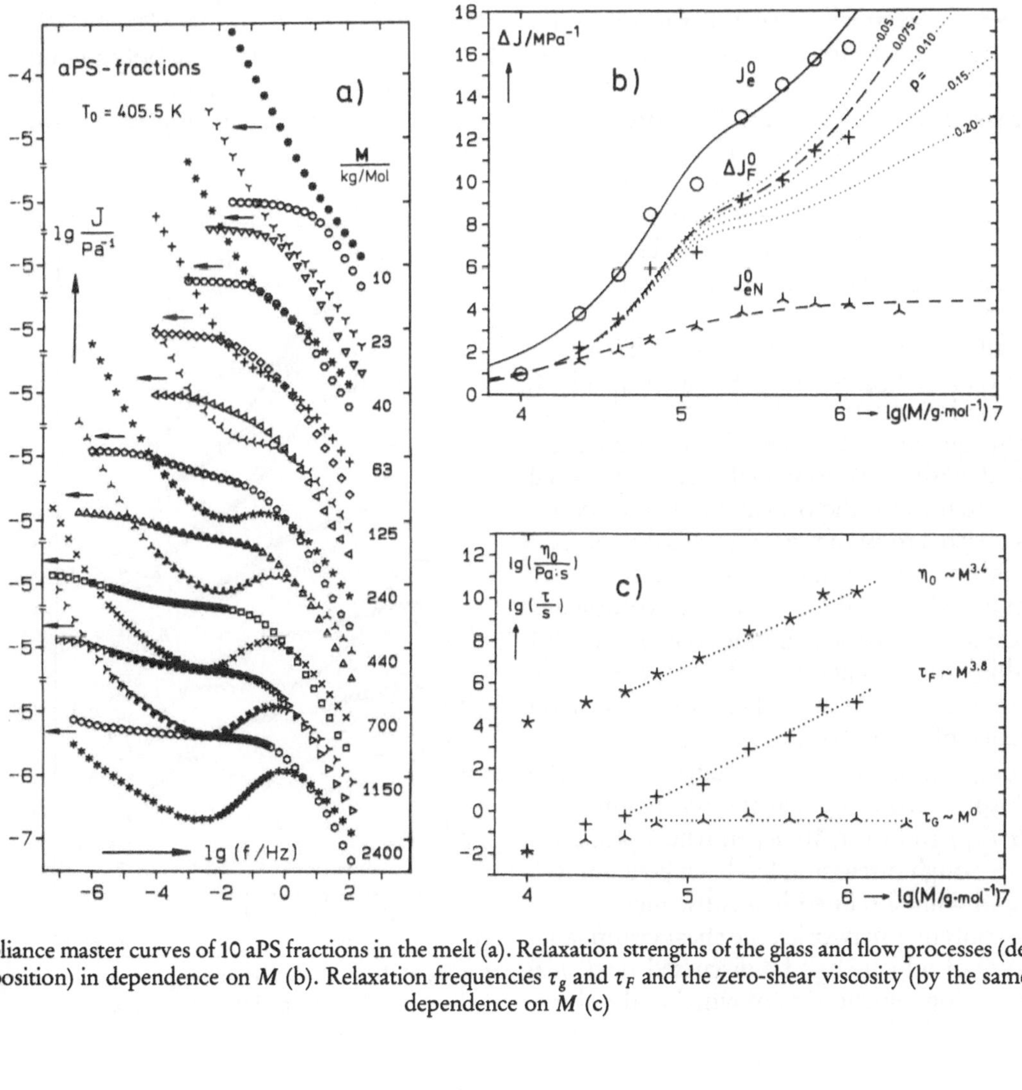

Fig. 17. Compliance master curves of 10 aPS fractions in the melt (a). Relaxation strengths of the glass and flow processes (determined by curve decomposition) in dependence on M (b). Relaxation frequencies τ_g and τ_F and the zero-shear viscosity (by the same analysis) in dependence on M (c)

turns out that a good fit can be achieved with two processes (glass relaxation and flow relaxation) and the viscous flow. This means that the shear band process between the glass and flow process is absent or lower in strength by at least a factor of 1/5 of the plateau compliance. Because all the chains in a fraction behave the same, it follows that $\phi = 1$. We may also say $\xi = 1$, because a rearrangement of strands would destroy the shearband at one position and a new band will be formed elsewhere in the melt. Therefore, we have to apply Equation (26a) instead of Equation (26) to estimate the shearband relaxation strength for fractions. Assuming for aPS: $-\Delta g_{rot}/kT = 2/3$, $\gamma_m = 9$, $d = 2s = 4l_o = 0.88$ nm, $x/r = 2$, $r_\infty/d = 6$, we get $\beta(1) \approx 0.018$ from Equation (25) and from Equation (26a) and Equation (14)

$$\Delta J_B^\infty = \beta(1) \frac{9}{10} \frac{s}{r} \gamma_m^2 \cdot J_{eN}^o \approx 0.11 \, J_{eN}^o \qquad (26b)$$

which is indeed much smaller than the plateau compliance and cannot be detected by decomposition of the master curves.

For discussion of the glass and flow process, the respective relaxation strengths $\Delta J_{eN}^o (= J_{eN}^o)$ and ΔJ_F^o are plotted (together with their sum, the steady state compliance J_e^o) versus lg M in Figure 17b and compared with the predictions of Equations (14) and (30), given by the dashed curves. q has been defined in Equation (8).

Additionally it was assumed that $-\Delta g_{rot}/kT = (2/3)(r_\infty/r)^{1/2}$; $f = 1$, $\Delta g_{def}/\Delta g_{g\infty}^* = -1$ and $n = M/M_o$ with $M_o = 104$ in Equation (10a); and finally $p_f^o = 1/250$ (cf. Table 1). The agreement between theory and experiment is good within experimental error. The small value of 0.07 to 0.10 for the probability p in Equation (27) points out that most of the tight folds do not cross the junctions but prefer to stay near them. Because it is still in dispute whether the steady state compliance does level off at high M or continue to increase [24], one could think that it depends on the polymer and its probability p which enters not only here but also in the M-dependence of the viscosity [1].

The decomposition of the master curves also yields the zero shear viscosity η_o and the relaxation times ($\tau_{g,F} = 1/2\pi f_{g,F}$) and their M-dependence, plotted in Figure 17c. A rough discussion has been given in Reference [1] and will be refined in a subsequent paper.

To lead to polymer networks, the shear compliance master curves of which shall be discussed next, it is worthwhile asking the question of whether the shearband process can be observed in polymer melts at all. The answer is definitely yes, if one inspects the master curves in Figure 18 as the fingerprints of binary aPS blends of different mixing ratios. Since M_2/M_1 is about 11.3, the shorter chains can rearrange (dynamically fractionate by reptation) within the shearband, gaining back their free energy of rotation. Therefore, Equation (26) must be applied with $\phi < 1$. With a more sophisticated application of the formulae above and a proper averaging over all chain distributions, the strength and relaxation frequencies of all three processes and their dependences on mixing ratio can be understood. Postponing the theory to a subsequent paper, it may only be mentioned here that a good fit to the M_w-dependence of the shearband relaxation strengths is not achieved unless about 40 files of meander cubes operate in parallel.

The shear compliance of polymer networks in dependence on crosslinking density

The shear compliance of polymer networks is characterized by two relaxation processes: the glass process and a lower frequency relaxation which finally

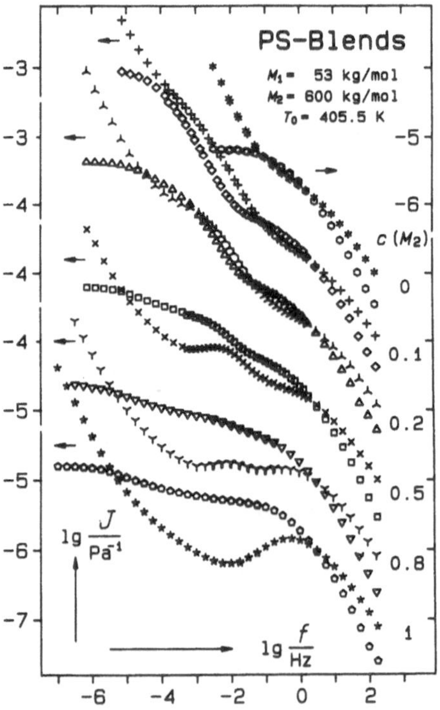

Fig. 18. Compliance master curves of 2 aPS fractions and 4 binary blends of them, the latter showing a rearranged shearband process in between the glass and flow relaxations

leads to the equilibrium compliance J_e. At low crosslinking, the low frequency process dominates in strength and is well separated from the glass process. With higher crosslinking it strongly decreases and accelerates until it seems to merge below the glass relaxation. This behaviour is common to end-linked model networks [25] as well as to statistical networks [26] and will be discussed in the last part of this paper. In Figure 19, shear compliance master curves for a synthetic isoprene rubber (IR)[11]) are presented [27].

Figure 19a shows the results of the radiation crosslinked samples (2 MeV-electrons and doses of 50 to 800 kGy) and Figure 19b similar measurements on three nonirradiated, thermally treated samples (i. e. as prepared, heated for 26 h and 3 days at $T = 353$ K, respectively).

Because there was no irreversible flow (the shear amplitudes used were less than $\gamma = 5 \cdot 10^{-3}$), most master curves (J', J''-pairs) could be perfectly decomposed into two relaxation processes, using Cole-Cole distributions[10]). The relaxation frequencies and relaxation strengths of these are plotted in Figures 20a and 20b, respectively.

The glass relaxation frequency is not influenced by the irradiation doses applied. The plateau compliance starts to decrease only at higher doses; a fact which has been interpreted by considering the effectiveness of the crosslinks present (p_c) on the hindering of intrameander shear (p_c^*)[12]).

The low frequency process is much more sensitive to crosslinking. We propose to attribute it to shearbands, if they are able to rearrange, thereby reducing their restoring forces[13]). The main experimental results are the dependence on the dose D or crosslinking density: $\Delta J_B^o \propto D^{-1}$, $f_B^o \propto D^{3.85}$, as is evident from Figures 20b and 20a[14]).

In the literature, the low frequency process is attributed to the slip of temporary entanglements and/or the onset of movement of pendant chains in the imperfect network [25–31]. Since, in the meander model of the bulk, the intrameander shear fluctuations determine the plateau compliance (and no entanglements are needed), the low frequency relaxation can only be caused by dangling ends or mobile strands during shearband deformation: if a fraction ξ of polymer segments has lost its orientation correlation $(x_2 - x_1)$ within the shearband, its relaxation strength should be increased from ΔJ_B^∞ to $\Delta J_B^\infty/\xi$ according to Equation (26) with $\phi = 1$. Such rearrangement is envisaged in Figure 15, (bottom) which might suggest that it will be a highly cooperative process of local tight fold reptation, provided its motion is not hindered by crosslinks.

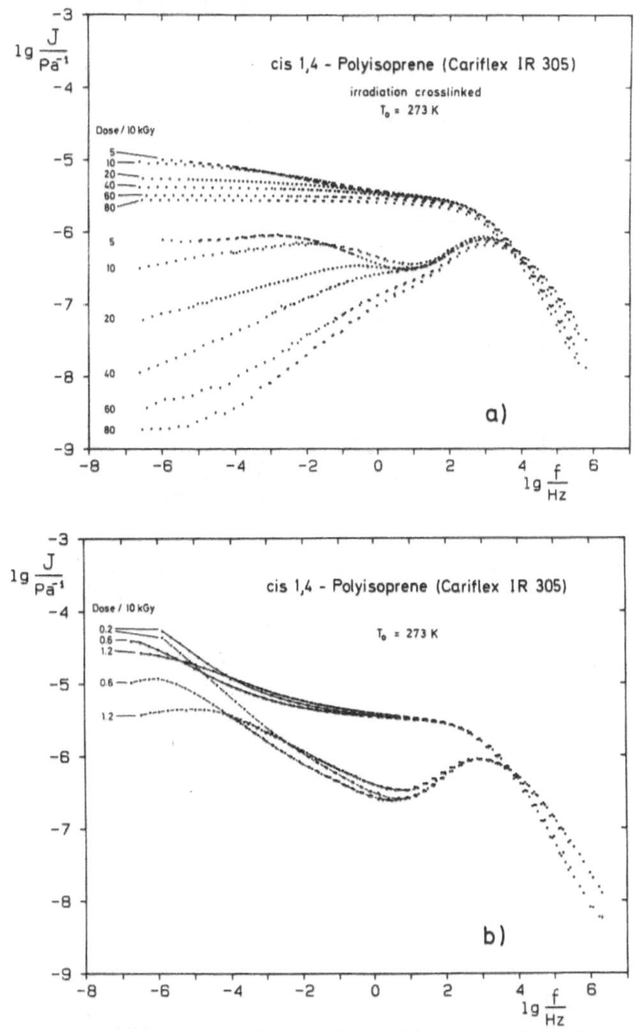

Fig. 19. Compliance master curves of 6 irradiation crosslinked IR-networks (a) and 3 thermally treated[14]) IR-samples (b) exhibiting the glass and shearband relaxation

[11]) All samples, the investigations of which will be discussed in the following, were prepared (and irradiation crosslinked) by Dr. U. Eisele, Bayer AG, Leverkusen.

[12]) Compare Equations (14a) and (15), and Figures 10 and 11.

[13]) At low doses there is still another more rigid component ΔJ_{BE}^o, f_{BE}^o, which we attribute to the shearbands when the chain ends only have rearranged. It can be detected because it is much faster than the main component but still separated from the glass process.

[14]) At this stage, it should be mentioned that the equivalent doses for the thermally treated samples were determined by a simultaneous extrapolation of both dependences.

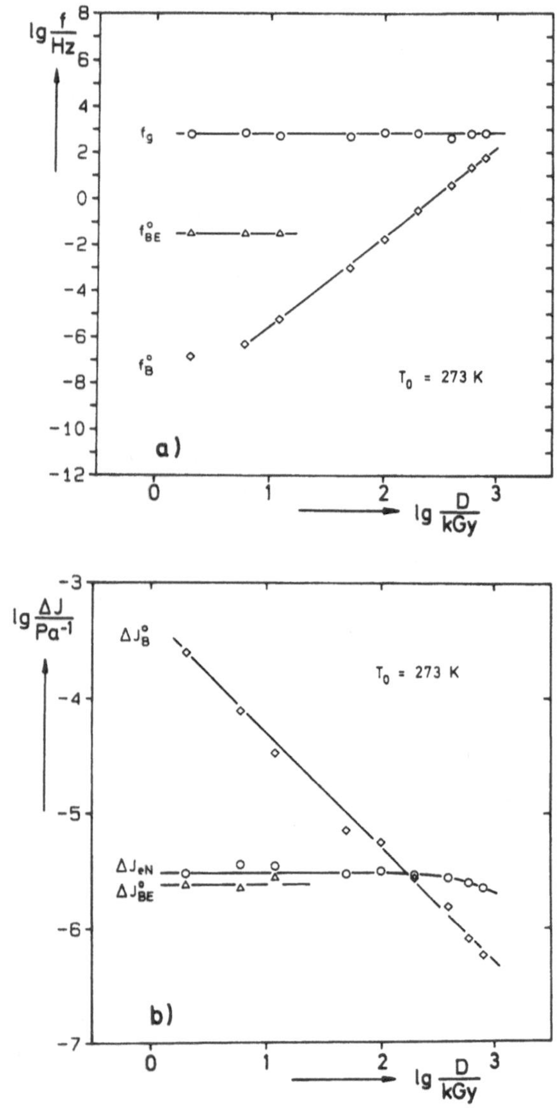

Fig. 20. Relaxation frequencies (a) and strengths (b), determined by the decomposition of the master curves in Figure 19. $f_B^o \propto D^{3.85}$ and $\Delta J_B^o \propto D^{-1}$ characterize the strong dependence of the shearband process on crosslinking

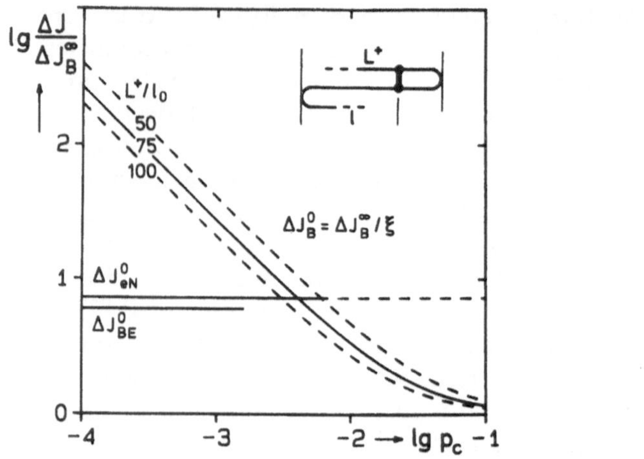

Fig. 21. Plot of the results of the model theory, to be compared with Figure 20b

To put it quantitatively, let us assume that only that part of a tight fold of length l, which is "open", may contribute to the rearrangement and lose its orientation correlation (cf. Fig. 21, inset). Then, ξ becomes approximately

$$\xi \approx 1 - \frac{l_o}{L^+} \int_0^{L^+/l_o} \exp\left(-\frac{l}{l_o} p_c\right) d\left(\frac{l}{l_o}\right) \tag{31}$$

($l_o$monomer length, L^+ average tight fold length, p_c crosslinking probability per monomer (= segment in the case of IR), and $\exp(-p_c l/l_o)$ the Poisson probability for l being free of crosslinks). In Figure 21, the ratio $\Delta J_B^o/\Delta J_B^\infty = 1/\xi$ is double-logarithmically plotted versus p_c for three L^+/l_o values. It can be quantitatively compared with Figure 20b, assuming the parameters: $s/d = 4/6$, $r/d \approx 7$, $d^3/kT_o = 0.058\ \mathrm{MPa}^{-1}$, $\gamma_m = 9$, $\beta \approx 0.02$ and therefore $\Delta J_B^\infty = 0.40\ \mathrm{MPa}^{-1}$ according to Equation (26a) and $\Delta J_{eN}^o = 2.9\ \mathrm{MPa}^{-1}$ according to Equation (14), if we also draw the ratio $\Delta J_{eN}^o/\Delta J_B^\infty$ into Figure 21. Figures 21 and 20b confirm that we have achieved a principal explanation of the relaxation strengths versus crosslinking. To finally obtain the experimental tightfold length L^+/l_o we read the abscissa of the crossover point in Figure 20b as 170 kGy and find the respective one in Figure 21, taking into account data from equilibrium swelling on these samples which gave $p_c \approx 2 \cdot 10^{-5}$/kGy. The so determined $p_c \approx 0.0034$ for the intersection point in Figure 21, is not far from that which would be expected if the tightfold length L^+/l_o agreed with $1/p_f^o = 75$ monomers (cf. Table 1), obtained from the radius of gyration.

Summarizing the interpretation of the relaxation strengths, one may state: ΔJ_B^∞, the strength of shearbands without any loss of segment orientation, cannot be detected (as in uncrosslinked polymer fractions) because it should be about 1/10 of the plateau compliance. A first step of rearrangement is attributed to the movement of chain ends and some dangling ends and can probably be observed as a minor component ΔJ_{BE}^o at low crosslinking. The main contribution to shearband rearrangement comes from the tightfolds not being fixed by crosslinks and decreases inversely with the crosslink density. The relaxation frequency of the main shearband process varies with $D^{3.85}$ which is not far from D^4 and may suggest a double diffusive kinetics during shearband deformation: at any stage of rearrangement, the shearband deformation will take place by the diffusion of dislocations and proceed much slower to equilibrium than the J_g-relaxation because of the larger deformation to be realized. But additionally, the equilibrium deformation at a given stress will change when the rearrangement proceeds at every shearband deformation cycle. A more quantitative treatment of this idea has still to be developed.

To conclude this part on isoprene rubber, two supplementary items of information gained from the analysis will be presented: (i) In Figure 22, the temperature dependence of the plateau compliance is plotted for all samples of different doses, showing a perfect $1/T$-behaviour suggested by Equations (14) and (14a).

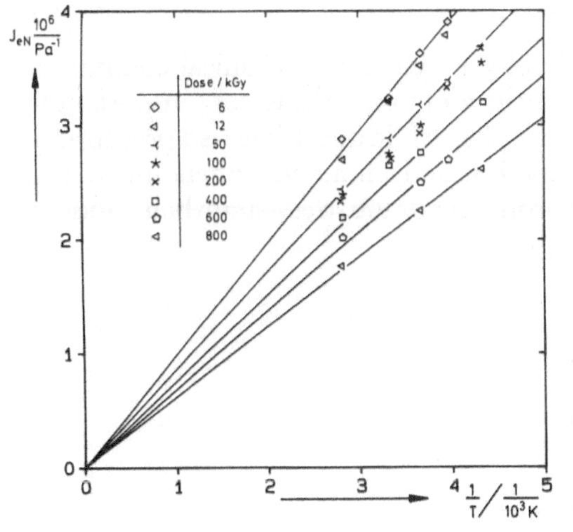

Fig. 22. The dependence of the plateau compliance J_{eN} of IR-networks on the reciprocal temperature

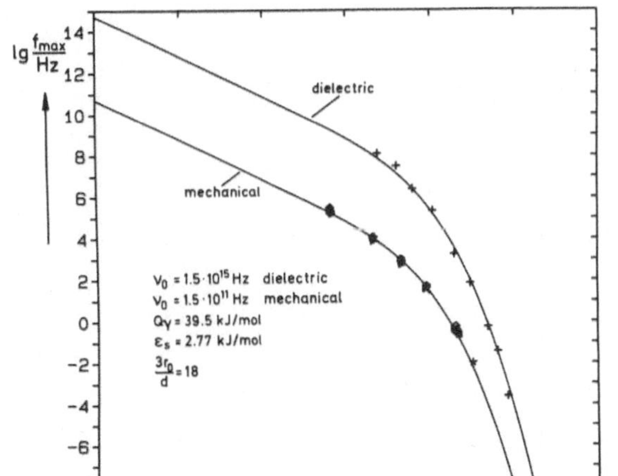

Fig. 23. The activation diagram of all IR-networks investigated including dielectric results on an uncrosslinked sample (+). The curves were drawn according to Equation (18)

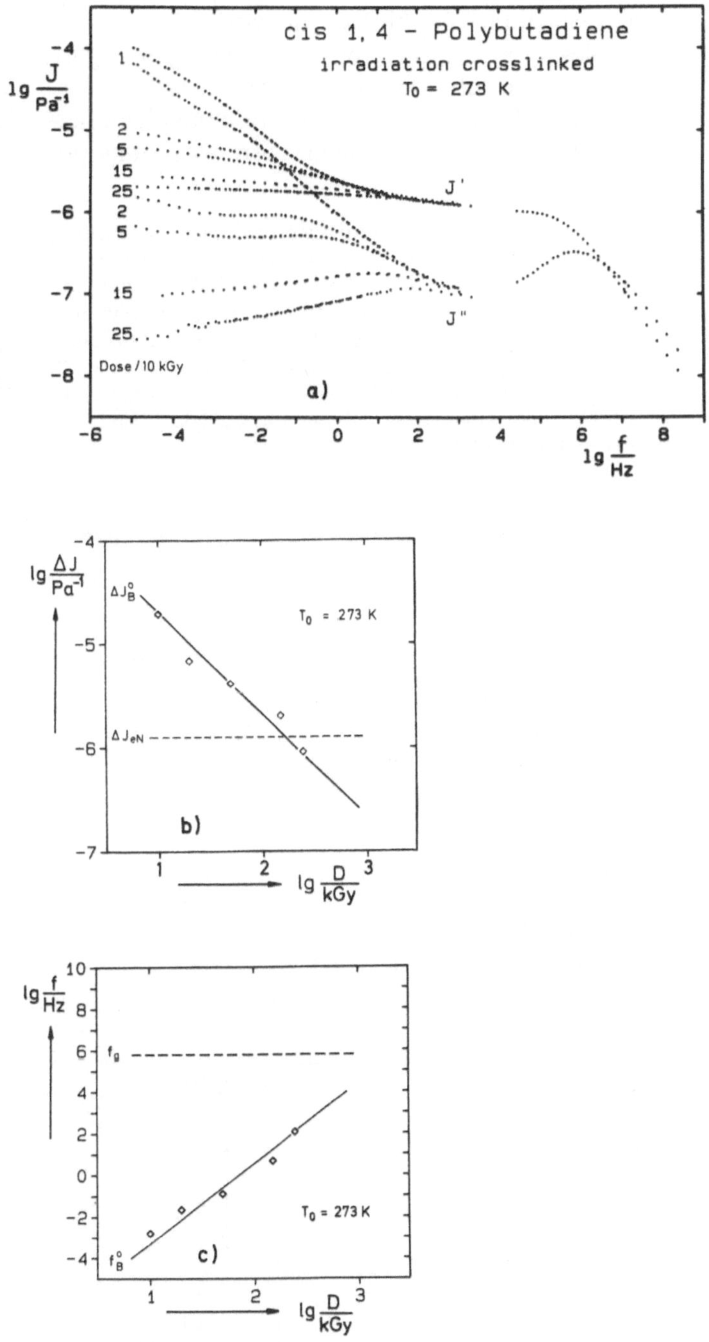

Fig. 24. Master curves of high cis-1,4-poly(butadiene) crosslinked with various doses, and the results of curve decomposition, similar to that of IR-networks

(ii) Figure 23 gives the activation diagram and its analysis with Equation (18) for glass relaxation. The relaxation frequencies taken from J''_g lie on a single curve for all samples investigated. The dielectric relaxation (ε''_g-maxima) of an uncrosslinked IR-sample appears shifted by 10^4 Hz to higher frequencies.

To emphasize the common nature of the relaxational behaviour of polymer networks, two more examples are presented at the end of this paper. Figure 24 shows the results of the dynamic mechanical measurements and its analysis on cis-1,4-polybutadiene networks[15]. To prevent crystallization we could not measure the whole master curves but used the activation diagram of the other polybutadiene (Fig. 26) to adjust the measurements above the crystallization temperature to those at low temperatures (after quenching) properly on the frequency scale. In the decomposition we did not take into account the small intermediate process due to chain ends in the shear-bands.

In Figures 25 and 26, results and their analysis from a noncrystallizable polybutadiene[15]) are shown. Only four samples with different crosslinking were investigated, but there is no doubt that the relaxational behaviour is very similar to the networks discussed above.

Conclusion

The study of the dynamic mechanical compliance of polymer melts and networks is very important to check molecular models in detail. This is a presupposition for its application to nonlinear phenomena, e. g. rheological properties or the stress-strain behaviour up

[15]) All network samples were kindly provided by Dr. U. Eisele, Bayer AG Leverkusen: (a) cis-1,4-poly(isoprene), Cariflex IR 305 (Shell). $M_w = 2 \cdot 10^6$, $M_n = 2.5 \cdot 10^5$ g/mol, anionically polymerized with Butyl-Li-catalyst, 93% cis-1,4, 7% 1,2. (b) cis-1,4-poly(butadiene), CB 11 (Bayer), $M_w = 2.8 \cdot 10^5$, $M_n = 1.4 \cdot 10^5$ g/mol bulk polymerized with Ziegler-catalyst, 93% cis-1,4, 3% trans-1,4, 4% 1,2. (c) poly(butadiene), Intene NF (Bayer), $M_w = 2.8 \cdot 10^5$, M_n $1.4 \cdot 10^5$ g/mol anionically polymerized with Butyl-Li-catalyst. 39% cis-1,4, 50% trans-1,4, 11% 1,2.

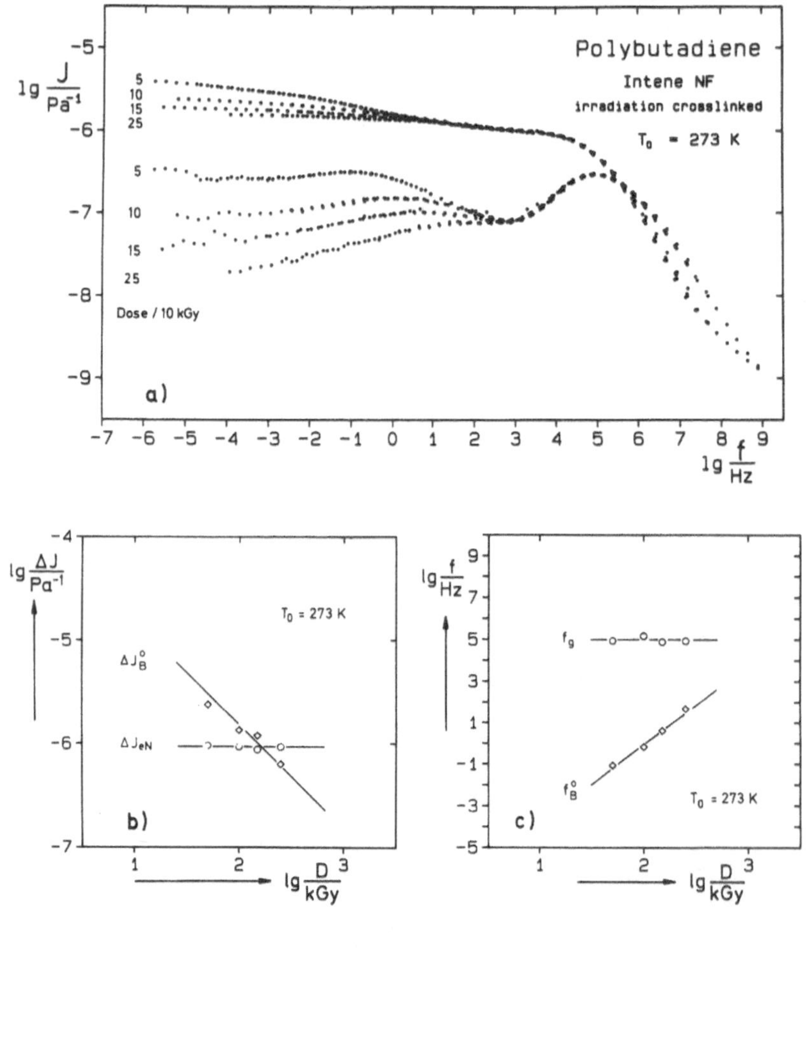

Fig. 25. Master curves of noncrystallizable poly-(butadiene)-networks and the results of curve analysis, similar to that of IR-and high-cis BR-networks

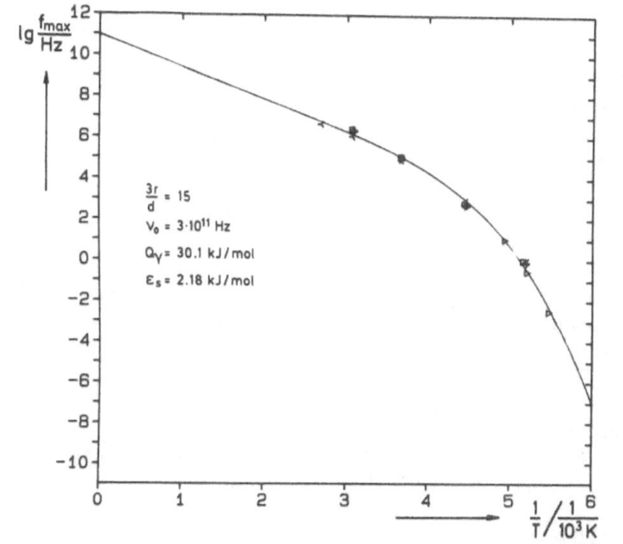

Fig. 26. Activation diagram of all BR-networks studied and discussed in Figure 25. The curve was drawn according to Equation (18)

to large strains in different deformation modes. The next step, which is under study, is that of the amplitude dependence of the relaxation strengths and frequencies and/or its dependence on superimposed stress, strain or shear rate [26].

Acknowledgement

The authors wish to thank Dr. U. Eisele for supplying the samples, the Deutsche Forschungsgemeinschaft and the Fonds der Chemie for financial support.

References

1. Pechhold W, v Soden W, Stoll B (1981) Makromol Chem 182:573
2. Pechhold W (1984) Makromol Chem, Suppl 6:163
3. Pechhold W, Sautter E, v Soden W, Stoll B, Großmann HP (1979) Makromol Chem, Suppl 3:247
4. Pechhold W, Stoll B (1982) Polym Bull 7:413
5. Pechhold W, Yeh GSY (1989) in preparation
6. Pechhold W, Grüner W, v Soden W (1988) Coll & Polym Sci, in preparation
7. Böhm M, Grassl O, Pechhold W, v Soden W (1987) Progr Coll & Polym Sci 75:62
8. Flory PJ (1953) Principles of Polymer Chemistry, Cornell University Press, Ithaca, New York
9. Bueche F (1962) Physical Properties of Polymers, Interscience Publications, Wiley, New York
10. Tanford Ch (1961) Physical Chemistry of Macromolecules, Wiley, New York
11. Pechhold W, Großmann HP (1979) Faraday Discuss Chem Soc 68:58
12. Pechhold W, Gross T, Großmann HP (1982) Coll & Polym Sci 260:378
13. Brandrup J, Immergut EH (1975) Polymer Handbook, sec Ed, Wiley, New York
14. Schmieg C (1986) Thesis, Ulm
15. Pechhold W (1980) Coll & Polym Sci 258:269
16. Daoud M, Cotton JP, Farnoux B, Jannink G, Sarma G, Benoit H, Duplessix R, Picot C, de Gennes PG (1975) Macromolecules 8:804
17. Pechhold W, Großmann HP, v Soden W (1982) Coll & Polym Sci 260:248
18. Heinrich W, Stoll B (1988) Coll & Polym Sci, in press
19. Winkler G, Stoll B, Sautter E, Pechhold W (1980) Coll & Polym Sci 258:289
20. Bakule R, Heinrich W, Pechhold W, Stoll B (1986) Coll & Polym Sci 264:185
21. Zentel R, Strobl GR, Ringsdorf H (1985) Macromolecules 18:960
22. Heinrich W, Stoll B (1985) Coll & Polym Sci 263:895
23. v Soden W (1987) Thesis, Ulm
24. Laun HM (1986) private communication
25. Havránek A, Ilavsky M, Nedbal J, Böhm M, v Soden W, Stoll B (1987) Coll & Polym Sci 265:8
26. Ferry JD (1980) Viscoelastic Properties of Polymers, 3rd ed, Wiley, New York
27. Böhm M (1987) Thesis, Ulm
28. Ferry JD (1976) Molecular Fluids, Gordon and Breach, London
29. Valles EM, Macosko CM (1979) Macromolecules 12:673
30. Havránek A (1982) Polym Bull 8:133
31. Bibbo MA, Valles EM (1984) Macromolecules 17:360

Received August 10, 1987;
accepted August 10, 1987

Authors' address:

W. Pechhold
Abt. Angewandte Physik
Universität Ulm
Oberer Eselsberg
D-7900 Ulm, F.R.G.

Discussion

RIGBI:

You have been speaking mainly of polyisobutylene and -isoprene polymers, and this sort of structure implies a long-range, crystalline-like structure, which we know does not exist at all in these polymers. Is there any other evidence you can use for this type of structure?

PECHHOLD:

I think, I made a remark in the very beginning, that I don't believe that the amorphous bundles have anything to do with crystalline regions. One may say they are small domains of a nematic type of order, limited to one meander cube size, because these cubes are destributed in orientation in the isotropic melt. On approaching the $T_{n/i}$-transition, the orientation correlation increases and enters a long-range order only below this temperature – which for PE has been extrapolated to be 200 K.

KELLER:

Are these structures meant to be equilibrium structures of the amorphous state, or do these structures just happen to be left there by incomplete melting or by contraction and polymerization and so on. Are these the equilibrium states of amorphous?

PECHHOLD:

We think that they are equilibrium structures and we have shown that the free energy is decreased by the superfolding and not increased, as in the first step of a nucleation process. Nevertheless, the meander model relies on two fundamental assumptions: the validity of the Cluster-Entropy hypothesis (CEH) and chain parallelism being favoured over interpenetrating coils. Our endeavour is to prove or disprove this model by direct and indirect experimental evidence.

KRÖNER:

Can these structures be considered amorphous similarly as we would perhaps call a polycrystal with small grains amorphous, too?

PECHHOLD:

I would say the identification of being amorphous is already inside these cubes, because the chains are highly disordered, intramolecularly by different conformations, as well as intermolecularly by dislocations. From X-ray or from any diffraction method, these structures are amorphous and the halo can be calculated from its short-range order. But there is some orientational anisotropy on a 100 Å scale, and if this could be ruled out, then these structures – you may call them nematics with very small grains – would be disproved.

KRÖNER:

Could you explain, perhaps in your model, exactly what is the difference between the network and the melt, because you have described both in terms of this meander?

PECHHOLD:

The meander superstructure of the melt is not changed by crosslinking, but the molecular motions I have described become more or less restricted.

First of all, the long-range motions, like the reptation of a molecule, drop out, as does the viscous flow. Next, the shear band motion gets reduced, which determines the extensibility of a rubber and at medium crosslinking the intrameander shear and therefore the plateau compliance starts to decrease, because according to our analysis only 1/30 of the total crosslinks are active here.

D. RICHTER:

You said that this local anisotropy should be seen on the scale of the order of 100 Å. I mean this is just the middle of small angle scattering. So why there is no evidence in small angle scattering for structures like that, or are there some explanations which you could bring forward?

PECHHOLD:

Small angle scattering in the melt cannot show up the meander size, because of lack of density differences. But in the first step of a cold crystallization, we recently found a Guinier scattering in this range.

Further evidence – even though a small effect – Dr. Sautter got from label experiments. Measuring the radius of gyration of n-alkanes in perdeuterated polyethylene or of oligostyrenes in perdeuterated polystyrene, he found [1, 2], that these radii are larger, by about 15 %, than those measured in the pure n-alkane or pure oligostyrene melt. We could account for this result by assuming chain parallelism, and simulating the scattering of the oligomer in the meander of the melt ($3r/d \approx 15$) and in its own environment (small blocks or the most tiny meanders, $3r/d = 6$).

[1] Pechhold W (1984) Makromol Chem Suppl 6:163
[2] Sautter E, Coll & Polym Sci, in press

LAUN:

In your model you have a one-dimensional coil arranged in a bundle that is superfolding such that you get the cube topology. Now, you told us that an entanglement, that rheologists or others normally talk about, would be, in your model, the junction of a bundle going from one cube to the other.

But what would be the picture of a raptation motion of a molecule, I think of tube-reptation, in your model? Would it be a reptation of the whole bundle or the reptation of a single molecule within the bundle?

PECHHOLD:

We assume the single molecules reptate within the superfolding bundles. You may look at them as topological tubes ($D \approx 5$ nm) in which chain reptation is confined mainly to its molecular tube ($d < 1$ nm). For viscous flow or flow relaxation to occur, the chains must reptate by about half their lengths to leave their initial junctions at the interface of shearbands and enter neighbouring ones. The relaxation strength ΔJ_F due to these spread junctions is derived in my paper. It suggests that cooperative reptation of several chains is not necessary in this model to account for the flow relaxation and also for viscous flow.

HAVRANEK:

Do you also have a model of entanglement with those blocks?

PECHHOLD:

On first sight, every cube-junction of four of these bundles could be thought to be an entanglement. But, according to model calculations – using spread junctions on the shearband interfaces – only the effectively bridging stems across the interface will act as temporary entanglements.

PAKULA:

What happens to those structures if you try to swell, for example, the samples, or if you try to mix two polymers?

PECHHOLD:

If it is a crosslinked system, swelling means an isotropic swelling of the cubes and each chain is retained in its initial bundle. The free energy of chain folding changes and must be introduced into the chemical potential of the mixture to give the equilibrium swelling, a similar procedure as Flory applied for the coil network. The shear compliance of a swollen network, as it depends on ϕ_p, will be discussed in a subsequent communication here. For compatible polymers, including mixtures of different molecular weights, the superstructure will not change very much but the molecular motions described may vary strongly in strength and kinetics.

For incompatible polymers, phase separation may take place either on the level of the meander-structure (e. g. block-copolymers) or at the coarse grain level.

PAKULA:

But it means that you introduce the solvent molecules into bundles?

PECHHOLD:

Yes, of course. The polymer molecules separate to leave a place for the solvent molecules and if the chains are crosslinked, the cross-section of the bundles must increase. If they are scarcely or not crosslinked, the superstructure can rearrange in solution and the swollen bundles may retain their initial diameter (e. g. in θ-solution). My most interesting problem is to describe the polymer solution from the very dilute regime up to the bulk. If you are interested, I can show you that we can account for the viscosity in dilute solution, which nicely gives the experimental values at θ-condition.

BOUÉ:

I was wondering if you get any molecular weight dependence of your bundle or of your meanders. To get the $M^{3.4}$-law, you have to add to the classical M^3-dependence another molecular weight dependence and I would like to know where you got it from.

PECHHOLD:

The flow relaxation time for unrestricted reptation would be $\tau_F = L^2/8D_1$ if one assumes curvilinear diffusion of a chain across half of its length L with the appropriate $D_1 = \Gamma_0 l_0^2 \lambda/2L$ ($\pm\, l_0$ displacement of the diffusing segment of length λ with jump rate Γ_0). The resulting M^3-dependence will be changed to M^4 if all the tight-folds were sessile at the cube junctions, to increase its shear fluctuations (because only one stem per molecule would cross the junction). Experimentally, we found $\tau_F \sim M^{3.8}$. The low shear viscosity can now be written $\eta_0 = \tau_F/J_e^0 \sim M^{3.4}$, because J_e^0 roughly varies with $M^{0.4}$.

WEYMANS:

Just a question concerning the two different length scales you have in the two different models. In the coil model, you have something like the persistence length that is, of course, interrelated to your meander length scale. Can you think of any experiment that can distinguish between the two length scales, I mean, all experiments so far referred to give the same results, or can be explained in both models. But can you think of any experiment that should be done, that can distinguish between the meander model and coil model?

PECHHOLD:

As I have shown, there is an interrelation of the persistence length in the coil model with the fold probability of tight-folded molecules within the meandering bundle, and may also be for stiffer molecules with the meander size. As to your question, I first of all must say that we have not by far looked at all experimental facts which have to be explained in our model (e. g. the SANS results on partially deuterated chains or networks, or the self-diffusion of labelled chains, rings or nanogels in topological networks). One experiment I would propose, is SANS on very high molecular weight linear labelled chains, for which we predict a maximum at low q in the Kratky-plot before it levels off. Direct evidence, of course, should come from detailed electron microscopy of freeze-fractured surfaces, thin sections and/or thin films spread from dilute solutions.

ILAVSKY:

I have roughly the same question. One can see that both models (the coil model and meander model) give the same results in many cases, and also in many cases the two maxima can be explained, by the random coil model. So, I would like to ask the question: what is the main goal, where is the main difference between these two approaches? Where is the large difference or what is new from only this model that cannot come from the coil model?

PECHHOLD:

This is a very difficult question. Our goal, or our approach, starting at Blasenbreys time, was to try to continue the solid state physics of crystals into amorphous materials. And in polymers we have a very exceptional case. You have to account for both crystalline and amorphous parts altogether, because you have two-phase structures. I think that a complete understanding of crystallization, of the long spacings observed at different quenching temperatures and of the memory effects will be achieved in the meander model. Also, the structure and rubber elastic properties of liquid crystalline polymers, including their transition data at isotropization, are easy to describe with the ordering of cube orientation. Further examples in favour of the meander model are its applications to swollen networks, thermoreversible gelation and other mesophases.

HAVRANEK:

I think that one thing may be in favour of this model. That is that there are some single points, like the dislocations in solid state physics.

PECHHOLD:

Yes, the meander structure of nearly parallel stems is predistined to introduce dislocations which mediate shear fluctuations and deformations, and are in equilibrium above T_g. I should mention that Dr. Havranek has got similar results to ours on networks, showing two relaxation processes in the compliance.

LAUN:

I would like to come back to a more practical point. Up to now, I was convinced that the steady state compliance of polymer melts, as long as the molecular weight distribution is similar and the average molecular weight is high enough, is more or less independent of molecular weight. In your graph, you have shown steady state compliances of polystyrene fractions and I could not see that the compliance became constant in the range of high molecular weights. Do you have an explanation for that behaviour?

PECHHOLD:

According to our measurements on PS, we cannot decide whether it levels off or not. PS-fractions of even higher M should be measured, but τ_F would increase beyond the experimental range. In my paper, I showed that the flow relaxation should increase in strength for high M, its slope depending on the concentration of effective bridges at the spread junctions.

ILAVSKY:

You decompose all original curves. If the decomposition does not work in the region of low molecular weights, this can, of course, be an artefact. If you look down you see that really in the region between 4 and 5 you have some little bend in the viscosity as well as for the relaxation time. I would rather consider it as an artefact.

PECHHOLD:

The bends in η and τ_F versus M are certainly no artefacts, but mark the critical molecular weight between 30 to 40 thousand. The decrease in J^0_{eN} at lower M we found in at least four systems. Sometimes it is necessary to have a little crosslinking, because then you prevent flow, and you can measure the plateau compliance and this we did, for instance, for Polybutadiene [3]. Of course, you are right to say that it is not firmly proved, but in our opinion it is highly probable, and it complies with the prediction of the meander model.

[3] Pechhold W (1984) Makromol Chem Suppl 6:163

ILAVSKY:

If you introduce crosslinking, you usually do this by irradiation, but then you are producing not only crosslinks but also chain scissions – that means pendent chains – into the network. This can affect the composition.

PECHHOLD:

Some chain scission is certainly there and has some (minor) influence on the shearband compliance, as well as on equilibrium swelling, as we know from the comparison with DCP-crosslinked Poly(isoprene).

Deuterium magnetic resonance on elongated networks: orientational order versus network structure

A. Dubault, B. Deloche, and J. Herz[1])

Laboratoire de Physique des Solides, CNRS-LA 2, Université de Paris-Sud, Orsay, France
[1]) Institute Charles Sadron (CRM-EAHP), CNRS-ULP, Strasbourg, France

Abstract: The orientational order generated in uniaxially strained (end-linked) polydimethylsiloxane rubbers is investigated as a function of the length of the precursor chains and of the polymer concentration at which the network was formed. The degree of orientational order was measured by means of the deuterium NMR technique performed on labelled network chains. The induced order increases with the density of crosslinking junctions and trapped entanglements. The observed effects are analysed versus the equilibrium swelling degree of the samples, in order to characterize the real network structure with respect to chain segment ordering.

Key words: rubber, polydimethylsiloxane, deuterium NMR, orientational order, trapped entanglements

Introduction

Different microscope investigations on constrained rubbers have already revealed some departures from the predictions of rubber elasticity based on volumeless gaussian chains. For instance, it has been observed that the stress optical coefficient does not remain constant with elongation, even for low deformations [1, 2]. These deviations are ascribed, in part, to the existence of interactions between chains which strongly perturb the over-simplified model of isolated chains. Different types of interaction may be considered. Firstly, short-range orientational correlations take place between chain segments, as was shown some years ago by one of us [3]. Such local interactions exist whatever the structure of the network, even for model networks having a well defined mesh size without imperfections. However, there may also be steric interactions arising from the interpenetration of the precursor chains during formation of the network. In contrast to the orientational correlations, these interactions are inherent to non-ideal networks and consequently depend on the type of crosslinking process. These topological constraints are commonly called trapped entanglements and can be viewed as physical knots that are more mobile than chemical crosslinks. Thus, it appears that the chemical mesh size is no longer a well adapted parameter to confidently account for the elastic behaviour of a real networks.

During recent years, the ^{2}H-NMR technique has emerged as a powerful tool for investigating the local behaviour of strained rubbers [4–8]. For instance, this technique has recently demonstrated that free homopolymer chains dissolved in a strained rubber are as well oriented as linked chains; this has produced indisputable experimental proof of the presence of strong orientational correlations. Herein, we present ^{2}H-NMR results related to the dependence of the induced orientational order with the network structure; this study was performed by changing the chemical mesh size or the polymer concentration during the crosslinking reaction.

Experimental conditions

Materials

All experiments were carried out on endlinked polydimethylsiloxane networks for which the chemical mesh size is defined by the length of the precursor. All networks are tetrafunctional and contain a known fraction (about 10%) of perdeuterated chains, statistically distributed inside the sample. The studied networks (referred as PDMS [D]) differ either by the molecular weight M_n of the pre-sor chains or by the concentration in the polymer, V_c, at which the crosslinking reaction was performed. Synthesis and all characteristics of the networks are fully detailed elsewhere [9, 10]; in Table 1 are reported only those characteristics necessary for understanding the present experimental results.

^{2}H-NMR method

The general features of deuterium magnetic resonance in anisotropic fluids have been largely developed in references [11, 12]. Due to its large sensitivity, this technique has been recently used successfully in the investigation of deformed rubbery materials. We may recall simply that in the case of rapid uniaxial orientations, the liquid-like NMR line is split into a doublet whose spacing $\Delta \upsilon$ may be expressed in frequency units as:

$$\Delta \upsilon = 3/2\ \upsilon_q P_2 [\cos \Omega] \cdot \langle P_2 [\cos \theta\ (t)] \rangle$$

where υ_q is the static quadrupolar constant of the C–D bonds, Ω is the angle between the stress axis and the steady magnetic field. Then, splitting $\Delta \upsilon$ gives access to the order parameter $\langle P_2 [\cos \theta\ (t)] \rangle$ of the C–D bonds relative to the uniaxial constraint.

^{2}H-NMR conditions

Samples were strips approximatively 1 mm thick, 6 mm wide and 40 mm long. Both ends of the strips were gripped by means of two jaws; one of these could be moved by means of a calibrated screw allowing adjustment of the elongation. Lengths were measured to within on accuracy of 0.02 mm using a micrometer fitted onto the microscope stage.

All ^{2}H-NMR spectra were recorded using FT NMR equipment: a CXP 90 Brucker operating at 2 T with a classical electromagnet. The steady magnetic field was perpendicular to the stretching direction (Ω = 90°).

Results and discussion

^{2}H-NMR spectra obtained for sample $A(M_n$ = 25 000 V_c = 0.7) at different elongation ratio λ are presented in Figure 1; these spectra illustrate the sensitivity of the method. We have to mention that we are in a favourable situation with the polydimethylsiloxane compound; in fact, its low glass transition temperature conjugated with the rapid rotation of methyl groups make the NMR line much narrower than those obtained from polybutadiene networks [7] for instance. Figure 2 shows the behaviour of the quadrupolar splitting $\Delta \upsilon$ in the small elongation range for the five networks studied. As earlier shown [3, 4], the data fit with the strain function $(\lambda^2 - \lambda^{-1})$ but the important experimental fact is the dependence on both the chemical mesh size and the crosslinking concentration V_c also. This clearly demonstrates that the induced orientational order is sensitive to the overall structure of net-

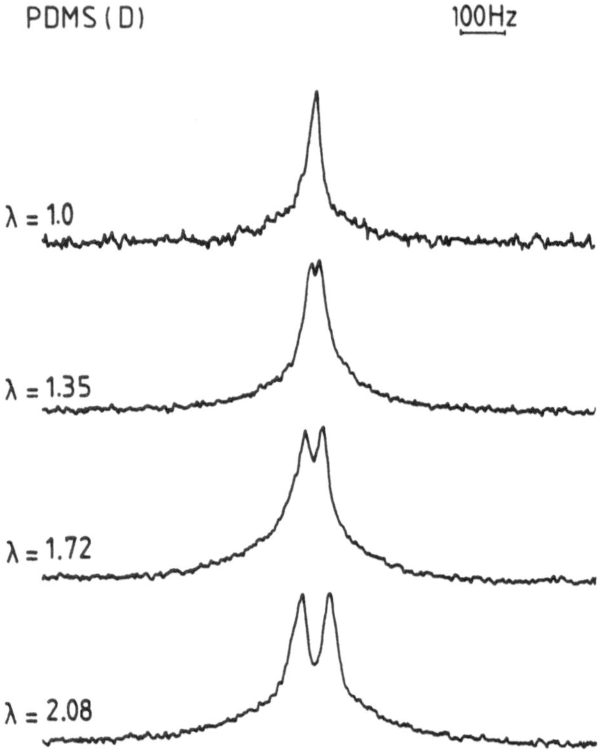

Fig. 1. 13 MHz ^{2}H-NMR spectra of perdeuterated chains of a uniaxially strained PDMS network (sample A in Table 1) for various extension ratios λ. The number of accumulation was typically 5 for λ = 1 and 250 for λ = 2.08

Table 1. Main characteristics of the PDMS [D] networks

Samples	C	B	B	A	A
M_n	3100	10500	10500	25000	25000
V_c	0.7	1	0.7	1	0.7
ϕ_e	0.185	0.167	0.135	0.1	0.067
Slopes p	43	42	37	28	17

M_n = molecular weight of precursor PDMS chains (g/mol); V_c = volume fraction of polymer at which network was formed in toluene; ϕ_e = volume fraction of polymer at swelling equilibrium in cyclohexane; slopes of the different straight lines in Fig. 2

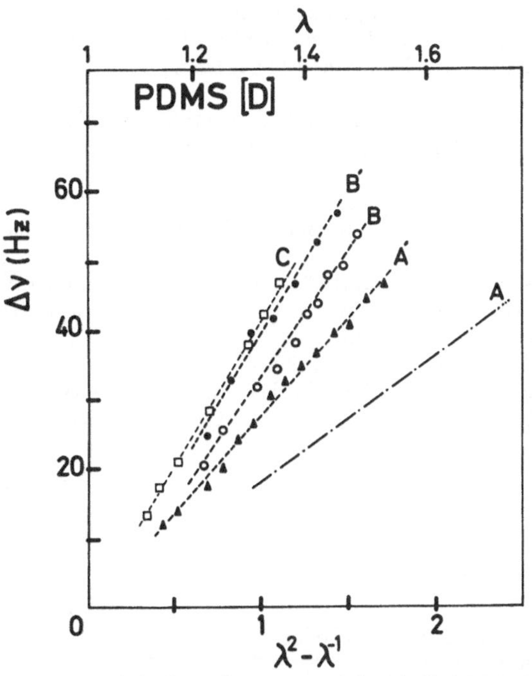

Fig. 2. Quadrupolar splitting $\Delta \upsilon$ of the labelled PDMS networks (referred to in Table 1) versus $[\lambda^2 - \lambda^{-1}]$. The numerous data for sample A have already been reported in Reference [5]: they fall along the line indicated

work (not only to M_n). This effect is quantified in Table 1 in which are reported the slopes $p = \frac{\Delta \upsilon}{\lambda^2 - \lambda^{-1}}$ of the observed straight lines; in fact, this parameter measures the efficiency of the stress on the local ordering.

At present, we have to relate these data to a variable specifying the structure of the network. As the chemical mesh size M_n does not completely define any real network, we are led to relate p to a macroscopic characteristic rather than a microscopic one. Figure 3 shows that the induced orientational order (p) varies as the polymer fraction ϕ_e at the swelling equilibrium in a good solvent (herein the cyclohexane). This signifies that the observed order parameter $\langle P_2(\theta) \rangle$ varies in the same manner as ϕ_e when either the chemical mesh size (M_n) or the crosslinking conditions (V_c) is changed. Moreover, in comparing the values of p obtained for A' with those obtained for A (V_c changed) and B' (M_n changed), we can conclude that a change in V_c has an effect on local ordering as important as a variation in M_n. These effects of V_c are commonly attributed to variations in the density of entanglements trapped during the crosslinking process; the data of Table 1 show that, with regard to segmental ordering, such physical constraints have a non-negligible efficiency compared to the chemical junctions. We can also add that the effect in V_c is less pronounced as the crosslinking density is reduced (comparison between $M_n =$ 10 500 and 25 000). Such an observation is consistent with the fact that the larger the size of the precursor chains, the larger the probability of trapping entanglements, and consequently, the higher the ratio of the physical constraints to the crosslinking junctions (for rubbers A and A').

From this study, it seems clear that the M_n^{-1} dependence deduced from the classical theory is difficult to test, even on model networks. In fact, there are always trapped entanglements whose density cannot be precisely evaluated. As perceived by ^{2}H-NMR measurements, these entanglements qualitatively act as chemical junctions in reducing the chemical mesh size. However, this apparent non-differentiation between the physical and chemical constraints does not hold in measuring the anisotropy of the overall chains, as has been done by small angle neutron scattering on similar PDMS networks [13]. At this scale of chain dimension, the trapped entanglements are not perceived as crosslinks; in fact, the anisotropy of polymeric coils under stress is sensitive to M_n only.

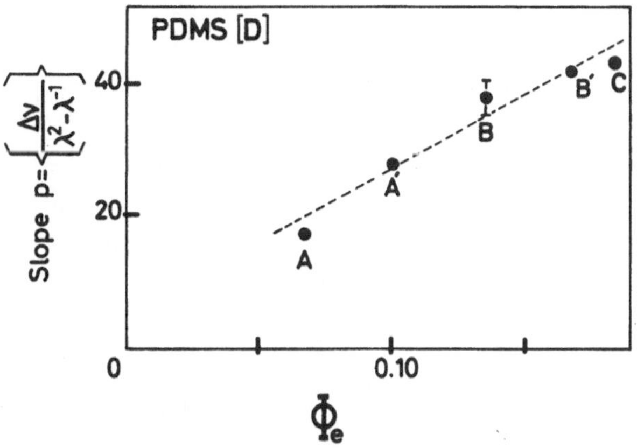

Fig. 3. Slopes p of the lines $\Delta \upsilon$ versus $[\lambda^2 - \lambda^{-1}]$ (data of Fig. 2) plotted versus the polymer fraction ϕ_e at swelling equilibrium in cyclohexane (uncertainties have been estimated from several experiments)

Conclusion

This work undoubtedly demonstrates that the segmental ordering induced in strained networks depends not only on the chain length between two adjacent crosslinks, but is also very sensitive to the experimental conditions during the network synthesis. In consequence, the elastic properties of rubbery materials would be better specified by macroscopic characteristics rather than the chemical mesh size which does not reflect the real structure of the network. As regards the orientational $\langle P_2 \rangle$, the equilibrium swelling degree ϕ_e appears well adapted to define accurately the topological structure of the network; in fact, the present results show that $\langle P_2 \rangle$ scales as ϕ_e whatever the densities of chemical junctions and trapped entanglements. In contrast, the existing neutron scattering measurements do not reveal a noticeable effect of trapped entanglements on the overall dimensions of distorted coils; such a comparison between the two techniques would be worth elaborating for a better understanding of the role of trapped entanglements. For instance, it may be important to test the validity of the sliplink notion [14] through the local (^{2}H-NMR) and the semi-local (SANS) influence of the trapped entanglements.

References

1. Jarry JP (1978) Thesis, Paris
2. Ong CS, Stein RS (1974) J Polym Sci Phys Ed 12:1599
3. Deloche B, Samulski ET (1981) Macromol 14:575
4. Dubault A, Deloche B, Herz J (1984) Polymer 25:1405
5. Deloche B, Beltzung M, Herz J (1982) J Phys Lett 43:763
6. Toriumi H, Deloche B, Samulski ET, Herz J (1985) Macromol 18:305
7. Gronski W, Stadler R, Jacobi MM (1984) Macromol 17:741
8. Deloche B, Dubault A, Herz J, Lapp A (1986) Europhys Lett 1 (12):629
9. Beltzung M (1982) Thesis, Strasbourg
10. Beltzung M, Picot C, Rempp P, Herz J (1982) Macromol 15:1594
11. Charvolin J, Deloche B (1979) In: Luckhurst GR, Gray GW (eds) The Molecular Physics of Liquid Crystals, chapter 15, Academic Press, London
12. Samulski ET (1985) Polymer 26:177
13. Beltzung M, Picot C, Herz J (1984) Macromol 17:663
14. Ball R, Doi M, Edwards SF, Warner M (1981) Polymer 22:1010

Authors' address

A. Dubault
Laboratoire de Physique des Solides
CNRS-LA 2
Université de Paris-Sud
F-91405 Orsay, France

Discussion

SAUTTER:

I want to ask: How large is the orientation factor as an absolute value?

DUBAULT:

It is a very small orientation, from 10^{-3} to 10^{-2}.

BOUÉ:

The radius of gyration is not a good quantity. I just want to say that the radius of gyration of one labelled mesh when you put, in addition, an entanglement, will not vary a lot. It will vary – depending on the model – but only by 5 % or 7 %. Did you exempt models?

DUBAULT:

Not exactly labelled. When you compare the anisotropy of this tube network you do not obtain a difference in neutron scattering. It is about the same variation in orientational order.

Progress in Colloid & Polymer Science Progr Colloid & Polymer Sci 75:49–54 (1987)

Stress-strain behaviour of model networks in uniaxial tension and compression

W. Oppermann and N. Rennar

Institut für Physikalische Chemie der TU Clausthal, Clausthal-Zellerfeld, F.R.G.

Abstract: Elasticity measurements performed on well-defined poly(dimethylsiloxane) networks (PDMS) reveal that there is a direct proportionality between the small-strain modulus, G, and the chemical network density, ν_{ch}, only at high network densities, whereas G is quite constant and in the order of the plateau modulus, G_N^o, at low network densities. This indicates that topological interactions, e. g. entanglements, contribute to the modulus in a certain range of network densities.

PDMS networks having well-defined topologies were prepared by endlinking fractionated PDMS chains ($\bar{M}_n$ ranging from 2000 to 62000 g mol^{-1}) with a pentafunctional cyclic siloxane. Generally, the sol fraction of the samples was below 1.5 % suggesting that the crosslinking reaction was quite complete.

Stress-strain isotherms in uniaxial tension and compression were measured at 333 K for these networks utilizing only one specimen in the same apparatus for the whole deformation range covered. At small and medium deformations, the reduced stress increases monotonically as a function of reciprocal elongation when going from extension to compression. A maximum in the Mooney-Rivlin plot may occur, if at all, in the compression range at $\lambda \leq 0.7$, in qualitative accord with some theoretical approaches.

Key words: Rubber elasticity, model networks, Mooney-Rivlin plot, entanglements.

Introduction

Considerable work has been done in recent years to test the applicability of molecular theories of rubber elasticity. Still, there are different opinions about the influence of trapped entanglements on the mechanical properties at equilibrium and about the shape of the curve of the reduced stress versus reciprocal elongation (Mooney-Rivlin plot) [1–9].

According to the classical statistical theory of rubber elasticity, the reduced stress σ_{red} of a polymer network deformed in simple elongation or uniaxial compression, is given by [10, 11]:

$$\sigma_{red} = \frac{\sigma_o}{\lambda - \lambda^{-2}} = A\,\nu\,k\,T\,\frac{\langle r^2 \rangle}{\langle r^2 \rangle_o}\,, \qquad (1)$$

where σ_o is the nominal stress (tensile or compressive force per cross-sectional area of the undeformed specimen) and the deformation ratio, λ, is the length of the deformed sample, L, related to the length at rest, L_o. A is a factor of the order of unity, ν the number of elastically effective network chains per unit volume, k the Boltzmann constant, T the temperature, $\langle r^2 \rangle$ the mean-square end-to-end distance of the chains in the undeformed network, and $\langle r^2 \rangle_o$ the same quantity for uncrosslinked chains. The structure factor A is unity for affinely transposed crosslinks [12–15], whereas A becomes equal to $(f-2)/f$ with the assumption of fluctuating crosslinks, f being the functionality of the crosslinks [16–18].

Most polymer networks do not conform precisely to Equation (1). The deviations have usually been described by the semi-empirical Mooney-Rivlin Equation [19, 20]:

$$\sigma_{red} = 2C_1 + 2C_2/\lambda\,, \qquad (2)$$

but this equation holds only in simple elongation and fails in uniaxial compression.

Several molecular models have been introduced which modify the classical statistical theory so as to come to a better agreement with experiments. One of these theories is based on the influence of entanglements which become permanently trapped in the network upon crosslinking. The elastic behaviour of these entanglements is assumed to be different from that of chemical crosslinks [3, 21].

On the other hand, it has been argued that the degree of fluctuation of the junctions depends on the magnitude of the deformation causing a strain dependence of the structure factor A [1, 22]. The upper limit of the reduced stress is given by the theory using affinely transposed crosslinks ($A = 1$). This should hold at small strains and in compression, whereas at infinitely large tensile strain the theory including fluctuating crosslinks applies.

Another explanation of the Mooney-Rivlin behaviour is based on a modification of the statistical theory, reconsidering the packing conditions in amorphous polymer networks [8]. From this it follows that the term $\langle r^2 \rangle_o$ in Equation (1) has to be replaced by one which depends on strain.

A more formal account makes use of the similarity of the result of the classical statistical theory of elasticity and the ideal gas law, treating the real rubber with a van der Waals type of approach. The parameters introduced are interpreted on a molecular level as finite chain extensibility and global interactions or crosslink fluctuations [9].

The results of the theoretical approaches differ in what they predict for the shape of the Mooney-Rivlin plot, especially in the vicinity of the undeformed state, and in the absolute value of σ_{red} at some specified strain. These predictions can be tested by experiments. The imperative basis for an experimental examination of the relationship between network structure and mechanical properties is the availability of networks of well-defined microstructure. Particularly, the network density, ν, and the functionality of the crosslinks, f, are of fundamental interest in the recent versions of theories. On this basis, the aims of the present study are:

1. The synthesis of networks whose topology is as simple and well-known as feasible,
2. The investigation of the mechanical behaviour in uniaxial tension and compression on one sample in one apparatus during the same experiment.

Experimental

Sample preparation

The method used for network formation was the crosslinking of vinyl-terminated precursor poly(dimethylsiloxane) chains (PDMS) by a pentafunctional crosslinker. The percursor polymers obtained from Wacker-Chemie GmbH, Burghausen, were fractionated using the solvent/nonsolvent system benzene/methanol primarily to remove low molecular weight cyclics and short chains. The molecular weight distribution was checked by GPC. M_w/M_n was below 1.5 in every case. The molecular weight of the fractionated material was determined by vapor pressure osmometry and by titration of the vinyl endgroups with mercury(II)-acetate. The results of the two methods agreed within 5 %, indicating that the percursor chains carried exactly two functional groups. A series of polymers ranging in molecular weight from 2000 to 62000 g mol^{-1} was employed to prepare the networks.

Crosslinking was performed in bulk by mixing the precursor chains with 1,3,5,7,9-pentamethyl-1,3,5,7,9-pentahydro-cyclopentasiloxane in stoichiometric ratio and adding a few drops of a solution of the catalyst, cis-dichlorobis(diethylsulfide)-platinum-(II), in toluene to give an overall Pt concentration of 5–10 ppm. The carefully stirred mixture of polymer, crosslinker, and catalyst was poured into suitably shaped molds and cured at $T_c = 333$ K under N_2 for seven days. The functionality of the crosslinker had been checked before use by titration with mercury(II)-acetate. The scheme of the crosslinking reaction is outlined in Figure 1.

The networks were extracted over 3 weeks with a mixture of hexane and benzene, then slowly deswollen and dried in vacuum. The weight loss due to extraction was less than 1.5 %. The chemical network density, ν_{ch}, was computed from the molecular weight of the precursor chains by $\nu_{ch} = \varrho/M_n$, ϱ being the density of the sample.

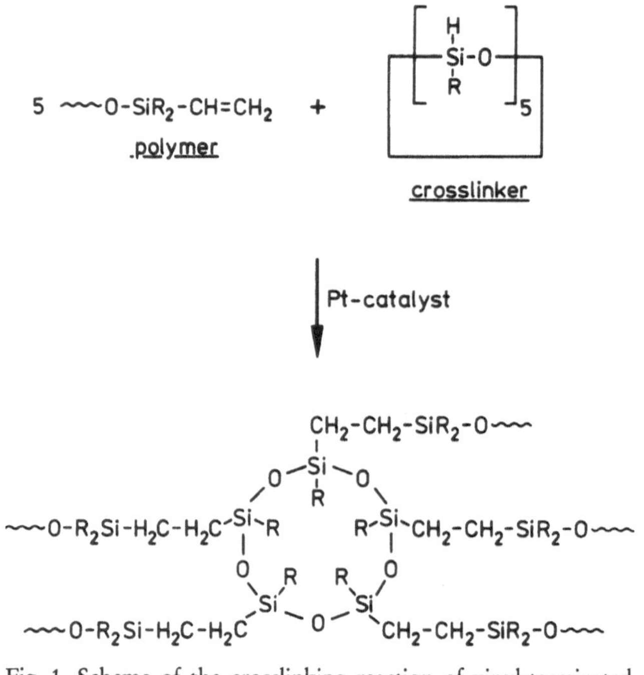

Fig. 1. Scheme of the crosslinking reaction of vinyl-terminated PDMS and pentafunctional cyclic siloxane (R ≙ CH_3)

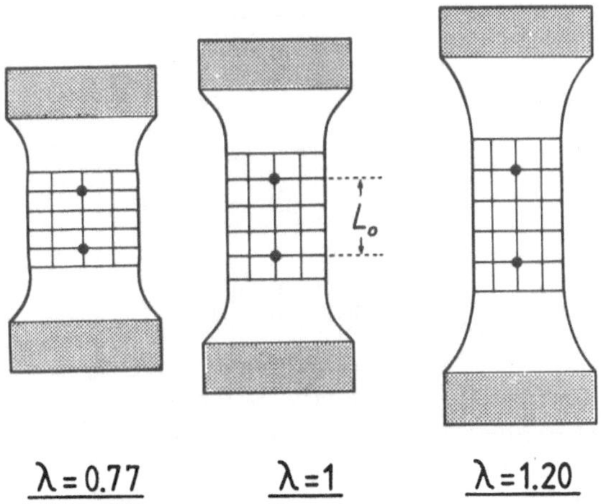

Fig. 2. Observed deformation of the specimen in compression ($\lambda < 1$) and elongation ($\lambda > 1$)

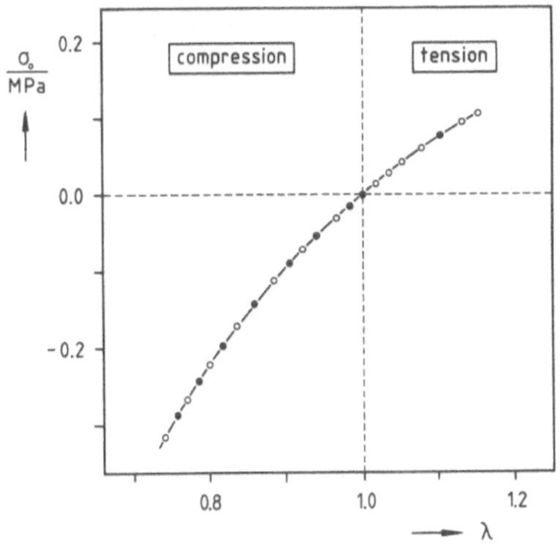

Fig. 3. Stress-strain curve in uniaxial tension and compression for a pentafunctional endlinked PDMS network with $\bar{M}_n = 18400$ g mol^{-1}; (○) increasing deformation; (●) decreasing deformation

Shape of the specimen, measuring apparatus

To comply with the above stated aims, the networks had to withstand compression and tension, both deformations being homogeneous throughout the main part of the sample. This requirement was fulfilled by using circular symmetric, dumbbell-shaped specimens. A check for the homogeneity of the deformation was made by drawing a grid on the surface of one of the samples and carefully observing the displacement of the intersections of the lines, as well as the contour of the sample, upon compression and elongation. The results of this test are depicted in Figure 2. It is seen that within the portion of the sample between the two marks the deformation is indeed homogenenous.

The ends of the dumbbell-shaped specimen were fixed by chucks to the apparatus which ensured a precise alignment of upper and lower clamps. This is compulsory for obtaining reasonable compressive deformations. The force necessary to deform the sample was measured by means of a high-precision load cell, while the deformation was obtained by determining the distance of two carbon black particles brought into the middle part of the sample about 20 mm apart. The approximate position of these two marks is also indicated in Figure 2.

In order to calculate the reduced stress, σ_{red}, at rather small deformations, it is necessary to pin-point the state of zero deformation. This was done by determining the state of zero birefringence, in the usual manner, by passing a laser beam through the sample, placing polarizers in crossed position before and behind the sample.

All measurements were performed at 333 K, i. e. the temperature of network formation, because under these conditions the memory term $\langle r^2 \rangle / \langle r^2 \rangle_o$ should be exactly unity.

Results and Discussion

Figure 3 shows a typical example for a stress-strain curve. It is for a PDMS network whose network chains have $M_n = 18400$ g mol^{-1} corresponding to $\nu_{ch} = 5.1 \cdot 10^{-5}$ mol cm^{-3}. It is seen that the deformation is completely reversible and there is a smooth transition from compression to extension. The deformation range accessible is usually about 20–30 % in both directions. At higher compression ratios, the sample will bend outwards although a lot of care had been taken to have as good an alignment as possible, whereas at higher extensions the samples break.

It is worth noting that the elongations at break become considerably larger when the stoichiometry is slightly unbalanced or the crosslinking reaction is incomplete. Stated otherwise, an ideal, clean model network shows bad properties for technical use.

The Mooney-Rivlin plots obtained for two of the samples differing about 30 % in network density, ν_{ch}, are shown in Figure 4. Three distinctive characteristics of this graph are to be noted. Firstly, the curves pass through the state of zero deformation ($\lambda = 1$) with a positive slope and without any indication of a discontinuity. Secondly, the reduced stresses measured for the two samples are virtually identical, although the network densities differ by approximately 30 %. Thirdly, the reduced stresses are far above the theoretical predictions even of the affine model, which yields 0.10 and 0.14 MPa, respectively, for the two samples.

The fact mentioned in the first statement clearly rules out such theoretical approaches which propose a maximum of the Mooney-Rivlin curve at $\lambda = 1$. The second and third points will be discussed in more detail on the basis of Figure 5, where the shear modulus, G, is

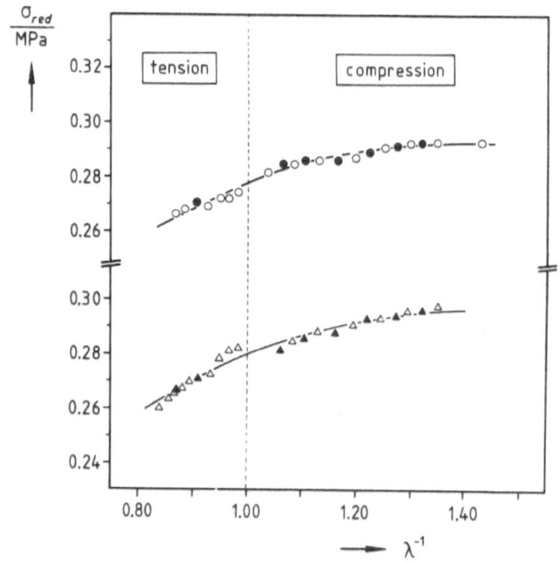

Fig. 4. Mooney-Rivlin plot for two pentafunctional endlinked PDMS networks. (○, ●) $\bar{M}_n = 18400$ g mol^{-1}; (△, ▲) $\bar{M}_n = 25200$ g mol^{-1}; (○, △) increasing deformation; (●, ▲) decreasing deformation

plotted versus the chemical network density making use of the relation

$$G = \lim_{\lambda \to 1} \sigma_{red} \,. \tag{3}$$

It is striking that over a wide range of network densities from $3 \cdot 10^5$ to $12 \cdot 10^{-5}$ mol cm^{-3} the modulus is fairly independent of network density, at variance with the predictions of the classical theory. This is also contradictory to Flory's approach, based on the concept of constrained fluctuations of crosslinks. If the network density exceeds about $12 \cdot 10^{-5}$ mol cm^{-3}, the modulus increases with further rising network density, the relationship between G and ν_{ch} in this range being apparently a direct proportionality. The last statement seems to be highly speculative because we have only two experimental points in that regime. It receives some support, however, when data from the literature are taken into account. Figure 6 is the same kind of plot as Figure 5 for results taken from various sources in the literature obtained on PDMS networks with tetrafunctional crosslinks [23–29]. The course of the curve is similar to that in Figure 5. Furthermore, the levels of the plateau are quite identical (around 0.3 MPa), and the mininum network densities from where the proportionality between G and ν_{ch} holds, are roughly identical, too.

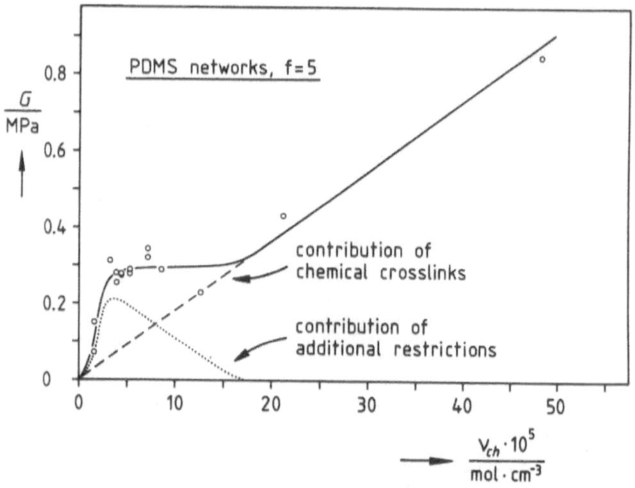

Fig. 5. Shear modulus G as a function of calculated network density ν_{ch} for *pentafunctional* endlinked PDMS networks at $T = 333$ K (our results)

It is our impression that this is enough evidence that the curves drawn in Figure 5 and 6 correctly reflect the real characteristics of the material properties, and we shall base our interpretation thereon.

The slopes of the curves in the proportionality range (at high network densities) correspond to structure factors $A = 0.67$ in both cases. This can be interpreted as having reduced fluctuations of crosslinks (intermediate state between the two extremes $A = 1 - 2/f$ for fully fluctuating crosslinks and $A = 1$ for nonfluctuating crosslinks). In the proportionality range, an effect

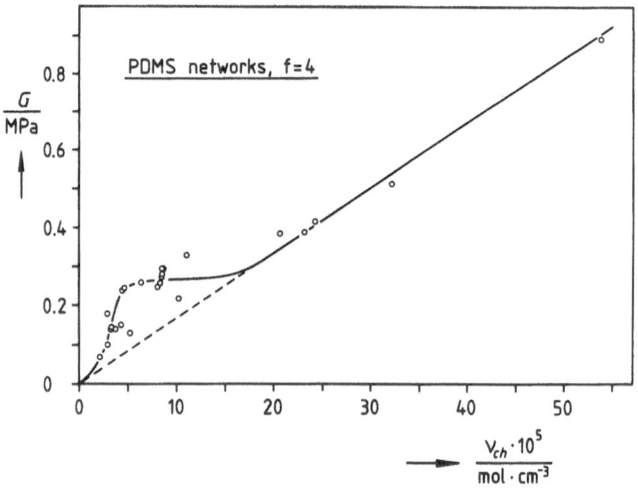

Fig. 6. Shear modulus G as a function of calculated network density ν_{ch} for *tetrafunctional* endlinked PDMS networks at $T = 298$–310 K (data from the literature [23–29])

of permanent entanglements need not be taken into account.

The height of the plateau in the range of network densities from $3 \cdot 10^{-5}$ to $12 \cdot 10^{-5}$ mol cm^{-3} is practically indentical to the plateau modulus G_N^o of uncrosslinked, high molecular weight PDMS [21]. It is, however, a true equilibrium quantity. (There are no appreciable relaxations observed on a time scale of one week at 333 K.) If one assumes that the degree of fluctuation of the crosslinks is independent of ν_{ch}, the extension of the proportionality lines (the dashed lines in Figs. 5 and 6) reflect the portion of the modulus due to permanent crosslinks. The differences between the solid lines and the dashed lines can then be treated as the contribution of additional restrictions, e. g. permanent entanglements. This difference has been depicted as a dotted curve in Figure 5.

According to this interpretation, the contribution of trapped entanglements to the equilibrium modulus is at most 3/4 of the plateau modulus. The entanglement contribution decreases when the chemical network density is increased (after a minimum chemical network density of $3 \cdot 10^{-5}$ mol cm^{-3}). This seems plausible because the mutual interspersion of the chains is less when the chains are shorter. The occurrence of the plateau is attributed to this fact.

From our data, the transition from the plateau to the proportionality range appears to be rather sharp. In many other publications this is not seen so distinctly. In our view this is due to the fact that the precursor polymers from which our networks were made, had been fractionated. The broader the molecular weight distribution of the network chains, the more the transition is smeared out over a certain range of network densities. When the networks are prepared by vulcanization, i. e. by statistical crosslinking of very long chains, the chain length distribution in the network is so wide that no plateau occurs, and the only indication of an entanglement contribution is given by an intercept in the graph of G versus ν_{ch} [30].

Some comment seems appropriate concerning one of our older papers where we found considerably lower moduli for some end-liked PDMS networks [31]. Those samples were prepared in solution. Under these conditions, many crosslinks are wasted to produce elastically ineffective cycles, and therefore the effective network density was much lower than supposed [32].

Acknowledgements

Financial support for this work by the Deutsche Forschungsgemeinschaft is gratefully acknowledged. We thank Wacker-Chemie GmbH, Burghausen, for the provision of some of the materials utilized in this study.

References

1. Flory PJ (1977) J Chem Phys 66:5720
2. Langley NR (1968) Macromol 1:348
3. Dossin LM, Graessley WW (1979) Macromol 12:123; Graessley WW (1982) Adv Polym Sci 47:67
4. Heinrich G, Straube E, Helmis G (1979) Plaste u Kautschuk 26:561
5. Flory PJ, Erman B (1982) Macromol 15:800, 806
6. Edwards SF (1977) Brit Polym J 9:140
7. Marrucci G (1981) Macromol 14:434
8. Schwarz J (1981) Polym Bull 5:151, 478
9. Kilian H-G (1983) Kautsch Gummi Kunstst 36:959; Kilian H-G Enderle HF, Unseld K (1986) Coll & Polym Sci 264:866
10. Treloar LRG (ed) (1975) The Physics of Rubber Elasticity, University Press, Oxford
11. Dusek K, Prins W (1969) Adv Polym Sci 6:1
12. Kuhn W (1934) Kolloid-Z 68:2; Kuhn W (1936) ibid 76:258
13. Hermans JJ (1947) Trans Farad Soc 43:591
14. Flory PJ (1950) J Chem Phys 18:108
15. Wall FT, Flory PJ (1951) J Chem Phys 19:1435
16. Duiser JA, Staverman AJ (1965) In: Prins JA (ed) Physics of Non-Crystalline Solids, North-Holland Pub Co, Amsterdam, p 376
17. Graessley WW (1975) Macromol 8:186
18. Edwards SF (1971) In: Chompff AJ, Newman S (eds) Polymer Networks, Structure and Mechanical Properties, Plenum Press, New York, p 83
19. Rivlin RS (1947) J Appl Phys 18:444
20. Mooney M (1940) J Appl Phys 11:582; Mooney M (1948) ibid 19:434
21. Ferry JD (ed) (1980) Viscoelastic Properties of Polymers, 3rd ed, Wiley, New York
22. Ronca G, Allegra G (1975) J Chem Phys 63:4990
23. Mark JE, Sullivan JL (1977) J Chem Phys 66:1006
24. Llorente MA, Mark JE (1980) Macromol 13:681
25. Kosfeld R, Heß M, Hansen D (1980) Polym Bull 3:603
26. Valles EM, Macosko CW (1979) Macromol 12:673
27. Macosko CW, Benjamin GS (1981) Pure Appl Chem 53:1505
28. Granick S, Pedersen S, Nelb GW, Ferry JD, Macosko CW (1981) Polym Sci Polym Phys Ed 19:1745
29. Meyers KO, Bye ML, Merrill EW (1980) Macromol 13:1045
30. Gleim W, Oppermann W, Rehage G (1986) Makromol Chem 187:1273
31. Oppermann W, Rehage G (1981) Coll & Polym Sci 259:1177
32. Oppermann W, Rose S, Rehage G (1985) Brit Polym J 17:175

Received February 2, 1987;
accepted March 16, 1987

Authors' address:

W. Oppermann
Institut für Physikalische Chemie
der TU Clausthal
Arnold-Sommerfeld-Str. 4
D-3392 Clausthal-Zellerfeld, F.R.G.

Discussion

ILAVSKY:

Did you also try to determine the distribution in functionality? I do not believe that you have exact functionality 5 for all your molecules. I mean for the small molecular weight prepolymer and a large molecular weight prepolymer. If you take it into account, the chemical contribution will change. But I don't believe that this will reduce its contribution of entanglements.

OPPERMANN:

When preparing the crosslinker we did some fractionation, so we are sure that the crosslinker we put into the sample really has functionality 5. Is that what you mean? These are small rings, small siloxane rings with hydrogen atoms on the ring. And you can, by high pressure chromatography, fractionate 5-6-7- membered rings and we took these fractions and we are pretty sure that we really have functionality 5. The other point is whether all functionalities react or not. But here we are also pretty sure, because we tried to unbalance this stoichiometry a little by using a small excess of crosslinker or excess precursor chains and then the modulus always drops down. So I think we are really in the range where we have the right stoichiometry.

ILAVSKY:

I saw the results which Hertz and also Macosko presented. They showed that even at the stoichiometry of SiH-group and double bounds, side reactions are going on, so that some part of the crosslinker is reacting between them.

OPPERMANN:

Mainly Gottlieb has done this analysis. This is important when having trifunctional crosslinkers. The effect drops down very rapidly, if you go to higher crosslink functionalities.

That is another reason why we used this 5-functionality which is rather uncommon, I think.

HAVRANEK:

What were the times of measurements of G and was it in the region of very small crosslinking density or not?

OPPERMANN:

It starts from a very high crosslinked density that corresponds to a molecular weight of about 4000 and goes down in crosslink density to a molecular weight corresponding to that between crosslinks of 60000. We did a torsional vibration experiment over three decades of frequency just to supplement these data. These are static data taken at measurement times of about 1 h or so. But one can see from the torsional vibration experiment, that there is practically no frequency dependence from 10-3-1 Hz, and we are pretty sure that we have the equilibrium modulus.

WINTER:

First one comment on the side reaction.

Careful investigations by Fisher and Gottlieb at Bersheva showed that there are no side reactions at very low temperature, let us say if you prepare your PDMS at room temperature. But if you prepare the samples at elevated temperatures side reactions are very severe. This is just one remark which I found also very interesting with respect to our experiments.

PICOT:

Have you done measurements on samples prepared at different concentrations in order to modulate the contribution of permanent entanglements?

OPPERMANN:

All the results I have presented here were crosslinked in bulk, and if you crosslink in solution, then you dry the sample and measure the dry sample – the modulus is lower. I think it is not only the effect of reduction of entanglements, but also the effect that as you dilute the system you get much more ineffective cycles in the samples. So we prefer to do it in bulk: it is done at 50 °C.

KILIAN:

May I ask a provocative question: How far are you sure that the samples are homogeneous, because they are so stiff and break at such small degrees of elongation?

OPPERMANN:

First we observed that the worse you make the sample, the better you can elongate. So we agree, if you argue the other way round.

Progress in Colloid & Polymer Science Progr Colloid & Polymer Sci 75:55–61 (1987)

General deformation modes of a van der Waals network

H. F. Enderle and H.-G. Kilian

Abteilung Experimental-Physik, Universität Ulm, Ulm, F.R.G.

Abstract: It is shown how multiaxial deformation modes can be described with the aid of a generalized van der Waals network model. By introducing the modified invariant $J = \beta I_1 + (1 - \beta) I_2$ the resulting van der Waals strain energy function allows in the case of biaxial homogeneous deformation to derive a set of equations of state. The calculated normal forces can be fitted to uniaxial elongation and compression experiments on the one hand and to data of general biaxial deformation modes on the other hand up to largest measures of strain.

Key words: Deformation modes, biaxial deformation, strain energy function, rubber elasticity, finite extensibility, Van der Waals theory.

Introduction

Taking account of finite chain extensibility involves an exclusion principle: macroscopic strain parameters of a real network cannot exceed a defined maximum value; all the states of deformation that are beyond these limits are not realizable [1, 2].

From the non-Gaussian theory of Kuhn [3] it appears that the energy density should run to infinitely large values in every state of maximum deformation. This constitutive singularity is the reason behind the well known upturn in the simple extension stress-strain pattern of rubbers [4, 8]. The asymptotics in the stress-strain pattern are different for different deformation modes; the upturn in the nominal forces of equibiaxially stretched rubbers is, for example, observed at smaller strains than in the mode of simple extension.

This phenomenon cannot be interpreted in terms of the Gaussian model because it does not show any mode dependence in the reduced representation. This is illustrated with the Mooney plot in Figure 1a, according to which the elastic response of the Gaussian network related to the "deformation function" of an incompressible elastic continuum ($D = \lambda - (\lambda)^{-2}$; λ = macroscopical strain), is identical for simple extension and uniaxial compression (uniaxial compression is a special realization of equibiaxial extension).

It is a significant finding that calculations with the aid of the Langevin phantom chain model of Kuhn [3] cannot be fitted to the experimental data in the range of small and intermediate strains (see Fig. 1b). Better correspondence is obtained in this range from calculations with the aid of Flory's phantom-network theory [5, 7, 8] which is, on the other hand, a failure at larger strains, due to its being based on the Gaussian chain model.

This short discussion throws light upon the fact that the individual elastic response obtained in each different deformation mode must be strongly related to finite chain extensibility. Crosslinkage fluctuation-effects are, on the other hand, existent, so they must also be considered if a full representation of the mechanical properties of real networks is being aimed at.

Kilian [4] has, for the first time, presented an approach in which both of these effects have been accounted for. Finite chain extensibility and "global interactions" were discussed in terms of a van der Waals approach. The maximum strain parameter, as one of the van der Waals parameters, can be directly related to the chain length in the network. The strain-energy relevant interaction between the chains is characterized with the aid of the second van der Waals parameter that is shown to be related to fluctuations in the crosslinks

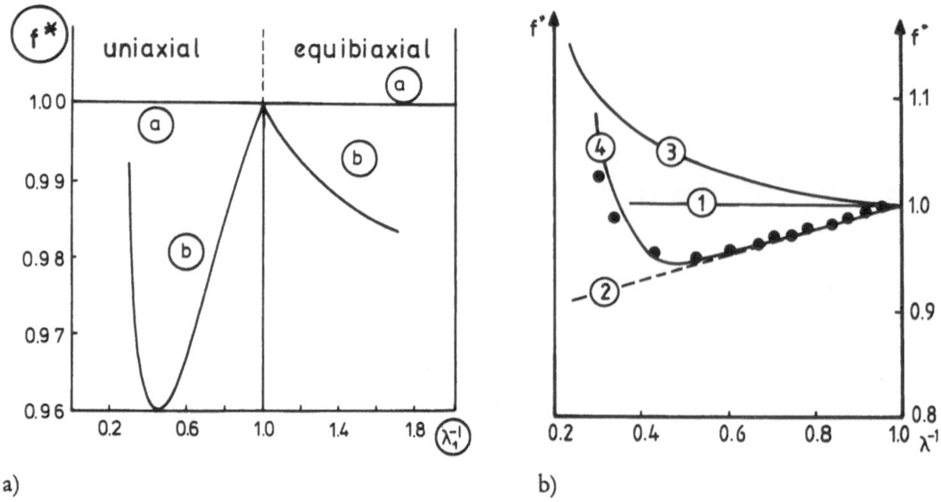

Fig. 1. a) Mooney plot of two different modes of deformation, simple and equibiaxial extension (the last realized by uniaxial compression): (a) Gaussian network, (b) van der Waals model; b) Reduced forces f^* of different models against the inverse stretching ratio: ① the Gaussian model; ② the Mooney – Rivlin approach (----); ③ single-chain phantom network (Langevin type of equation of state); ④ van der Waals model calculated with the parameters $y \equiv \lambda_m{}^2 = 53$, $a = 0.26$; (●) experimental data obtained for SBR

[5–7]. A satisfactory representation of stress-strain measurements is possible within the total range of strains (see Fig. 1).

In discussing different deformation modes, then, like simple extension, simple shear and equibiaxial extension, geometrical conditions have been proposed. The maximum strain in the directions of interest should be related to the chain length parameter of the representative chain in the real network [9].

These very first ideas should now be put into a general formulation, so as to allow for an interpretation of every multiaxial deformation mode. The description should, of course, be confined to a generalization of the van der Waals model to arrive at a quantitative description with the use of its "structure parameters", the actual chain length and the interaction parameter.

Theory

We will consider incompressible à priori isotropic mechanically stable systems deformed under equilibrium conditions. Since the elastic system is characterized by its strain energy function [1, 2, 10], it is clear that we have to seek the most general formulation of the elastic potential of a van der Waals network.

The system is believed to be represented as a "quasi-continuum" comprised of a defined number of energy-equivalent subsystems of deformation, in the case of a perfectly homogeneous network represented by the chains themselves. For an incompressible and isotropic elastic body of this type we are led, for the principle of material objectivity [11], to the simplest form of the strain energy function comprising two of the three strain invariants (I_1, I_2) of the Cauchy-Green deformation tensor $\underset{\sim}{C}$ [1, 2, 10, 11]. Taking account of the van der Waals corrections, the universality in the formulation of the strain energy function must necessarily become reduced. This should be indicated by the more distinct

$$W = W(I_1, I_2; y, a) \tag{1}$$

whereby the strain invariants are written as [1, 11]

$$I_1 = tr(\underset{\sim}{C}) \tag{2}$$

$$I_2 = 1/2\,\{(tr(\underset{\sim}{C}))^2 - tr(\underset{\sim}{C}^2)\}\,. \tag{3}$$

The length parameter y is defined as the number of statistical segments per chain [3, 4] while "a" is the phenomenological "fluctuation parameter" in the van der Waals version.

The van der Waals parameters

If the chains themselves are not too short, the maximum extensibility of a single entity can be ap-

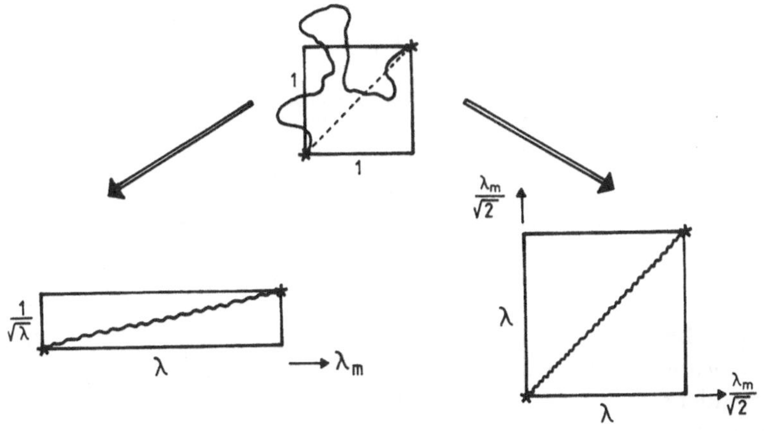

Fig. 2. Uniaxial and equibiaxial deformation in heuristical limit states

proximated by the use of the "Gaussian measure" [4,7–9]

$$(\lambda_{\text{max-chain}})^2 = y \tag{4}$$

y is the number of Kuhn-segments.

If the totality of the crosslinks is now transformed according to the law of affinity [8,9] we are led to the significant consequence

$$y = (\lambda_{m1})^2 + (\lambda_{m2})^2 + (\lambda_{m3})^2 \,. \tag{5}$$

The maximum strain parameters in the three directions of the coordinate frame, are assumed to be uniquely determined by the single chain structure parameter y. The consequences of Equation (5) are illustrated in Figure 2. In the mode of simple elongation (left side of Fig. 2) the maximum strain $\lambda_m^{(1)}$ is estimated to be roughly identical with

$$\lambda_m^{(1)} \simeq y^{1/2} \,. \tag{6}$$

In the equibiaxial mode (right side in Fig. 2) the maximum macroscopic strain in both coordinate axes is reduced to the value

$$\lambda_m^{(2)} = \lambda_m^{(1)}/\sqrt{2} \,. \tag{7}$$

In swelling a dry network, this process corresponds to the mode of equitriaxial extension, such that the maximum strain in all directions should now become equal to

$$\lambda_m^{(3)} = \lambda_m^{(1)}/\sqrt{3} \,. \tag{8}$$

Hence, it is a consequence of the symmetry condition as defined in Equation (5) that the geometrical constraints produced by finite chain extensibility of the single chains are behind the mode dependence of the maximum strains.

The size of the interaction parameter "a" characterizes global interaction, due to diffusive fluctuations of the crosslinks [6,12]. Many plausible reasons could be given for defending the assumption that the interaction energy via diffusive crosslinkage motion should be mode independent [6,13]. Let us accept in the present paper, without discussion, that the interaction parameter "a" of van der Waals networks does not depend on the mode of deformation.

The strain-energy function

It is the object of this section to find a generalized formulation of the strain energy function, to find infinitely large energy densities when the rubber is strained to its maximum. This singularity has to be met irrespective of the shape the rubber is forced into.

Anticipating later results, let us define the generalized strain invariant by [14]

$$J = \beta I_1 + (1-\beta)\, I_2 \,. \tag{9}$$

The energy-density pole at the heuristic state of maximum strain is indeed, uniquely defined by bringing the van der Waals strain-energy function into the generalized form [15]

$$W = G[-2y\{\ln(1-\eta) + \eta\} - (2a/3)\{(J-3)/2\}^{3/2}] \tag{10}$$

where

$$\eta = \{(J-3)/y\}^{1/2} \,. \tag{11}$$

The modulus is explicitly expressed as [16,17]

$$G = \varrho RT\,(\xi/M_c) = \varrho RT\,(\xi/yM_s) = \varrho RT/yM_o\,. \quad (12)$$

Where ϱ is the density, T the absolute temperature, R the gas constant. The molecular weight of the chains M_c can be written as

$$M_c = yM_s \quad (13)$$

with M_s as the molecular weight of the statistical Kuhn segment [3]. But the statistical segment is by no means stretching invariant. To accept strain dependent changes of diffusive conformational freedoms per statistical segment, is equivalent to defining the actual stretching invariant unit with respect to the system's entropy by [3,16–18].

$$M_o = M_s/\xi\,. \quad (14)$$

We have a simple possibility of proving the quality of this approach by determining the maximum strain parameter λ_m from the relative course of simple extension experiments and then to find out whether the moduli of a set of differently crosslinked rubbers can be computed with an invariant M_o [17,43].

The biaxial deformation modes

To now describe biaxial stress-strain experiments, we need to know the two independent Cauchy-stress components as a function of the macroscopic strain variables in the two directions of interest (see Fig. 3).

From the condition of incompressibility

$$I_3 = \det(\underset{\sim}{C}) = 1 \quad (15)$$

we are led to

$$\lambda_3 = 1/(\lambda_1\lambda_2)\,. \quad (16)$$

With the knowledge of the strain energy function, we can then derive the biaxial elastic material constitutive equations (the biaxial mechanical equations of state) [19] the most general formulation of which reads

$$\sigma_i = \{(\lambda_i)^2 - (\lambda_i)^{-2}\,(\lambda_j)^{-2}\}\{dW/dI_1 + (\lambda_j)^2 dW/dI_2)\} : i, j = 1, 2 \quad (17)$$

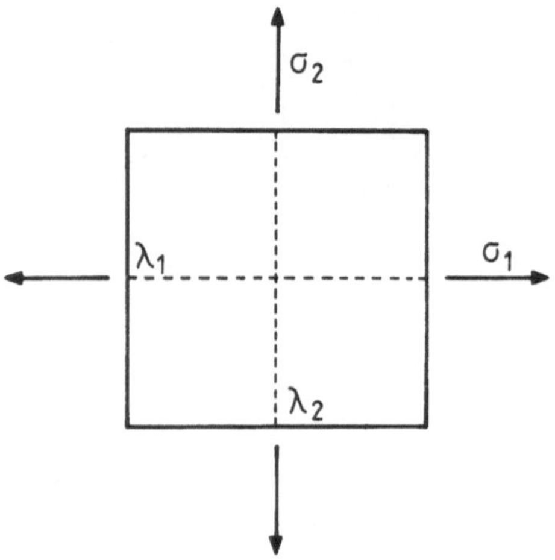

Fig. 3. Homogeneous biaxially streched sheet of rubber

whereby

$$dW/dI_1 = \beta dW/dJ$$

$$dW/dI_2 = (1-\beta)\,dW/dJ \quad (18)$$

and

$$dW/dJ = (1-\eta)^{-1} - a\,\{(J-3)/2\}^{1/2}\,. \quad (19)$$

Comparison with experiments

A distinct example of a biaxial stress-strain measurements selected from extensive investigations published by Kawabata et al. [20,21] are depicted in Figure 4b. After the equibiaxial stretch up to $\lambda_1 = \lambda_2 = 1.9$, λ_2 was decreased stepwise while λ_1 was left unchanged. Finally, λ_2 was brought to that reduced value where the normal stress σ_2 disappears. This means that the system is then brought into the mode of simple extension. All the intermediate states of deformation cover, therefore, a large set of biaxial deformation modes, including the mode of pure shear. The experimental nominal forces f_1 and f_2 can be fairly well described with the aid of the Equations (17), whereby that set of elementary parameters (G_o, y and a) has been employed, with was previously adjusted by fitting equibiaxial measurements (Fig. 4a).

The quality of our fit to simple elongation and compression measurements of Flory and Pak [22] can be seen by evidence from the plot depicted in Figure 5. We have to notice here that within the scope of the theory involved, the representation could only be made per-

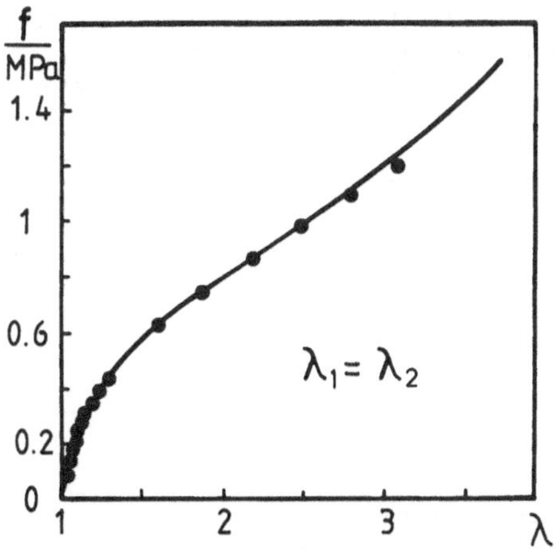

Fig. 4a. Normal force $f_1 = f_2 = f$ of an equibiaxially stretched rubber sheet according to [20] and [21]. Solid line calculated with $G_o = 0.49$ MPa, $y = 150$, $\beta = 1$

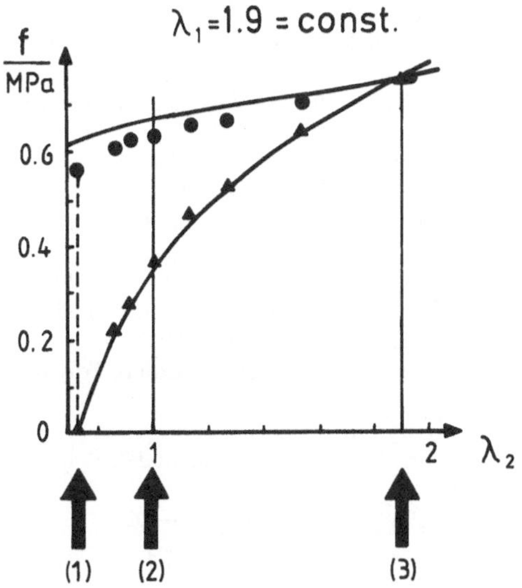

Fig. 4b. Normal force f_1, f_2 of a biaxial stretched sheet of Polyisoprene versus strain ratio λ_2, according to [21]. Solid lines calculated with the set of parameters: $G_o = 0.49$ MPa, $y = 150$, $a = 0.24$, $\beta = 1$. "One-dimensional modes" of deformations are indicated with arrows: (1) simple extension (2) pure shear (3) equibiaxial extension

fect by assigning the parameter β to a value smaller than one.

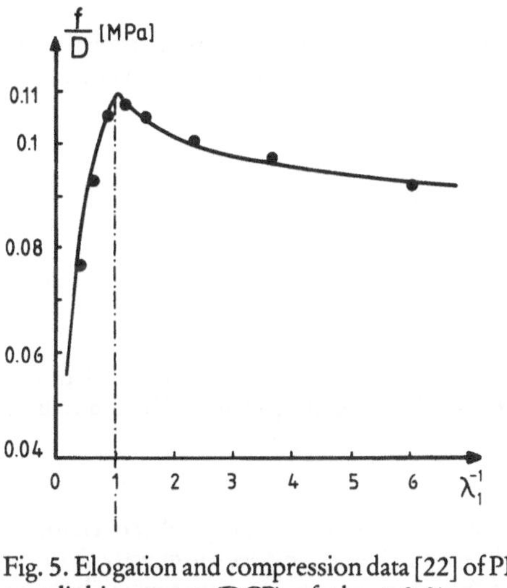

Fig. 5. Elogation and compression data [22] of PDMS samples with crosslinking agent (DCP) of about 3 % mapped in "Mooney" fashion. Solid line calculated with the aid of the parameters: $G_o = 0.11$ MPa; $y = 380$, $a = 0.24$, $\beta = 0.94$

Final Remarks

It has been shown that the elastic response of rubbers to different deformation modes can be fully represented using a generalized van der Waals description. What is evidenced is that the mode dependence of the normal stresses is a plausible consequence of having the real network constituted by chains of finite length. The maximum value of the first strain invariant is then simply assigned equal to the chain length parameter y (Eq. (5)). This ensures that the strain-energy function of van der Waals networks goes to infinitely large values for each state of maximum strains, irrespective of the shape the sample is forced into. This constitutive singularity fixes, for real networks, the accessible volume in the λ_i-space to a sphere of the radius y.

The van der Waals interaction parameter "a", which characterizes the strain-energy modification due to fluctuations of the crosslinks, indeed seems not to show any mode dependence.

The initial slope of the reduced force in the Mooney plot depends on both the van der Waals parameters, thus illuminating the necessity to account for finite chain length and global interaction, even at the smallest strains.

There a need for a finer, at least plausible, interpretation of the parameter β.

References

1. Adkins JE, Green AE (eds) (1970) Large Elastic Deformation, 2nd ed, Clarendon Press, Oxford
2. Truesdell CA (ed) (1966) The elements of continuum mechanics, Springer, Berlin
3. a) Kuhn W (1936) Kolloid Z 76, 3:258
3. b) Kuhn W, Grün F (1942) Kolloid Z 101:248
4. Kilian H-G (1981) Polymer 22:209
5. James HM, Guth E (1947) J Chem Phys 15:669

6. Vilgis Th (1984) Ph D Thesis, University of Ulm
7. Flory PJ (ed) (1953) Principles of Polymer Chemistry, Cornell, Ithaka
8. Treloar LRG (ed) (1975) The Physics of Rubber Elasticity, 3 rd ed, Clarendon Press, Oxford
9. Smith TL, Dickie RA (1969) J Polym Sci A2 7:635
10. Rivlin RS (1948) Phil Trans R Soc A241:379
11. Truesdell CA, Toupin RA (eds) (1966) The Classical Field Theories, Handbuch der Physik III/1, Springer Berlin
12. Vilgis Th, Kilian H-G (1986) Coll & Polym Sci 264:137
13. Kilian H-G, Schenk H, Wolff W (19??) to be published
14. Larson RG, Monroe K (1984) Rheol Acta 23:10
15. Kilian H-G, Vilgis Th, Enderle HF (1984) Coll & Polym Sci 262:696
16. Flory PJ (1985) Polym J 17:1
17. Kilian H-G, Unseld K, Enderle HF (1986) Coll & Polym Sci 264:866
18. Dusek K, Prins W (1969) Adv Polym Sci 6:1–102
19. Rivlin RS, Saunders DW (1951) Phil Trans R Soc A243:251
20. Kawabata S (1973) J Macromol Sci B8 3:605
21. Kawabata S, Matsuda M, Tei K, Kawai H (1981) Macromolecules 14:154
22. Pak H, Flory PJ (1979) J Polym Sci Polym Phys Ed 17:1845
23. Pietralla M (1982) Habilitationsschrift, Ulm

Received March 18, 1987;
accepted April 14, 1987

Authors' address:

H.-F. Enderle
Universität Ulm
Abteilung Experimentelle Physik
Oberer Eselsberg
D-7900 Ulm, F.R.G.

Discussion

ILAVSKY:
If y is the finite chain-length parameter it is proportional to the molecular weight between crosslinks. Hence, the modulus G_0 and y should be directly related. Is that so or not?

ENDERLE:
Yes, they are directly related.

ILAVSKY:
In that case, y is not an adjustable parameter.

ENDERLE:
It should not be an adjustable parameter, but in our calculation we had G, the parameter of the modulus, fitted too. Theoretically, it should be related.

ILAVSKY:
The modulus is directly related to the new number of elastic active chains. From the modulus, one can calculate directly the value of y, so that it is not necessary to have three parameters but only two.

KILIAN
This is not straightforwardly possible in any case: the magnitude of the parameter M_0 implicates the knowledge of the size of those units which are stretching invariant. Additional problems come about due to "prefactor problems". For networks prepared under the same conditions – hoping that the prefactor correction terms are not then changed, so that they are accounted for by adjusting M_0 – the two van der Waals parameters are left for every fit of experiments. Since the interaction parameter, a, seems not to depend on the degree of crosslinking, the maximum strain parameter alone should describe the stress-strain curves of differently crosslinked networks. This was found to hold for NR-networks, as measured by Mullins.

ILAVSKY:
I would expect that there should also be something like a statistical length.

KILIAN:
This is a deeper problem. When calculating the maximum strain parameter, the statistical Kuhn segment should be appropriate. Yet this segment seems not to behave like a rigid rod without any changes of its "intrinsic" entropy. To assign M_0 to smaller numbers is equivalent to assuming that there are distinct conformational changes within the statistical segments accompanied by an entropy change that depends uniquely on the strain.

TOMKA:
I think the problem is that in a rubber we never know what the entropy elastically effective element of the system is from the chemical structure.

KILIAN:
This is correct. The concrete molecular interpretation should be derived from the discussion of M_0.

TOMKA:
We say very easily that we have the modulus and the number of chains and these seem to relate. But even if we know the prefactor, we do not know the correlation to the chemical structure, and what acts as the elastically active element. It is very easy to use the concept of an elastically active element, but nobody can define it chemically.

KILIAN:
I would not totally agree with your statement. M_0 is an average measure for the size of the stretching invariant unit, the size of which gives the order of magnitude of stretching invariant chain-

standing units (mostly found to comprise 2 to 4 chain-standing units).

TOMKA:

But you can get the segment number and this does not correlate with the structure of the segment.

ILAVSKY:

From the other measurements we know for large amount of polymers, of course, how large the statistical segment is. Whenever we took the values of Flory (determined for polymer solutions) for the network, it worked!

PIETRALLA:

There is the geometric statistical segment. With that you can calculate, using a statistical theory like that of Kuhn and Grün, the mean square length and other geometric parameters. But when measuring the force or the heat exchange, for example, you have to deal with entropically equivalent segments, and these are different because each equivalent geometric statistical segment has some entropy contribution by its internal conformational abilities when performing the next step.

The contribution to the entropy change may even depend on the direction of the step just in contrast to the geometrical segment.

STADLER:

The statistical segment you determine from dilute solution has nothing to do with a statistical segment you determine from birefringence or from NMR or something like that. It is specific to the method.

Progress in Colloid & Polymer Science Progr Colloid & Polymer Sci 75:62–69 (1987)

Dynamic shear compliance of swollen networks and its dependence on crosslink density*)

M. Böhm, O. Grassl, W. Pechhold, and W. v. Soden

Abteilung Angewandte Physik, Universität Ulm, F.R.G.

Abstract: Some results on irradiation crosslinked isoprene networks, together with data from the literature on other networks are used to study the general behaviour of the reduced shear compliance $J(\phi)/J(1)$ in dependence on the polymer volume fraction ϕ up to the equilibrium swelling. Within the limit of the uncrosslinked melt, the ϕ^{-2}-law for the plateau compliance is interpreted by multiple-step intra meander shear. Its other limit, a $\phi^{-2/3}$-law, already reached at medium crosslinking, appears simply as the (one-step) intrameander shear of the isotropically swollen blocks.

Key words: Shear compliance, swollen networks, meander model.

Introduction

The relaxation processes controlling the dynamic shear-compliance of polymer melts and networks have been recently discussed within the framework of the meander model [1]. In that paper, the properties of the bulk were of interest in dependence on molecular weight and crosslinking density. For further checking of the model we started dynamic mechanical measurments on the same isoprene networks swollen to different polymer volume fraction ϕ in n-nonane and n-dodecane. The first results are presented here and compared with the predictions of the meander model of the isotropically swollen state.

Experimental

The preliminary results reported here were obtained on radiation crosslinked poly(isoprene), Cariflex IR 305. From the same samples used in [1–3] to measure shear compliance mastercurves of the bulk in dependence on crosslinking density, specimens were cut and swollen to different ratios ϕ^{-1} in n-dodecane or n-nonane. The weight of each test piece was determined before adjusting it in the double sandwich holder[1]) and immediately after the measurement – usually 10^{-3} to 10^{2} Hz at 1 to 3 temperatures. In the case of n-nonane as swelling agent, strips of cotton soaked with n-nonane were placed in the temperature controlled cell to generate the respective vapour pressure.[2]) Using this method it was possible to keep the weight of the specimen constant within 1 p.c..

Results

In Figure 1, a representative set of mastercurves is shown which was obtained on eleven 50 kGy irradiated IR-samples swollen in n-nonane to different degrees, from the dry state ($\phi_p \equiv \phi = 1$) up to $\phi = 0.16$ which is not far from equilibrium swelling ($\phi = 0.13$) in this solvent. The ordinate for the real part J' of the compliance has been shifted by two decades against that of the imaginary part of J'', to avoid crossing over of the two sets of curves. The reference temperature for all swelling ratios was chosen to be $T_o = 273$ K, as it was in Reference [1].

Most interestingly, it follows from Figure 1 and its analysis, that the mastercurves do not change form with the polymer volume fraction ϕ, but appear shifted to higher frequencies and higher compliances with

*) Dedicated to Professor Dr. Helmut Dörfel on the occasion of his 60th birthday.

[1]) The measurements with the mechanical spectrometer (10^{-4} to 200 Hz) including the double sandwich sample holder and other details of the device are described in [2, 4].

[2]) In spite of the thin open edges of the specimen, evaporation would have been enhanced by the weak flow of dry nitrogen used in all experiments below room temperature, to prevent condensation of water or deposition of ice at the cell opening for the driving shaft.

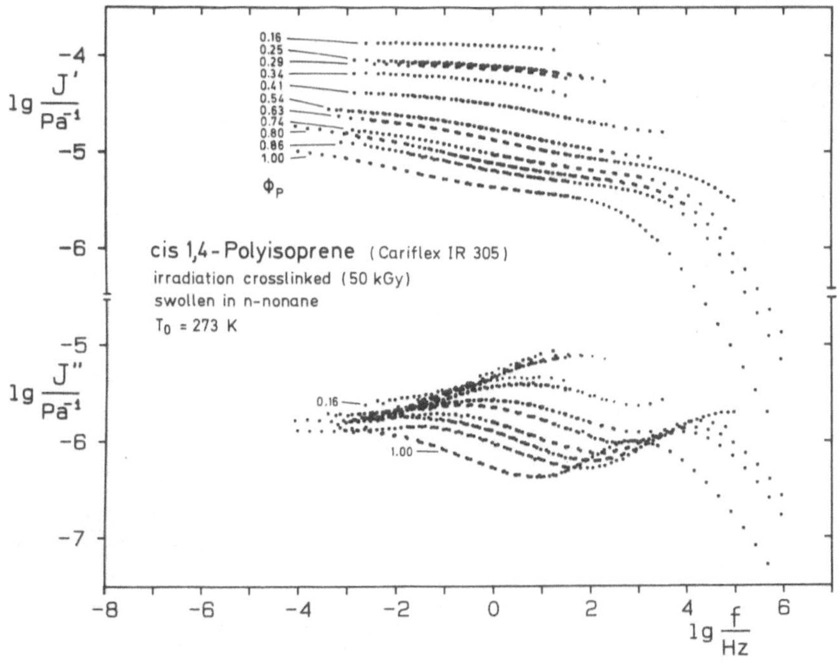

Fig. 1. Shear compliance mastercurves of irradiation crosslinked Poly(isoprene), swollen to different polymer volume fractions ϕ

an increasing degree of swelling. This constant shape – being a kind of fingerprint of the network formed by crosslinking – can be assured from this investigation only up to $\phi \approx 0.6$, corresponding to a temperature range of 293 to 218 K[3]), but will be assumed to be true also for lower ϕ. This implies that the two relaxation processes, the *glass process* (at higher frequencies) and the *rearranged shearband process* (at lower frequencies) [1] remain at constant ratios of relaxation frequencies, strengths and widths, independent of the degree of swelling.

Figure 2 gives the activation diagram for the glass process, deduced from the measurements in Figure 1: the positions of the symbols were calculated from the frequency shift factors used to obtain the mastercurves and the position of the glass process at the reference temperature $T_0 = 273$ K. At lower ϕ – where this process would be outside the frequency and temperature[3]) range available – the above assumption of constant shape of the mastercurves was applied to get its position in Figure 2. The activation curves fitted to the experimental points were drawn after the meander theory [13], to be discussed below. The calorimetric glass temperature of the dry sample was $T_g = 209$ K.

Similar measurements of the dynamic shear compliance at different swelling ratios were performed on samples irradiated with 100, 200, 400, 600 and 800 kGy. All the results so far obtained will now be evaluated and compared with selected data from the literature.

Evaluation of the compliance shift factors

Whereas the frequency shift of the mastercurves (in Fig. 1) with decreasing ϕ can be discussed in the activation diagram, it is appropriate to plot the vertical shift factors, i. e. lg $[J(\phi)/J(1)]$ versus lg ϕ in a separate diagram. In Figure 3 such a plot is shown for all samples investigated in this work.

Even though the scattering of the data[4]) is somewhat large, there is no doubt that for slight crosslinking, the slope of the shift factor versus ϕ in the double logarithmic plot is not far from -2 and tends to -0.6 at medium crosslinking. For higher swelling ratios (at low crosslinking) the slope decreases into the range -0.6 to -0.3, but more, as well as more accurate,

[3]) This temperature limit is due to the melting point of n-nonane. More comprehensive measurements will be done on a series of DCP-crosslinked IR-samples, using Methylcyclopentane (T_m = 131 K) as swelling agent.

[4]) This scattering must be due to uncertainties of the shift factors (geometry, coupling) and/or of the swelling ratios, and should be reduced in subsequent measurements.

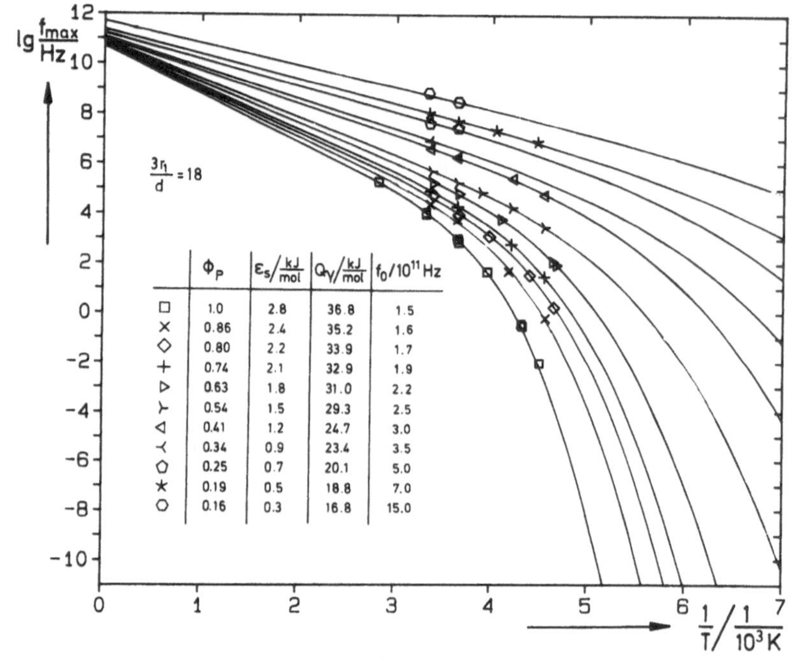

Fig. 2. Double logarithmic plot of the glass-relaxation frequency f_{max} versus $10^3/T$ from the frequency shift factors of Figure 1. The curves are drawn according to the dislocation model theory, Equation (13). The fit parameters are given in the inset

data are needed to achieve a more rigorous conclusion. The shift factors for those samples swollen to equilibrium stay on the lower limit line of slope $-2/3$ for medium doses and approach a final slope in the range -1.5 to -2 with decreasing crosslinking.

Because of the (probable) constant shape of the mastercurves and because only shift factors (relative to the dry state) are to be discussed, it is worth-while to select corresponding data from the literature, irrespective of applied frequency or deformation mode and plot them in a similar diagram to Figure 3. This has been done in Figure 4, with symbols indicating the systems investigated together with the applied methods.

All the data fit into the scheme already described in the context of Figure 3: The diluted melt (PMMA) shows a nearly exact ϕ^{-2}-law, whereas the other limit (for strongly or medium crosslinked samples) appears to be a $\phi^{-2/3}$ dependence.

It should be mentioned that recent data by Ilavský and Dusek [12] on POPT (M_n = 710 g/mol)/MDI model networks prepared in different amounts of xylene show a $\phi^{-1/3}$ (or even less) dependence of the compliance ratio to the respective dry states. First measurements by our group [3] on similar samples prepared in the bulk and swollen afterwards in Diethyleneglycoledimethylether tend in the same direction. (We think that inner syneresis into solvent blocks takes place at the expense of the isotropic swelling of the meander cubes).

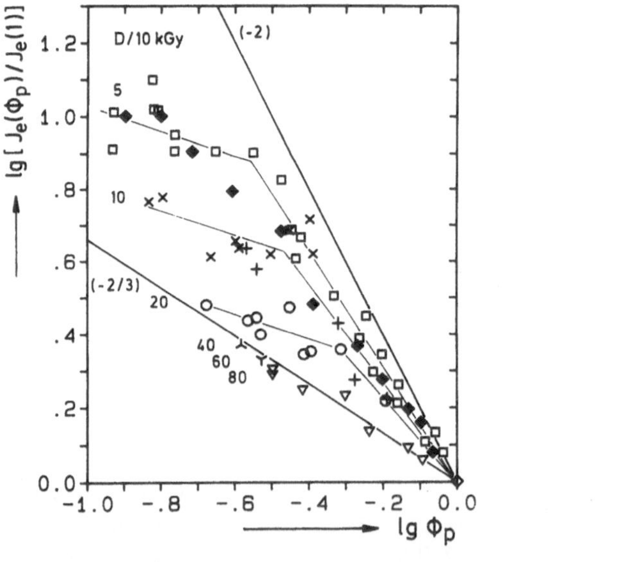

Fig. 3. Double logarithmic plot of the compliance shift factors obtained from all measurements, including those in Figure 1. The symbols indicate the irradiation dose and the swelling agent. They refer to following IR-samples: ◆ 50 kGy in n-nonane [3]; □ 50 kGy in n-dodecane [2]; + 100 kGy in n-nonane [3]; × 100 kGy in n-dodecane [2]; ○ 200 kGy in n-dodecane [2]; ⅄ 400 kGy in n-dodecane [2]; Y 600 kGy in n-dodecane [2]; ▽ 800 kGy in n-dodecane [2]. The limiting slopes are obviously -2 for the uncrosslinked melt and about $-2/3$ for medium or strongly crosslinked networks

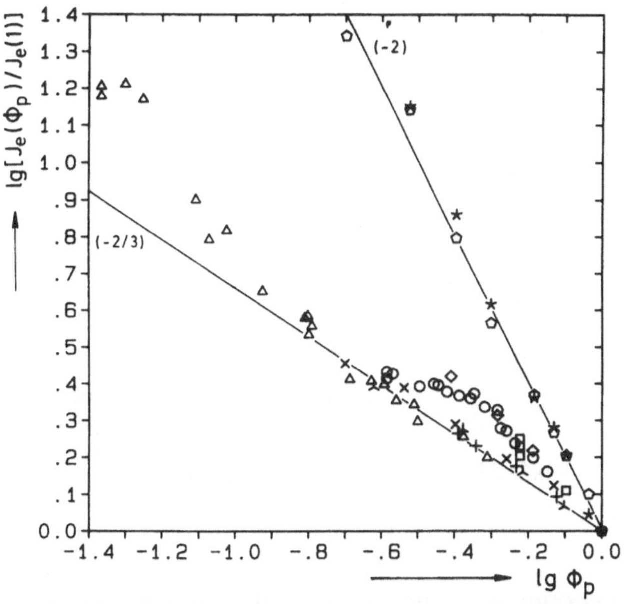

Fig. 4. Selected data from literature plotted similarly as in Figure 3. The symbols indicating networks and references are as follows: + NR, DCP-vulcanizate/n-decane, different ϕ, stress-strain [5]; × NR, DCP-vulcanizate/benzene, different ϕ, stress-strain [6]; ◇ NR, DCP-vulcanizate/n-decane, different ϕ, stress-strain [7]; □ PDMS, irradiation crosslinked/different solvents, different ϕ, stress strain [8]; ○ trans-1.5-poly (pentenamer), DCP-vulcanizate/CCl_4, different ϕ, stress-strain [9]; △ Epoxy-amine networks/DMF, equilibrium swelling, Rheovibron 110 Hz [10]; ⬠,★ PMMA, uncrosslinked/Diethylphthalate, different ϕ, concentric cylinder type rheometer and capillary viscometer [11]

Discussion

In this final section we are not yet in the position to present a complete model theory for the dynamic shear compliance of swollen networks in dependence on crosslink density. Instead we will try to understand the limiting laws for the compliance shift factor $J(\phi)/J(1)$ at zero and medium crosslinking as well as the changes of the activation diagram with the polymer volume fraction ϕ.

In the frame of a simplified coil model the shear compliance is usually written [13–15]

$$J = (R_o^2/R^2)/\nu kT \qquad (1)$$

where R is the actual and R_o the ideal root mean square end-to-end distance (or radius of gyration) of network strands between crosslinks and/or entanglements, and ν their number per unit volume. Using this formula, the dependence of the shear compliance J on the swelling ratio ϕ^{-1} can be estimated for the different regimes:

a) The plateau compliance of uncrosslinked polymer melts should read $J_{eN}^o \approx \phi^{-2}/\nu_e kT$, if it is intuitively assumed that the number of entanglements per unit volume $\nu_e(1)$ in the bulk state decreases proportional to the probability of polymer/polymer contacts, i.e. $\sim \phi^2$, and R, the average length of the entanglement network strands, does not deviate from that of an ideal coil. A similar explanation, fitting the experimental results, has been given in Reference [16].

b) For slightly crosslinked melts, this might also be true for J_{eN}, but the equilibrium compliance J_e (at low frequencies, c.f. Fig. 1) should be mainly determined by the number of (chemical) crosslinks per unit volume ν_{ch} because of the temporary nature of the entanglements or its slipping. Therefore J_e should read $J_e \approx \phi^{-1/3}/\nu_{ch}(1)kT$ after Equation (1), if one assumes $\nu_{ch} = \nu_{ch}(1) \cdot \phi$ and $R = R_o\phi^{-1/3}$ for isotropic swelling. This prediction does not agree with experiment (Fig. 1), at least for low and medium swelling (Fig. 3).

c) For medium or strongly crosslinked networks – where equilibrium and network compliance are about equal – the latter reasoning should be valid and $J_e \approx J_{eN} \approx \phi^{-1/3}/\nu(1)kT$. This limiting law does not fit the experimental result $J(\phi)/J(1) = \phi^{-2/3}$ (c.f. Figs. 3, 4).[5]

Since the simplified coil model cannot account for the majority of experimental facts, we now turn to the *meander model* to derive its predictions, at least for the limiting laws:

a) The plateau compliance of bulk polymer melts has been attributed to 1d-step shear fluctuations between adjacent layers of molecules in the meander cube [1, 20], yielding

$$J_{eN}^o(1) = d(r+x)^2/9kT = dr^2/kT \qquad (2)$$

after Reuss-averaging. The reason for restricting its relative displacement to one chain-distance d is that larger displacements would increase the interfacial energy of the (crossed) interfaces between rotated neighbouring cubes. In the diluted melt multiple, i.e. id-steps should be admitted, provided the (small) solvent molecules stay at the i-edges and are capable of filling them out by stacking under appropriate angles.

[5]) The $\phi^{-1/3}$-dependence has been derived from the coil model by several authors [17, 18]. The derivation by Kilian [19] – yielding a $\phi^{-2/3}$-law – is possibly not self-consistent.

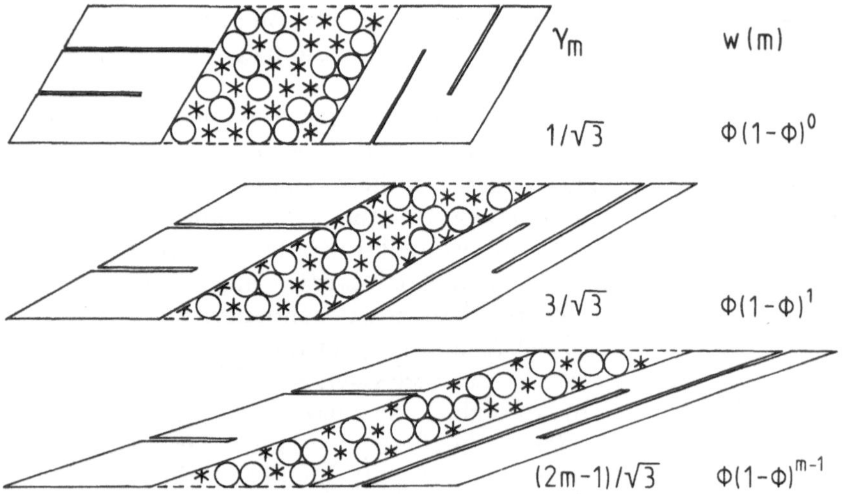

Fig. 5. Multiple-step intrameander shear deformation as it is assumed to occur in diluted melts. (○) polymer stems; (∗) files of solvent molecules

The polymer segments, although flexible, are much more restricted to the internal angles of the main chain and cannot fill any free space. As visualized in Figure 5, we assume a more or less cylindrical packing of polymer stems and files of solvent molecules within the cross-section of the meander cube[6]), and admit shear deformations up to md-displacements to occur between adjacent layers with probabilities

$$W(m) = \phi(1-\phi)^{m-1} \; (m = 1,2,3\ldots 3r/d). \quad (3)$$

The local shear deformations of a m-layer read

$$\gamma_i = \pm\,(2i-1)/\sqrt{3} \; (i = 1,2,3\ldots,m). \quad (4)$$

To calculate the plateau compliance of the diluted melt, we write a first approximation to the free energy of shear deformation per m-layer of diluted stems

$$\frac{\Delta g_{\mathrm{def}}}{kT} = \sum_{i=1}^{m} (z_i \ln z_i + \bar{z}_i \ln \bar{z}_i) + \frac{d(r+x)^2}{kT}\sigma_{21} \times \left[\gamma_{21}^{(m)} - \frac{1}{\sqrt{3}}\sum_{i=1}^{m}(2i-1)(z_i-\bar{z}_i)\right] + \mu\left[1-\sum_{i=1}^{m}(z_i+\bar{z}_i)\right] \quad (5)$$

denoting the concentrations of $\pm\, id$-steps within the m-layer by $z_i, \bar{z}_i$, and the paraelastic shear of an average m-layer by $\gamma_{21}^{(m)}$. The last two terms comprise both conditions to be satisfied, with Lagrangian multipliers as front factors. Since the definition of σ is the derivative of the free energy per unit volume for γ, the volume of a representative layer $d(r+x)^2 = 9dr^2$ appears in the first multiplier.

The minimization of Equation (5) for $z_i, \bar{z}_i$ yields

$$z_i, \bar{z}_i = \frac{\exp\left[\pm\,(2i-)\sigma^*\right]}{2\sum_{i=1}^{m}\cosh(2i-1)\sigma^*} \quad (6)$$

with the abbreviation $\sigma^* = d(r+x)^2\sigma_{21}/\sqrt{3}\,kT$.

From the first condition one concludes using Equation (6)

$$\gamma_{21}^{(m)} = \frac{1}{\sqrt{3}} \frac{\sum_{1}^{m}(2i-1)\sinh(2i-1)\sigma^*}{\sum_{1}^{m}\cosh(2i-1)\sigma^*} \quad (7)$$

$$\approx \frac{\sigma^*}{3\sqrt{3}}(4m^2-1) \text{ for } \sigma \to 0\,.$$

The paraelastic shear $\gamma_{21}^{(m)}$ of the m-layers contributes to the total shear deformation with the probability $W(m)$, yielding

$$\gamma_{21} \approx \sum_{m=1}^{3r/d} \frac{d(r+x)^2}{9\,kT}(4m^2-1)\phi(1-\phi)^{m-1}\sigma_{21}\,. \quad (7a)$$

[6]) Its size $(r+x) = 3r$ will be nearly equal to that in the bulk melt, at least at θ-condition [21].

We further assume that multiple-step displacement can only occur in one of two perpendicular directions, the other being restricted to $1d$-steps. Therefore the anisotropic shear compliance for small amplitudes becomes

$$2(S_{2121} + S_{1212}) = \partial\gamma_{21}/\partial\sigma_{21} + \partial\gamma_{12}/\partial\sigma_{12}$$

$$\approx \sum_{m=1}^{3r/d} \frac{d(r+x)^2}{9\,kT}\,(4m^2+2)\phi(1-\phi)^{m-1}\,. \tag{8}$$

Without proof we transfer the result for the shear compliance in chain direction – as derived for the $1d$-step in [22] –

$$2(S_{3131} + S_{3232}) = \frac{2}{3}\,2(S_{1212} + S_{2121}) \tag{8a}$$

and obtained by Reuss-averaging the result of this simple model theory for the plateau compliance of diluted melts

$$J_{eN}^{o}(\phi) = \frac{1}{5}\,(S_{2121} + S_{1212} + S_{3131} + S_{3232})$$

$$\approx \sum_{m=1}^{3r/d} \frac{d(r+x)^2}{9\,kT}\,\frac{2m^2+1}{3}\,\phi\,(1-\phi)^{m-1}\,. \tag{9}$$

Equation (9) reduces to Equation (2) for the bulk state and describes the limiting law of the shear compliance shift factor for diluted (uncrosslinked) melts

$$\frac{J_{eN}^{o}(\phi)}{J_{eN}^{o}(1)} \approx \phi\left[1 + 3(1-\phi) + \frac{19}{3}\,(1-\phi)^2 + \ldots + \frac{2m^2+1}{3}\,(1-\phi)^{m-1} + \ldots\right] \tag{9a}$$

which is approximately equal to

$$\phi^{-2} \approx \phi\left[1 + 3(1-\phi) + 6\,(1-\phi)^2 + \ldots + \frac{m(m^2+1)}{2}\,(1-\phi)^{m-1} + \ldots\right] \tag{10}$$

and fits the experiment (in Fig. 4) within the measured range.

b) For slightly crosslinked melts, two facts have to be explained: (i) the slope of $\lg\,[J(\phi)/J(1)]$ versus $\lg\phi$ (Fig. 3) starts to deviate from -2. Qualitatively this is due to the onset of a restoring force of bundles which are stretched during shear deformation (c. f. Fig. 5). A quantitative treatment must introduce an equivalent free energy term into Equation (5) and will be postponed until more accurate experimental data are available. (ii) The low frequency relaxation strength behaves similarly to the plateau compliance – i. e. $\sim\phi^{-2}$ [7]) and not $\sim\phi^{-1/3}$, as predicted by the coil model – as is pointed out by the constant shape of the mastercurves in Figure 1. In the meander model this relaxation is attributed to the rearrangement of shear bands, and its strength ΔJ_B has been derived in Reference [1], Equation (26):

$$\Delta J_B = \frac{\beta(\phi)}{\beta(1)}\,\frac{\Delta J_B^{\infty}}{\phi\xi}\,. \tag{11}$$

ΔJ_B^{∞}, the relaxation strength without any loss of segment orientation, and ξ, the fraction of polymer segments which are hindered to reorient by crosslinks, will not depend on ϕ in a first approximation. Because $\beta^{-1}(\phi) = 1 + \exp\left[-\phi\cdot 3\frac{d}{s}\Delta g_{rot}/kT\right]$ from [1], Equation (25), ΔJ_B should be roughly proportional to ϕ^{-2}, if this function is approximated by a power law, and if

$$3(d/s)\,\Delta g_{rot}/kT \approx 2.$$

c) For medium or strongly crosslinked networks – where equilibrium and plateau compliance are about equal – stretching of bundles (during shear) becomes much more difficult and finally restricts the intrameander shear to $1d$-displacements only (in which no stretching of bundles is involved). The respective plateau-compliance should be that of the crosslinked melt (c. f. [1], Equation (14a)), the only difference being the factor $\phi^{-2/3}$ which is due to the larger area of the cube layers in the swollen state, i. e.

$$J_{eN} = P(p_c^*)\,d(r+x)_1^2\,\phi^{-2/3}/9\,kT \tag{12}$$

yielding the limiting law of the shear-compliance shift factor for medium or strongly crosslinked networks

$$J(\phi)/J(1) = \phi^{-2/3} \tag{12a}$$

in agreement with experiment (Figs. 3, 4).

---

[7]) At very low crosslinking.

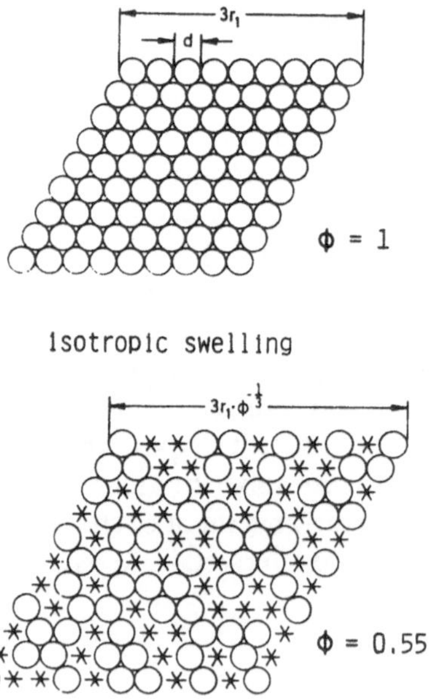

Fig. 6. Isotropic swelling of a meander cube, in cross-section

Finally, it may be of interest to discuss the geometry of the isotropically swollen meander cube at medium or strong crosslinking (when each chain is retained in its initial bundle), and to have a drawing for a special case ($\phi = 0.55$ in Fig. 6):

If $3r_1/d$ is the number of parallel stems in the dry cube layer (or segments per cube segment line) one counts in the isotropically swollen state:

Sites/segment line	$\phi^{-1/3} 3r_1/d$
Sites/cube cross section	$\sigma^{-2/3}(3r_1/d)^2$
Segments/segment line	$\phi^{2/3} 3r_1/d$
Segments/cube cross section	$\phi^{1/3}(3r_1/d)^2$.

The first two relations follow from geometry of isotropic swelling, the latter from its consequence, a stretching (partial defolding) of the tightfolded molecules in the chain direction by $\phi^{-1/3}$. The statistical element for the intrameander shear, the cube layer of diluted stems, becomes $\phi^{-2/3}(3r_1)^2 \cdot d$ which has been used to derive Equation (12). For Equation (12a), d was assumed equal to that of the melt.

Turning to the activation diagram (Fig. 2), the curves were drawn according to the dislocation model theory of the glass relaxation which yields

$$f_{\max} = \frac{f_o}{\pi} \exp\left(-\frac{Q_y}{kT}\right) \times \left[1 - \left(1 - e^{-\frac{\varepsilon_s}{kT}}\right)^{\frac{3r_1}{d}\phi^{2/3}}\right]^{3\frac{d}{s}\left(\frac{3r_1}{d}\right)^2 \phi^{-2/3}} . \quad (13)$$

For the dry state ($\phi = 1$) this formula was derived in References [1, 23, 24]. It must be modified for the swollen state in three respects:

i) The exponent of the inner bracket denotes the number of polymer segments/ segment line and has to be multiplied by $\phi^{2/3}$.

ii) The exponent of the outer bracket counts the average number of segment lines per cube (each of which must contain at least one segment dislocation). This number increases with isotropic swelling by the factor $\phi^{-2/3}$.

iii) ε_s, Q_y and f_o have to be fitted anew for every swelling ratio ϕ^{-1}, because ε_s depends solely, Q_y and f_o partly, on intermolecular interaction which is changed by the solvent.

For a rigorous check of the modified Equation (13), relaxation frequencies in a much wider temperature range must be available[3]). First measurements on POPT/MDI-networks support strongly Equation (13).

Acknowledgement

Support by the Deutsche Forschungsgemeinschaft and the Fonds der Chemie are gratefully acknowledged.

References

1. Pechhold W, Böhm M, v Soden W (1987) Coll Polym Sci, this volume
2. Böhm M (1987) Thesis, Ulm
3. Grassl O (1986) Diplomarbeit, Ulm
4. v Soden W (1987) Thesis, Ulm
5. Mullins L (1959) J Appl Polym Sci 2:257
6. Gumbrell SM, Mullins L, Rivlin RS (1953) Trans Farad Soc 49:1495
7. Allen G, Kirkham MJ, Padget J, Price C (1971) Trans Farad Soc 67:1278
8. Yu CU, Mark JE (1974) Macromolecules 7:229
9. Gebhardt G (1978) Thesis, Clausthal
10. Ilavský M, Bogdanova LM, Dusek K (1984) J Polym Sci Phys Ed 22:265
11. Masuda T, Toda N, Aoto Y, Onogi S (1972) Polymer J 3:315
12. Ilavský M, Dusek K (1986) Macromolecules 19:2139
13. James HM, Guth E (1949) J Polym Sci 4:153
14. Tobolsky AV (1962) Properties and Structure of Polymers, Wiley, New York
15. Hoffmann M, Krömer H, Kuhn R (1977) Polymeranalytik I, Thieme, Stuttgart

16. Graessley WW (1974) Adv Polym Sci 16:1
17. Flory PJ (1967) Principles of Polymer Chemistry 6, Cornell, Ithaca
18. Treloar LRG (1958) The Physics of Rubber Elasticity 2, Clarendon Press, Oxford
19. Kilian HG, Enderle HF, Unseld K (1986) Coll & Polym Sci 264:866
20. Pechhold W (1984) Makromol Chem, Suppl 6:163
21. Pechhold W, Grüner W, v Soden W (1988) Coll & Polym Sci, in preparation
22. Pechhold W (1980) Coll & Polym Sci 258:269
23. Pechhold W, Sautter E, v Soden W, Stoll B, Grossmann HP (1979) Makromol Chem Suppl 3:247
24. Heinrich W, Stoll B (1988) Coll & Polym Sci, in press

Received September 5, 1987;
accepted September 7, 1987

Authors' address:

W. Pechhold
Universität Ulm
Abt. Angewandte Physik
Oberer Eselsberg
D-7900 Ulm, F.R.G.

Progress in Colloid & Polymer Science Progr Colloid & Polymer Sci 75:70–82 (1987)

Thermomechanics of polymer networks

Yu. K. Godovsky

Karpov Institute of Physical Chemistry, Moscow, U.S.S.R.

Abstract: Calorimetric determinations of the total energy exchange in polymer networks provide information about both thermodynamic and molecular quantities characterizing the deformation process and, therefore, have a fundamental importance in investigating rubber elasticity. The thermomechanical behaviour of the chemically crosslinked polymer networks, filled networks, rubberlike thermoelastoplastics and crystalline networks are discussed. Thermomechanics of the crosslinked networks is considered from the point of view of the interchain entropy and energy contributions to the free energy of deformation and the temperature coefficient of the unperturbed chain dimensions. The comparison with the results obtained on isolated macromolecules demonstrates that the classical Gaussian theory of rubber elasticity quantitatively predicts the intrachain entropy and energy contributions at simple deformations of the networks and their independence of the deformation (at small and moderate deformations). The interchain changes of internal energy, vibrational entropy and volume resulting form the deformation are also considered and it has been concluded that they support the theory only at small deformations. Analysis of the entropy and energy effects resulting from the simple extension of the stress-softened networks filled with different fillers shows that, in many cases, the entropy and energy contributions are dependent on the concentration of the fillers, which contradicts the classical theory of rubber elasticity. Some reasons for the dependence are considered. Thermomechanical studies of SBS ands SIS block copolymers with a hard block content of below 40 % show that the energy contributions accompanying uniaxial extension are independent of the hard block content and degree of deformation. The energy contributions for diene blocks coincide well with the results for chemically crosslinked diene networks. On the other hand, the thermomechanical behaviour of the segmented polymers with the small molecular weight of the soft blocks and the large content of the hard block is determined not only by intrachain conformational changes but by intermolecular changes, both in the soft and hard blocks. Some possible deformation mechanisms which lead to such thermomechanical behaviour are considered. Although it is widely accepted that the free energy of the uniaxial deformation of the two-phase crystalline networks is purley intrachain, our calorimetric investigations show that the thermodynamics of the deformation of these networks is controlled by interchain changes in the amorphous regions. To support this conclusion some thermomechanical results for oriented and unoriented crystalline networks are considered.

Key words: Thermomechanics, networks, deformation calorimetry, energy contribution, block copolymers.

Introduction

The thermomechanical study of rubber elasticity has a very long history and has thrown much light on the physical nature of rubber elasticity. It dates back to the beginning of the last century when Gough published his qualitative observations (made with his lips) on the behaviour of unvulcanized natural rubber [1]. Over half a century later Joule measured the temperature changes accompanying the uniaxail adiabatic stretching of vulcanized rubber and demonstrated that the temperature decreased for small extensions but increased for extensions above ca. 14 %, which is referred to as the adiabatic inversion point for natural rubber. As a result of these measurements, Joule supposed that rubberlike elasticity has to be entropic in origin. In 1932, Meyer, Susich and Valko suggested that the decrease in entropy with elongation is a consequence of orientation of the molecular chains of the rubber. The simplest statistical theory, developed at the same time, locates these entropy changes in the number of configurations accessible to flexible chain molecules. Both intrachain and intermolecular contributions from changes in internal energy are considered to play no role. Force-temperature measurements at constant pressure and length (such experiments are often referred to as thermoelastic studies) carried out from 1930–1950 confirmed Joule's thermodynamic analysis and the statistical theory. This type of measurement was the main source of quantitative experimental information on thermoelasticity of polymer networks for some decades [1–3].

Another experimental approach to the thermoelasticity of polymer networks was then developed. It consisted of direct calorimetric measurements of heat effects resulting from deformation, and simultaneous measurements of the mechanical characteristics. Müller [4] was the first to use stretching calorimetry for thermodynamic study of polymer deformation. He demonstrated that this approach is very usefull for characterizing thermomechanical properties of polymer networks. In his early studies, Müller used a gas calorimeter which was, unfortunately, not sensitive enough to study very small heat effects. This became possible after the calorimeter was introduced at the end of the 1960's by our group in Moscow [5, 6] which was based on the Calvet principle and possesses a very high sensitivity. Today this thermomechanical approach is used in studying not only rubberlike elasticity but also thermomechanical behavoiour of crystalline and glassy polymers, as well as polymer blends, block copolymers and composites. The modern state of this area of polymer physics has been considered recently [6].

The presentation deals with the main ideas of the thermomechanics of polymer networks, with emphasis on the intrachain and interchain changes resulting from the deformation.

Thermomechanics of chemically crosslinked polymer networks

Calorimetric experiments involve direct measurements of the heat and work resulting from the uniaxial deformation under constant pressure and temperatures

$$W_{P,T} + Q_{P,T} = (\Delta H)_{P,T}\,. \quad (1)$$

The molecular processes responsible for rubber elasticity are usually interpreted in terms of constant volume changes of entropy and internal energy. Hence, enthalpy and entropy terms in Equation (1) must be expressed as the sums of contributions arising from a constant volume process and a strain-induced volume dilation

$$(\Delta H)_{P,T} = (\Delta U)_{V,T} + (\partial U/\partial V)_{P,T}\, d\,V + P\, d\,V \quad (2)$$

$$T\,(\Delta S)_{P,T} = T\,(\Delta S)_{V,T} + T\,(\partial S/\partial V)_{P,T}\; d\,V\,. \quad (3)$$

Stretching of a rubber is accompanied by, as a rule, a very small increase in volume, which means that the term $P\,d\,V$ is negligibly small in comparison with other terms. However, the energy and entropy terms are large, both of the order of 400 J/cm^3. Thus, while the volume changes have a very small effect on mechanical work, they make an important contribution to each of its constant pressure components $(\Delta H)_{P,T}$ and $(\Delta S)_{P,T}$ [7]. Unfortunately, it is very difficult to use the rigorous thermodynamic approach to obtain products $(\partial U/\partial V)_{P,T}\; d\,V$ and $(\partial S/\partial V)_{P,T}\; d\,V$ and, therefore, the majority of thermoelastic and thermomechanical studies include some approximations in the treatment of thermodynamic results. Many years ago, Flory et al. [8] proposed using the Gaussian network theory to obtain an expression for the dilation term and, since then, this approach is widely used to treat thermomechanical data. In our studies we followed this approximation.

At constant volume and temperature we arrive at

$$W_{V,T} = (\Delta U)_{V,T} - T\,(\Delta S)_{V,T}\,. \quad (4)$$

The following relations are called the entropy and energy contributions to the free energy of deformation (or to the elastic force f), respectively

$$\text{entropy contribution} - T\,(\Delta S/W)_{V,T} = f_S/f \quad (5)$$

$$\text{energy contribution} \quad - (\Delta U/W)_{V,T} = f_U/f\,. \quad (6)$$

The thermodynamic quantity of primary interest is the energy contribution exhibited by a deformed polymer network. Early molecular theories of rubber elasticity supposed that the elasticity of a polymer network is exclusively entropic. But it is now well known that deformation of polymer networks is accompanied by intramolecular energy changes which are closely related to the conformational energies of macromolecules. The classical theory of rubber elasticity for a Gaussian polymer network takes into account the change of conformational ennergy by incorporating the front factor into the equation of state for simple elongation of compression [3, 6, 8]

$$f = \nu k T L_i^{-1} (\langle r^2 \rangle_i / \langle r^2 \rangle_0) \cdot (\lambda - \lambda^{-2}) . \qquad (7)$$

In this equation ν is the number of elastically active chains in the network, $\langle r^2 \rangle_i$ is the mean square end-to-end distance of a network in the undistorted state, $\langle r^2 \rangle_0$ that of the corresponding free chains, L_i the length of the undistorted sample. The other values have their usual meaning. The temperature dependence of $\langle r^2 \rangle_i$ is related to the thermal expansion of the sample, but $\langle r^2 \rangle_0$ is a characteristic of the chemical structure of chains.

From Equation (7) one may derive Equation (8), which relates the intramolecular energy changes to the mean square end-to-end distance of unperturbed chains

$$(\Delta U/W)_{V,T} = T d \ln \langle r^2 \rangle_0 / dT . \qquad (8)$$

The temperature dependence of the unperturbed chain dimensions may be both positive and negative and, consequently, the simple deformation of polymer networks may be accompanied by both an increase and decrease in the internal energy. Thus, Equation (8) is of considerable importance for two reasons. Firstly, it permits comparison of the results of thermomechanical measurements on polymer chains in bulk, that is in the network structure, with results of viscosity measurements on chains of the same polymer, essentially isolated, in dilute solution. Secondly, it also establishes the relationship between the purely thermodynamic quantity with its molecular conterpart, which can be interpreted in terms of the rotational state theory of chain configuration. Finally, it is very important to note that in Gaussian statistical theory the energy contribution is independent of strain, which is a direct consequence of its intrachain nature.

Equation (7) also leads to Equation (9) which allows us to determine from calorimetric experiments the energy contribution, at constant pressure, by invoking the statistical theory, as modified by Flory et al. [8]

$$(\Delta U/W)_{V,T} = 1 + (Q/W)_{P,T} - 2\alpha T/(\lambda^2 + \lambda - 2) \qquad (9)$$

where α is the volume thermal expansivity of the undeformed sample. The last term on the right side of Equation (9) corresponds to the relative intermolecular change of internal energy which arises as a result of a volume change during the course of deformation. It is strongly dependent on deformation. The difference between energy changes under P and T constant and V and T constant is very important at small deformations. As this term characterizes the relative internal energy change resulting from the volume change, it must be identical to the corresponding expression for solids. For quasihookian solids, the relative energy change is expressed by the equation [6]

$$(\Delta U/W)_{\Delta V} = 2\beta T/\varepsilon + 1 \qquad (10)$$

where β is the linear thermal expansivity, and ε is the strain. For networks using Equation (9) we arrive at the equation

$$(\Delta U/W)_{\Delta V} = 2\alpha T/(\lambda^2 + \lambda^{-2}) \approx 2\beta T/\varepsilon . \qquad (11)$$

As is seen from Equations (10) and (11) these terms are identical which means that Flory's approximation for intermolecular changes is a very good approximation at small network deformations.

Calorimetric results for typical polymer networks at small and moderate deformations (Fig. 1) [6, 9] can be treated using the above expressions (Fig. 2). The results indicate that the free energy contains a significant energy contribution, independent of strain. In some studies published in literature, a dependence of energy contribution on deformation is found at small deformation [10]. Although various attempts have been made to explain this dependence, Shen [11] emphasized some years ago that, due to the presence of the intermolecular term, a small experimental error, which is experimentally unavoidable, will greatly affect the calculation of the energy contribution. Allen and Price [7, 12] studied thermoelasticity of natural rubber at constant volume and found that the value of energy contribution is in good agreement with values obtained at constant pressure and does not show any

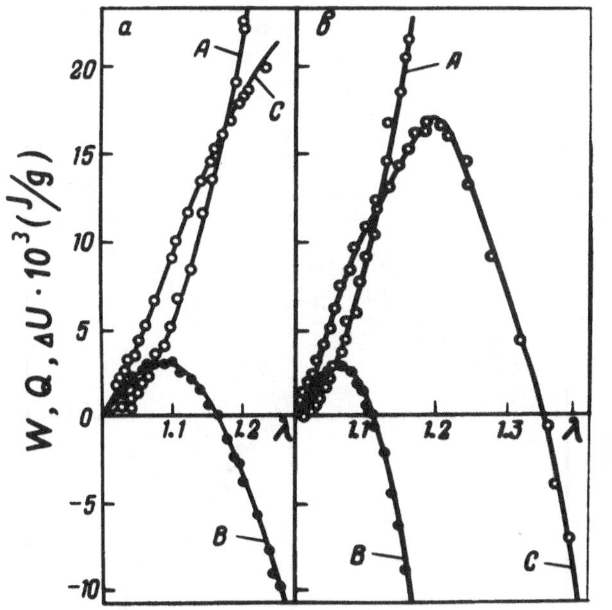

Fig. 1. Mechanical work (A), heat (B) and internal energy (C) on stretching samples from the unstrained state to λ at room temperature [6]. a) Natural rubber; b) ethylene-propylene rubber

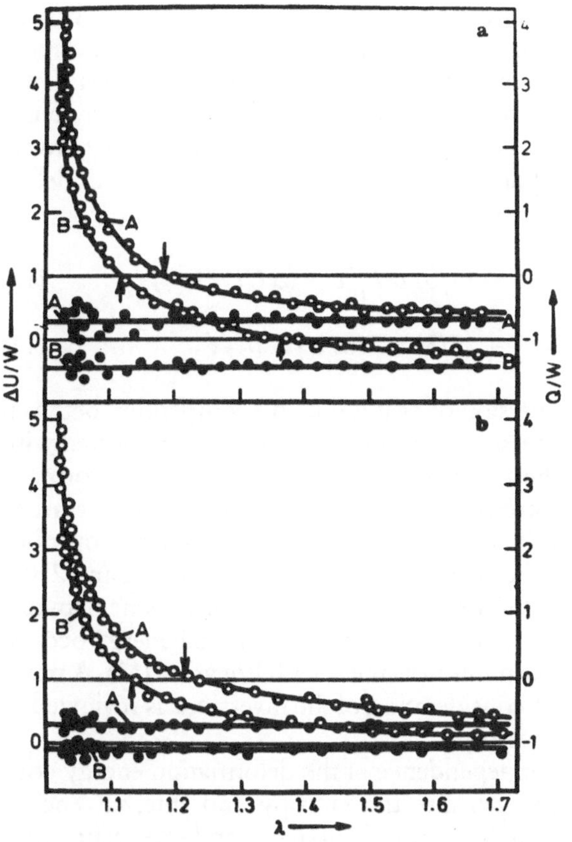

Fig. 2. Calorimetrically determined relative entropy and energy contributions as a function of extension ratio λ [6]. (○) = $(Q/W)_{P,T}$ and $(\Delta U/W)_{P,T}$; (●) = $(Q/W)_{V,T}$ and $(\Delta U/W)_{V,T}$. a) Natural rubber (A), ethylene-propylene rubber (B); b) polydimethylsiloxane rubber (A), polychloroprene rubber (B). Arrows indicate the inversion points

significant dependence on strain. Hence, one may conclude that, in good agreement with the statistical theory, thermomechanical measurements lead to an energy contribution independent of deformation.

Table 1 demonstrates a comparison of the energy contributions obtained in thermoelastic and thermomechanical measurements, as well as values for the temperature dependence of chain dimensions calculated from the energy contributions and from the viscosity-temperature measurements on isolated chains. A very good agreement is quite obvious. Hence, the current statistical theory of rubber elasticity quite satisfactorily predicts the values of intramolecular changes and their independence of deformation at elongation.

Simple elongation and compression at constant pressure must be accompanied by interchain effects resulting from volume change on deformation [2, 6, 7]. The correct experimental determination of these effects is difficult because of very small absolute values of volume changes. These measurements, however, are important for understanding the molecular mechanism of rubber elasticity and checking the validity of the statistical theory. Interchain effects in polymer net-

Table 1. Energy contributions and values of $d \ln \langle r^2 \rangle_0 / dT \times 10^3$

Polymer	$(\Delta U/W)_{V,T}$	f_U/f [3]	$d \ln \langle r^2 \rangle_0 / dT \times 10^3$ f_U/f or $(\Delta U/W)_{V,T}$	Viscosity-temperature
Natural rubber	0.21	0.18	0.66	0.6 [3, 11]
Polydimethylsiloxane	0.27	0.24	0.86	0.71 [3]
Ethylene-propylene rubber	− 0.42	− 0.45	− 1.48	− 1.4
Polychloroprene rubber	− 0.10	− 0.10	− 0.34	–
Polyethylene	–	− 0.45	− 1.53	− 1.2 [3, 11]

works are reflected, first of all in thermomechanical inversions at low strains, which arise from a competition between intra- and intermolecular changes. Calorimetric results demonstrate this fact very obviously (see Fig. 1). According to the statistical theory, inversions of heat and internal energy must occur at deformations [6]

$$\lambda_Q = 1 + (2/3)\,\alpha T/(1 - (\Delta U/W)_{V,T}) \quad (12)$$

$$\lambda_U = 1 - (2/3)\,\alpha T/(\Delta U/W)_{V,T}\,. \quad (13)$$

The inversion of heat is due to a competition between the increase of the vibrational entropy connected with the volume change and the decrease of the conformational entropy. The inversion of internal energy arises from different signs of inter- and intrachain contribution to the energy. As is seen from Figures 1 and 2, the energy contribution in a ethylene-propylene network is negative and an inversion of internal energy occurs, which is in full agreement with Equation (13). A more detailed consideration of the inversions is reviewed in Reference [6].

The independence of the deformation energy contribution permits the entropy and internal energy changes to be resolved into intra- and interchain components. Typical results of this treatment are shown in Figure 3. According to the statistical theory, the absolute values of the intermolecular changes in the internal energy and entropy associated with the volume dilation may be expressed by the equation

$$T(\Delta S)_{\Delta V} = (\Delta U)_{\Delta V} = \nu k T L_i^{-1}(\langle r^2\rangle_i/\langle r^2\rangle_0) \cdot \alpha T(\lambda - 1)/\lambda\,. \quad (14)$$

We recognize from this equation that the internal energy change resulting from the volume change of a Gaussian network is exactly balanced by the equivalent change of entropy and, thus, this volume change does not contribute to the free energy of deformation. It can be seen from Figure 3 that the prediction of Equation (14) coincides with the experimentally-obtained interchain energy changes only in the region of low strains, i. e. below approximately 30%. Such deviations are also found for force-temperature measurements [11].

Using Equation (14) it is easy to arrive at an equation for strain-induced volume dilation

$$\Delta V/V = \nu k T L_i^{-1}(\langle r^2\rangle_i/\langle r^2\rangle_0)\,\varkappa\,(\lambda - 1)/\lambda \quad (15)$$

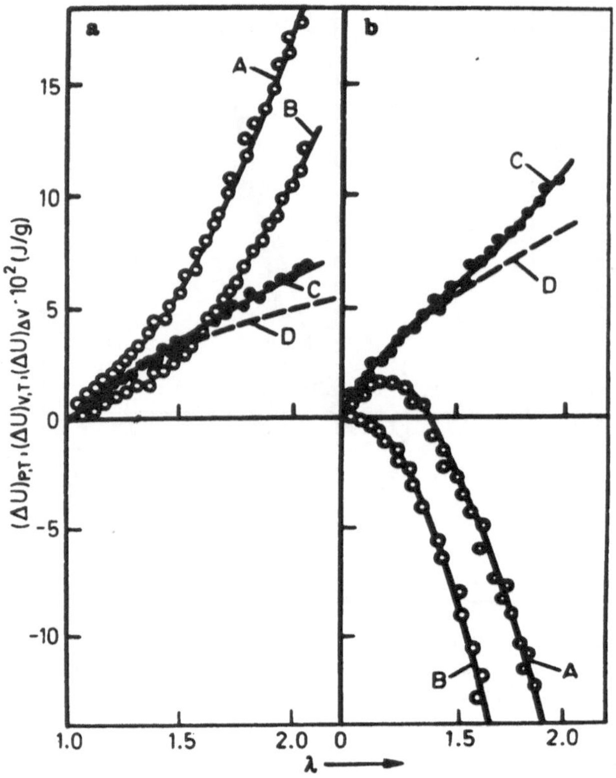

Fig. 3. Inter- and intrachain energy changes on stretching samples from the unstrained state to λ at room temperature [6]. a) Natural rubber; b) ethylene-propylene rubber. (A) = $(\Delta U)_{P,T}$; (B) = $(\Delta U)_{V,T}$; (C) = $(\Delta U)_{\Delta V}$; (D) = according to Equation (14)

where $\varkappa$ is the isothermal compressibility. Typical results obtained with deformation calorimetry are shown in Figure 4. Calorimetric and dilatometric results agree very well. The statistical theory again underestimates the volume changes on deformation.

Thus, in contrast to the intrachain changes which follow the statistical theory, not only at small but also at moderate deformations, the interchain changes of internal energy, vibrational entropy and volume only follow the statistical theory at small strains.

A number of attempts have been made in the literature to describe the interchain changes, first of all the volume dilation, resulting from the extension of a network. Treloar [13] attempted to consider the volume change, making use of various functional relations for the stored energy including the phenomenological Mooney-Rivlin equation and Ogden type equation. As a result of the analysis Treloar concluded that neither the statistical theory, nor the phenomenological equations of the Mooney-Rivlin or Ogden type can correctly predict the volume changes accompanying

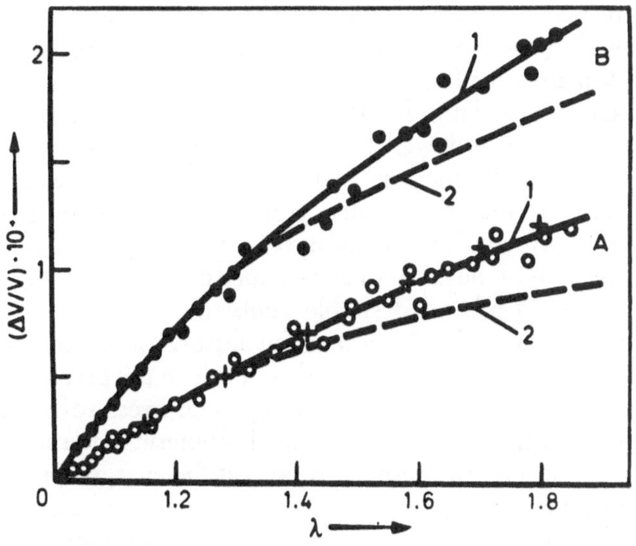

Fig. 4. The relative strain-induced volume dilation of polymer networks as a function of extension ratio λ [6]. (A) = natural rubber; (B) = ethylene-propylene rubber. (O) and (●) = calorimetric data; (+) = dilatometric data [29]. Curves 2 correspond to Equation (15)

simple extension of polymer networks. Thus, the apparent agreement between experimental behaviour of volume dilation and the prediction based on the Mooney-Rivlin form of the free energy of deformation, which has been claimed in some published studies [13], seems to be fortuitous. On the other hand, Kilian [14] recently described the strain-induced volume dilation for some polymer networks using his van der Waals equation of state. He demonstrated that these volume changes are dependent not only on the compressibility of the network but also on the van der Waals parameters and the pressure coefficient of the interaction parameters.

During the last decade, the classical theory of rubber elasticity has been carefully reconsidered and a number of theoretical models have been formulated to describe the behaviour of real networks with steric restrictions (for references, see Ref. [6]). According to the models, the free energy of deformation in the networks with constrained chains contains an additive contribution to that describing the phantom network. Therefore, the thermoelasticity of the new models is considerably more complex than that of the phantom networks. Unfortuantely, the majority of the new models contain temperature-dependent parameters which are difficult to relate directly to molecular characteristics of real polymer networks. Therefore, these models are still awaiting comparison with the experimental results.

Filled networks

In spite of their long history, reinforcement mechanisms and elastic properties of filled elastomers remain the subject of numerous experimental investigations, but there have been only a small number of investigations concerned with thermomechanical measurements.

It is well known that in the presence of reinforcing fillers the elasticity modulus of the elastomers at small and moderate deformation increases in the first approximation, according to the Guth-Smallwood equation

$$E = E_o\,(1 + 2.5\,\varphi + 14.1\,\varphi^2) \qquad (16)$$

where E_o is the modulus of elasticity of the unfilled elastomers and φ is the volume fraction of the filler. Thus, for filled elastomers the mechanical work of deformation, the change of entropy and internal energy should include parameter φ. Nevertheless, because the energy contribution has the relative form, it must not be dependent on the presence of the filler. This is a consequence of the intrachain character of the contribution in elastomers. It has been suggested that the role of the filler is increasing the active chains due to the network-filler links. Therefore, the first and most important problem in the thermoelasticity of filled elastomers is the determination of the energy contribution and compared to the values obtained for unfilled networks.

The results found in the literature are shown in Figure 5. Both silica and carbon black of different particle size and surface characteristics were used as fillers. In all studies it was found that, as in unfilled networks, the energy contribution is independent of deformation, but its magnitude is dependent on the physical characteristics of the filler, used in such a way that the more effective the filler as a reinforcing agent, the larger the value of energy effect. Indeed, even a small amount of active filler content considerably increases the energy contribution, which is in contradiction to the assumed increase only in the concentration of active network chains caused by the filler. For a less active filler the energy contribution is independent of the filler content at small and moderate amounts and only at higher content does it increase. It is very important to note that regardless of the sign of the intrachain energy changes in an unfilled network, the reinforcement

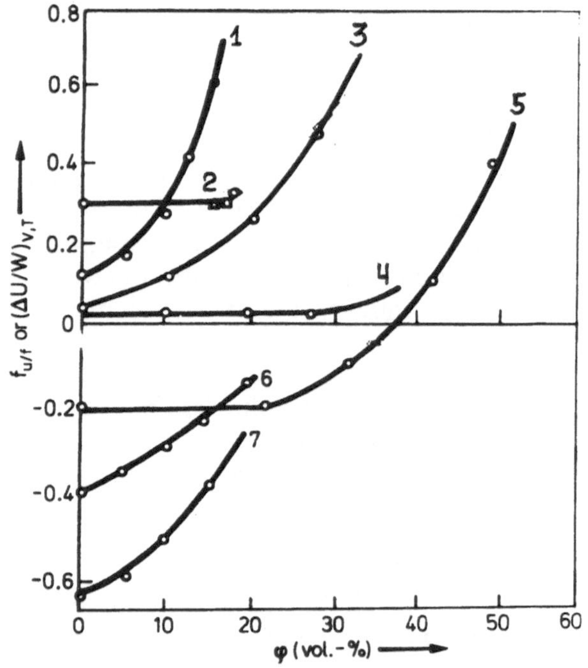

Fig. 5. Dependence of the energy contribution on the filler content. (1) = filled silicon rubber [15]; (2) = filled silicon rubber [16]; (3, 4) = butyl rubber [17]; (5) = styrene-butadiene rubber [18]; (6) = ethylene-propylene rubber [19]; (7) = plasticized PVC [20]

always contributes positively to the energy. This may be due only to intermolecular changes. Hence, these results demonstrate that changes in internal energy rather than entropy, as sometimes supposed, are important in reinforcement. The dependence of the energy contribution on the filler amount and its reinforcing ability demonstrates that in filled elastomers the energy contribution seems to lose its obvious meaning as only a measure of the intrachain effects. Thermomechanical measurements are very much concerned with the thermodynamic quantities responsible for reinforcing of filled elastomers and provide an excellent method of probing such interactions. However, to date, very few such measurements have been reported and we are, therefore, still at the very beginning of understanding the thermodynamic aspects of the reinforcement process of filled elastomers.

Rubberlike block copolymers

Rubberlike block copolymers commonly consist of the dominant soft (rubbery) component and a rigid component. The physical reason for the appearence of rubberlike elasticity in such block copolymers is connected with immiscibility and microphase separation, which leads to the formation of domain structure with domain size of the order of 100 Å. These rigid domains act as crosslinks for the rubbery matrix. The most important and well known example of elastomeric block copolymers is that of styrene-butadiene-styrene, triblock copolymers commonly containing 20–40% styrene. The glassy styrene blocks are located at the ends of the linear or star-shaped chains. This arrangement leads to desirable rubberlike properties of these materials. An important feature of rubberlike block copolymers observed, as a rule in the first loading cycle, is the stress softening. Such stress-softened block copolymers are typical elastomeric materials.

Calorimetric study showed that the energy contributions both for polybutadiene and polyisoprene block copolymers are independent of deformation up to very high extensions [6, 9]. The energy contributions obtained for polybutadiene and polyisoprene are in good agreement with the results for usual chemically crosslinked networks [6]. This seems to indicate that there are no considerable intermolecular changes in the rubbery matrix even at very large deformations. The hard block content also has no influence on the energy contribution.

Other typical representatives of rubberlike block copolymers are the so-called segmented polyurethanes and polyesters. Their block lengths are generally much shorter in comparison with polystyrene-polydiene block copolymers which tends to limit their extensibility. The segmented block copolymers, with a high content of hard block, exhibit a remarkable thermomechanical behaviour. First of all, their stress-softening is accompanied, as a rule, by a considerable residual deformation, which may reach approximately 50% of the total deformation. The residual deformation is a consequence of plastic deformation of rigid domains and orientation of the domains in the stretching direction.

Figure 6 shows the thermomechanic characteristics of segmented polyurethanes with butadiene soft block [21]. The energy changes in the sample containing less than 50% hard block differ principally from those samples with 50% content or more. In the sample with 42% hard block, the internal energy begins to decrease at some elongation and becomes negative. Such a behaviour is similar to that of typical SBS block copolymers and is controlled by the negative intramolecular energy contribution of polybutadiene blocks. In samples of more than 50%, the internal energy increases considerably with elongation.

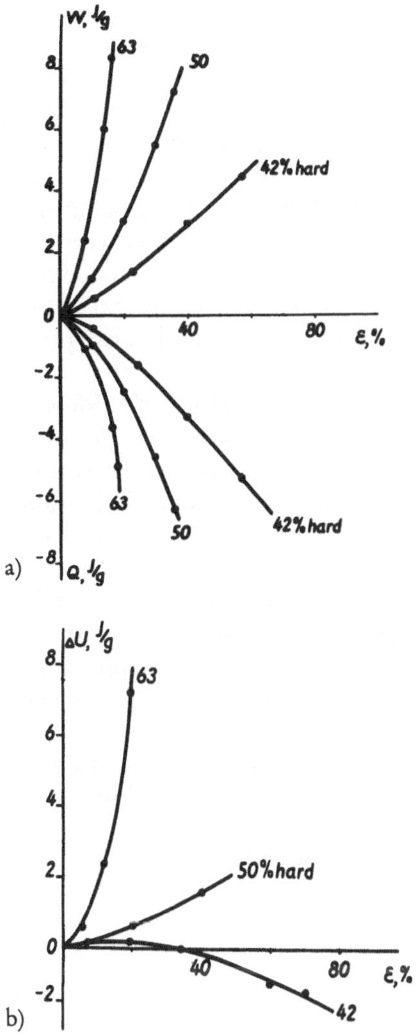

Fig. 6. a) Dependence of work and heat on strain for polyurethanes with various hard block contents [21]. b) Internal energy changes as a function of strain for polyurethanes with various hard block contents [21]

Considerations of the possible reasons for such behaviour lead to the conclusion that the considerable internal energy increase can arise only as the result of the intermolecular changes in the rubbery component. This means that intermolecular changes dominate the thermomechanical behaviour of these samples. Of course, the relatively large reversible deformations are the result of conformational changes of the soft blocks, but we can say nothing definite concerning the energy of the conformational changes from their thermomechanical properties. This conclusion is quite different from that arrived at concerning the chemically crosslinked networks and tri- and multiblock copolymers with the hard block content of less than 50 %. Such a behaviour is very characteristic of not only this particular type of segmented polyurethanes. Similar behaviour has also been observed in block and graft copolymers with polydimethylsiloxane as a rubbery component [6].

Perhaps the most striking example of a dramatic change in thermomechanical behaviour with an increasing hard block content has been observed in segmented polyesters. Typical thermomechanical behaviour of two segmented polyester-type multiblock copolymers with 40 and 55 % of the hard (polybuthylenetherephthalate) block content is shown in Figure 7. Polytetramethyleneoxide, which is the rubbery block in these systems, possesses a rather high negative energy contribution -0.47^3. Following this value one can estimate the change of the intrachain energy as a function of deformation. The results of the estimation, together with the calorimetrically determined internal energy changes, as well as the changes according to the statistical theory, are shown in Figure 8. It is seen that for the sample with 40 % hard block content the experimental and estimated data are close only in the region of small and large extensions; in the intermediate region there is a large discrepancy indicating that in this region there is a considerable positive energy contribution, which seems to arise as a result of the intermolecular changes in the rubbery matrix. Nevertheless, the conformational changes dominate.

The situation changes dramatically in the sample with 55 % hard block. The energy changes revealed calorimetrically do not follow the theoretical predictions even qualitatively. Indeed, there is a thermo-

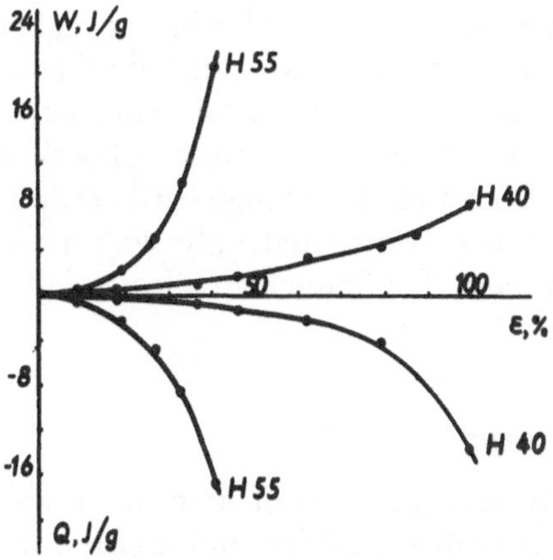

Fig. 7. Dependence of work and heat on strain for polyester-type block copolymer "Hytrel" with various hard block contents [21]

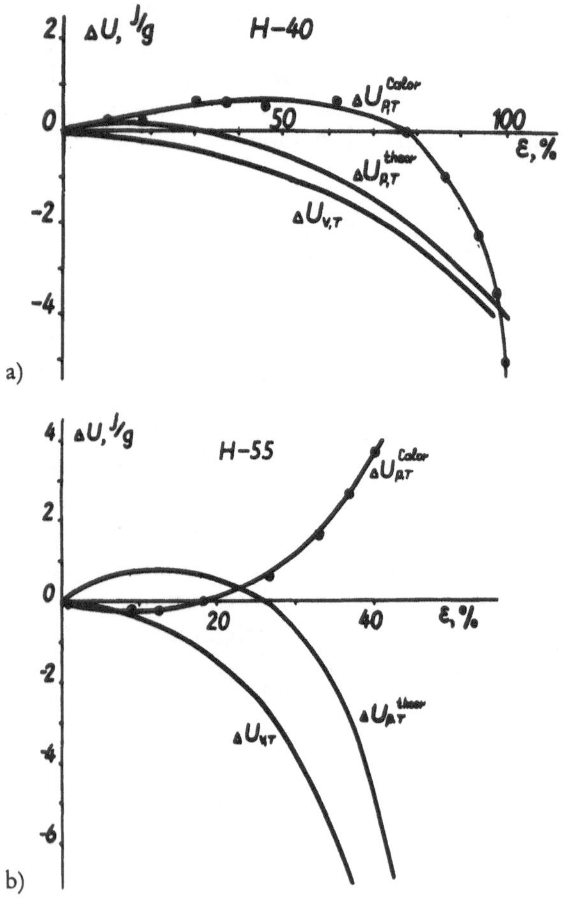

Fig. 8. Internal energy changes as a function of strain for polyester-type block copolymer "Hytrel" with 40% (a) and 55% (b) hard block content [21]. $\Delta U_{V,T}$ and $\Delta U^{\text{theor}}_{P,T}$ are computed according to the theory of rubber elasticity with the corresponding set of parameters

mechanical inversion behaviour (which is very unusual for networks): at small strains, the internal energy decreases and then increases considerably at higher extensions. Summing up, one may conclude that at present there is sufficient experimental evidence to indicate that the free energy of a strained rubberlike block copolymer with a relatively high hard component content cannot be originated only within the chains of the network. The interchain effects play a considerable role.

Crystalline networks

Crystalline polymers are, as a rule, two-phase systems consisting of both crystalline and amorphous regions, and therefore, above their glass transition temperature, they can be considered as networks in which crystallites are taken to be the solid filler and act as multifunctional crosslinks. It has been suggested that the elastic properties of such crystalline networks should be related to the conformational changes in the amorphous regions [22, 23]. The chains in the amorphous phase are thought to be in a highly oriented state even in the absence of external stress. The change in the enthalpy on deformation of the amorphous regions is thought to be small and can be neglected. According to this approach, the free energy of deformation of two-phase crystalline polymers above T_g is also purely intrachain.

This approach was postulated in the early 1960s and since then has become widely used for considering the elastic properties of crystalline polymers. As one can see from publications from Ulm University, dealing with polymer networks, scientists there also follow this approach [24]. This idea is usually used for considering the mechanical behaviour of semicrystalline polymers and estimating the modulus of elasticity. Let us consider whether this approach can be applied successfully to the thermomechanical behaviour of crystalline networks.

According to the thermomechanics of solids, work is a parabolic function of strain and heat is a linear function of strain; therefore, the heat to work ratio is a hyperbolic function of strain [6]

$$W = E\varepsilon^2/2 \tag{17}$$

$$Q = \beta T E \varepsilon \tag{18}$$

$$Q/W = 2\beta T/\varepsilon . \tag{19}$$

The only phenomenological parameter which controls solely the heat to work ratio and, accordingly, the internal energy, is the macroscopic linear coefficient of thermal expansion of an undeformed sample. Solid polymers follow these relations quite satisfactorily, as one can see from the data for polystyrene and amorphous and crystalline polyethylenetherephthalate at room temperature (Fig. 9). Our results are very similar to those obtained in other studies [25, 26]. Calorimetrists from Ulm University demonstrated that the heat to work ratio of defined in Equation (19), also describes completely the data for a steel wire [27]. Hence, in spite of the provocative simplicity of these relations, they do correctly describe the thermomechanical behaviour of various solids.

Figure 10 shows typical results for crystalline polymers above their T_g. The undrawn sample absorbs

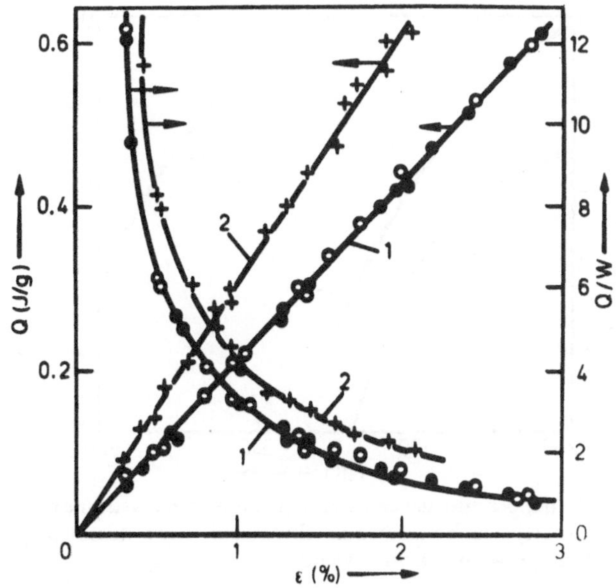

Fig. 9. Heat and heat-to-work ratio as a function of strain at 20 °C [6]. (1) = polyethylenetherephthalate ((○) = amorphous, (●) = crystalline); (2) = polystyrene

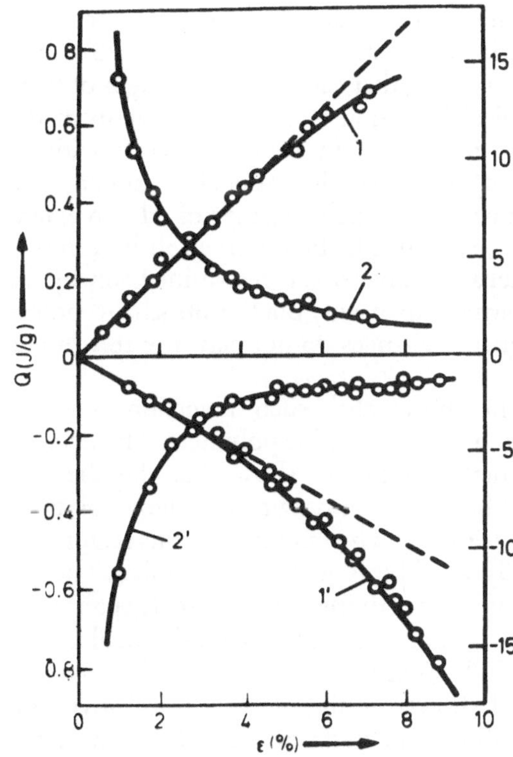

Fig. 10. Heat (1, 1′) and heat-to-work ratio (2, 2′) for reversible deformation of low density polyethylene at 20 °C [6]. 1, 2 = isotropic sample; 1′, 2′ = cold-drawn sample ($\lambda = 4$, deformation along orientation axis). Dotted lines correspond to Equation (18)

heat under extension, similarly to the polymers in Figure 9. This means that it has a positive coefficient of thermal expansivity. The heat to work ratio also follows the theoretical curve quite satisfactorily. These results demonstrate that undrawn crystalline polymers behave similarly to typical solids. This would suggest that the presence of crystallites in these networks prevents the amorphous chains from deforming, exclusively due to conformational rearrangements. The deformation is accompanied by a volume change. This conclusion is quite consistent with the Poisson ratio for typical crystalline polymers (0.30–0.35), i. e. the volume increases under extension. Hence, at least for undrawn crystalline polymers, the thermomechanical properties cannot be related to the conformational changes of the extended tie molecules in the non-crystalline regions.

The situation turns out to be less obvious and more complex for drawn polymers. Cold drawing of crystalline polymers leads to a change in the sign of thermal effects accompanying the reversible stretching of a drawn sample, i. e. a reversible extension of drawn polymers is accompanied by the evolution of heat. It could be suggested that this heat evolution is really the consequence of the stretching of highly oriented tie molecules in amorphous regions, as has been supposed for crystalline networks. According to up-to-date models, the drawn crystalline polymers possess such tie-molecules in both the intrafirbrillar and interfibrillar amorphous regions [6]. The drawn polymers are able to deform reversibly at 10–30 % and such deformations can be related to the intramolecular conformational changes. However, as is seen from the heat to work ratio (Fig. 10) in this case it also has a parabolic form typical for solids with a negative thermal expansivity along the extension direction.

It is a well known fact that above T_g most drawn crystalline polymers often exhibit a shrinkage on heating. Although the shrinkage seems to include two molecular mechanisms, the conformational elasticity of oriented tie molecules and the shrinkage of oriented crystallites which is known to possess negative expansivity, the dominant contribution to the macroscopic shrinkage produces the amorphous regions.

To understand the thermal shrinkage features of drawn crystalline polymers, it is very interesting and important to compare typical data for stretched rubbers and drawn crystalline polymers. According to the statistical theory of the isoenergetic chains, the linear thermal expansion coefficient of a stretched network

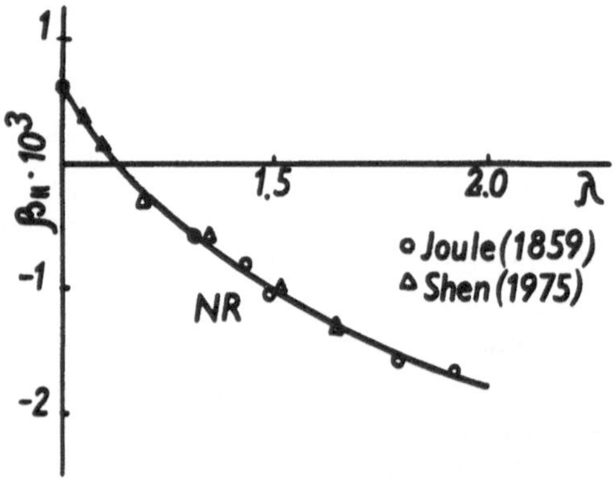

Fig. 11. Dependence of the linear thermal expansion coefficient of natural rubber on extension ratio λ

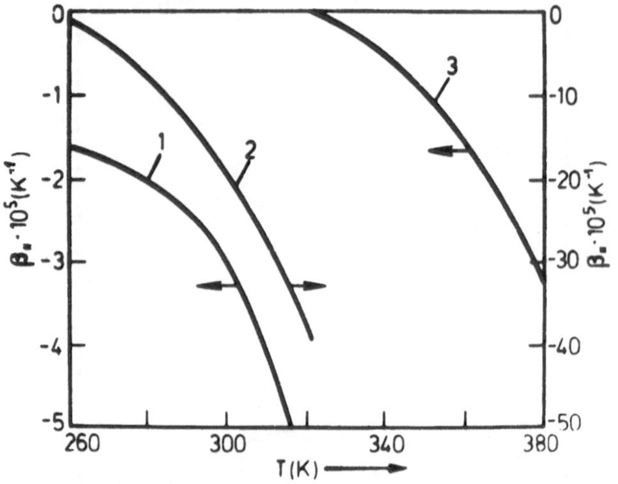

Fig. 12. Temperature dependence of β_{II} for drawn crystalline polymers [6]. (1) = high density polyethylene; (2) = low density polyethylene; (3) = polyoxymethylene

along the stretching direction β_{II} depends on the deformation, according to the following equation [6]

$$\beta_{II}(\lambda) = -1/T(\lambda^3 - 1)/(\lambda^3 + 2) + \alpha/(\lambda^3 + 2). \quad (20)$$

As seen from this equation, an extended polymer network possesses a negative thermal expansivity along the orientation axis, similar to that of a gas, that is more than two orders of magnitude higher than that of macromolecules incorporated in a crystalline lattice. Experimental results for natural rubber (Fig. 11) from both Joule [1] and a more recent [11] study follow this equation quite satisfactorily up to moderate extensions. According to Equation (20) shrinkage is a very weak function of temperature.

Data for typical drawn crystalline polymers (Fig. 12) show that their shrinkage is about one to two orders of magnitude lower than that for stretched rubbers. The temperature dependence of β_{II} for drawn polymers reveals its very large increase with temperature. All these findings concerning the thermal shrinkage of drawn crystalline polymers above T_g, enable us to conclude that although, the amorphous regions are solely responsible for this shrinkage, the interchain interaction prevents oriented chains from free shrinkage. With increasing temperature, the interchain interaction decreases and the shrinkage increases. Hence, two-phase drawn crystalline networks are an extremely interesting class of polymer solids, with a very large negative thermal expansivity. Their large thermal shrinkage is controlled by conformational changes strongly restricted by interchain interactions.

If drawn crystalline polymers are really solidlike (from the thermodynamic point of view), then according to the thermoelasticity of solids, an inversion of the internal energy must occur on elastic extension [28]

$$\varepsilon_{inv} = -2\beta T \quad (21)$$

due to a different dependence of the work spent and heat evolved during strain (see Eqs. (17), (18)). It is worth emphasizing that the molecular origin of thermomechanical inversion is quite different from the internal energy inversion in classical networks with the negative temperature coefficient of chain dimensions. In classical networks, such a change of ΔU on extension cannot, in principle, be observed. It is seen that thermomechanical inversion is determined only by the value of thermal expansivity but the intrachain conformational energy changes do not play any role in this case.

Figure 13 shows that such inversion is really observed in drawn crystalline polymers. The data for polyethylenetherephthalate show that the thermomechanical behaviour of drawn crystalline polymers above and below T_g, when the amorphous regions are undoubtedly solidlike, is qualitatively similar. The difference is the value of the negative thermal expansivity. For PET, it is of an order of magnitude smaller than that for HDPE and PP.

Figure 14 shows the thermomechanical inversion for LDPE and segmented polyester. The similar behaviour of both these materials support the suggestion that in rubberlike block copolymers with a large hard component content the conformational elasticity of the soft block is hardly restricted by intermolecular in-

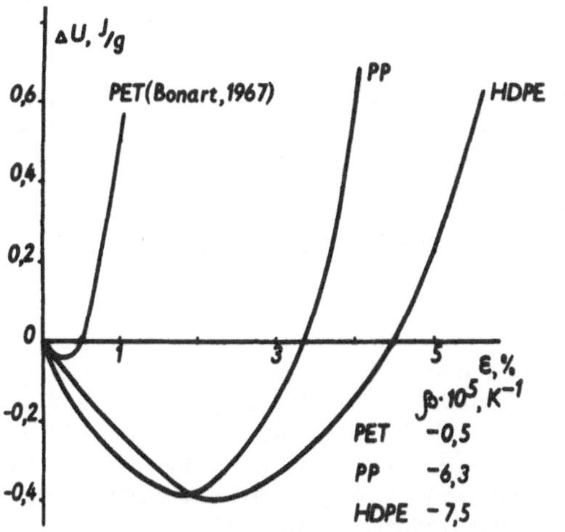

Fig. 13. Internal energy changes ΔU as a function of strain for drawn high density polyethylene, polypropylene and polyethylenetherephthalate [6]

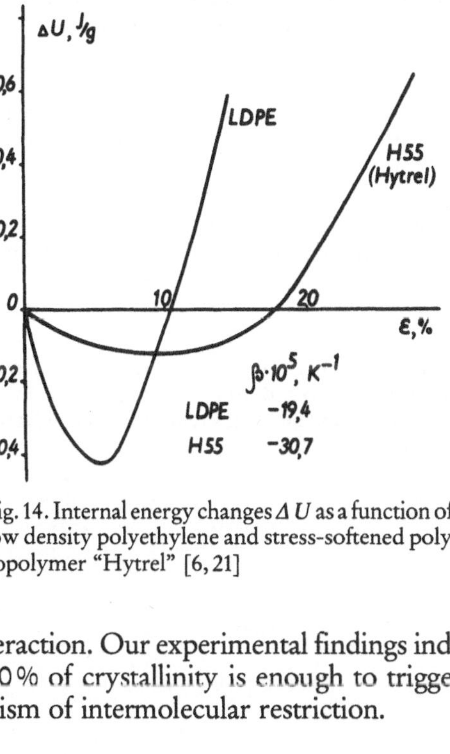

Fig. 14. Internal energy changes ΔU as a function of strain for drawn low density polyethylene and stress-softened polyester-type block copolymer "Hytrel" [6, 21]

teraction. Our experimental findings indicate that 30–40% of crystallinity is enough to trigger the mechanism of intermolecular restriction.

Conclusion

In typical unfilled networks only conformational elasticity occurs and the free energy of deformation is exclusively intramolecular. A small intermolecular volume, energy and entropy changes always accompanying deformation are attendent effects and do not contribute to the free energy. As revealed by thermomechanical studies, a real reinforcement of networks takes place when the free energy of deformation includes both the intrachain conformational changes and some intermolecular contribution. Unique thermomechanical properties of crystalline networks, especially in a drawn state, as well as some oriented block copolymers are the consequences of the simultaneous action of the conformational and intermolecular elasticity.

References

1. Flory PJ (1953) Principles of Polymer Chemistry, Ithaca
2. Treloar LRG (1975) The Physics of Rubber Elasticity, 3 Ed, Clarendon Press, Oxford
3. Mark JE (1973) Rubber Chem Technol 46:593; (1976) J Polym Sci Macromol Rev 11:135
4. Müller FH (ed) (1969) Thermodynamics of Deformation, Calorimetric Investigation of Deformation Processes, Rheology 5, Academic Press, New York, p 417
5. Godovsky YK, Slonimsky GL, Alekseev VF (1969) Vysokomol Soedin A11:1181
6. Godovsky YK (1986) Adv Polym Sci 76:31
7. Price C (1976) Proc R Soc Ser A 351:331
8. Flory PJ, Hoeve CAJ, Ciferri A (1959) J Polym Sci 34:337
9. Godovsky YK (1981) Polymer 22:75
10. Krigbaum WR, Roe R-J (1965) Rubber Chem Technol 38:1039
11. Shen M, Kroucher M (1975) J Macromol Sci C12:287
12. Allen G et al (1971) Trans Farad Soc 67:1278
13. Treloar LRG (1978) Polymer 19:1414
14. Kilian H-G (1980) Coll & Polym Sci 258:489; (1981) 259:1084
15. Galanti AV, Sperling L (1970) Polym Eng Sci 10:177
16. Papkov VS et al (1975) Mechan Polym N 3:387
17. Zapp RL, Guth E (1951) Ind Eng Chem 43:430
18. Oono R, Ikeda H, Todani I (1971) Angew Makromol Chem 46:47
19. Romanov A, Marcincin K, Jehlar P (1982) Acta Polym 33:218
20. Godovsky YK, Bessonova NP, Guzeev VV (1983) Mechan Polym N 4:605
21. Godovsky YK, Bessonova NP, Konjuchova EV, Tarasov SG (1984) Book of preprints, Vol A2, paper A48, Rubber 84, Moscow
22. Krigbaum WR, Roe RJ, Smith KJ (1964) Polymer 5:533
23. Lohse DJ, Gaylord RJ (1978) Polym Eng Sci 18:512
24. Heise B, Kilian H-G, Pietralla M (1977) Progr Coll & Polym Sci 62:16
25. Morbitzer L, Hentze G, Bonart R (1967) Kolloid Z Z Polym 216–217:137
26. Schmid JBM, Wohlrab J, Goritz D (1986) Coll & Polym Sci 264:236
27. Kilian H-G (1982) Coll & Polym Sci 260:895
28. Godovsky YK (1982) Coll & Polym Sci 260:461
29. Christensen RC, Hoeve CAT (1970) J Polym Sci A1, 8:1503

Author's address:

Yu. K. Godovsky
Karpov Institute of Physical Chemistry
Moscow 107120, U.S.S.R.

Discussion

KILIAN

Have you checked the behaviour of semicrystalline networks at elevated temperatures? Did you then obtain the same results?

GODOVSKY:

Well, the same, only the level of effects was somewhat smaller. But when the crystallites exist, the result is the same, quantitatively I mean.

ILAVSKY:

We saw this morning that the deviation from these various theoretical models depends on the molecular weight between the crosslinks. The energy contribution *U* or *F* should not depend on crosslinking, it should be the same for all crosslinking densities, but I wonder if this was proved by systematic calorimetric measurements?

GODOVSKY:

Well, as far as I know, there are no direct calorimetric results in literature. Nevertheless, Mark and Shen analyzed this problem using results obtained by means of thermoelastic measurements, not by direct calorimetry. They came to the conclusion that the energy contribution is independent of crosslinking density.

ILAVSKY:

Segmented polyurethanes or segmented polyester are not classical networks.

Is there any model which considers something like interphase interactions, because I don't think that you have two distinct phases in these systems, so that if you deform, you also deform some interphase. Do you have something, like a 2- or 3-phase model for the thermodynamic description?

GODOVSKY:

Well, I have demonstrated here only the results for a rather large hard block content, but we have the data for a small content of a hard block. There is no such type of behaviour: they behave rather classically. When you increase the content of the hard block, these dramatic changes appear.

KILIAN:

Have you identified the "solid-like phenomena" at small strains only?

GODOVSKY:

It is very difficult to have a large reversible deformation in crystalline polymers in undrawn state, because anelasticity and cold drawing appear so suddenly. I have demonstrated only the data concerning the reversible deformation. In the drawn state, which indicates a totally necked sample, you have a reversible deformation in low density polyethylene of about 30 %.

RIGBI:

I would like to ask about your experiments on the drawn materials. You stretched the drawn materials only once. What would happen if you allowed them to come back to zero position and stretched them again? Would you get the same picture?

GODOVSKY:

The deformation is mechanically reversible but thermodynamically irreversible, and there are dramatic changes in the heat effects.

KELLER:

When you say, you get intermolecular energy on stretching the drawn polymer, do you mean that you can get reversible induced crystallization?

GODOVSKY:

Oh no, I am afraid not a crystallization because if you have a crystallization, the heat involved during this stretching is much higher than the work spent, and here this situation is quite different.

Progress in Colloid & Polymer Science Progr Colloid & Polymer Sci 75:83–103 (1987)

Polymer network structure as revealed by small angle neutron scattering

C. Picot

Institut Charles Sadron (CRM-EAHP), Strasbourg, France

1. Introduction

The molecular theories of rubber elasticity relate chain statistics and the chain deformation mechanism to the deformation of macroscopic materials [1–6]. Until recently, the tests of validity of the theories of polymer networks were usually made by investigations of bulk properties, such as stress strain behavior and swelling equilibria. Such type of measurements are quite straightforward but their interpretations within the framework of molecular theories can be, in many cases, a matter of controversy. In this sense, measurement at the molecular level of geometric changes in a polymer network is fundamental for the theory. It constitutes a direct check of the models and an unambigous test of the consistency between theory and experiment. During the last decade, the advent of small angle neutron scattering (SANS) has allowed us to measure molecular dimensions unambiguously and to characterize the conformation of the chains in bulk [7–8] and semi-dilute polymer systems [9]. Therefore it was in principle possible to apply this technique to the study of the molecular structure of polymer networks and thus to test the molecular theories of rubber elasticity directly at the molecular level.

The advantages of the SANS are twofold [10–11]:

i. The range of scattering vectors accessible on low angle spectrometers ($10^{-3} - 1$ Å^{-1}) matches particularly well with the characteristic polymer dimensions.

ii. Neutron scattering is the result of neutron-nucleus interaction and hence different isotropic species may interact with neutron differently. In particular, hydrogen and deuterium have scattering lengths ($b_H = -0.374\ 10^{-12}$ cm, $b_D = 0.667\ 10^{-12}$ cm respectively) that differ in both sign and magnitude. As a consequence, a deuterated molecule in a protonated matrix will be visible and vice versa.

This last property allows specific labeling by deuteration of different sites of a polymer network in order to analyze its structure. Different types of labeling have then been used to investigate the topology of a network, and are schematically represented in Figure 1:

1. Networks labeled at the cross links (Fig. 1a): this type of labeling provides informations about the spatial distribution of the junction points of the network
2. Networks with labeled elementary chains (Fig. 1b): in this case, the conformation of chain at the level of the mesh can be characterized
3. Networks with labeled paths (Fig. 1c): such labeling allows one to analyze the conformation of a chain trajectory through several junction points
4. Networks H in a solvent D or the reverse (Fig. 1d): in this situation, the neutron scattering results from the spatial correlations between all the elements of the network and leads to the evaluation of that is called the full correlation function.

The aim of this paper is to present an overview of the main results obtained in this field. We will firstly give a brief introduction to small angle neutron scattering as applied to the study of polymer networks. The salient features of the experimental results will be then presented and discussed in the framework of existing molecular theories of rubber elasticity.

2. Small angle neutron scattering (SANS) by polymers and networks – general background

Detailed reviews devoted to the application of SANS to the study of polymer systems are available in the literature [12–15]. These reviews provide information on the experimental technique as well as on the fundamental theories related to scattering by polymers.

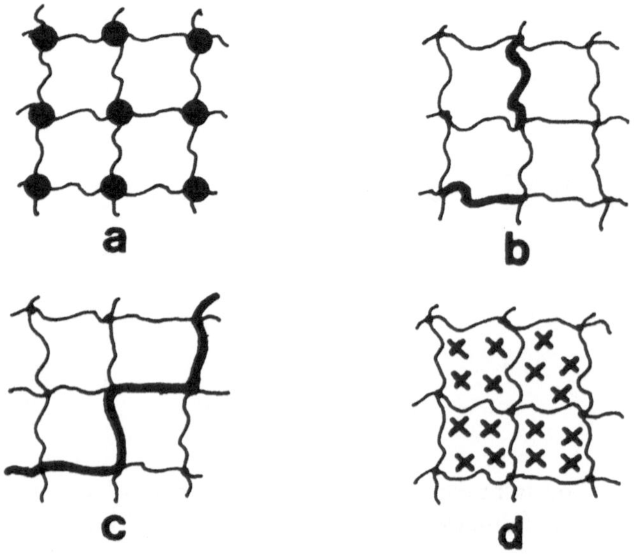

Fig. 1. Labeling of polymer networks for SANS (a) Labeled crosslinks; (b) Labeled elementary chains; (c) Labeled paths; (d) Network (H) in solvent (D) or the reverse

We will restrict ourselves to describing some salient points, as a background, to enable a better understanding of the experimental results that will be presented.

2.1 Experimental

The intensity of neutrons scattered by any molecular system can be written in a very general form:

$$I(q) = K\,(d\Sigma(\underline{q})/d\Omega)_{\text{tot}} \quad (1)$$

where K is an experimental constant. The scattering vector $\underline{q}$ has, in the case of purely elastic scattering, a magnitude given by:

$$|q| = (4\pi/\lambda)\sin\theta/2$$

where λ is the neutron wavelength and θ the angle of scattering.

In an isotropic measurement, such as for a bulk polymer in the undeformed state, a polymer solution or a swollen network, the scattered intensity is recorded on a bidimensional counter and then averaged for a given scattering angle on all azimuthal angles. For an anisotropic measurement, such as polymer samples under uniaxial deformation, the principal components parallel and perpendicular to the stretching directions are considered. These two components are obtained by selecting appropriate sectors or strips on the multicounter ($q_{\parallel}$ and $q_{\perp}$).

The term $d\Sigma(q)/d\Omega$ appearing in Equation (1) is the total differential scattering cross section per solid angle $d\Omega$. This term is the sum of a coherent and an incoherent contribution:

$$(d\Sigma(q)/d\Omega)_{\text{tot}} = (d\Sigma(q)/d\Omega)_{\text{coh}} + (d\Sigma/d\Omega)_{\text{inc}}\,. \quad (2)$$

The first term (coherent), which is phase dependent, contains all the physical information about the structure while the second term (incoherent) is a flat signal. This last term can be substracted experimentally by using an appropriate measurement of incoherent flat background.

For a multicomponent system, the coherent scattering cross-section is generally expressed by:

$$(d\Sigma(q)/d\Omega)_{\text{coh}} = \left\langle \sum_{ij}^{N} b_i\, b_j \exp\left(i\,\underline{q}\cdot(\underline{r}_i - \underline{r}_j)\right)\right\rangle \quad (3)$$

where N is the number of scattering centers, $\underline{r}_i$ and $\underline{r}_j$ denote the positions of centers i and j with a scattering length density b_i and b_j. The phase term appearing in Equation (3) is the Fourier transform of the spatial pair correlation function of the scattering centers calculated on the ensemble average (represented by $<>$) taken over all configurations. Thus the coherent scattering term gives direct information about the spatial distribution of the scattering centers.

It is not the purpose of this paper to give a detailed analysis of the scattering functions corresponding to each specific case of labeling. We will limit ourselves to only a brief presentation of the basic concepts used in the calculations and discuss the parameters gained from the SANS measurements.

2.2 Models for SANS by labeled networks

2.2.1 The scattering function of networks with labeled crosslinked

Neutron coherent scattering experiments carried out on this type of networks will provide information about the spatial distribution of crosslinks of the network. Such labeled networks are usually synthesized by endlinking chains terminated at both their extremities by short perdeuteriated sequences. The interesting question which arises is how the scattering correlation function is modified by building a three-dimensional network from these endlabeled chains.

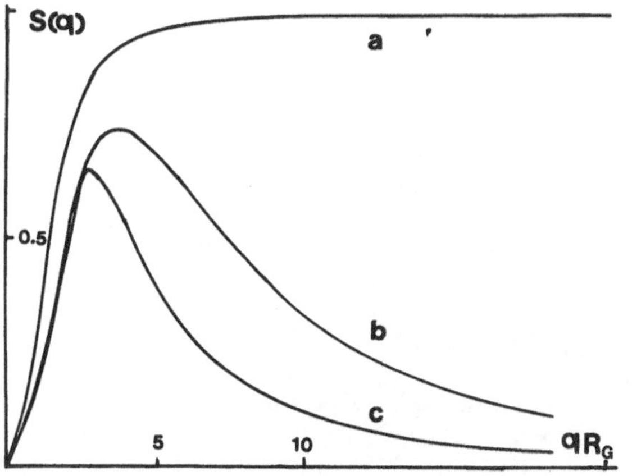

Fig. 2. Scattering correlation function of DHD "pseudo copolymers" α is the D fraction: (a) $\alpha = 0.999$; (b) $\alpha = 0.9$; (c) $\alpha = 0.5$

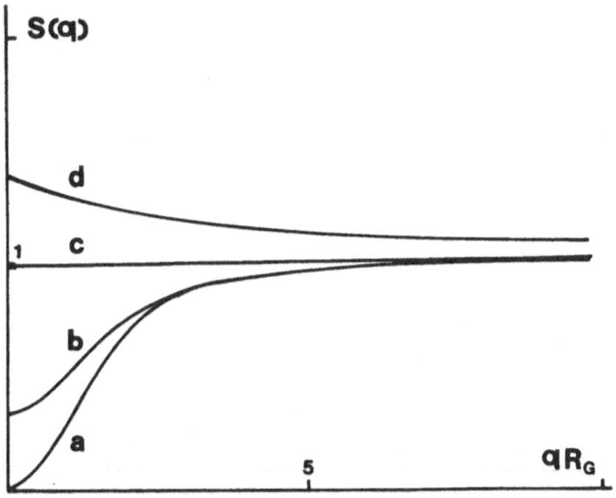

Fig. 3. Scattering correlation function of end-labeled chains ($\alpha = 0.999$). Effect of the molecular weight polydispersity: (a) $M_w/M_n = 1$; (b): 1.2; (c): 2; (d): 3

Let us first recall some basic results concerning melts of endlabeled chains. A first approach, based on the random phase approximation, has been formulated by de Gennes [16–17]. This author shows that although each coil is considered as Gaussian, the total scattering function is very different from that of a Gaussian. For a simple case, where endlabeling is punctual, the scattering function exhibits, in the small q range, a "correlation hole" of size comparable with the overall dimension R_g of one chain. This correlation hole results from the depletion of the concentration of monomers of other chains in the vicinity of a reference chain.

At large q value (smaller distances than R_g), the interference effects between different labels drop out. The intensity reduces to the value for single units. The resulting scattering correlation function is displayed in the Figure 2 as curve a.

For chains labeled over a finite fraction (α) of their length, (D-H-D pseudo copolymer) the same arguments prevail as previously in the small q range ($q < 1/R_g$). On the other hand, in the domain of large q ($qR_D > 1$, R_D being the size of the D portion) the scattered intensity is convoluted by interference effects between D units belonging to a sequence of the same chain, leading to a q^{-2} decrease of the intensity. As a consequence, this type of labeling leads to a scattering correlation function represented in Figure 2 curves b and c. The intensity exhibits a maximum at a given intermediate scattering vector q_m. It must be pointed out that such a peak does not correspond to any local order or segregation of deuterated species to which it was initially often ascribed.

Detailed theoretical calculations based on random phase approximation have been developed by Leibler [18–19] in order to define the exact shape of the scattering correlation function of various partly labeled linear polymer melt systems.

More recently, Benmouna and Benoît [20] have proposed a more intuitive approach (based on a generalization of Ornstein-Zernicke theory) to polymer systems which leads to identical results. Some typical behaviours related to such triblock pseudo-copolymers are represented in Figure 2.

Furthermore, calculations [19] show that the scattering function is strongly altered by the polydispersity of the system (molecular weight and composition polydispersity). This effect results in a rise in intensity in the small q range, leading, for some extreme situations, to the cancellation of the correlation peak (see Fig. 3).

As a further step, recent calculations have been carried out by Benoît [21] for linear multiblock copolymers with a large number N of labeled sequences (HDHDH...). The results show that the shape of the scattering correlation function is only slightly modified when N increases. As seen in Figure 4, for $N > 10$, the peak position and amplitude do not depend on N.

Benoît et al [21] have recently extended these calculations to a linear succession of labeled elements of various structures (comb-like polymers, stars etc.). The main conclusion is that the correlation function is

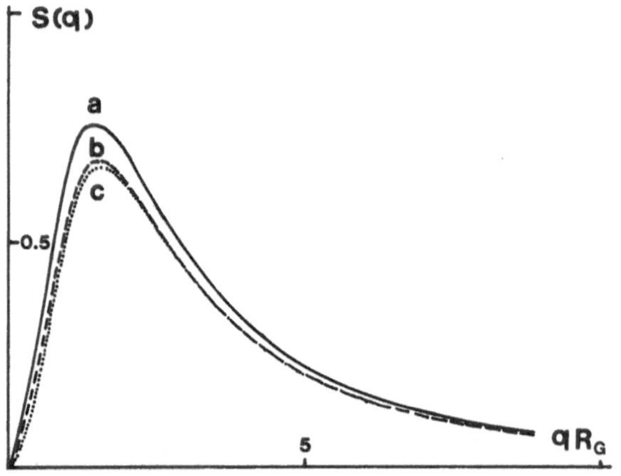

Fig. 4. Scattering correlation function of a multiblock copolymer (HDHDH . . .). N is the number of repeating units (α = 0.1). (a) $N = 1$; (b) $N = 10$; (c) $N = 1000$

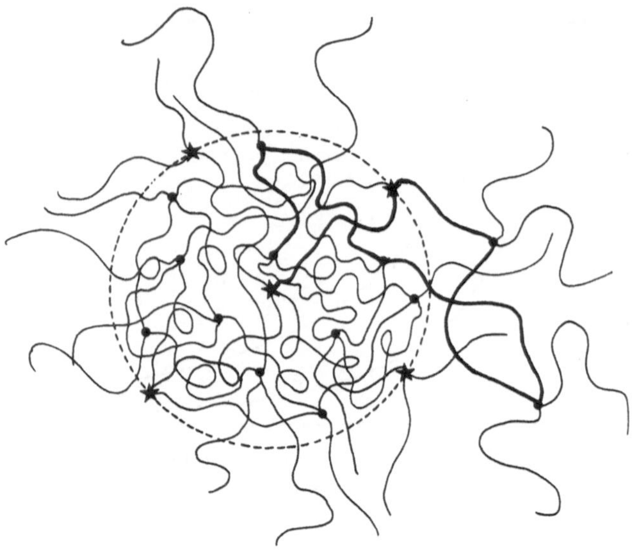

Fig. 5. Schematic representation of a dry network. The reference sphere has a radius equal to the end-to-end distance of an elementary chain. Topological (★) and spatial (●) neighbours of the central cross link. The thick line represent the shortest path between two spatial neighbours (From Ref. [24])

not strongly affected by the architecture of the pseudo copolymer.

The ultimate stage, to derive the correlation function of endlabeled chains linked together in a Gaussian three-dimensional network, is somewhat more tricky to evaluate. The scattering of the network is equivalent to the scattering of an assembly of stars (labeled at their centers). One can assume that the effect of the elastic chains on each junction of the network is the same as creating a potential well in which the crosslinks are located. This creates a non-random distribution of these crosslinks which can be characterized by a correlation function $H(q)$ that defines their positions. Benoît et al. [22] have shown, using this approach, that the shape of the scattering function is only slightly affected by the influence of $H(q)$ and, consequently, mainly dominated by the scattering of the basic labeled units (as for a linear succession of labeled structures). One could hope to extract $H(q)$ from the experimental scattering curves if the exact molecular parameters of the labeled units were known and if strong deviations from this model were present in the system (inhomogeneities, deviation to Gaussian statistics, etc.).

As it will be seen later in the experimental section, the first results obtained on this type of labeled networks exhibit, as expected, a sharp peak which has been interpreted as a first approach in terms of a spatial mean distance h between junctions. In fact, this interpretation can be very hazardous due to the fact that neighbouring crosslinks are not necessarily connected by elementary elastic chains of the network. As was originally remarked by Flory [23] and discussed more recently by Bastide [24], in a loose dry network the elementary chains are widely interspersed. The distinction has to be made between topological neighbouring crosslinks (linked by an elementary chain) and spatial neighbours (connected by a longer path through the network). As a consequence, a given crosslinks has two kinds of first neighbours (cf. Fig. 5). The interpretation of the correlation peak in terms of a distance then leads to an average value which does not discern between the two types of neighbouring junctions. This point will be developed more quantitatively in the experimental section.

2.2.2 *The scattering function of chain-labeled networks*

Let us first recall that for the simple case of a bulk made of deuterated chains, among identical hydrogeneous polymers it has been shown both theoretically [25, 26] and experimental [26–28] that the coherent scattering contribution makes $P(q)$ appear exclusively, the single-chain scattering function (normalized to 1 for $q = o$) which characterizes the chain conformation

$$(d\Sigma(q)/d\Omega)_{\mathrm{coh}} \propto (b_D - b_H)^2\, C_D(1 - C_D)P(q) \quad (4)$$

where C_D is the concentration of labeled species.

It has also been shown [26, 29] that for a ternary system made of two polymers differing only in their scattering lengths (H and D) in solution in a solvent of scattering length b_s equal to the average scattering length of the mixture of the two polymers, the coherent contribution can be expressed in a form identical to Equation (4). This situation allows one to determine the single chain scattering function in situations like semi-dilute solutions or swollen networks independently of the intermolecular interactions. For dilute solutions of polymer chains obeying Gaussian statistics, the single chain scattering function is given by the classical Debye expression:

$$P(q) = 2/(qR_g^2)^2 \left(\exp(-q^2R_g^2) + q^2R_g^2 - 1\right) \quad (5)$$

where R_g is the radius of gyration of a labeled chain.

In the small angle region (Guinier range: $qR_g < 1$) this reduces to:

$$P(q) = 1 - q^2R_g^2/3\,. \quad (6)$$

Most of SANS data can be interpreted in this range of scattering vectors by the Zimm representation.

In the intermediate range of q (submolecular range: $R_g^{-1} < q < a^{-1}$ (a being the statistical unit length), the scattered intensity reflects the conformation of the chains at shorter distances than the overall mean dimension of the chain. For Guassian chains, $P(q)$ varies like q^{-2}; for excluded volume conformations like $q^{-5/3}$, for rod-like particles like q^{-1}.

Let us now turn to the question of scattering by a labeled chain linked into a network. The ideal network consists of locally mobile chains that are bonded together a cross-linkages into a random unbreakable network. The scattering function will reflect the constraints imposed on the labeled probe by the network (either at rest or deformed). The mechanism of network chain deformation has been the subject of much debate for more than 40 years. They are wide latitudes within which the theorists may make models. A good review by Eichinger [5] gives a critical overview of the models which have been proposed since the pionnering works of Kuhn up to the more recent works in this field. We will limit our discussion to describing some basic mechanisms of chain deformation which have been proposed and for which scattering functions have been derived.

The calculation of the scattering function of deformed chains has been devoted to that of elementary elastic chains [30,33] but could be extended to models of multilinked chains [31–35]. Many experimental results deal with networks labeled at the level of the elementary elastic chain; thus, we will restrict ourselves to some quantitative predictions concerning this case. Regarding the results reported for labeled multilinked chains, we will compare them directly with theoretical predictions derived from references [31] to [35]. The theoretical predictions can be then discussed in the framework of the following classical models:

1. The affine deformation [30]

Among the simplest models, the affine deformation model assumes that all the segments of the network chain are oriented as if embedded in an affinely deforming continuum. The resulting deformation is affine at all levels of segmental distances along the chain. For a uniaxially stretched network (elongation ratio Λ), the molecular deformaton α_i is characterized by the ratios of the radii of gyration $R_{g\parallel}$ and $R_{g\perp}$, parallel and perpendicular to the stretching direction divided by R_{go} the radius of gyration corresponding to the unstretched state

$$\alpha_\parallel = R_g/R_{go} = \Lambda \text{ and } \alpha_\perp = R_g/R_{go} = \Lambda^{-1/2} \quad (7)$$

where the network is assumed to be incompressible. The molecular and macroscopic deformations are equivalent.

For an isotropic swelling (swelling ratio Q), the affine deformation leads to:

$$R_g \infty Q^{1/3}\,.$$

2. The junction affine model or end-to-end pulling model [33]

This model assumes that the vector connecting the ends of the linked chain is deformed affinely. The crosslink junctions are fixed in space. The resulting molecular deformations are:

$$\alpha_\parallel = \left(\frac{1+\Lambda^2}{2}\right)^{1/2} \quad \text{and} \quad \alpha_\perp = \left(\frac{\Lambda+1}{2\Lambda}\right)^{1/2}. \quad (8)$$

This model can be considered as a limiting case of the phantom network behavior which is discussed below.

3. The phantom network model [31–35]

This model takes into account the spatial fluctuations of the crosslinking junctions; it is the average position of these crosslinks which is transformed affinely when the network is strained. The fluctuations of a junction about its average position is supposed to be independent of the strain. These assumptions lead to molecular deformations α_i, directly derivable from the Equation (8) where Λ is replaced by $\Lambda_{ph}^2 = \Lambda^2 (f-2) + 2/f$. In that case, α_i is dependent on the functionality f of the network. For instance, the parallel component of the molecular deformation is given by:

$$\alpha_{\parallel} = \left(\frac{1+\Lambda^2 - 2/f(\Lambda^2-1)}{2}\right)^{1/2}. \qquad (9)$$

For tetrafunctional networks

$$\alpha_{\parallel} = \frac{(\Lambda^2+3)^{1/2}}{2}. \qquad (10)$$

If f becomes infinite (which corresponds to the theoretical case in which the crosslink fluctuations are frozen) the result corresponding to the end-to-end pulling is obtained again.

4. Intermediate and phenomenological models [34]

In a series of publications, Ullman [33–35] has calculated the scattering functions for different model networks. Starting from the phantom network described as an ensemble of volumeless chains interacting only at their junction points, two basic ideas, resulting from recent modifications of the theory of rubber elasticity, have been introduced in these calculations:

i. The junction fluctuations are damped by repulsive interactions of neighbouring chains in a real elastomer (36). The severity of restrictions on junction fluctuations (with respect to those allowed in the phantom model) is characterized by a parameter k ($k = o$: phantom; $k = \infty$: fixed junction approximation).

ii. As mentioned earlier [23–24], junction points can be divided into two classes: topological neighbours, which are directly connected by an elementary chain, and spatial neighbours, connected by longer paths (several elementary chains). Upon deformation (swelling or stretching) distances between topological neighbours are supposed to be modified less than distances between spatial neighbours. Network deformation results in a chain unfolding without significant modification of the dimension of the elementary chains. This process of chain rearrangement is taken into account in the calculations, by introducing a deformation Λ^* (at the level of the elementary chain coordinates) related to the macroscopic deformation Λ, according to:

$$\Lambda^{*2} = (1-\alpha)\,\Lambda^2 + \alpha. \qquad (11)$$

Note that $\Lambda^* = \Lambda$ if $\alpha = 0$ corresponds to the affine model and $\Lambda^* = 1$ if $\alpha = 1$ describes a process in which only network unfolding takes place, without chain deformation.

The resulting expression of molecular deformation can be written within this framework as:

$$\alpha_{\parallel} = \left[\frac{1+\Lambda^{*2}(1-(2/f(1+k)) + 1/f(1+k)\Lambda^{*2})}{2}\right]^{1/2}. \qquad (12)$$

Thus, the adjustment of the parameters α and k should allow a phenomenological description of the chain behavior as revealed by SANS.

As mentioned before, the calculations given above deal with the scattering by an elementary chain of the network. A series of experiments have been performed on multilinked labeled chains. Several calculations [31–35] based on the phantom model have been devoted to this problem. The theoretical results given in these references allow one to predict also the behavior of the radius of gyration of the multilinked chain as the response of the scattering function in the whole range of scattering vectors. As will be seen in the experimental section, SANS measurements in the submolecular range of q on labeled paths provide much information about the process of chain deformation at different scale of distances.

Let us now consider some questionable points concerning measurements of swollen networks [4]. According the calculations presented above, the molecular deformation can be generalized by simply replacing Λ by $(Q/Q_o)^{1/3}$ leading, for instance, in the phantom network model, to:

$$R_g/R_{go} = \left(\frac{(Q/Q_o)^{2/3}+3}{4}\right)^{1/2} \qquad (13)$$

where Q is the swelling ratio of the network during the measurement. Q_o and R_{go} are the initial values of swelling and chain dimension of the network.

This generalization can be considered as non-rigorous, taking into account the following considerations:

Prediction of the molecular deformation depends critically on the choice of the reference state concerning both Q_o and R_{go}. Q_o can be measured macroscopically but does not necessarily correspond to the dimension at the molecular scale. On the other hand,

R_{go} can be directly measured by a scattering experiment. If the above calculations are applied crudely they would predict that the elastic chain of a gel, prepared in a good solvent, undergoes a supercoiling during deswelling, leading to an end-to-end distance of the elastic chains of a dry network smaller than the imperturbed dimensions.

Swelling is considered an "external field" which produces a purely geometric change of chain dimensions. Therefore, the modification of solvent/polymer interactions (excluded volume effect) is not considered.

This approach does not account for presumed changes of the fluctuation of junction points during the swelling process.

For all these reasons, measurements of network chain dimensions in the swollen state must be discussed very carefully.

2.2.3 *The full scattering function*

When a polymer network (H) is swollen at equilibrium in a deuterated solvent, a q dependent scattering signal is observed due to the strong contrast between the H and D species. The same observation is made for a D network swollen in an H solvent.

The coherent scattering function $S_T(q)$, measured:

$$(dZ/d\Omega)_{\mathrm{coh}} \propto (b_D - b_H)^2\, S_T(q) \tag{14}$$

is called the full correlation function because it results from the Fourier transform of the pair correlation function of all the scattering segments of the network chains.

An important contribution to the understanding of networks swollen at equilibrium has been obtained from the analogy with semi-dilute polymer solutions [16]. In a swollen network, the polymer chains are connected together by chemical crosslinks, while in semi-dilute solutions, the polymer chains interact by transient junctions (entanglements). From this view, semi-dilute polymer solutions should display different behavior only in the very low-frequency dynamic properties (scale of time $\gg$ disentanglement time in solution).

As a consequence, the static correlation function of a swollen gel observed by SANS in principle should be identified as that produced by an equivalent semi-dilute solution of linear polymer.

A semi-dilute solution of high molecular weight linear polymer is characterized by correlation lengths [16, 37]. The correlation length can be identified as the average distance between two entanglements. For distances smaller than ξ, the pair correlation function is dominated by excluded volume interaction between segments from the same chain leading to [16]:

$$S_T(q) \propto \frac{1}{q^{5/3}} \qquad (q\xi > 1)\,. \tag{15}$$

For distances $r > \xi$, the internal single chain correlations are screened out by all pairs of monomer correlations.

This total pair correlation function has been derived by de Gennes [16] and Edwards [37] leading by Fourier transform to a Lorentziantype scattering law:

$$S_T(q) \propto \frac{1}{q^2 + \xi^{-2}} \qquad (q\xi < 1)\,. \tag{16}$$

In order to obtain experimentally the correlation length ξ, these relations can be used in two ways: first by the determination of the crossover in q behavior from q^2 to $q^{5/3}$ and from the Lorentzian broadening of the scattering correlation in the range of q where $q\xi < 1$.

The variation of ξ with the polymer concentration has been theoretically predicted by Edwards [32] using a self-consistent field approach and by de Gennes [16] through the relationship between polymer statistics and phase transition problems.

More recently, Schaefer et al. [38] have developed a new theory to reconcile the scaling and mean field approaches. Briefly, the calculations show that ξ should vary for semi-dilute solutions according $c^{-3/4}$, c^{-1} and $c^{-1/2}$ for good, θ, and marginal solvents, respectively.

For a polymer network swollen at equilibrium the relation with scaling laws of semi-dilute solutions is made, according de Gennes [16], through the "c^* theorem", c^* being the concentration at which the chain begin to overlap. This theorem says that a gel swollen by a good solvent automatically maintains a concentration c_e proportional to c^*. Therefore, apart from proportionality factors, a swollen networks should display correlation length behaviors (as revealed by SANS) similar to those observed in semi-dilute solutions. It must be pointed out that the ξ value observed in gels is not necessarily the distance between two chemical crosslinking points, due to the presence of physical crosslinks (entanglements) which can be present in the network. It can, for instance, be dependent on the conditions of the network synthesis (concentration during the crosslinking process etc.).

Finally it should be mentioned that all the behaviors described above correspond to ideal networks with a uniform spatial distribution of the crosslinks.

Practically, most of the real networks do not correspond to this ideal view. Long distance fluctuations of crosslinking densities can result from the network's structure (see, for instance, the model of self-similar nets proposed by Boué et al. [54], from inhomogeneities introduced during the crosslinking process or from the molecular weight polydispersity of the meshes of the network. In the swollen state, such inhomogeneities will result in large concentration fluctuations giving rise to a strong forward small angle scattering. The mean square amplitude and the mean square correlation distance of these fluctuations can, in principle, be evaluated from the absolute scattering measurements of $I(o)$ and from the initial slope of the scattering function, according the theories of scattering by statistical fluctuations.

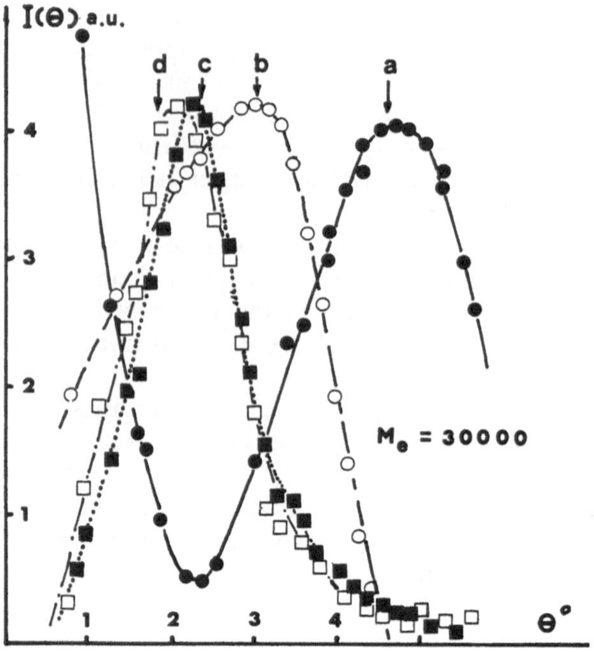

Fig. 6. Experimental scattering function of polystyrene networks with labeled crosslinks: (a) Dry; (b,c,d) swollen in cyclohexane, carbone disulfide, benzene (From Ref. [39])

3. Experimental results

For the sake of clarity, we will first report some of the most important and characteristic SANS experimental results obtained as a function of the different types of labeling. Some of them have enabled us to answer basic questions concerning the network structure but others have created questions which have not yet been fully solved. In part 4, the observed experimental features will be discussed altogether.

3.1 Networks with labeled crosslinks

The pioneering SANS measurements on this type of networks were carried out by Duplessix et al. [39] on polystyrene networks obtained by endlinking bifunctional linear endlabeled chains with divinylbenzene (DVB). As seen in Fig. 6, the scattered intensity exhibits a sharp peak of correlations. The position of this peak is strongly dependent on the swelling ratio Q which was varied by using solvents of different thermodynamic quality (C_6H_{12},CS_2,C_6H_6). As a first approach to interpreting this maximum, these authors used a model of pair correlation diffraction by a pseudolattice to calculate a mean pair separation distance h between junction points.

We have seen in the previous part of this paper that this interpretation is very questionable, but at the time when these results were published, the RPA approach was not yet available. Nevertheless, if this experimental parameter h is plotted as a function of $Q^{1/3}$ (linear macroscopic dilatation of the network (see Fig. 7), a linear variation is observed up to a given swelling ratio ($Q \approx 10$) which corresponds exactly to the concentration during the crosslinking process ($c_p = 0.1$ g cc^{-1}). As far as the meaning of this parameter h can be accredited, this result shows that up to the crosslinking concentration, the mean distance between junction points varies affinely with the macroscopic deformation of network. Above this value, the increase of h is considerably reduced. The information obtained using this approach is somewhat tricky to interpret. If h is considered as an average distance between neighbouring crosslinks, it does not make the distinction between spatial and topological neighbours. From Bastide's evaluations based on molecular conditions of packing the chains in a network [24], the ratio n/f of the total number of neighbours over the number of topological neighbours of a given referenc crosslink, can be estimated. This ratio varies as $M^{1/2}/f^2$. Concerning the series of networks under discussion, it can be estimated as between 10 and 25. Thus, in these samples pre-

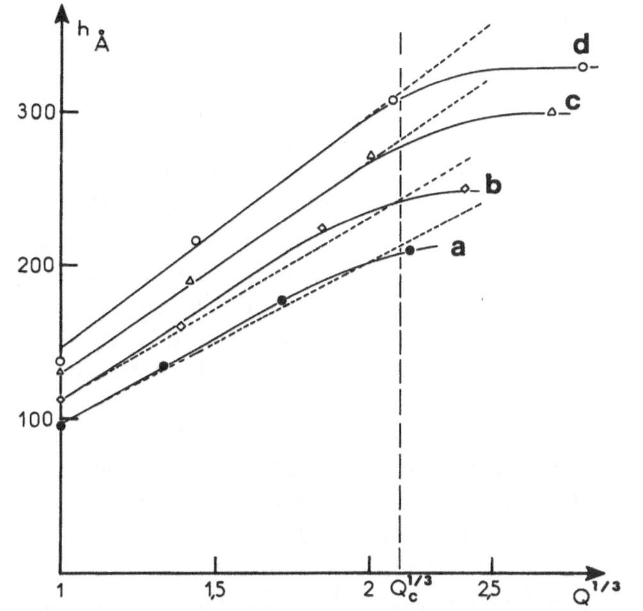

Fig. 7. Variation of the mean separation distance h between crosslinks as a function of $Q^{1/3}$ for different mesh lengths: (a) M_e = 21 000; (b) M_e = 30 000; (c) M_e = 44 000; (d) M_e = 50 000 (From Ref. [39])

pared from dilute solutions, the elastic chains are highly intertwined; spatial neighbours are predominant with respect to the topological ones.

The $Q^{1/3}$ behavior of that observed for $Q < Q_c$ could be interpreted by a process of chain disinterspersion with a volumic affine displacement of function points. For $Q > Q_c$, the network effect could take place and, the neighbouring crosslinks being mainly of topological type, h is mainly governed by solvent chain interactions.

Despite the low precision of the reported data, some comments can be made using the information provided by the RPA. According to the theoretical calculations, a system of multilinked labeled species of equivalent composition should exhibit a correlation peak in a q position corresponding to $qR_g \approx 2$ (R_g being the radius of gyration of the repeating unit, i.e., the mesh of the network). From this relation, R_g can be calculated for the dry networks and the samples swollen in cyclohexane and in benzene (or CS_2). This leads to the following values: 29, 46 and 65Å, which can be compared to the radii of gyration of the equivalent free chain (M_w = 30 000) in the equivalent situations: 49, 49 and 62 Å. One can see that for swollen gels the agreement between R_g and the value of the free chain dimensions is quite reasonable, while, on the other hand, R_g obtained for the dry network is considereably smaller. This would suggest that in this case the system does not correspond to its equilibrium state and could be interpreted in terms of chain interspersion resulting from the deswelling of the network from its state of crosslinking (Q_c = 10). Thus, in order to support this interpretation, further precise measurements on this type on network, crosslinked at different dilutions, would provide useful information on the spatial distribution of the crosslinks. Some experiments have been carried out recently on PDMS networks by Oeser et al. [40]. Unfortunately, the results obtained have not provided the expected information due to the flat scattering profile of the correlation function resulting from the rather high molecular weight polydispersity of the labeled crosslinked chains (M_w/M_n = 1.8). As has been mentioned earlier, the correlation hole effect is strongly smoothed by the composition or molecular weight polydispersity effect.

3.2. Chain labeled networks

As mentioned before, the main problem underlying statistical theories of rubber elasticity is the knowledge of the conformation of the chains in relation to the deformation of the network. One of the basic points is the description of chain conformation in the dry network at rest, which will be discussed in a first part. The problem of the chain conformation when the network is swollen at equilibrium (or deswollen) will then be presented. Finally, the behavior of the chains in uniaxially stretched networks will be described. For the sake of clarity, labeled elementary chains (L.E.C.) and labeled paths (L.P.) be discussed successively.

3.2.1 Networks in the dry undeformed state

Without dwelling on the complex problem of network topology, the basic premise of rubber elasticity theories is that the distribution of network chain configurations is unaffected both by the interlinking process and by the possible formation of entanglements that impede fluctuations of the junctions. Let us recall that this hypothesis basically concerns the elementary chains, for which the mean square end-to-end distance remains the same as before crosslinking.

i. L.E.C.: Measurements of the mean dimensions of labeled elementary chains have been carried out by Duplessix et al. [39] on model polystyrene networks

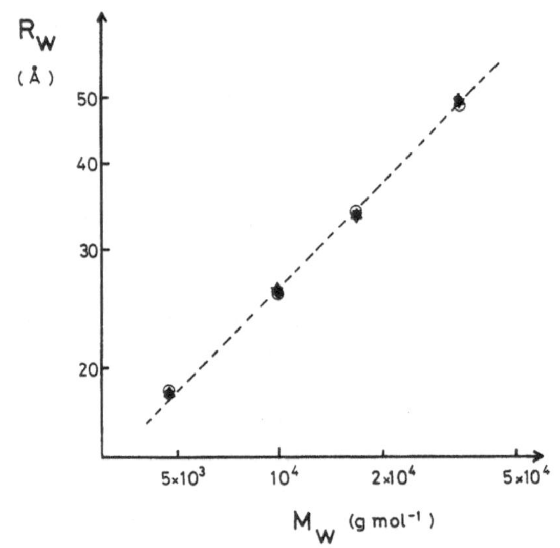

Fig. 8. Log-log plot of the variation of the radius of gyration of elastic chains in bulk PDMS networks as a function of their molecular weights (★); equivalent free chains (O). (From Ref. [41])

and on polydimethylsiloxane (PDMS) networks by Beltzung et al. [41]. Both series of measurements evidence no measurable departures from the unperturbed conformation after the crosslinking has been achieved. As an example, Figure 8 shows a comparison of the mean dimensions of free chains in the bulk state with the equivalent interlinked chains in PDMS.

These results raise some comments: PS networks have been crosslinked in dilute solutions ($c_p = 0.1$) while PDMS crosslinked samples have been synthesized in the bulk state or at very small dilution ($0.6 < c_p < 1$). A consequent modification of the dimensions could be expected, particularly in PS networks, resulting from deswelling of the crosslinking stage to the dry state. This efect can be evaluated from the theoretical predictions of Ullman [33]. On the basis of the phantom model network and assuming a functionality of the order of 5, deswelling should result in a decrease of the radius of gyration of less than 5% with respect to the unperturbed dimensions. This effect is within the range of experimental precision. For PDMS networks, the upper value of the chain contraction should be of the order of 2%. Thus, with regard to the overall dimensions, as seen by SANS, the elementary chains of a network do not exhibit any detectable difference from the equivalent free chains in the bulk state.

ii. L.P.: Most of the SANS measurements published in the literature have dealt with the determination of the dimensions of labeled multilinked chains. Clough [42] and Richards [43] have found that for PS networks obtained by γ-irradiation in the bulk state, the radius of gyration of labeled paths is the same as for an equivalent chain in the uncrosslinked material. Furthermore, the results of Richards show a q^{-2} dependence of the intensity scattered by the labeled crosslinked chains over the intermediate range of the scattering vector, confirming their Gaussian nature. These last results are independent of crosslink density along the labeled path. A similar conclusion has been obtained by Yu [44] and Sperling [45] from measurements of the mean dimensions of labeled paths in polybutadiene networks crosslinked in the bulk state or close to it.

Recently, quite an important series of experimental results on this topic have been reported by Bastide et al. [46]. These authors investigated a series of PS gels with the same crosslinking density and a mesh size estimated at between 45000 and 50000. These samples differed only by the molecular weight of the labeled path multilinked to the network ($1.5\,10^{-5} < M_{LP} < 1.6\,10^6$). The cross-linking was performed by γ-irradiation in solution ($\sigma_p = 0.1$) under theta conditions. It was confirmed that the chain conformation was not modified after the cross-linking had been achieved. The next step of the measurements was to investigate this series of samples in the dry state (deswollen by a factor 10). In Fig. 9, the radii of gyration of the multilinked labeled path are plotted as a function of their molecular weights. For comparison, the dimensions of the same deuterated free chains dispersed in a polystyrene (H) melt are also plotted. These latter follow the classical $M^{1/2}$ variation law corresponding to Gaussian conformation and can be considered, within a good approximation, as the reference chain dimensions during the crosslinking stage. Concerning the labeled paths in the deswollen networks, it can be observed that this M dependence is no longer followed.

The results indicate that the longer the paths, the more important the chain contraction induced by the deswelling is. A tentative extrapolation to the path size, corresponding to a nonvariation of the dimension during, deswelling, would lead to $M_{LP} \approx 50000$, which corresponds, perhaps fortuitously, to the mesh size of the network. If this value is not a pure coincidence, this result would corroborate the observation made by Duplessix on PS networks with labeled elementary chains.

These results can be compared more quantitatively to the theoretical predictions of the classical deformation models. The hatched areas of Fig. 9a and 9b

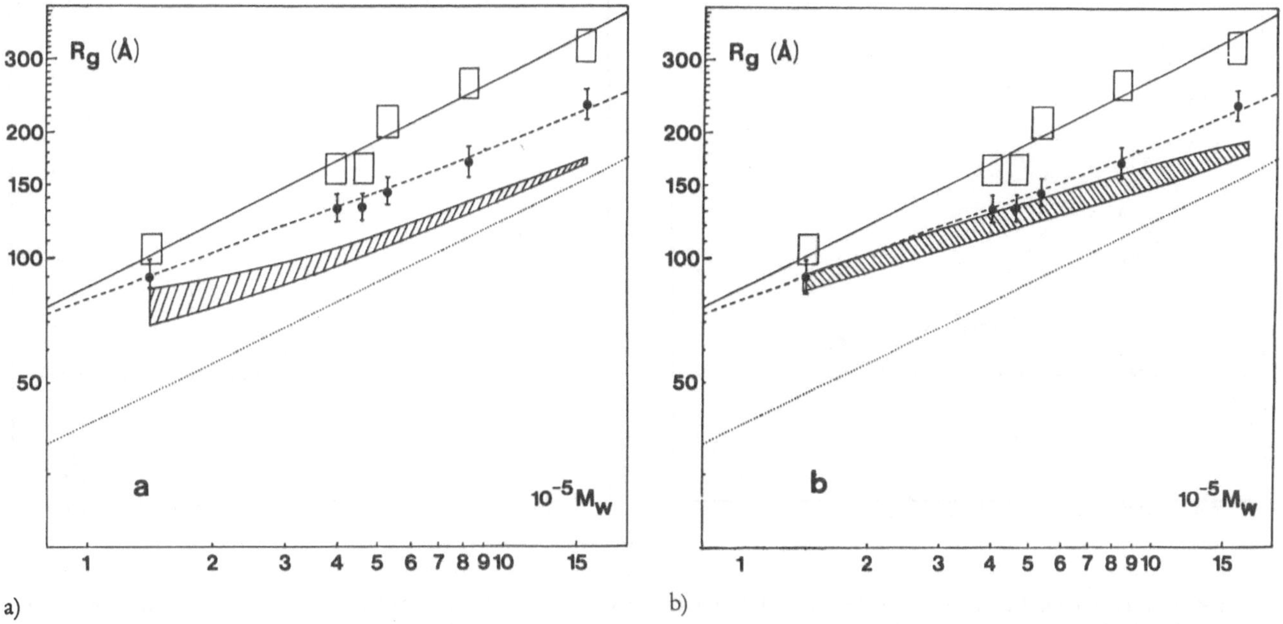

Fig. 9. Log-log plot of the variation of the radii of gyration of labeled paths in PS dry networks as a function of their molecular weights (●). Comparison with uncrosslinked chains (□) and with theoretical models: (a) junction affine, (b) phantom network. (From Ref. [46])

represent, respectively, these predictions for the affine deformation of the junctions and for the phantom network model. The width of these areas results from the uncertainty on the mesh size of the network. Both models overestimate the deformation of the chain leading to a disagreement with classical theories. The information provided by the measurement of R_g is limited: it gives only an estimation of the overall dimension of the chains. A more precise description of chain behavior is provided by the analysis of the scattering correlation function $S(q)$ in the intermediate range of q.

Let us again refer to the results of Bastide on networks with labeled paths, prepared as before. In the present case, the path is the longer ($M \approx 2.8\ 10^6$). The most appropriated representation for investigating the chain behavior in the submolecular range is the Kratky-Porod representation, which consists in plotting q^2 $S(q)$ versus q [57]. For Gaussian unperturbed chains, this representation leads to plateau-shaped curves, the height of the plateau being proportional to the mass of the chain per square unit length. Therefore, in Bastide's experiment (drying the gel from the swollen state) if the collapse of the labeled path was homogeneous at any scale of distance, the scattering function would be simply shifted to the upper direction (referring to the unperturbed chains). As seen in Figure 10, the resulting curve exhibits a hump which indicates the existence of a loss of affineness; the more important scale of observation is local (large q). This observation is not qualitatively in contradiction with the predictions of rubber elasticity theories. However, a more quantitative comparison with these theories has been made by Bastide using the "junction affine" and the "phantom" models (Figs. 10a and 10b, respectively). One observes that no agreement can be obtained between the theoretical and experimental curves. Similar results have been observed for networks with different degrees of crosslinking and different length of labeled paths. The loss of affineness seems therefore to proceed according to topological chain rearrangements which are not included in the classical theories of network deformation. Figure 11 shows that a fairly good agreement between experiment and theory can be obtained within the framework of the junction affine model by introducing into the calculations an apparent deswelling ratio ($Q_{ap} = 4$) smaller than the real one ($Q = 10$). The same conclusion holds for networks with different lengths of labeled paths, i.e., the shorter the path, the smaller Q_{ap} to be introduced into the calculation in order to obtain the theoretical fit. Henceforth, the multilinked chain deformation can be described by an

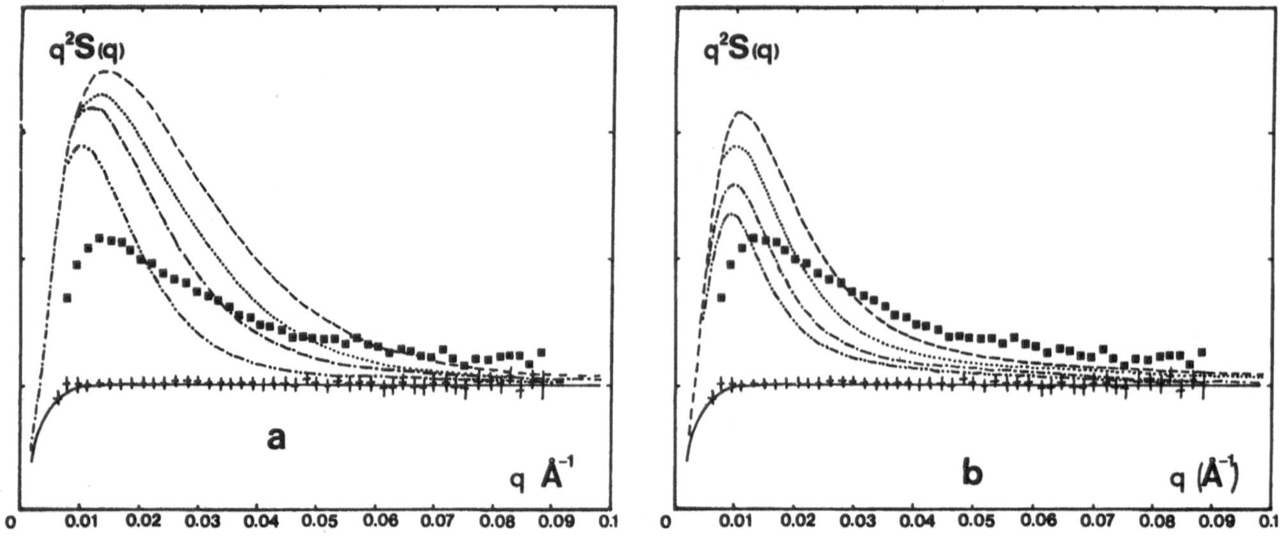

a) b)

Fig. 10. Kratky-Porod plot for the scattering function of deuterated chains ($M_w = 2.8 \cdot 10^6$). (+) Uncrosslinked chains in melt matrix; (■) labeled path linked to a network (deswollen by a factor 10); 35 000 < M (mesh) < 50 000. Calculated intensities scattered by multilinked chains (deswelling = 10). (----) M (mesh) = 35 000; (···) M (mesh) = 50 000; (-·-·-) M (mesh) = 75 000; (-···-) M (mesh) = 150 000. (a) Junction affine model; (b) phanttom network model (From Ref. [46])

affine displacement of the junctions smaller than the macroscopic one and depending on the length of the path. As emphasized by Bastide, this last comment is rather difficult to be admitted. It can be only asserted that the affine deformation of the crosslinks takes place at curvilinear distances, along a given path in the network, which are much longer than the mesh size. Such behavior would support the existence of a non-affine reorganization of the crosslinks resulting from the chain interspersion. This description is particulary valid for loose networks.

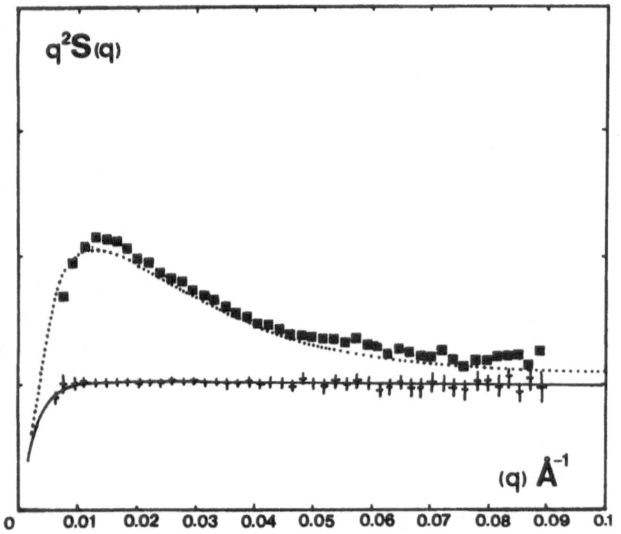

Fig. 11. The same as Figure 10, except the dotted curve is calculated for the junction affine model, a mesh size of 35 000 and a deswelling ratio equal to 4 (the exp. value is 10). (From Ref [46])

Thus, from these examples it can be seen that the chain conformation of a labeled path of a network in the quiescent dry state is dependent on the state of the system during the crosslinking process (bulk or dilute). For networks prepared in the bulk state, the conformation is not modified by the crosslinking. When the chains are linked together in the presence of a large amount of diluent, the crosslinking does not affect the conformation in the corresponding state but, as the network is deswollen, the chain trajectory (through several crosslinks) does not collapse uniformly. The scattering correlation function reveals a threshold of affineness. As a consequence, the memory effect from the stage of cross-linking appears to be a long scale distance effect which results in a rather low variation at the level of the mesh size.

3.2.2 Networks swollen at equilibrium, deswelling

In the Introduction, we raised some points questioning the comparison of experimental data with theoretical predictions. One of the key points was the choice of the reference state for theoretical calculations of the mean dimensions in the swollen state. In most results reported on labeled elementary chains, the reference which has been adopted is the chain dimension in the dry quiescent state. Indeed, it has been shown that in this case, the elementary chains as seen by SANS do not exhibit dimensions different from the unperturbed Gaussian coil. The same holds for labeled paths, for which most network samples have been prepared in the bulk (except Bastide's samples). Thus, the expected values of the mean dimensions of an elementary chain in a network swollen at equilibrium are currently expressed in the framework of different models by:

– $R_g = Q^{1/3}\, R_{g,\,dry}$ for the pure affine model

– $R_G = \left(\dfrac{Q^{2/3}+1}{2}\right)^{1/2} R_{g,\,dry}$ for the junction affine model

– $R_G = \left(\dfrac{Q^{2/3}+3}{2}\right)^{1/2} R_{g,\,dry}$ for the phantom model

Corresponding expressions can derived for multi-linked chains and are available in the literature [35].

i. L.E.C. Pioneering measurements on the dimension of the elementary chains in a swollen network were carried out by Duplessix et al. [39]. The result obtained on polystyrene networks swollen at equilibrium by solvents differing in their quality show that the variation of the radius of gyration of the elementary elastic chain is much lower than that predicted from the end-to-end pulling model.

For most of the samples investigated, the measured dimensions are at an even lower level than the phantom model values. It must be noted that the elastic chain dimensions of the polystyrene network swollen in cyclohexane at room temperature are the same as in the bulk, even though the network was swollen by a factor of 3. This result shows that the solvent-polymer local interaction is predominant compared to the geometric effect (change of volume after swelling) regarding the dimensions of the elementary elastic chains.

More recently, Beltzung et al. [47] have reported results obtained on a series of PDMS networks swollen at equilibrium. These networks were differed in their mesh size, functionality and polymer concentration during the crosslinking process.

Although some of the results were in agreement with the phantom or the end-to-end pulling models,

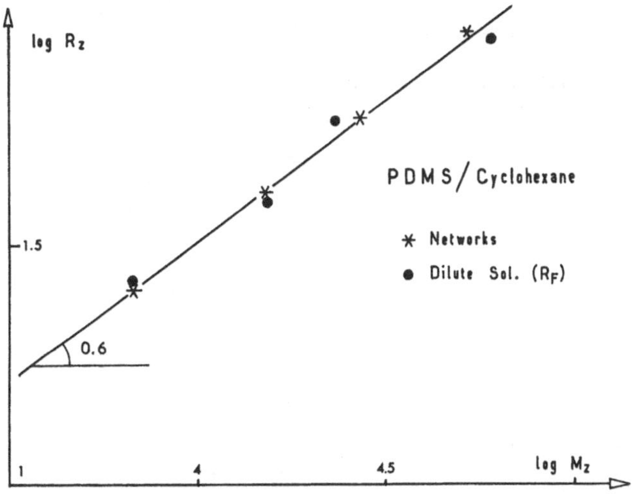

Fig. 12. Log-log plot of the variation of the radii of gyration of elastic chain in swollen (cyclohexane) PDMS networks as a function of their molecular weights (∗); equivalent free chains in dilute solution (●). (From Ref. [47])

these authors showed a remarkable agreement between the conformation of the elementary network chain and that of the free chain in dilute solution, whatever the functionality and concentration during the crosslinking process. This is illustrated by Fig. 12, in which the radii of gyration of the network chains and those of the corresponding free chains are plotted vs. the molecular weight on a double-logarithm scale. The exponent of the molecular weight dependence is quite close to 0.6, typical of excluded volume behavior. These experiments allow one to conclude again that conformation of elementary chains is mainly governed by the local polymer solvent interaction, whereas the equilibrium swelling degree of the gels depends on the volume fraction at which the network was prepared, as well on its functionality. Furthermore, this study strongly supports the de Gennes approach [16] of the behavior of a network swollen at equilibrium in a good solvent. This approach makes the analogy between a swollen gel and a semi-dilute solution of high molecular weight polymer chains. The mesh size of the network corresponds to the "screening length" of the polymer solution. As a consequence, the equilibrium polymer concentration c_e of the swollen gel is related to the overlap concentration c^* of a solution of macromolecules of molecular weight corresponding to the elementary chains of the network.

In order to investigate more deeply the relation between macroscopic and microscopic properties of swollen gels Bastide et al. [48] have carried out a quite

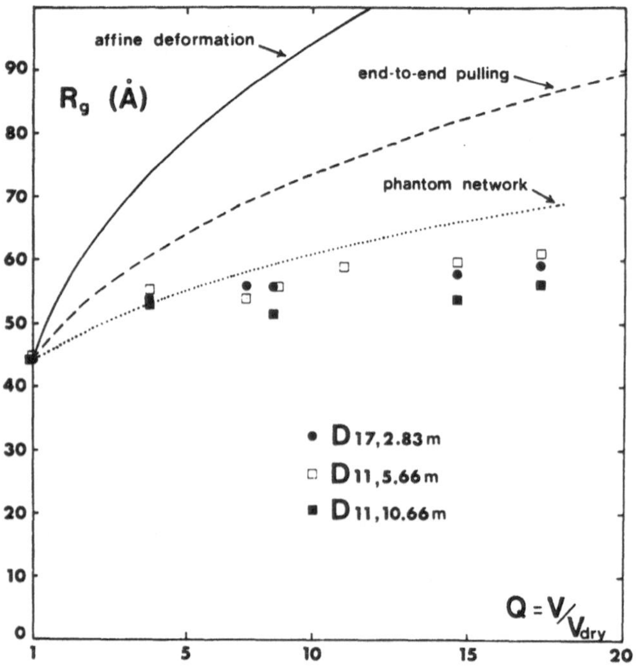

Fig. 13. Osmotic deswelling of PS chain labeled network. Variation of the radius of gyration of the labeled elementary chains as a function of the degree of swelling (From Ref. [48])

instructive experiment, which consisted of measuring the radius of gyration of the elastic chains of a polystyrene network as a function of the swelling degree in benzene. The network was first swollen at equilibrium in pure benzene, then the gel was osmotically deswollen by the addition of large polymer chains.

Figure 13 displays the dependence of the radius of gyration on the degree of swelling. It can be noticed that R_g does not exhibit any appreciable dependence for swelling degrees ranging from 5 to 18. On the same figure are plotted the theoretical predictions for the three different models for which the reference state has been chosen as the dry state.

It must be noted that the experimental data lie close to the phantom network model and that a better agreement with this model may be attained if Q_o (Eq. 13) is taken as an adjustable parameter. Under this condition, a better fit would require $Q_o < 1$, implying a slight supercoiling of the chains in the dry state. However, this procedure looks arbitrary and unrealistic, taking into account the experimental precision for the evaluation of the degree of supercoiling and the fact that this would involve the fluctuations of the crosslink positions being independent of the degree of swelling of the network. It has been proposed by Bastide et al. that during these deswelling experiments, the topology of the network is mainly governed by chain rearrangements, the gel behaving like a three-dimensional accordion for which large macroscopic deformations do not result in a significant change of molecular dimensions at the level of the mesh size. The dimensions of the labeled chains are once again mainly governed by the nature of the local polymer solvent interactions.

ii. L.P. Let us first recall Bastide's experiments [46] showing that the chain conformation of labeled paths in polystyrene networks, prepared by γ-crosslinking a semi-dilute solution of linear chains in the "theta conditions", is equivalent to that of the free chain in the same condition of dilution.

More recently, Richards et al. [43] have investigated quite extensively the conformation of multilinked chains in polystyrene networks (crosslinked in the bulk state) swollen at equilibrium in cyclohexane and in toluene. The number of crosslinks per chain was approximately ranged between 2 and 6 for labeled path of molecular weight between 100 000 and 170 000. For networks swollen in cyclohexane at 308 K (θ condition), although the gels were swollen by a factor of 3–6 times, depending on the crosslinking density, the values of the radii of gyration were indistinguishable from the bulk value within experimental precision. Concerning the networks swollen in toluene, depending on the crosslinking density, the swelling ratio of the investigated networks varied from 8 to 24. Despite this wide range of variation of the macroscopic dimensions, a remarkable constancy of the mean dimensions of the labeled paths was observed. However, the value obtained was slightly larger (18 %) than the interpolated value of an equivalent free chain in a good solvent, extracted from the literature. The molecular deformation of the labeled chains was then compared with the predictions for multilinked chains in the limiting cases of phantom and end-to-end pulling models. Strictly speaking, classical molecular theories cannot account for the constancy of the mean dimensions of the networks swollen in cyclohexane, taking into account the significant macroscopic swelling. This observation could again strongly support the concept of an unfolding mechanism without deformation of the elementary network chains. Concerning the behavior of networks swollen in toluene, Richards et al. have noticed that the ability of the phantom model, albeit better than the end-to-end pulling model for predicting the molecular deformation of the multilinked chain, is far from convincing, even allowing for the large experimental error. These authors have considered the possibility of threshold for the affineness of the deformation of the crosslinking points and compared experimental data to theoretical predictions where deformation is applied to the extremities of the labeled paths. Using this approach, they have shown that the model of affine displacement of the extremities leads to a reasonable agreement with the data relative to the more highly crosslinked networks, while for networks of low crosslink density the phantom behavior gives a more satisfactory description. This trend is at least qualitatively consistent with Flory's constrained function model [36]. Similar but punctual results on polybutadiene networks appear in a paper by Fernandez et al. [45]. In their paper, Richard et al. have also reported some measurements in the intermediate range of q, which exhibit a Gaussian behavior for networks swollen in cyclohexane and an excluded volume behavior for gels swollen with toluene, providing a further indication that the local polymer-solvent interaction is predominant with regards to the chain conformation. A very striking result, also reported in this paper, concerns the variation of the radius of gyration of labeled paths in networks of different crosslinking density in cyclohexane as a function of different crosslinking density in cyclohexane as a function of temperature over the range 308–338 K. While the swelling ratio increases monotically as a function of temperature, the values of the radius of gyration display a maximum around 320 °K. For the highest temperature, R_g decreases catastrophically below the unperturbed dimension. The cause of this feature remains unresolved at the present time.

All these results illustrate the difficulty of interpreting the molecular deformation taking place during the swelling (or deswelling) process of a polymer network in the framework of classical models of rubber elasticity. New processes of molecular deformation or concepts must be invoked (chain desinterspersion, threshold of affineness etc.).

3.2.3 *Uniaxially stretched networks*

Measurements of uniaxially chain labeled networks have been devoted both to samples with labeled elementary chains and to samples with labeled paths. As already mentioned, these two types of labeling allow us to probe the molecular deformation at the level of the mesh size and at curvilinear distances through several crosslinks, respectively. Therefore, we will report separately the experimental results obtained for these two types of labeling.

L.E.C.: The first experiments were reported by Hinkley et al. [44] on polybutadiene networks. The results were shown to be in qualitative accord with the prediction of the junction affine deformation model.

More recently, Beltzung [49] et al. have given quite an illustrative experimental description of the chain behavior in PDMS networks prepared by endlinking. The molecular deformation has been investigated as a function of the mesh size, the functionality and the concentration of polymer during the crosslinking process of the network. The main conclusions emerging from the experimental results are the following:

For short chains, the molecular deformation may be approximately described by the end-to-end pulling model, while for the longest chains, deformation is lower than the phantom model prediction.

Molecular deformation increases with the functionality of the network.

No significant change in the chain deformation process is observed by varying the concentration during the crosslinking process.

None of the classical models have allowed to interpret the whole set of experimental data, but a strong correlation has been evidenced between the molecular deformation and the ratio n/f of spatial to topological neighbours (which characterizes the degree of interspersion of the chains): the molecular deformation continuously decreases with increasing values of the ratio n/f. This result strongly supports the concept of an unfolding mechanism, occuring mainly for networks of large mesh size and low functionality (see Fig. 14).

In a recent work, the molecular behavior of bimodal PDMS networks was investigated by Oeser et al. [50]. Such networks, made of mixtures of short chains (S.C.: $M \approx 3000$) and long chains (L.C.: $M \approx 25\,000$), are of considerable interest because they have ultimate properties which are much better than those of usual (unimodal) PDMS elastomers [51]. The advantage of SANS is that both behaviors of S.C. and L.C. can be observed by labeling selectively one or other of these chains. Figures 15a and 15b represent respectively the molecular deformation in the direction parallel to sketching for the L.C. and S.C. of bimodal networks of different compositions submitted to an uniaxial extension. The results illustrate that for a given composition, the molecular deformation of the L.C. is higher than that of the S.C., except for the highest S.C. content. This behavior can be qualitatively interpreted by associating with the molecular deformation an equilibration of the local constraints, involving the highest

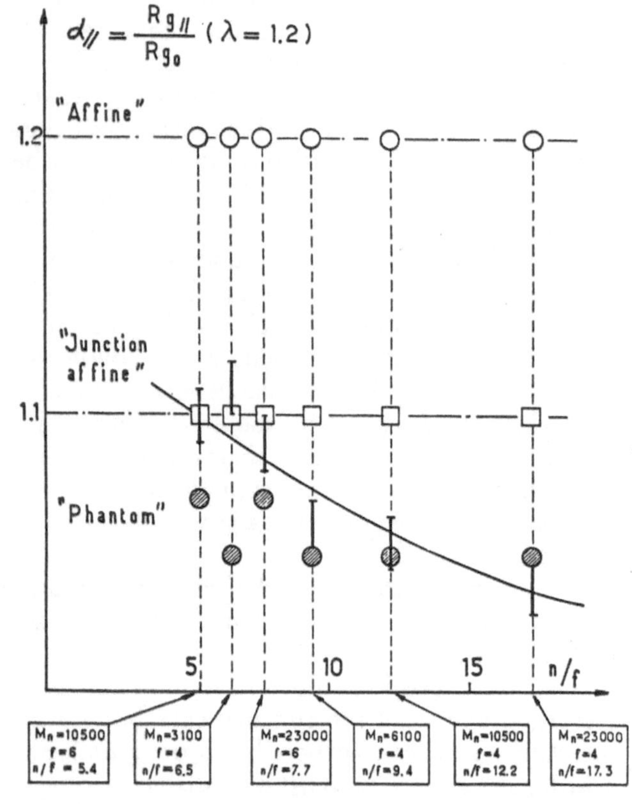

Fig. 14. PDMS networks: uniaxial stretching. Variation of the elementary chain deformation as a function of the ratio n/f of spatial to topological neighbouring crosslinks and comparison with models chain deformation. (From Ref. [49])

modulus of the S.C. and their limited extensibility. Theoretical calculations carried out by Vilgis et al. [52] based on a non-Gaussian theory of rubber elasticity are in good semi-quantitative agreement with these experimental results.

L.P.: The experimental measurements on networks with labeled paths are not manifold. Clough et al. [42] have reported results on stretched polystyrene networks prepared by γ-irradiation. By analyzing the variation of the radius of gyration as a function of the stretching ratio, they have shown a molecular deformation dependence following the prediction of the phantom network model. The conclusion from these results remains questionable, however, taking into account the fact that the samples were not necessarily at stress equilibrium and the limited information provided by the evaluation of the radius of gyration.

A more detailed analysis of the molecular deformation of labeled paths in stretched polystyrene networks has been recently undertaken by Bastide and Boué [53, 54]. The details of the experiment are described in Reference [54], so we will limit ourselves to some comments about the results. Their experiments were mainly concerned with comparison of the chain relaxation mechanism in melts and rubbers and we will focus our discussion on the results obtained for crosslinked systems stretched (at 150 °C) than relaxed at long times (30 min) far behind the maximum Rouse time of the free part of chains between two crosslinks. In principle, equilibrium under a constant strain should be obtained, but the authors have pointed out that an important relaxation can still be observed at these large times. This point is already a matter of discussion. Comparison of the experimental results with the scattering form factors calculated from models has led to several unexpected observations. For the smaller stretching ratio ($\lambda = 1.46$), the fit of the experimental

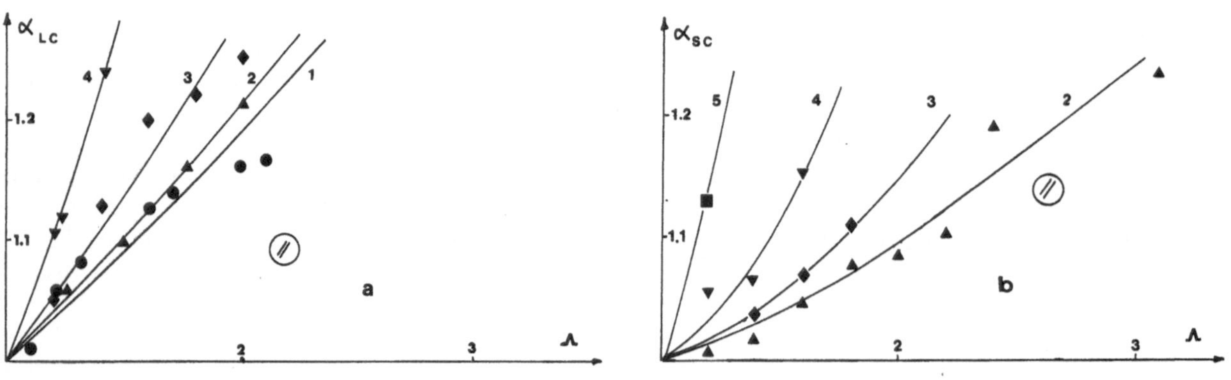

a) b)

Fig. 15. Bimodal PDMS networks: uniaxial stretching. Variation of chain deformation as a function of the stretching ratio for different weight fractions (wt %) of short chains: (1) wt = 0; (2) wt = 10; (3) wt = 50; (4) wt = 90; (5) wt = 100; (a) Long-chain behavior ($M_w = 25\,000$); (b) short-chain behavior ($M_n = 3\,000$)

data could be obtained in the parallel and transverse directions, but by using different models: phantom network in the parallel direction and junction affine in the perpendicular direction, leading to the conclusion that the labeled path is less deformed (more relaxed) in the parallel than in the transverse direction to the sample (more deformed, less relaxed). For a sample stretched at a higher stretching ratio ($\lambda = 4.6$), the results exhibit a similar tendency but an enhanced one. However, no satisfactory fit with calculated scattering form factors could be obtained.

These results can be compared to those of Beltzung et al. [49] on PDMS networks with labeled meshes. For most of the samples investigated, there was a general tendancy to a more pronounced molecular deformation in the direction transverse to the stretching, than in the parallel direction.

These results illustrate the complexity of the mechanism of chain deformation in a rubbery network and the "unfitness" of classical molecular theories to give a realistic picture of the molecular behavior as observed by SANS. Some attempts have been made by Boué and Bastide [54] to get a better understanding of the molecular behavior in such crosslinked systems. The ideas are first to reconsider the nature of the topological structure of a network and, secondly, to analyze separately the contribution of the chain deformation and of its end-to-end vector orientation. The incidence of these two considerations is largely commented on in the paper by these authors, published in this volume.

3.3 The full correlation function of network swollen at equilibrium

As dicussed in Section 2.2.3, the analysis of the full correaltion function $S_T(q)$ of networks swollen at equilibrium is based on their analogy with semi-dilute solutions.

The first measurements of $S_T(q)$ were carried out by Bastide et al. [4] on fully deuterated polystyrene networks swollen at equilibrium in CS_2 and by Geissler et al. [55] on polyacrylamide gels swollen in D_2O. With regard to the polystyrene gels, the results exhibit a $q^{-5/3}$ dependence of $S_T(q)$ in the $q\xi > 1$ regime, similar to the equivalent semi-dilute solution, but a large excess of scattering was observed in the $q\xi < 1$ domain attributed to the presence of inhomogeneities in the gel. Because of this disturbing forward scattering, the correlation length ξ, attainable for $q\xi < 1$, could not be determined. Concerning the polyacrylamide gels, despite the presence of inhomogeneities in the gel structure, the Lorentzian behavior given by Equation (16) was observed, allowing the determination of ξ for gels prepared at various concentrations and various crosslinking densities. The correlation length was found to vary with $c^{-3/4}$ for gels in the crosslinking state as well as for gels at swelling equilibrium. The results are similar to those obtained for semi-dilute solutions and compatible with the prediction of scaling analysis in the good solvent situation.

Quite recently, a detailed study of a measure of the correlation length in randomly crosslinked polystyrene gels swollen at equilibrium in toluene and in cyclohexane ($35 < T < 60$) was reported by Davidson et al. [56]. Therefore, the measurements have allowed us to characterize the behavior of the swollen gels in a variety of thermodynamic environments.

In the situation of a good solvent (toluene) the crossover between the q^2 and $q^{5/3}$ behavior could be observed, as in semi-dilute solutions. The Lorentzian broadening of $S_T(q)$ in the $q\xi < 1$ domain was used to extract the values of ξ. The results show a particularly good agreement with the measures of equivalent semi-dilute solutions ($\xi\, c^{-3/4}$) except for the highest concentrations ($c > 0.06$), where the ξ variation exhibits an important departure from the scaling predictions. It has been suggested by the authors that this deviation could be attributed to the existence of permanent crosslinks, but the mechanism by which this violation could take place is not yet very clear. For gels swollen in cyclohexane in θ conditions ($T = 35$ °C), the power-law exponent of the variation of ξ with c was found to be slightly lower than the predicted value of unity measured for solutions ($\xi \propto c^{-0.85 + 0.06}$). For temperatures above the θ point, the power-law exponent decreases and for $T = 60$ °C, the highest temperature investigated, $\xi \propto c^{-0.56+0.05}$. This behavior corresponds to the marginal solvent situation. This last result is in contrast to the behavior of semi-dilute solutions in the same situation, which corresponds to strong excluded volume interactions. The authors suggest that the presence of permanent crosslinks introduces restrictions on the excluded volume effect as compared to the semi-dilute solutions of linear polymers.

Therefore, the measurements of $S_T(q)$ on polymer gels provide some support for the analogy between semi-dilute solutions of linear polymers and highly swollen networks. The power laws predicted and measured for solutions could be reproduced in most cases, but some differences have also been evidenced. It appears particularly that the excluded volume regime is strongly restricted for swollen gels, as compared with semi-diluted solutions. A major difference arises from the excess of scattering observed on a swollen gel at small angles which attests the presence of inhomogeneities.

4. Conclusion

This review shows that small angle neutron scattering has provided new and valuable information about the polymer network structure at the molecular level. This information, besides questioning the complex topology of a network and the mechanism of molecular deformation, has led to important conclusions.

First, it has been shown that for a network in the bulk state the conformation of the elementary elastic chains is Gaussian whatever the concentration at which the crosslinking process takes place. Non supercoiling could be observed. This result contradicts the concept of a "memory term" generally used in the classical theories of elasticity. If the network in the bulk state keeps a memory from the crosslinking stage, the resulting effect can be observed at curvilinear distances along a path in the network much larger than several mesh sizes. This reveals the high degree of interspersion of the chains, particulary in the loose networks, which has been the subject of most investigations.

It was shown that the large macroscopic deformation resulting from swelling a network does not result in an important chain deformation at the level of the mesh. This behavior cannot be interpreted in the framework of classical affine or phantom network models. The main features can be described, on the basis of topological considerations (chain desinterspersion, loss of affineness etc.) and a predominant effect of local polymer-solvent interactions. The analogy between swollen gels and semi-dilute solutions of linear polymers supports this scheme.

This analogy has also been evidenced by analysis of the full scattering correlation function of swollen networks, which exhibits (with some limitations) scaling behaviors comparable to semi-dilute solutions. Furthermore, this last type of experiment has increased the importance of long-range fluctuations of segment densities which can be inherent to the network structure (trees, self-similar gasket; cf. Ref. [54]) or to the network preparation. Obviously, SANS leads to parameters averaged on the overall structure.

The loss of affineness resulting from the chain disengagement has been cited for interpreting the weak local molecular deformations observed in samples deformed in the bulk state. The results could not be interpreted within the framework of a unique molecular model. It has been shown that the relevant characteristic governing the mechanism of molecular deformation is the local connectivity, which is related to the number of neighbouring cross-links directly connected by an elementary chain.

The whole set of the molecular information obtained by SANS has contributed to a better understanding of the molecular behavior of rubbery networks and should provide new concepts for further theoretical developments.

References

1. Treloar LRG (1975) The Physics of Rubber Elasticity, 3rd ed Clarendon, Oxford
2. Flory PJ (1953) Principles of Polymer Chemistry, Cornell University, Ithaca, New York
3. Dusek K, Prins W (1969) Adv Polym Sci, 6:1
4. Candau S, Bastide J, Delsanti M (1982) Adv Polym Sci 44:27
5. Eichinger BE (1983) Ann Rev Phys Chem 34:359
6. Mark JE (1982) Adv Polym 44:1
7. Cotton JP et al (1974) Macromolecules 7:863
8. Kirste RG, Kruse WA, Ibel K (1975) Polymer 16:120
9. Daoud M et al (1975) Macromolecules 8:804
10. Bacon GE (1967) Neutron Diffraction, Clarendon Press, Oxford
11. Gurevich II, Tarasov LV (1968) Low Energy Neutron Physics, North-Holland Publishing Co, Amsterdam
12. Maconnachie A, Richards RW (1978) 19:739
13. Richards RW (1978) Dev Polym Charact 1
14. Ullman R (1980) Ann Rev Mater Sci, 10:261
15. Picot C (1982) Pethrick RA, Richard RW (eds) Static and Dynamic Properties of the Polymeric Solid State, Nato ASI, C4, Reidel, Dordrecht
16. de Gennes PG (1979) Scaling Concepts in Polymer Physics, Cornell University, Ithaca, New York
17. de Gennes PG (1970) J de Phys 31:235
18. Leibler L (1980) Macromolecules 13:1602
19. Leibler L, Benoit H (1981) Polymer 22:195
20. Benoit H, Wu W, Benmouna M, Mozer B, Bauer B, Lapp A (1985) Macromolecules 18:986
21. Benoit H, Hadziioannou G, Macromolecules, to be published
22. Benoit H, Picot C, Krause S, Macromolecules, to be published
23. Flory PJ (1976) Proc R. Soc Lond, Ser A, 351:1666
24. Bastide J, Picot C, Candau S (1951) J Macromol Sci Phys B19(1):13
25. Cotton JP et al (1974) Macromolecules 16:120
26. Akcasu AZ et al (1980) J Polym Sci 18:863
27. Wignall GD et al (1981) Polymer 22:886
28. Fischer EW et al (1979) Polym Prepr Ann Chem Soc, Div Polym Chem 20(1):219
29. Williams CE et al (1979) J Polym Sci, Polym Lett Ed 17:379
30. Benoit H et al (1975) Macromolecules 8:451
31. Pearson DS (1977) Macromolecules 10:696
32. Warner M, Edwards SF (1978) J Phys A, A-11, 1649
33. Ullman R (1979) J Chem Phys 71:436
34. Ullman R (1982) Macromolecules 15:582
35. Ullman R (1982) Macromolecules 15:1395

36. Flory PJ (1979) Macromolecules 12:119
37. Edwards SF (1966) Proc Phys Soc Lond 88:265
38. Schaefer DW (1984) Polymer 25:387
39. Benoit H et al (1976) J Polym Sci, Polym Phys Ed 14:2119
40. Oeser R, Picot C (1986) ILL Exp Rep 427
41. Beltzung M, Picot C, Rempp P, Herz H (1982) Macromolecules 15:1594
42. Clough SB, Maconnachie A, Allen G (1980) Macromolecules 13:774
43. Davidson SN, Richards WR (1986) Macromolecules 19:2576
44. Hinkley JA, Han CC, Mozer B, Yu H (1978) Macromolecules 11:836
45. Fernandez AM, Sperling LH, Wignall GD (1986) Macromolecules 19:2572
46. Bastide J (1985) Springer Proc Phys, 5, Phys of Finely Divided Matter, Boccara and Daoud Ed
47. Beltzung M, Herz J, Picot C (1983) Macromolecules 16:580
48. Bastide J, Duplessix R, Picot C (1984) Macromolecules 17:83
49. Beltzung M, Picot C, Herz J (1984) Macromolecules 17:663
50. Oeser R, Lapp A, Herz J, Picot C (1986) ILL Exp Rep 434
51. Mark JE, Tang MY (1984) J Polym Sci, Polym Phys Ed 22:1849
52. Vilgis T et al (1986) Macromolecules 19:1212
53. Bastide J, Boué F (1986) Physica 140 A, 251, North Holland, Amsterdam
54. Boué F, Bastide J, Buzier M, Collette C, Lapp A, Herz J (1987) Progr Coll & Polym Sci 75:152–170
55. Geissler E, Hecht AM, Duplessix R (1982) J Polym Sci, Polym Phys Ed 20:225
56. Davidson NS, Richards RW, Maconnachie A (1986) Macromolecules 19:434
57. Bastide J, Herz J, Boué F (1985) J Phys 46:1967

Received November 19, 1987;
accepted November 20, 1987

Author's address:

C. Picot
Institut Charles Sadron (CRM-EAHP)
6, rue Boussingault
F-67083 Strasbourg, France

Discussion

ILAVSKY:

You presented the results on polystyrene networks for the swelling experiment which were done from the equilibriuim swelling to the dry state. Of course, if you are drying samples you are coming from the rubbery to glassy states. Have you taken this fact into account and if so, how? If you go to glassy states, you can have much free volume frozen in, and then the changes in the chain dimension will be smaller than you have already found. I would also like to say that we were trying to collect results from the literature on swelling and mechanical behaviour of networks and we wanted to get an agreement between the polymer solution at different concentrations, and networks at the same volume fraction of polymer. We had to take into account the elasticity term for the swelling equilibrium.

PICOT:

Concerning the first point, the important feature concerns the regime far above the dry state. For the highest Q values, the polymer-solvent interactions are predominant and the polymer chains are in a typical excluded volume regime. Close to the dry glassy state, the interpretation of variation of the dimensions is more complicated; the number of contacts between chain segments increases and one tends to a regime of semi-dilute or concentrated solutions. Concerning the second point, the comparison between polymer solutions and swollen gels will be model-dependent.

RIGBI:

Why did you develop this osmotic deswelling system, and have you considered deswelling at very low temperatures well below Tg in order to maintain the conformation?

PICOT:

The basic idea for the deswelling experiment was to vary the macroscopic dimensions of the gels without modifying the thermodynamic properties of the solvent. We have not considered deswelling at very low temperatures.

D. RICHTER:

At the beginning of your talk you showed a strong correlation between the crosslinks in this polystyrene system. Is this still a valid statement or is this an artificial result, due to size of the crosslinks? As is well known in new experiments of the PDMS, one does not see anything like that.

PICOT:

This point was not very clear at the time at which these results were published. Now, according to the calculations made by Benoit et al. on linearly multilinked labelled stars, it appears that the peak observed on PS network systems (see Fig. 6) is a pure correlation hole effect, as observed on end-labelled chains. Recent theoretical calculations on three-dimensional assemblies of labelled stars tend to confirm this explanation. The fact that this correlation hole effect has not been observed on PDMS samples is a pure consequence of the elastic chains polydispersity.

PIETRALLA:

I would like to evoke two simple physical rules which may help to understand some of your results. The first is: when we have a equilibrium force of extension, we must have some rate of transfer of momentum. The momentum can only be transferred by the chains at the crosslinks, and this means that in any network the crosslinks must fluctuate if they have not too great a mass. Thus, a trifunctional network must have the greatest fluctuations of the network points. With higher functionality, the fluctuations will be squeezed down, so you can have the puzzling effect that a system crosslinked at, e.g., particles, will show nearly ideal behaviour in the sense that the crosslinks do not fluctuate. And so, regarding Prof. Kröner's question, the measure of mesh size over functionality gives some hint of the mass of the crosslink, and this is the reason why you always find the results between the ideal junction and the phantom behaviour. The second rule is simply the equilibrium of force in a system. Regarding your bimodal networks, you can find in the literature some arguments that very short chains will have full extension and will purely orientate. But we are of the opinion that this is not possible, because short chains coming to the limiting extensibility will exert an infinitely increasing force. Since you must have an equilibrium of the average forces between the chains, this means that in your bimodal system the long chains must have a longer elongation, because the short chains are too close to the limit; they are on the upturn already.

PICOT:
I agree with you.

KILIAN:
On swelling, the chains of an initially dry network become extended. It is possible to compute the entropy decrease by considering the deformation as equi-triaxial. Finite-chain extensibility effects come into play. In your experiments, the networks were prepared in the swollen state under equilibrium conditions. My problem now is what happens on deswelling? There are no simple reasons for believing that the radius of gyration should become altered. The set of network might operate like a "three-dimensional telescope", allowing increasing interpenetration of the chains without marked changes of their conformation.

PICOT:
This is a complicated topological question. Our picture is that in the deswelling experiment the predominant mechanism taking place is a rearrangement of the elementary chains (interspersion), for which the chain extensibility does not play any role.

KILIAN:
It is a compression, so you strictly think that you are compressing your sample down to a vanishing volume. But this can't be so. So a lot of trouble can arise in this experiment, to my understanding.

ILAVSKY:
I have only one minor comment. If one can calculate the number of topological and spatial neighbours, this should be done with all your results. If there is some theory of network formation, one can then come to a reasonable number of elastically effective crosslinks. And then, of course, if you have elastically effective crosslinks, you can calculate whether they can be in the volume. And if you now assume that all elastically active crosslinks should also be spatial neighbours, then you come to the conclusion that the pressure is very large and this is impossible. In that case I think it is possible, at least in the first approximation, to approach how spatial and topological crosslinks should differ.

PICOT:
We have calculated the ratio of spatial to topological neighbours as an average value from the molecular parameters of the networks (mesh, size, functionality). It would be interesting to take your comments into consideration.

WEYMANS:
Just a question concerning the bimodal networks' strange behaviour which we observed. If you mix low molecular weight components with high molecular weight components, at certain concentration ratios, a kind of demixing might occur and agglomerates are built. Agglomerates of crosslinked low molecular weight deuterated chains might reduce the crosslink density of your material, and since I have to count the whole agglomerate as one crosslink, in a way, would you agree that such an easy explanation might be the reason for the peculiar findings?

PICOT:
Peculiar observations have been made, in fact, concerning the behaviour of the transverse component of the short chains molecular deformation. In most of the cases, the data analysis does not exhibit the presence of agglomerates of pure ***d*** species at least. But this does not necessarily reject the possibility of having short chains (*h* + *d*) clusters.

WEYMANS:
I'm just referring to some results published by J. Mark. It was observed that if you add a certain concentration of low molecular weight PDMS, you find a very strange energetic effect on the mechanical properties of the network.

KILIAN:
Have you discussed the entropy changes that can uniquely be related to the density of the crosslinks (not discussing at the moment the pre-factor problems)? This result is based on the principle of having the kinetic conformational energy equiparted. It is therefore interesting to know whether it is possible to understand, for example, simple extension experiments in terms of the new model.

PICOT:
We have not so far considered this approach.

STADLER:
We take the advantage of deuterium NMR to be that there are nearly no approximations in the case in which the system is above Tg. It is found that the mobility of the chain is much higher than the time scale of the NMR experiment; you get a single line of the order of about 500 Hz. If you stretch the sample, you now get a quadrupolar splitting and this quadrupolar splitting is directly correlated to the second Legendre polynomial with the angle between CD bond vector and the magnetic field vector. So this is what we observed in the fully deuterated network and this is what we observed in the case of the oligomers. The oligomers are not attached to the chain. Similar things were done by Deloche – the free chains in the network keep an orientation in the equilibrium state. There is no relaxation of the quadrupolar splitting. In the case of the polybutadienes, we even get a splitting into 4 lines corresponding to the cis and trans double bonds. This means these oligomers keep a residual orientation. If we extrapolate the quadrupolar splitting to the oligomer fractions zero (that means to the pure network) this orientation is comparable to the orientation of the weight average in the deuterated chain. And the question is, why is there a residual orientation of these networks which is fairly high?

PICOT:
The d-NMR is a very sensitive technique for probing local orientation. As suggested by Deloche and Dubault, the observed orientation could be attributed to some very local orientational coupling between the elementary segments of the chains.

BOUÉ:

You can do a comparison between the NMR experiment and the neutron experiment very easily because you can use exactly the same sample. In both cases, you use a deuterated mesh or a deuterated free chain inside a non-deuterated matrix, so it is very easy to do the same experiment at the same time. So here we present the experiment in neutron scattering. We use the same plot (Kratky-plot) and you see the big circles – they are for an isotropic chain. This is a stretched network. The little points (quite dark), are from a labelled mesh. That means the labelled species is really linked to the network and the middle curve is the isotropic sample. The upper curve is the perpendicular direction for the stretched sample. (all for a labelled mesh). The other curve, with open circles, is for a free chain in a strained network. It can be in a strained network or in an unstrained network, you get exactly the same form factor. So in NMR you get a doublet for labelled mesh and also for a free chain, but when you look at neutron scattering you don't see the same thing for the labelled mesh in a strained network, you see an anisotropy, and for a labelled free chain in the strained network you don't see any anisotropy. You can interpret that quite easily if you say that this orientation (that you see in NMR) is a kind of nematic orientation which acts only at the level of a segment at small orientation. If you make it quantitative you will get something like $(3\langle\cos^2\theta\rangle - 1)/2 \approx 10^{-3}$, because DMR is very sensitive. But in neutron scattering it should give 10^{-3} in may be the plateau value, or, if you measure the radius gyration, it should be 10^{-3} in the variation of the radius of gyration. This is not something that you can detect. (This is if you accept a kind of nematic orientation as a totally affine deformation.) So it changes affinely at all scales from 1 to 1.0001. This you cannot see in neutron scattering. When you have labelled meshs linked to the network, it is very different: you also get $(3\langle\cos^2\theta\rangle - 1)/2 = 10^{-3}$ at the level of one monomer, so you see the same thing in DMR; but in neutron scattering you see at larger scales and then the deformation is not of a totally affine type, it is a end-pulling type. So when you go to small angles, you see a larger and larger deformation. That is the reason why you see a difference between the strained and unstrained labelled mesh.

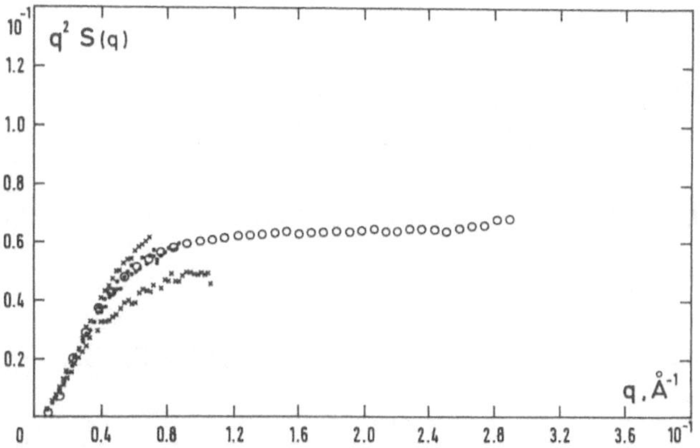

Kratky-plot $q^2S(q)$ versus q. (●) Isotropic network with labeled mesh; (×) Deformed network with labeled mesh; deformation ratio $\lambda \sim 1.3$. Upper curve is perpendicular and lower curve is parallel to the stretching direction; (O) Free labeled chain dissolved in a deformed network with $\lambda = 1.3$. (From Boué et al. Europhys Letters 1(12), 637–645 (1986)

Progress in Colloid & Polymer Science Progr Colloid & Polymer Sci 75:104–110 (1987)

Transient Networks

Evolution of rheology during chemical gelation

H. H. Winter

Max-Planck-Institut für Polymerforschung, Mainz, F.R.G.

Abstract: Rheology is a sensitive measure of the evolving molecular structure in a crosslinking polymer. Dynamic mechanical experiments in small amplitude oscillatory shear give the storage modulus $G'(\omega, p)$ and the loss modulus $G''(\omega, p)$ as a function of frequency ω. The extent of crosslinking, $p(t)$, changes with reaction time. Dynamic mechanical experiments allow detection of the gel point (GP) and give a macroscopic description of the *critical gel* state (network polymer at GP). This critical gel state is used as a reference for describing the entire evolution of rheology. The most surprising discovery of these experiments was that critical gels exhibit stress relaxation in a power law, i. e. the relaxation modulus is given as $G(t) = St^{-n}$. The relaxation exponent, n, depends on network structure. The power law behavior is an expression of mechanical self similarity (fractal behavior). The range of self similarity is defined between an upper and a lower frequency limit. The lower frequency limit (reciprocal of characteristic relaxation time) corresponds to an upper scaling length, the *correlation length*, which is of the order of the linear size of the largest molecular cluster (of pre-gel) or of the largest remaining percolation cluster (of post-gel). High frequencies probe relaxation within single chains. The upper frequency limit corresponds to a lower scaling length, the *glass length*, which is given by the dimension of the molecular network units responsible for glassy behavior. The correlation length and, hence, the characteristic relaxation time increase in the approach of the gel point, diverge to infinity at the gel point, and then decrease again with increasing extent of crosslinking. The critical gel has no characteristic length or time scale. All observations are restricted to polymers at a temperature above the glass transition temperature and at frequencies much below the glass frequency.

Key words: Network polymers, gel point, polyurethane, fractal, critical phenomenon.

Introduction

Chemical gelation of polymers has been studied extensively, for applied and for basic scientific reasons. Applications are governed by the change of properties as a function of reaction extent. Basic interest comes from the fact that gelation is a critical phenomenon [26] and that very simple properties are expected at the critical point (gel point).

This study is mostly concerned with the rheological expression of chemical gelation. The initial material is, in our case, of low molecular weight and, therefore, behaves as a Newtonian liquid. The molecular weight grows with the proceeding crosslinking reaction, consequently the viscosity increases, and elasticity sets in after reaching a critical molecular weight. The crosslinking polymer is still a liquid and it exhibits viscoelastic behavior similar to that of a polymer melt or solution. The molecular weight increases further and most dramatic changes of properties occur close to the gel point. The gel point (GP) is defined as the instant at which the weight average molecular weight diverges to infinity (infinite sample size) or at which the largest molecular cluster extends across the sample (finite sample size). At GP, phase transition occurs to a viscoelastic solid. The equilibrium modulus of the solid grows with higher and higher crosslink density until the reaction stops.

The process may be called chemical gelation to distinguish it from physical gelation, in which the network is formed by reversible association mechanisms between macromolecules. For model networks at GP, the critical extent of crosslinking, p_c, was predicted by Flory [17,18], and Stockmayer [28] and later Gordon [19] and Macosko and Miller [22] as a function of crosslink functionality and stoichiometric ratio. The predictions have been found to be in close agreement with experiment [5] in spite of differences between model networks and real polymer samples. This suggests that GP can be detected by monitoring the degree of crosslinking and waiting until it has reached the theoretical p_c-value. However, the degree of crosslinking is difficult to measure with sufficient accuracy. Difficulties arise from side reactions which parallel the crosslinking reaction [16,23]. The critical reaction time, t_c, for reaching GP can be measured directly. However, t_c is not a material parameter. It depends on both molecular properties and processing history.

Knowing the instant of chemical gelation is important for processing of crosslinking polymers, since shaping has to occur before GP while the polymer is still able to flow and stress can relax to zero. The most common rheological tests for detecting GP measure the appearance of an equilibrium modulus [14,1] or the divergence of the steady state shear viscosity [1,2,6,21]. Measurement of the equilibrium modulus is extremely difficult since its value is zero at GP and it remains below the detection limit for a considerable time. Measuring the diverging steady state shear viscosity has the advantage that the experiment is extremely simple. This method, however, has servere disadvantages which make it less suitable for characterizing gels [31]:

1. Near GP the relaxation time becomes very large and steady shear flow can no longer be reached.
2. The network structure near GP is very fragile. It gets broken by the large deformation of the 'steady' shear flow, causing an apparent delay of GP.
3. In the approach of GP, the gel is a viscoelastic liquid which behaves in a shear thinning way.
4. GP is found by extrapolation. The actual experiment ends some time before GP when the stress increases beyond the limit of the instrument or the strain exceeds the deformability of the sample.

Dynamic mechanical experiments in small amplitude oscillatory shear are better suited, not only for detecting GP but for measuring the entire evolution of rheology.

The following discussion is based on the experimental studies of Chambon and Winter [9], Chambon, Petrovic, MacKnight, and Winter [10], Chambon and Winter [11], and Winter, Chambon, and Morganelli [30] on endlinking polydimethylsiloxane and polyurethane samples near the gel point and at temperatures much above the glass transition temperature. Details of the experiments can be found in these publications and will not be repeated here. Experimental difficulties arise from the fact that any large strain (during sample preparation or transfer, during mechanical measurements) breaks the molecular structure of chemical gels irreversibly. The samples of this study have therefore been prepared directly in the rheometer (no transfer). Large strain has been avoided during the mechanical measurements. Numerous solution swelling experiments [12] confirmed that the rheologically observed GP (by dynamic mechanical measurement) coincides with the transition from a completely soluble state to an insoluble state.

Linear viscoelasticity at the gel point

The transition through GP occurs gradually. A limiting behavior at GP exists jointly for the liquid and the solid. This limiting behavior is the property of the 'critical gel' or the 'polymer at GP'. It should, however, be emphasized that no independent state of matter exists directly at GP and it is impossible to make a polymer directly at GP. A real polymer sample is either still before GP (viscoelastic liquid) or it is already beyond GP (viscoelastic solid).

The rheological behavior of the critical gel is well understood. A simple power law was found to govern the dynamic mechanical behavior. The storage modulus, G', and the loss modulus, G'', are related as

$$G'(\omega) = \frac{G''(\omega)}{\tan(n\pi/2)} = \frac{\pi}{2\Gamma(n)\sin(n\pi/2)} S\omega^n$$

where $\Gamma(n)$ is the Legendre Gamma Function. The 'gel strength' S depends on the mobility of the chain segements (given by persistency length and crosslink density, for instance). The relaxation exponent n may have values in the range

$$0 < n < 1 .$$

Stoichiometrically balanced endlinking networks were found to relax with $n = 1/2$ [10,29] while stoichiometrically imbalanced networks follow

$n = 1/2$ for excess of crosslinker and
$n > 1/2$ for lack of crosslinker

[11,30].

It is interesting to note that, independently of these experiments, power law relaxation was predicted for the critical gel [7, 8]. These predictions are based on the assumption that an analogy exists between dielectric and mechanical properties of percolating systems [13].

The power law behavior is an expression of self similar structure [24, 25]. For the critical gel it was found to extend over a frequency range of more than five decades [9] which was the entire experimental range. A *lower* frequency limit is given by the correlation length of the self similar structure. This correlation length diverges at GP and the lower frequency limit could theoretically be equal to zero. However, a practical lower frequency limit is given by the finite sample size, i.e. at a scale of observation which exceeds the size of the sample in the rheometer. The *upper* frequency limit of the dynamic mechanical behavior is given by the molecular structure. At high frequency the scale of observation decreases below the lower scaling length first of the network (distance between junctions points) and then of the polymer chain, which one might call the 'glass length'. The glass length is given by the size of the network element which determines the glassy behavior at low temperature. At this small length scale, vitrification becomes important and deviation from the self similar behavior is expected. This transition is not the subject of the immediate study. For the following, we will tacitly assume that the scale of observation is sufficiently larger than the glass length. We obviously neglect the details of the molecular structure by neglecting the high frequency glass behavior.

Analysis of the dynamic mechanical data shows that stress relaxation at GP occurs in a power law [11, 30]

$$G(t) = S\,t^{-n}; \quad p = p_c\,.$$

The relaxation modulus can be introduced into a general constitutive equation for linear viscoelasticity [4, 15]. This results in the 'Gel Equation' [29]

$$\tau(t) = S \int_{-\infty}^{t} (t - t')^{-n}\, \dot{\gamma}(t')\, dt', \quad p = p_c$$

which describes every known linear viscoelastic phenomenon at GP. With the gel equation, many new rheological experiments may be invented for detecting GP.

Evolution of linear viscoelasticity at increasing degree of crosslinking

The evolution of rheological properties during crosslinking may be observed through the dynamic viscosity

$$|\eta^*(\omega, p)| = \sqrt{G'^2 + G''^2}/\omega$$

using the extent of crosslinking, p, as parameter. The dynamic viscosity evolves gradually at the crosslinking reaction proceeds, see Figures 1 and 2. Each of the

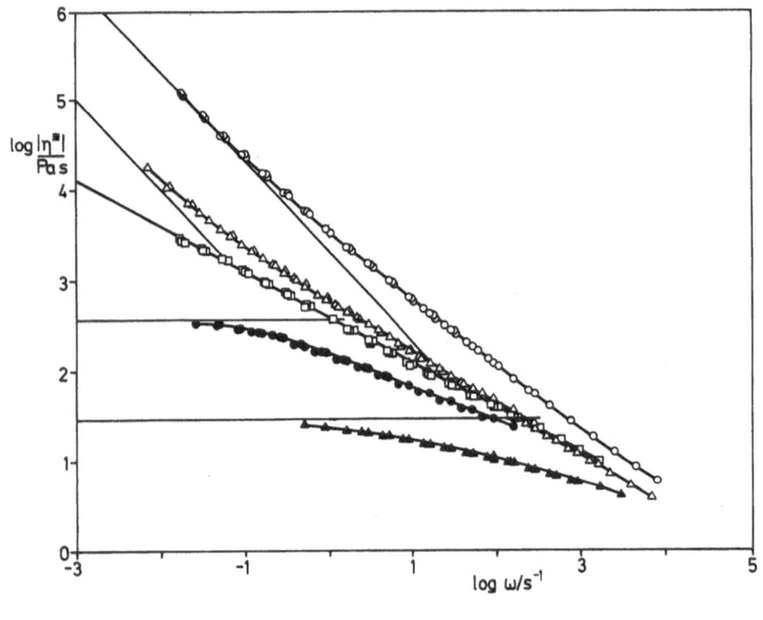

Fig. 1. Dynamic viscosity, as measured in a small amplitude oscillatory shear experiment (data of Chambon and Winter, [9]). Parameter is the degree of crosslinking, p. The power law exponent of the critical gels is $n = 1/2$. Parameter is the distance from the gel point −260, −120, 0, 120, 260 s (starting from lowest curve)

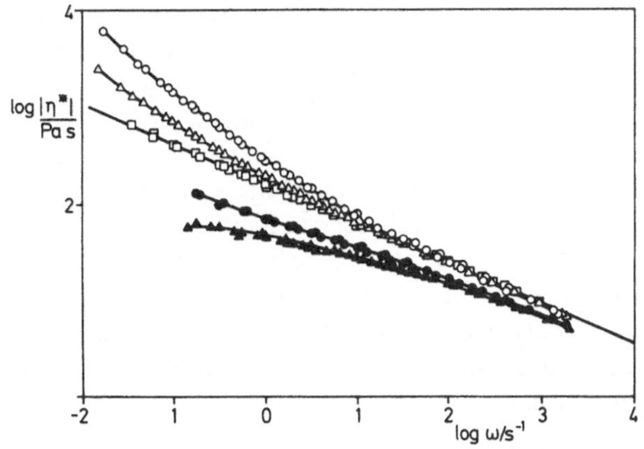

Fig. 2. Dynamic viscosity as measured in a small amplitude oscillatory shear experiment (data of Chambon and Winter [11]). Parameter is the degree of crosslinking, p. The power law exponent of the critical gel is $n = 0.58$. Parameter is distance from GP $-180, -108, 0, 12, 150$ s

samples has stable molecular structure since the reaction was stopped by poisoning the catalyst after predetermined reaction times.

The components of our samples are initially of low molecular weight. This results in a low viscosity, independent of frequency, and no elasticity. With increasing conversion p, the viscosity increases (and some elasticity sets in). At low frequency, the viscosity is constant (zero slope as expected by a viscoelastic liquid) and equal to the zero shear viscosity $\eta_o(p)$. Shear thinning (frequency dependence) is observed at high frequencies, with an onset frequency which shifts to lower and lower values as GP is approached ($p \to p_c$). At GP, the dynamic viscosity follows a power law

$$\eta^*(\omega, p_c)| = a S (i\omega)^{n-1},$$

with

$$a = \pi / (\Gamma(n) \sin(n\pi)).$$

S is a material constant and n is the network specific relaxation exponent. The onset of shear thinning has shifted to the zero frequency limit, $\omega \to 0$, i. e. the material behaves as shear thinning at all frequencies. Beyond GP, the dynamic viscosity grows further, with a dramatic increase at low frequencies (slope of -1, as expected from a solid) which is associated with the appearance of an equilibrium modulus $g_\infty(p)$.

Near GP, typical asymptotic behavior for the dynamic viscosity at low and at high frequency is found:

$$\text{low } \omega \text{ asymptote: } |\eta^*| = \begin{cases} \eta_o(p) & \text{for pre-gel} \\ g_\infty(p)/\omega & \text{for post-gel} \end{cases}$$

$$\text{high } \omega \text{ asymptote: } |\eta^*| = a S \omega^{n-1}.$$

The high ω asymptote is the same for the pre-gel and the post-gel. These asymptotes are drawn in Figures 1 and 2 and values are given in Table 1.

Intersects of the asymptotes with the power law of the critical gel define characteristic relaxation times of the crosslinking polymer, see Figure 3. The characteristic relaxation time of the pre-gel is defined as the reciprocal of the frequency at which the zero viscosity asymptote intersects with the power law of the critical gel:

$$\lambda(p) = [\eta_o(p)/(aS)]^{1/(1-n)}; \quad p < p_c.$$

Equivalently, the characteristic relaxation time of the post-gel is defined by the reciprocal of the frequency at which the low frequency asymptote g_∞/ω intersects with the power law of the critical gel:

$$\lambda(p) = [aS/g_\infty(p)]^{1/n}, \quad p_c < p.$$

The characteristic relaxation time of the crosslinking system undergoes an interesting evolution. The initial relaxation time is very short, since the molecular weight of the reaction components is low. Near GP,

Table 1. Asymptotic values of dynamic mechanical data of crosslinking samples of Figures 1 and 2

n	$t - t_c$ (s)	η_o (Pa s)	g_∞ (Pa)	S (Pa s^n)	λ (s)
1/2	− 260	35	–	–	$7.4 \cdot 10^{-3}$
	− 120	370	–	–	0.83
	0	–	–	226	–
	120	–	100	–	16.0
	260	–	2000	–	0.042
0.58	− 180	71	–	–	0.15
	− 108	156	–	–	1.0
	0	–	–	73	–
	72	–	28	–	19.0
	150	–	89	–	2.6

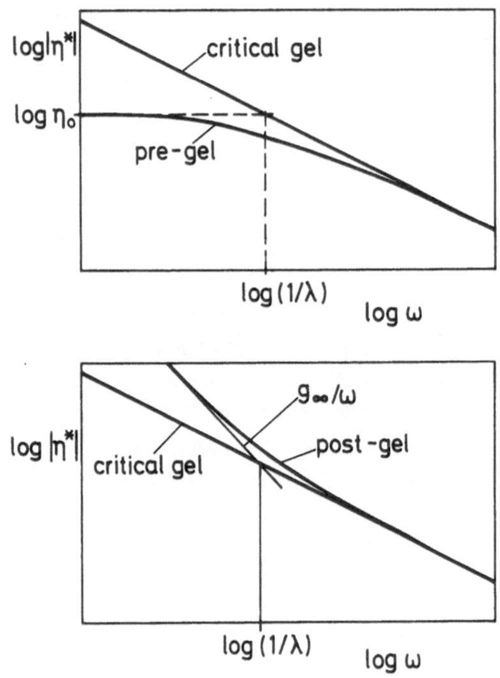

Fig. 3. Definition of the characteristic relaxation time of the pre-gel and the post-gel

the relaxation time rises sharply. The evolution is so dramatic that it could not yet be measured in detail. At GP, it diverges to infinity and the relaxation spectrum does not contain a characteristic time any more. Beyond GP, the relaxation time decreases. The maximum relaxation time of the final network is again very short, provided that the network has reached a high degree of perfection. The final relaxation time is expected to be of the same order as the initial one.

With these reference parameters, the viscosity curves are shifted to a form which allows comparison of the shape of the curves, (see Figs. 4 and 5). The high frequency slope is nicely seen in Figure 4, while data do not agree as perfectly in Figure 5. The viscosity curves of the pre-gel and the post-gel seem to curve much more sharply near GP then away from GP. A universal shape of the entire viscosity curve near GP does *not* seem to exist.

Scaling approximation near gel point

Gelation is a critical phenomenon and the evolution of properties near GP has been described by power laws [27] with critical exponents s and z

$$\text{pre-gel:}\quad \eta_o(p) \sim (p_c - p)^{-s}$$

$$\text{post-gel:}\quad g_\infty(p) \sim (p - p_c)^{z}\,.$$

Combination with the preceeding equations gives power law relations for the characteristic relaxation times in the pre-gel and the post-gel, however, within the critical range ($|p - p_c| \ll 1$):

$$\text{pre-gel:}\quad \lambda(p) \sim (p_c - p)^{-s/(1-n)}$$

$$\text{post-gel:}\quad \lambda(p) \sim (p - p_c)^{-z/n}\,.$$

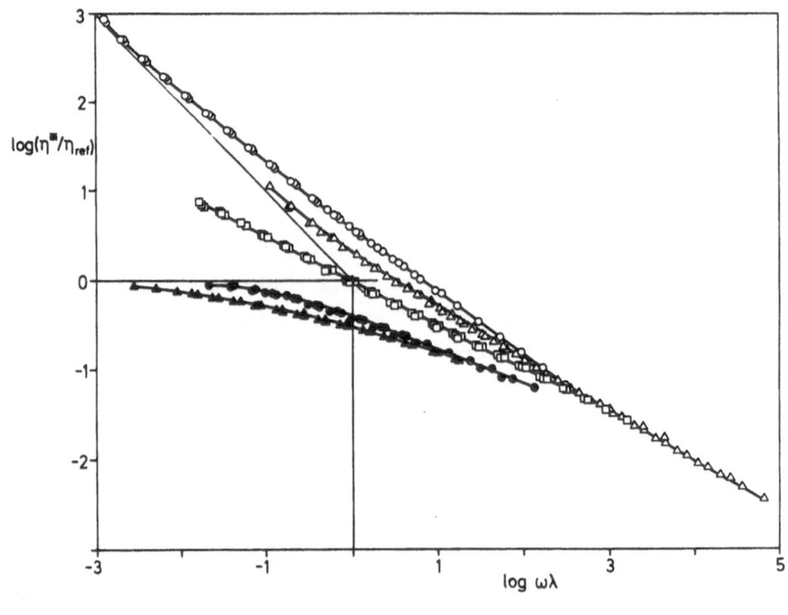

Fig. 4. Shifted viscosity curves of Figure 1 for gel with $n = 1/2$. Parameters for the shift are given in Table 1. The reference viscosity η_{ref} was chosen as η_o in the pre-gel and $g_\infty \lambda$ in the post-gel

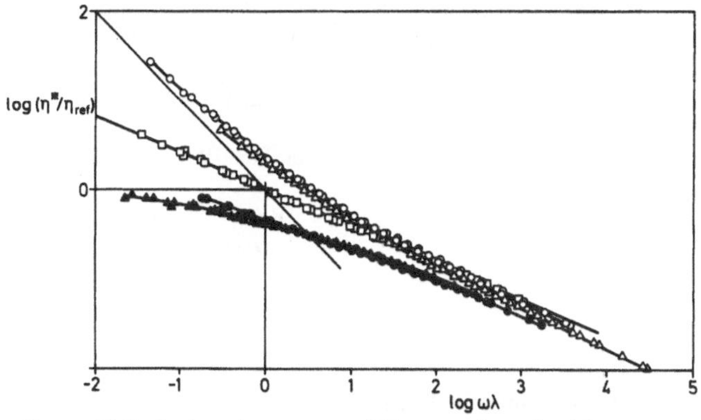

Fig. 5. Shifted viscosity curves of Figure 2 for gel with $n = 0.6$. Parameters for the shift are given in Table 1. The reference viscosity η_{ref} was chosen as n_o in the pre-gel and $g_\infty \lambda$ in the post-gel

The available experimental data is insufficient for determining the exponents. Adam et al. [1] found for polyurethane gels that z is about four times larger than s. Their values seem to depend on network structure. The gel was broken during the flow experiment and it is not yet clear whether it is possible to identify the power laws of broken gels with critical exponents.

It is interesting to consider a special case in which the characteristic relaxation time of the pre-gel and the post-gel follow the same power law (symmetry at $p = p_c$). This is the case if the exponents in the two equations above are set equal and a relation between s and z is calculated

$$n = \frac{z}{z + s}.$$

Such a relation has already been proposed from theoretical arguments by Clerc et al. [7, 8] for precolating systems. With $n = 1/2$, this relation would predict that the critical exponents are equal, $s = z$.

Conclusions

Gelation is a continuous process with a limiting relaxation behavior at the gel point (GP). The limiting behavior is valid for both the liquid and the solid in the approach of GP ($p \rightarrow p_c$). Singular behavior is only expected in the limit of zero frequency or zero time, i. e. outside the experimentally accessible range. The measured linear viscoelastic behavior of a network polymer at GP (*critical gel*) is simple, as expected for a critical state: stress relaxes in a power law, t^{-n}. This power law behavior of the critical gel is useful as a reference state for describing the evolution of rheology during gelation and, especially, during the transition through the gel point.

A characteristic relaxation time is defined by the intersecting asymptotes of the viscosity curves. This relaxation time increases in the approach of GP, diverges at GP, and then decreases again. There does not exist a characteristic time for the critical gel.

Detection of GP is made relatively easy by the fact that stress relaxation in a critical gel occurs in a power law. Equivalently, the dynamic moduli and the dynamic viscosities follow a power law in frequency which can be detected in a dynamic mechanical experiment.

It should be emphasized again that the entire study is restricted to polymers at temperatures much above the glass transition. The dynamic mechanic behavior would be considerably more complicated if vitrification interfered with the chemical gelation [20]. Physical entanglements are assumed to be of no importance in this study, since the molecular weights of the prepolymers were chosen as below the entangelment limit. Furthermore, the power law relaxation behavior has only been observed with endlinking systems. It might also occur with other systems and the range of validity will have to be studied in further experiments.

Acknowledgements

Financial support through NSF Grant MSM-860 1595 and helpful discussions with Dr. T. Vilgis at the MPI/Mainz are gratefully acknowledged.

References

1. Adam M, Delsanti M Durand D (1985) Macromolecules 18:2285
2. Apicella A, Masi P, Nicolais L (1984) Rheol Acta 23:291
3. ASTM D 1638-74
4. Bird RB, Armstrong R, Hassager O (1977) Dynamics of Polymeric Liquids, J Wiley, New York
5. Bistrup SA (1986) PhD Thesis, University Minnesota
6. Castro JM, Macosko CW, Perry SJ (1984) Polym Com 25:82
7. Clerc CP, Tremblay AMS, Albinet G, Mitescu CD (1984) J Phys Lett 45:L913
8. Clerc CP, Gireau G, Laugier JM, Luck JM (1985) J Phys A 18:2565
9. Chambon F, Winter HH (1985) Polym Bull 13:499
10. Chambon F, Petrovic ZS, MacKnight W, Winter HH (1986) Macromolecules 19:2146
11. Chambon F, Winter HH (1987) J Rheol 31:683
12. Chambon F (1986) Ph D Thesis, University Massachusetts
13. de Gennes PG (1979) Scaling Concepts in Polymer Physics, Cornell University Press, Ithaca, New York

14. Farris RJ, Lee C (1983) Polym Eng Sci 23:586
15. Ferry JD (1980) Viscoelastic Properties of Polymers, J Wiley, New York
16. Fisher A, Gottlieb M (1986) Proc of Networks 86, Elsinore Denmark, Aug 1986
17. Flory PJ (1941) J Am Chem Soc 63:3083, 3091, 3096
18. Flory PJ (1953) Principles of Polymer Chemistry, Cornell University Press, Ithaca, New York
19. Gordon M (1962) Proc R Soc, London Ser A 268:240
20. Harran D, Laudouard A (1986) J Appl Polym Sci
21. Lipshitz S, Macosko CW (1976) Polym Eng Sci 16:803
22. Macosko CW, Miller DR (1976) Macromolecules 9:199, 206; (1979) Polym Eng Sci 19:272
23. Macosko CW, Saam JC (1986) The Hydrosilation Cure of Polyisobutene, to be published
24. Muthukumar M (1985) J Chem Phys 83:3161
25. Muthukumar M, Winter HH (1986) Marcromolecules 19:1284
26. Stanley HE (1985) Introduction of Phase Transition and Critical Phenomena, 2nd Ed, Oxford University Press, New York
27. Stauffer D, Coniglio A, Adam M (1982) Adv Polym Sci 44:74
28. Stockmayer WH (1943) J Chem Phys 11:45; (1944) 12:125
29. Winter HH, Chambon F (1986) J Rheol 30:367
30. Winter HH, Morganelli P, Chambon F (1987) Macromolecules, submitted
31. Winter HH (1987) Polym Eng Sci, Dec 1987, in press

Received January 21, 1987;
accepted March 16, 1987

Author's permanent address:

H. H. Winter
Department of Chemical Engineering
University of Massachusetts
Amherst, MA 01003, U.S.A.

Discussion

D. RICHTER:

First question: I am surprised that you said that your fractal dimension is equal to 2 (for the stoichiometrically balanced network). This is just the fractal dimension of a Gaussian chain.

WINTER:

That is indeed an amazing result. The linear chain has the same fractal dimension as the stoichiometrically balanced network at the gel point. If the stoichiometry is not balanced, we can produce critical gels with any dimension between 2 and 3. We have not yet found critical gels with a dimension below 2.

D. RICHTER:

Second question: You introduced just the Hausdorff dimension or the geometrical dimension of the fractals into the equation for the dynamic viscosity. I think it should be the spectral dimension of the fractals.

WINTER:

One might expect that all three fractal dimensions should be considered: the Hausdorff dimension, the spectral dimension and the walk dimension. This situation is simplified by the Alexander-Orbach conjecture, which states that for percolating systems, such as critical gels, the spectral dimension is equal to 4/3, independent of the geometric dimension of the problem. We therefore cannot use the spectral dimension for describing the range of relaxation exponents. The remaining fractal dimensions are interconnected, so that we are only free to use one of them for describing our experiments. For convenience, we related the relaxation exponent n to the Hausdorff dimension. You could equally well rearrange the result and express it in terms of the walk dimension.

ILAVSKY:

You mentioned that the structure dimension is equal to 2 and so is, as you will recall, the dimension of the Gaussian chain.

WINTER:

This does not mean that the critical gel has the structure of a Gaussian chain. It forms a highly imperfect network out of interconnected Gaussian chains, swollen in other Gaussian chains, but it certainly is not a Gaussian chain itself. The value of 2 might be just a coincidence. We actually expect to find critical gels with fractal dimensions between 1 and 3. This would correspond to relaxation exponents between 1/3 and 3/5.

REINECKER:

Can you directly verify the relation between the mass and the radius of the gel so that you have an independent verification of this fractal structure?

WINTER:

Such data would be good to have. As a first step, we are planning to measure the linear size of the largest clusters by light scattering. Major difficulties arise from the lack of contrast in our samples and from impurities in the sample. We do not know how to measure the mass of these largest clusters. However, even if we knew the mass-volume relation of the largest cluster, our information would still be insufficient, since the selfsimilar behaviour of the critical gels seems to originate from the entire distribution of macromolecular clusters and not only from the largest one.

Orientation of macromolecules and elastic deformations in polymer melts. Influence of molecular structure on the reptation of molecules*)

H. M. Laun

Kunststofflaboratorium, BASF Aktiengesellschaft, Ludwigshafen am Rhein, F.R.G.

Abstract: Melt elasticity has a strong impact on both the processing behaviour of polymers and end use properties of fabricated parts. This paper compiles in the first part relations describing time-dependent and steady-state orientations (flow birefringence) as well as elastic strains for different deformation histories. Analytical expressions based on the relaxation time spectrum and rubber-like liquid theory are obtained for small shear or elongational strains. An at least approximate description is possible for high deformation rates and strains. Some fundamental theoretical predictions are compared with experimental results obtained on polystyrene and polyolefine melts of different molecular structure.

In addition, the second part presents fundamental experimental results on the influence of average molar mass and molar mass distribution on dynamic moduli, viscosity functions, normal stress coefficients, recoverable shear strains, extrudate swell, entrance pressure losses, and flow instabilities. The kind of side groups of the C-C-backbone as well as the type and number of chain branches in polyolefines affects the viscosity level and the temperature dependence (flow activation energy) of the rheological quantities. Long chain branching causes deviations from a thermorheologically simple behaviour. The experimental results are discussed in simple model images, taking into account the reptation motion of the molecules.

Key words: Polymer melt, elastic properties, molecular orientation, molar mass distribution, branching.

1. Introduction

Macromolecular fluids like the melts of thermoplastics exhibit noticeable non-Newtonian properties. For instance, a pronounced viscosity decrease with increasing shear rate is observed [1, 2]. This shear-thinning behaviour is in most cases of advantage for processing, since, e. g. in extrusion, the energy consumption increases less than proportional to the flow rate.

Processing of thermoplastics in the molten state causes deviations from the equilibrium conformation of the macromolecules, viz. the isotropic distribution of chain segment orientations at rest is converted to an oriented state. The degree of orientation is dependent on the rate, magnitude, and type of deformation. If the polarizabilities of segments parallel and perpendicular to the main chain are different, the second moment of the segment orientations can be detected by means of flow birefringence [3, 4].

Orientations can be frozen in by a rapid decrease of temperature during or immediately after moulding. They are of technical relevance since they influence the mechanical properties of the manufactured parts [5, 6].

Another important phenomenon of macromolecular fluids are elastic strains. Orientations that are still present in the material give rise to reverse deformations when the molecules tend to re-establish their equilibrium conformation after unloading [7]. The amount of recoverable strains depends on processing conditions and the molecular structure of the material. A well known elastic effect is the extrudate swell after

*) Extended version of a paper presented at: Makromolekulares Kolloquium, Freiburg, March 1985. Dedicated to Professor Dr. F. R. Schwarzl on the occasion of his 60th anniversary.

flow through a die. This change in cross-section has to be taken into account, e.g., in blow-moulding and extrusion of profiles.

Frozen-in orientations due to rapid cooling before complete recoil may cause changes of the shape of moulded parts when the mobility of the molecules is increased at elevated temperatures [8]. An example is the shrinkage of annealed blown films and spun fibres. Non-uniform frozen-in orientations in injection moulded parts give rise to distortions when the temperature is increased.

It is the aim of this paper to put together the basic relations that describe time-dependent orientations and elastic deformations of a macromolecular network and to compare these with experimental results obtained on polymer melts of different molecular structure. Simple model images taking into account the reptation motion will be used to visualize the microscopic origin of the observed structure-dependent melt elasticity.

2. Prediction of stresses and elastic strains

Here, the fundamental relations that connect stresses, recoverable strains, and orientations for an ideal rubber, a rubber-like liquid, and a nonlinear polymer melt are compared. Since thermoplastic polymer materials behave in a rubber-like way at high deformation rates, the starting point will be the ideal rubber.

a) Ideal rubber

The flexible macromolecules are attached to one another by permanent chemical crosslinks (Fig. 1a). A macroscopic strain gives rise to an affine deformation of the network. The stretching of strands causes a tensile stress which is proportional to the increase of the square of the crosslink distances [7,9].

Simple shear of magnitude γ (cf. Fig. 1b) from the equilibrium state (depicted by broken lines) into the cube shape (full lines) gives rise to shear stress components $\sigma_{12} = \sigma_{21}$ which are proportional to the total shear strain:

$$\sigma_{21} = G\,\gamma\,. \tag{1}$$

According to rubber theory the shear modulus G is given by

$$G = \frac{\varrho RT}{M_c} \tag{2}$$

M_c being the length of network strands, T the absolute temperature, R the gas constant, and ϱ the density [3]. In simple shear, normal stresses are also observed [7]. The directly measurable quantity is the primary normal stress difference N_1:

$$N_1 = \sigma_{11} - \sigma_{22} = G\,\gamma^2\,. \tag{3}$$

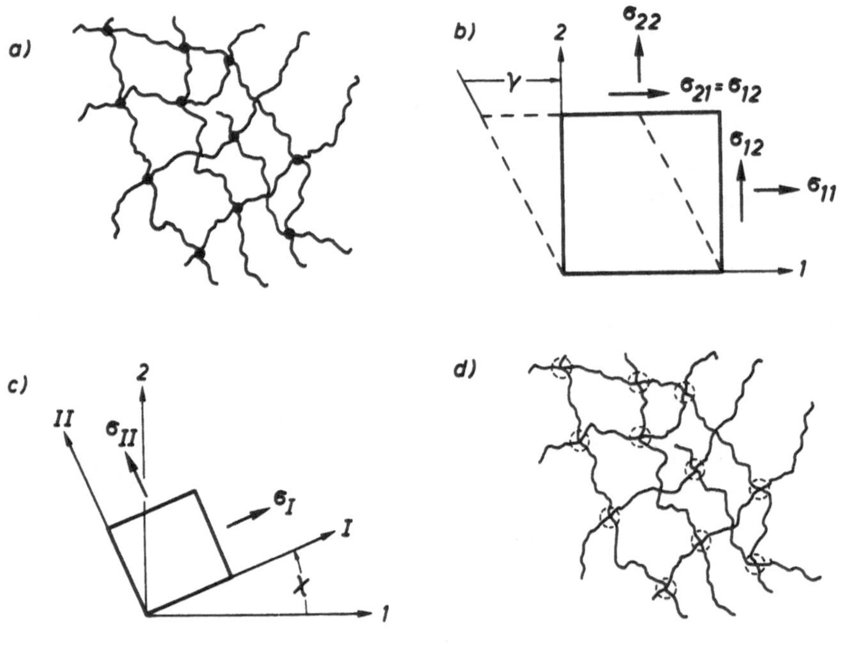

Fig. 1. a) Model of an ideal rubber with permanent chemical crosslinks, b) Components of stress for simple shear in 1- direction Equilibrium state (----), c) Directions of principal normal stresses, d) Model of a rubber-like liquid with temporary physical crosslinks

Since the deformation of an ideal rubber is fully reversible the recoverable shear strain γ_r is equal to the total strain γ. It can be expressed by means of Equations (1) and (2) by the ratio of primary normal stress difference and shear stress:

$$\gamma_r = \gamma = \frac{\sigma_{11} - \sigma_{22}}{\sigma_{21}} . \tag{4}$$

In an orthogonal co-ordinate system I, II rotated by an angle χ (cf. Fig. 1c) the state of stress is described merely by principal normal stresses σ_{I} and σ_{II}. The transformation equations read [4]

$$2\,\sigma_{21} = (\sigma_{\mathrm{I}} - \sigma_{\mathrm{II}}) \sin 2\chi \tag{5a}$$

and

$$\sigma_{11} - \sigma_{22} = (\sigma_{\mathrm{I}} - \sigma_{\mathrm{II}}) \cos 2\chi . \tag{5b}$$

It is interesting to note that the angle χ is given by the ratio of N_1 and σ_{21}, viz. the recoverable strain γ_r (cf. Eq. (4)):

$$2 \cot 2\chi = \frac{\sigma_{11} - \sigma_{22}}{\sigma_{21}} . \tag{6}$$

As shown by Kuhn and Grün [10] the stress optical law is valid for a Gaussian network. That means the differences Δn of the refractive indices in the directions of the principal normal stresses (e. g. directions I and II in Fig. 1c) are proportional to the differences $\Delta\sigma$ of the corresponding principal stresses:

$$\Delta n = C\,\Delta\sigma . \tag{7}$$

The stress-optical coefficient C in Equation (7)

$$C = \frac{2\pi}{45\,kT} \frac{(n^2 + 2)^2}{n} (\alpha_1 - \alpha_2) \tag{8}$$

is proportional to the difference of the polarizabilities α_1 and α_2 of a chain segment parallel and perpendicular to the direction of the chain [3]. Here, k is the Boltzmann constant and n the average refractive index. The chain segments are considered as rigid, such that contributions to the birefringence due to changing valence angles (stress birefringence) can be neglected and the orientation birefringence is dominating.

In simple shear, both the extinction angle χ and the birefringence $\Delta n = \sigma_{\mathrm{I}} - \sigma_{\mathrm{II}}$ can optically be measured [4]. The corresponding stresses σ_{21} and $\sigma_{11} - \sigma_{22}$ follow from Equations (5a), (5b).

The stress optical law is valid independent of the kind of deformation. For uniaxial elongation, which will also be considered in the following, rubber theory gives for the tensile stress $\sigma = \sigma_{11} - \sigma_{22}$:

$$\sigma = G\,(\lambda^2 - 1/\lambda) . \tag{9}$$

Here, one principal stress is parallel to the draw direction. The stretch ratio $\lambda = l/l_o$ is the quotient of stretched length l and initial length l_o.

b) Rubber-like liquid

Following Lodge [7] the concept of a 3-dimensional network can be applied to the melts of macromolecules if the assumption is made that the network points are now temporary physical junctions called entanglements (Fig. 1d) which are created and decay by statistical motions of the chains (cf. [11]). The time average of the entanglement number, however, remains constant.

Rubber-like liquid theory yields a viscosity which is independent of the shear rate (see below). Its great importance for real melts is due to the fact that it correctly describes the asymptotic behaviour of the stresses in the limiting case of small deformation rates or small deformations [4]. The stress optical law is fully valid for the rubber-like liquid. This means that flow-birefringence is able to detect the time-dependent state of stress.

For a step-shear experiment of magnitude γ_o at time $t = 0$ we now obtain

$$\sigma_{21}(t) = \mathring{G}\,(t)\,\gamma_o \tag{10}$$

and

$$N_1\,(t) = \mathring{G}\,(t)\,\gamma_o^2 . \tag{11}$$

As for an ideal rubber, the shear stress is still proportional to the shear γ_o and the primary normal stress difference proportional to the square of the shear step. However, the stresses are now time dependent, which is described by a time-dependent relaxation modulus $\mathring{G}\,(t)$. The modulus decays with increasing time since the deformed network is gradually replaced by a relaxed network.

Using a discrete relaxation time spectrum with relaxation times τ_i and relaxation strengths g_i the relaxation modulus may be expressed by

$$\mathring{G}(t) = \sum g_i \exp(-t/\tau_i) . \qquad (12)$$

The relaxation time spectrum of a given material has in general to be determined experimentally [12, 13]. In some cases semimolecular theories also enable us to calculate spectra from the molecular mass distribution (cf. [14–18]).

The generalization of Equations (10), (11) for arbitrary shear histories yields

$$\sigma_{21}(t) = \int_{-\infty}^{t} \frac{\partial \mathring{G}(t-t')}{\partial t'} [\gamma(t) - \gamma(t')]\, dt' \qquad (13)$$

and

$$N_1(t) = \int_{-\infty}^{t} \frac{\partial \mathring{G}(t-t')}{\partial t'} [\gamma(t) - \gamma(t')]^2\, dt' . \qquad (14)$$

Here $\gamma(t) - \gamma(t')$ is the relative shear strain between the current time t and the past time t'. These relative strains, weighted by the time derivative of the relaxation modulus, the so-called memory function [7, 11], are summed up over the total past time.

Of special interest in the following are the differences in magnitudes of the recoverable strains of an ideal rubber and of a rubber-like liquid. It can be shown that the total recoil in simple shear under lateral shape constraint (constraint recovery) after unloading is given by [13, 15, 19]:

$$\gamma_r(t) = \gamma(t) - \frac{1}{\eta_o} \int_{-\infty}^{t} \sigma_{21}(t')\, dt' \qquad (15)$$

η_o being the constant viscosity of the rubber-like liquid:

$$\eta_o = \int_0^{\infty} G(t')\, dt' = \sum g_i \tau_i . \qquad (16)$$

The shear stress in Equation (15) has to be calculated according to Equation (13).

For a step shear deformation of magnitude γ_o at time $t = 0$ and unloading of the sample at a time $t > 0$ the total recoverable shear strain γ_r follows, from Equations (10), (15), as

$$\gamma_r(t) = \gamma_o \left[1 - \frac{1}{\eta_o} \int_0^t \mathring{G}(t')\, dt'\right]$$

$$= \gamma_o \frac{\sum g_i \tau_i \exp(-t/\tau_i)}{\sum g_i \tau_i} . \qquad (17)$$

It is easy to show that immediately after the step, the shear deformation is completely reversible ($\gamma_r = \gamma_o$). With increasing time of shape constraint, however, the recoverable part of the shear strain decays to zero. Obviously, directly after the step shear all temporay entanglements from the initial stress-free state are still present such that the memory to the unsheared state is perfect. With increasing time, more and more entanglements of stretched chains are replaced by those of relaxed chains, until the memory of the initial state has completely vanished. It is also interesting to note that the decay of the recoverable shear strain with increasing time of relaxation (shape constraint) occurs more slowly than that of the shear stress (Eqs. (10), (12)) due to the factors τ_i in the numerator of Equation (17).

A comparison of the time-dependent shear stress and primary normal stress coefficient during stress relaxation shows that, similar to that for the the ideal rubber, the validity of the Lodge-Meissner-relation

$$\frac{N_1(t)}{\sigma_{21}(t)} = \gamma_o \qquad (18)$$

is found. However, as given by Equation (17) the stress ratio N_1/σ_{21} is only equal to the recoverable shear strain for $t = 0$. It is higher than γ_r for $t > 0$. It follows that for a rubber-like liquid, the ratio of primary normal stress difference and shear stress is not a measure of elastic deformations!

For a step shear rate test, $\dot{\gamma}_o$ being the constant shear rate for $t > 0$, the relevant time-dependent stresses calculated from Equations (13), (14) follow as

$$\sigma_{21}(t) = \dot{\gamma}_o \sum g_i \tau_i (1 - \exp(-t/\tau_i)) \qquad (19)$$

and

$$N_1(t) = \dot{\gamma}_o^2\, 2\sum g_i \tau_i^2 (1 - (1 + t/\tau_i) \exp(-t/\tau_i)). \qquad (20)$$

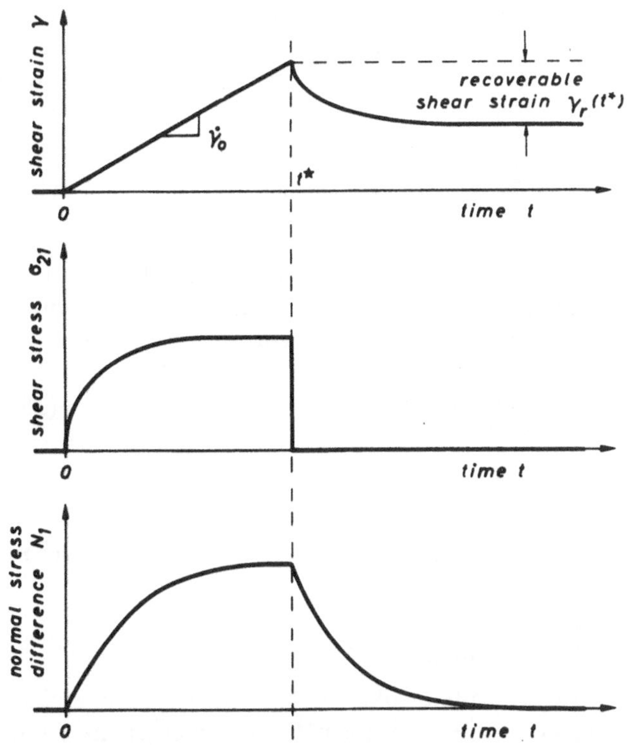

Fig. 2. Time dependence of the shear strain, shear stress, and primary normal stress difference in a stressing test at constant shear rate $\dot{\gamma}_o$ (until a steady state is reached) and subsequent recoil under lateral shape constraint due to unloading ($\sigma_{21} = 0$) at time t^* for a rubber-like liquid (schematic)

Equations (15), (19) yield a total recoverable strain γ_r for unloading the sheared sample at a time t of:

$$\gamma_r(t) = \dot{\gamma}_o \frac{\sum g_i \tau_i^2 (1 - \exp(-t/\tau_i))}{\sum g_i \tau_i} . \tag{21}$$

The time dependence of the quantities σ_{21}, N_1, and γ_r during a stressing test at shear rate $\dot{\gamma}_o$ and subsequent unloading ($\sigma_{21} = 0$) at time $t = t^*$ is schematically depicted in Figure 2. Due to the lateral shape constraint, N_1 gradually decays to zero for $t > t^*$. Equation (21) gives the total recovery $\gamma_r(t^*)$ if t is set equal to t^*, but not the time dependence of the recoil.

Two limiting cases are contained in Equation (21). For a short duration of the stressing test, such that t is smaller than the shortest relaxation time of the spectrum in Equation (12), an expansion in series of the exponential functions gives

$$\gamma_r(t) = \gamma(t) = \frac{N_1(t)}{\sigma_{21}(t)} \text{ for } t < \tau_i$$

as for an ideal rubber with complete reversibility of the strain.

For a long duration of the stressing test, such that the time of shear is bigger than the longest relaxation time of the spectrum, a steady-state shear flow is achieved. Viz, the stress components and the recoverable shear strain attain time-independent values. These steady-state properties, characterized by the subscript s, are

$$\sigma_{21,s} = \dot{\gamma}_o \sum g_i \tau_i = \dot{\gamma}_o \eta_o , \tag{22}$$

$$N_{1,s} = \dot{\gamma}_o^2 \, 2 \sum g_i \tau_i^2 = \dot{\gamma}_o^2 \theta_o \tag{23}$$

and

$$\gamma_{r,s} = \dot{\gamma}_o \frac{\sum g_i \tau_i^2}{\sum g_i \tau_i} = \frac{\dot{\gamma}_o \theta_o}{2 \eta_o} . \tag{24}$$

The steady-state primary normal stress difference related to the square of the shear rate is the primary normal stress coefficient θ_o which is independent of the shear rate for a rubber-like liquid. Besides θ_o and the constant viscosity η_o the so-called steady-state compliance J_e^o is commonly used as a third characteristic rheological quantity:

$$J_e^o = \frac{\gamma_{r,s}}{\sigma_{21,s}} = \frac{\sum g_i \tau_i^2}{(\sum g_i \tau_i)^2} = \frac{\theta_o}{2 \eta_o^2} . \tag{25}$$

J_e^o represents the recoverable shear strain related to the shear stress in the steady-state of the preceeding stressing test and can be expressed by θ_o and η_o.

If one compares the total recoil from the steady-state of shear flow with the ratio of primary normal stress difference and shear stress, the famous relation [7]

$$\gamma_{r,s} = \frac{N_{1,s}}{2 \sigma_{21,s}} \tag{26}$$

is obtained. It indicates that the steady-state recoverable shear strain of a rubber-like liquid is only half of the recoverable strain of an ideal rubber (Eq. (4)) at the same stress ratio! In fact, the reduced recoil of a temporary network compared to the ideal rubber is not surprising. The recoil takes some time. During the recoil process part of the stressed network is gradually replaced by relaxed chains which impede the still stressed part of the network in the reverse deformation.

A third kind of deformation to be considered here is oscillatory shear at small shear amplitude $\hat{\gamma}$ and low angular frequency ω. Written in complex notation as

$$\gamma(t) = \hat{\gamma} \exp(i\,\omega\,t) \, . \tag{27}$$

the amplitude and phase shift of the resulting shear stress

$$\sigma_{21}\,(t) = \hat{\gamma}\, G^* \exp(i\,\omega\,t) \tag{28}$$

is characterized by the complex dynamic modulus G^*

$$G^* = G' + iG'' \, , \tag{29}$$

G' being the storage modulus and G'' the loss modulus [14].

We can now make use of the fact that for any shear deformation, the total shear strain γ can be decomposed into a recoverable (elastic) shear strain γ_r and a viscous (irreversible) shear strain γ_v:

$$\gamma = \gamma_r + \gamma_v \, . \tag{30}$$

The decomposition can experimentally be verified (cf. e.g. [20, 21]. In the time derivative of Equation (30) the viscous and elastic fractions of the shear rate $\dot{\gamma}_v$ and $\dot{\gamma}_r$, respectively, can for slow deformations, be expressed by the steady-state values of viscosity (Eq. (22)) and compliance (Eq. (25)). This yields the differential equation

$$\dot{\gamma} = \dot{\gamma}_v + \dot{\gamma}_r = \sigma/\eta_o + \dot{\sigma} J_e^o \, . \tag{31}$$

By using Equation (27) for the shear strain and Equation (28) for the shear stress and separation of real and imaginary parts, the components of the dynamic modulus G^* for small angular frequencies follow as

$$G'(\omega) = \frac{\theta_o}{2} \frac{\omega^2}{1 + \omega^2 \tau_R^2} \, , \tag{32a}$$

and

$$G''\,(\omega) = \eta_o \frac{\omega}{1 + \omega^2 \tau_R^2} \, . \tag{32b}$$

Here, the quantity τ_R is a characteristic retardation time

$$\tau_R = \eta_o J_e^o = \theta_o/(2\eta_o) \, , \tag{33}$$

the meaning of which will become more clear in the following section. Equations (32a), (32b) are valid for $\omega < 1/\tau_R$. For the limiting case of $\omega \ll 1/\tau_R$ the well known relations

$$G' = \frac{\theta_o}{2} \omega^2 \text{ and } G'' = \eta_o\, \omega \tag{34a, b}$$

are obtained. They enable us to directly determine the basic rheological quantities η_o and J_e^o from the frequency dependencies of the loss and storage moduli:

$$\eta_o = \lim_{\omega \to o} G''\,(\omega)/\omega \tag{35a}$$

and

$$J_e^o = \lim_{\omega \to o} G'\,(\omega)/G''\,(\omega)^2 \, . \tag{35b}$$

c) Shear-thinning polymer melts

Lodge's rubber-like liquid theory yields steady-state viscosities and normal stress coefficients which do not depend on the shear rate (compare Eqs. (22), (23)). The reason for this discrepancy compared to real melts is the assumption of a constant time average entanglement density. Only if one allows the network points to decrease in number with increasing deformation or deformation rate, then a shear-thinning behaviour will be obtained.

The most simple approach is the extension of the relaxation modulus as a product of the already introduced time-dependent modulus $\mathring{G}\,(t)$ and of a solely deformation-dependent function $h(\gamma)$, called damping or attenuation function [11, 22–24]. For a step shear deformation, Equations (10), (11) now have to be replaced by

$$\sigma_{21}(t) = G(t, \gamma_o)\, \gamma_o \tag{36}$$

and

$$N_1(t) = G(t, \gamma_o)\, \gamma_o^2 \, , \tag{37}$$

where the time and strain dependent relaxation modulus is given by

$$G\,(t, \gamma_o) = \mathring{G}\,(t)\; h\,(\gamma_o) \, . \tag{38}$$

The justification for the separability assumption in Equation (38) has been experimentally verified (e.g.

[12, 25] for a variety of polymer melts, by direct measurements of the relaxing shear stress and primary normal stress difference. A simple one-parameter approximation of the attenuation function is a single exponential

$$h(\gamma) = \exp(-n\gamma), \tag{39}$$

the damping constant n having values in the range of $n = 0.18$ to $n = 0.33$. As shown elsewhere [12] the factorized modulus and Equation (39) yield analytical expressions for the steady-state viscosity η_s

$$\eta_s(\dot{\gamma}) = \sum \frac{g_i\,\tau_i}{(1 + n\,\dot{\gamma}\,\tau_i)^2} \tag{40}$$

as well as for the steady-state primary normal stress coefficient θ_s

$$\theta_s(\dot{\gamma}) = 2\sum \frac{g_i\,\tau_i^2}{(1 + n\,\dot{\gamma}\,\tau_i)^3}\,. \tag{41}$$

The constant values of η_s and θ_s in the limit of small shear rates ($\dot{\gamma} \ll 1/n\tau_i$) are consistent with the viscosity and normal stress coefficient of a rubber-like liquid.

Only recently has it been shown [25] that the steady-state recoverable shear strain $\gamma_{r,s}$ can be analytically expressed by

$$\gamma_{r,s}(\dot{\gamma}) = \dot{\gamma}\,\frac{\sum g_i\,\tau_i^2/(1 + n\,\dot{\gamma}\,\tau_i)^2}{\sum g_i\,\tau_i/(1 + n\,\dot{\gamma}\,\tau_i)^2}$$

$$= \frac{N_{1,s}(\dot{\gamma})}{2\,\sigma_{21,s}(\dot{\gamma})} + \frac{n\,\dot{\gamma}^3}{\sigma_{21,s}(\dot{\gamma})}\sum \frac{g_i\,\tau_i^3}{(1 + n\,\dot{\gamma}\,\tau_i)^3}\,. \tag{42}$$

Equation (42) predicts for the shear-thinning range that the recoverable shear strain is bigger than the value of primary normal stress difference over twice the shear stress [25].

So far, only shear properties have been considered. Several consequences of a factorized strain and time dependent modulus on the nonlinear behaviour of the elongational viscosity μ and the recoverable elongational strain ε_r have already been discussed in [21]. Here we only have to keep in mind that the Hencky measure of strain ε, commonly used in rheology, is defined by

$$\varepsilon = \int_{l_0}^{l} \frac{dl}{l} = \ln \lambda\,. \tag{43}$$

From Equation (9) follows the approximation for small total strains

$$\sigma = G\,[\exp(2\varepsilon) - \exp(-\varepsilon)] \approx 3\,G\,\varepsilon\,. \tag{44}$$

This linear viscoelastic limiting case, if generalized for a nonpermanent network, yields analogous relations to Equations (19), (21). For a given deformation history $\varepsilon(t) = \gamma(t)$ the tensile stress is just three times the shear stress. As a result, the time-dependent recoverable strain for a constant (Hencky) strain rate $\dot{\varepsilon}_o$ follows as [13]:

$$\varepsilon_r(t) = \dot{\varepsilon}_o\,\frac{\sum g_i\,\tau_i^2\,[1 - \exp(-t/\tau_i)]}{\sum g_i\,\tau_i}\,. \tag{45}$$

Figure 3 shows a comparison of measured recoverable elongational strains ε_r (solid symbols) and recoverable shear strains γ_r (open symbols) of an LDPE melt [21, 25]. Details of the experiments are to be found in [25–27]. The sample has been deformed at constant deformation rates up to different total deformations and was subsequently unloaded. The recoverable strains measured after total recoil are related to the preceeding deformation rates and are plotted versus the time of deformation at a given constant rate.

We find that the experimental results in shear and elongation coincide on a single curve for the beginning of the stressing tests, independent of the deformation rate. The solid line in Figure 3 is the reduced recoverable shear strain of a rubber-like liquid (Eq. (21)) and linear viscoelastic fluid calculated by means of a discrete relaxation time spectrum given in [12]. This curve is identical to the reduced recoverable elongational strain of a linear viscoelastic fluid (Eq. (45)). However, if the original nonlinear strain function [exp (2ε) − exp (−ε)] of a rubber-like liquid is used, instead of the approximation in Equation (44), this would give partially higher values of ε_r than predicted by Equation (45). In that case an $\dot{\varepsilon}_o$ – invariant master curve cannot be obtained.

It is interesting to note that the solid line correctly describes the pronounced elastic behaviour at relatively high deformation rates over a wide time range. Deviations from the asymptote only become significant at large total deformations. This is the range where a decrease of the entanglement density with increasing strain becomes relevant. For comparison, the behaviour of an ideal rubber (fully reversible deformation) is depicted in Figure 3 by the broken straight line. Additional examples for the predictability of reco-

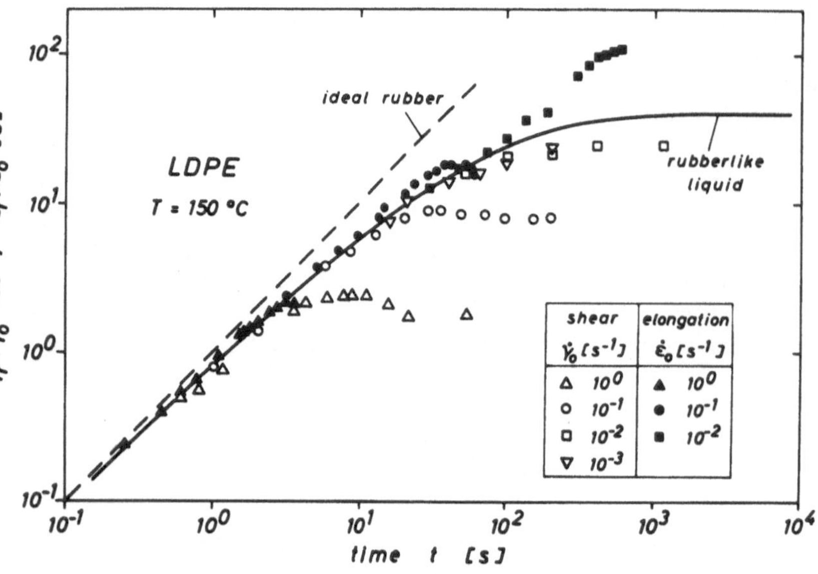

Fig. 3. Recoverable shear strains (open symbols) and recoverable elongational strains (solid symbols) related to the constant shear rates $\dot{\gamma}_o$ and elongational strain rates $\dot{\varepsilon}_o$, respectively, versus time. The full line represents the reduced recoverable shear strain of a rubber-like liquid (Eq. (21)), the broken line denotes the fully reversible deformation of an ideal rubber

verable strains of other polymers may be found in [13, 25].

d) Validity of the stress optical law

The validity of the stress optical law for shear-thinning melts, too, has been intensively investigated by Janeschitz-Kriegl [3, 4] and coworkers on different thermoplastics. As an example, Figure 4 shows results obtained from a polystyrene melt [28].

The polarizability of a polystyrene chain segment parallel to the chain is lower than that in the transverse direction. That is why the stress optical coefficient C of polystyrene melts is negative. By independent measurements of the birefringence (Δn and χ) as well as of the stress components (σ_{21} and N_1) in the steady-state of shear flow, it was found that the stress optical coefficient is independent of the shear rate (upper part of Fig. 4). The investigated $\dot{\gamma}_o$-range also includes pronounced shear-thinning.

In addition, by evaluating the time-dependence of the optical and mechanical quantities for a step shear rate test with a constant shear rate of $\dot{\gamma}_o = 0.015\ s^{-1}$, it was also found that C does not depend on the time of shear in the investigated range (lower part of Fig. 4).

The time and deformation rate invariance of the stress optical coefficient of real polymer melts means that the second moment of segment orientations detectable by flow birefringence, only reflects the state of stress in the material. Therefore, mechanically measured principle normal stress differences can

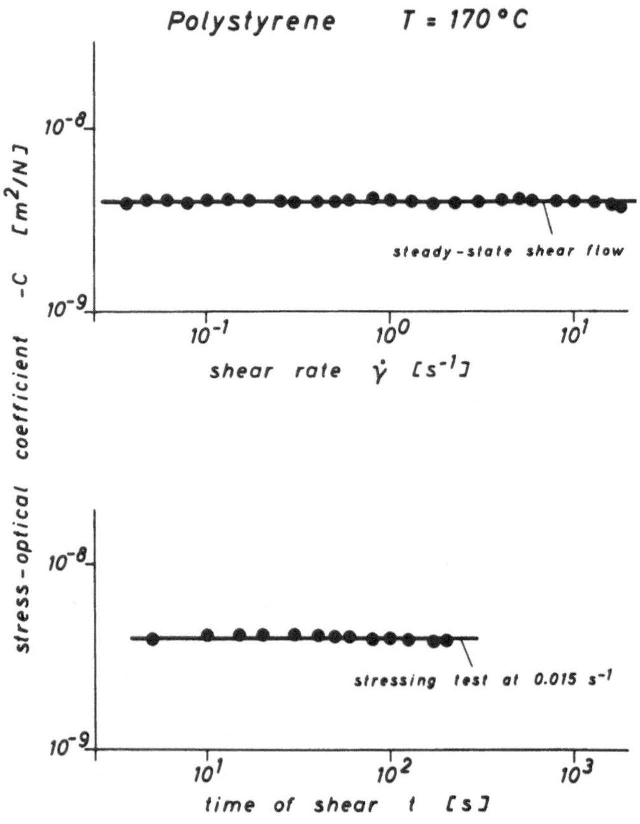

Fig. 4. Experimental examination of the validity of the stress optical law, after [28]. The top diagram shows the stress optical coefficient determined on a polystyrene melt in steady-shear flows at different constant shear rates. The lower diagram represents the time dependence of the stress optical coefficient of the same melt in a stressing test at shear rate of $\dot{\gamma}_o = 0.015\ s^{-1}$

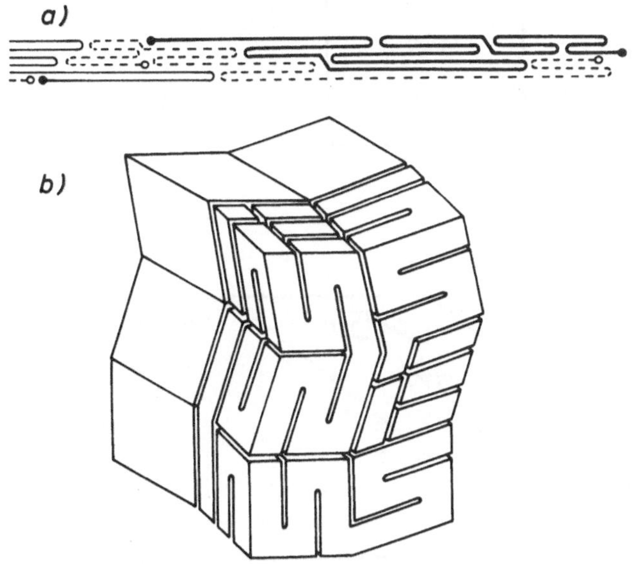

Fig. 5. a) Bundle of molecules with tight folds of single chains; b) Arrangement of meander-like folded bundles in a fluctuating cube topology (after [30])

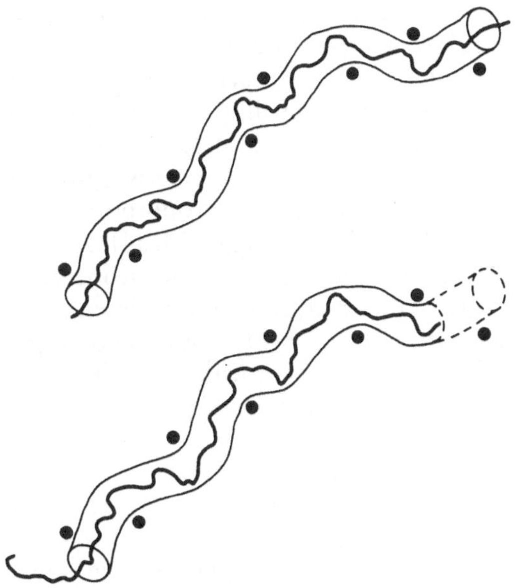

Fig. 6. Reptation motioni of a molecule out of its tube, limited by adjacent molecules (after [33, 34])

also be regarded as a measure of the molecule orientations.

We have the somewhat surprising result that flow birefringence and the state of stress, both dependent on the same moment of segment orientations, behave differently in time compared to the recoverable strains, although melt elasticity is also due to orientations of the macromolecules. Equations (5a) and (15) formally interrelate the shear stress history, flow birefringence, and recoverable shear strain.

3. Model consiterations

a) Reptation motion

So far the nature of the temporary network points of a rubber-like and shear-thinning melt has not yet been defined. In the literature quite different interpretations may be found. The original idea of the random coil model [29] are entanglements caused by a crossover of chain segments that move into opposite directions. The image of an isolated network point ignores the presence of adjacent molecules and does not give an explanation how the chains can disentangle.

With respect to a detailed description of the mutual arrangement of molecules, the meander model of Pechhold [30] is most advanced. Here the molecules of a melt are combined in bundles which are arranged in fluctuating cubes (Fig. 5). Tight folds within the bundles cause a random coil-like neutron scattering behaviour [31]. In this model the entanglements correspond to the locations where a bundle passes from one to an adjacent cube [32].

It is far outside the topic of the present paper to compare advantages and limitations of coil and bundle models. Two comments, however, should be made here. Firstly, although both approaches represent extreme cases, mutually penetrating random coils and a priori parallelized but fluctuation chain segments, they give comparable predictions for a variety of rheological phenomena observed on polymer melts. Secondly, the molecular process essential for viscous flow is a reptation motion of the molecules in both models, at least if the ideas of de Gennes [33] and Doi-Edwards [34] are accepted for random coils.

Due to the presence of neighbouring molecules (depicted by dots in Fig. 6) the lateral mobility of a molecule is reduced. It can only move freely within a tube. If the melt is sheared or stretched, this will cause an affine deformation of the tube and, within the tube, of the molecule. A thus obtained shape does not, in general, represent an equilibrium conformation. Therefore, within a short time, the chain segments that have been partially stretched or compressed, will be redistributed uniformly along the tube. The time scale for this process is normally shorter than the time resolution of common rheological measurements above the glass transition temperature [4].

For viscous flow, another much slower process is essential: by diffusion in the chain direction (reptation) out of the deformed tube, the molecule may find a new equilibrium conformation. The characteristic time for the diffusion of the centre of mass is proportional to the third power of the molecule length [33, 34]. A more detailed evaluation, taking into account the increased lateral mobility of the chain ends and the fluctuations of the tube length [35–37] yields a noticeably higher exponent. This is in agreement with the well known molar mass dependence of the zero shear rate viscosity η_o [14],

$$\eta_o = K\, M_w^{3.4}\,, \tag{46}$$

since η_o is proportional to the reptation time. Here, M_w is the weight average of the molecular mass distribution.

When a polymer melt is deformed, the fast deformation of the tube, and with it the orientation of the molecule, competes with the reptation out of the deformed tube. During flow an average degree of orientation will be the result. The orientations will be more pronounced the faster the rate of strain. Without reptation a given deformation will remain completely reversible. This explains, e. g., why melts with very long molecules – so-called thermoplastic elastomers [38] – the reptation time of which is in the range of several days, behave like a crosslinked rubber although they are, in fact, fluids. In the case of most thermoplastics, however, the reptation times at commonly applied processing temperatures are considerably shorter. Therefore, a noticeable rubber-like behaviour is observed at short deformation times and high deformation rates only.

Figure 7 shows a rough model to visualize the competition of orientation and reptation. Here, viscous flow is represented by a mutual slip of two adjacent melt elements. An elastic deformation is depicted by an affine deformation of the elements. A molecule that crosses the slip plane prevents slip and is deformed in its tube. Only after a reptation of that very molecule out of one element is slip possible and the elastic deformation is now replaced by a viscous (irreversible) displacement. After that procedure, no elastic deformation is present in the volume elements. Of course, the real behaviour can only be approximated by a large number of such simplified processes occuring in parallel and in series.

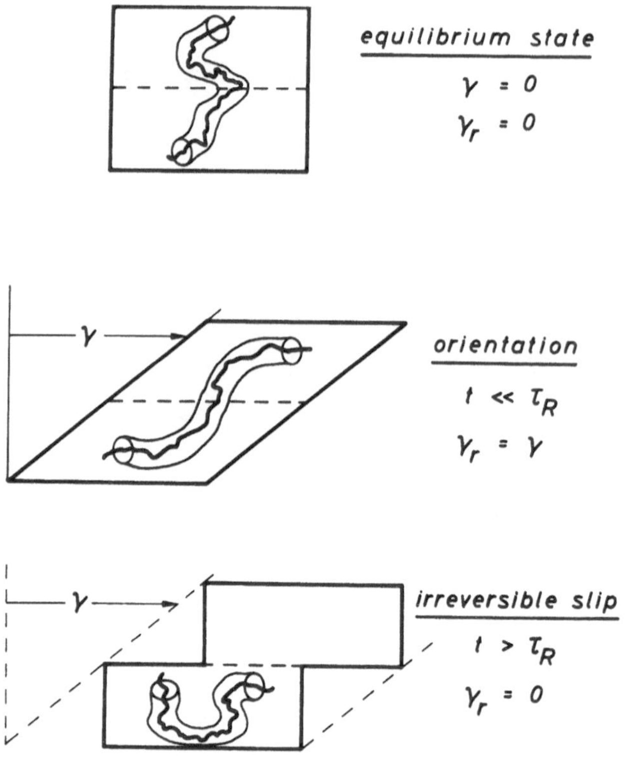

Fig. 7. Rough model to visualize elastic strains by a deformation of the tube and of irreversible flow by a slip of volume elements

As an example, Figure 8 shows recoverable elongational strains of an LDPE melt [21] versus total strain ε for different Hencky strain rates $\dot{\varepsilon}_o$. The experimental details are described in [26].

At the lowest stretch rate ($10^{-3}\ s^{-1}$) the deformation of the melt is slow enough to give the molecules enough time to reduce their orientation by reptation processes. As a result, the recoverable strain ε_r remains very small. With increasing strain rate, however, the reversible part of the elongation becomes more and more dominant because now the reptation is too slow to enable a full decay of the repeated tube orientations. For $\dot{\varepsilon}_o = 1\ s^{-1}$ and $\varepsilon = 2$ more than 80 % of the total strain is reversible. The recoverable part is still higher at the start of stretching. For comparison, a fully reversible deformation is depicted by the broken line.

It is also evident from Figure 8 that for high elongation plateau values of ε_r are obtained. Here an equilibrium (steady-state) between build up and decay of orientations is reached. The maximum value of ε_r = 2.2 measured corresponds to a shrinkage of the stretched and subsequently unloaded melt by a factor of 9!

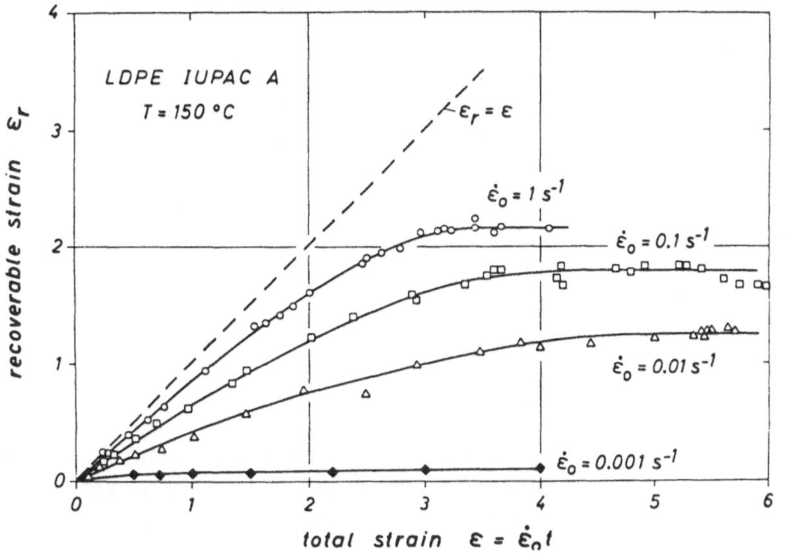

Fig. 8. Recoverable elongational strains of an LDPE melt versus total strain for different constant Hencky strain rates after [21]. Limiting case of a fully reversible deformation (- - - -)

The value of the characteristic retardation time τ_R for the decay of elastic deformations can be calculated from basic rheological quantities introduced above. We assume that a temporary network is repeatedly step-sheared by $\Delta\gamma$ and subsequently kept under shape constraint for a time period Δt (Fig. 9). This corresponds to an average shear rate of $\dot{\gamma}_o = \Delta\gamma/\Delta t$. If the step height $\Delta\gamma$ is chosen equal to the steady-state recoverable shear strain $\gamma_{r,s}$ at the shear rate $\dot{\gamma}_o$ and the time of rest equal to the characteristic time τ_R, such that the elastic deformations have almost completely decayed after the rest period, we obtain

$$\Delta t = \tau_R = \frac{\Delta\gamma}{\dot{\gamma}_o} = \frac{\gamma_{r,s}(\dot{\gamma}_o)}{\dot{\gamma}_o} . \tag{47}$$

It follwos with Equation (24):

$$\tau_R = \eta_o J_e^o = \frac{\sum g_i \tau_i^2}{\sum g_i \tau_i} . \tag{48}$$

The same quantity has already been obtained for oscillatory shear (Eq. (33)). The characteristic retardation time is proportional to the zero shear rate viscosity and to the steady-state compliance. According to Equation (47) a recoverable strain greater than 1 will be obtained in a constant shear rate test if the reciprocal of the deformation rate is smaller than τ_R. This is true for shear and elongation.

Of considerable practical interest is the influence of molecular properties, such as the weight average molar mass M_w, the type and breadth of the molar mass distribution MMD, and the number and type of branches, schematically depicted in Figure 10, on the elastic properties of polymer melts. Based on the reptation model several useful conclusions can be drawn.

b) Influence of average molar mass and molar mass distribution

For melts having a uniform length of the molecules the tube model yields a steady-state compliance J_e^o which is independent of the molar mass M as long as M

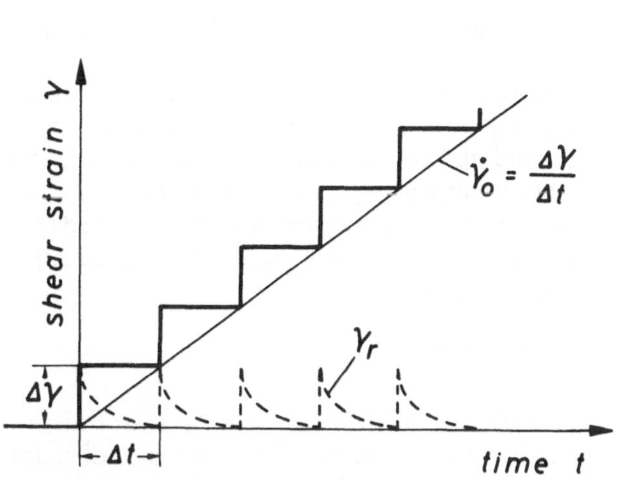

Fig. 9. Schematic diagram to explain the interrelation between the characteristic retardation time τ_R, shear rate $\dot{\gamma}_o$, and recoverable shear strain $\gamma_{r,s}$

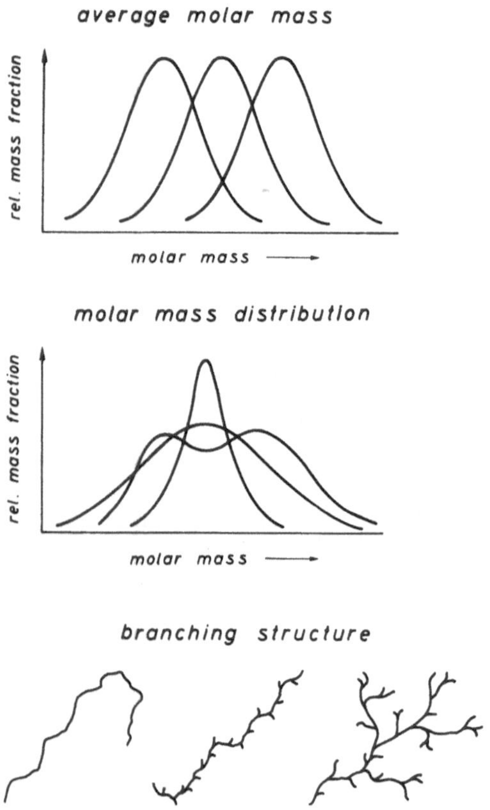

Fig. 10. Important molecular properties influencing the viscoelasticity of polymer melts

exceeds a critical value M_c' [39]. From Equations (46), (48) follows an increase of the characteristic retardation time τ_R according to

$$\tau_R = K_R\, M^{3.4}\,. \tag{49}$$

For comparable molar mass distributions versus log M (e.g. log-normal distributions at constant ratio of weight and number average molar mass M_w/M_n) we expect the compliance J_e^o to be independent of M_w. In fact, such a behaviour is experimentally observed [40–43] for $M_w > M_c'$. For that special case only Equation (49) can be generalized to

$$\tau_R = K_R'\, M_w^{3.4}\,.$$

A variation in the shape of the molar mass distribution, for instance, by adding a few very long molecules, may give rise to a dramatic change in the elastic properties of melts. Qualitatively this can be demonstrated for the blend of two narrow distributed components if one assumes that, as a first approximation, the relaxation processes are simply superimposed.

Let $\tau_1 = \tau$ and $g_i = g$ be the single relaxation time and relaxation strength of the low molar mass component 1. Similarly the high molar mass component 2 is characterized by $\tau_2 = x\tau \gg \tau_1$ and $g_2 = g$. Here $g_2 = g_1$ is a necessary condition to achieve the same compliance J_e^o in Equation (25) for both components. The factor x follows from the ratio of the molar masses or the viscosities:

$$x = \frac{\eta_{o2}}{\eta_{o1}} = \left(\frac{M_2}{M_1}\right)^{3.4}\,. \tag{50}$$

The addition of a small volume fraction $c \ll 1$ of component 2 to component 1 yields, for the zero shear rate viscosity η_{ob} of the blend,

$$\eta_{ob} \approx g\tau + c\,gx\tau = (1 + cx)\,\eta_{o1}\,. \tag{51}$$

From Equation (25) the compliance of the blend J_{eb}^o follows as

$$J_{ob}^o \approx \frac{g\tau^2 + cg(x\tau)^2}{(g\tau + cgx\tau)^2} = \frac{1 + cx^2}{(1 + cx)^2}\, J_e^o\,. \tag{52}$$

For simplicity, we shall restrict ourselves to $cx < 1$ and $cx^2 \gg 1$ (e.g., $M_2/M_1 = 5$, $\eta_{o2}/\eta_{o1} = 238$ and $c = 0.1\,\%$). Under that condition the blend viscosity is less than twice the viscosity of component 1. However, the blend compliance attains values which are considerably higher than the compliance of each component. It follows that the characteristic retardation time of the blend τ_{Rb} is considerably increased, compared to the value of component 1, but still remains lower than the characteristic time of the pure high molar mass component. (The numeric example gives: $\eta_{ob} = 1.24\,\eta_{o1}$, $J_{eb}^o = 37.6\,J_e^o$, and $\tau_{Rb} = 46.6\,\tau_{R1} = 0.196\,\tau_{R2}$).

Figure 11 represents an attempt to visualize the reason for the increased compliance of a binary blend. We can consider one long molecule as surrounded by significantly shorter ones. The slip of volume elements is only marginally impeded, since the shorter molecules can make way for the long one by means of their fast reptation motion. In that case long molecule will have tube walls which are partially permeable for lateral movements! Therefore, the viscosity increase caused by the long molecule remains small. For a more quantitative discussion of the so-called tube renewall process, see [65, 66].

In contrast to the low molar mass surrounding, the long chain is significantly oriented since its decay of

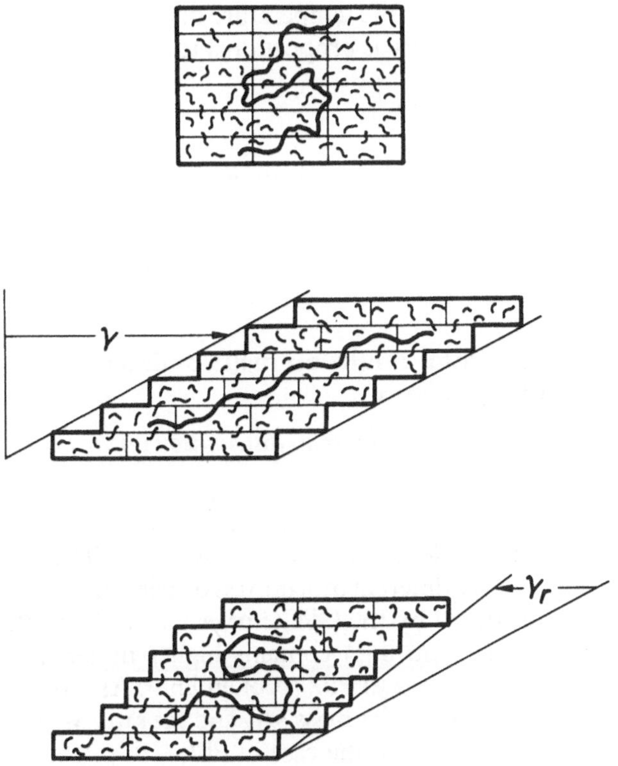

Fig. 11. Increase of melt elasticity due to low concentrations of a high molar mass component in a matrix of significantly shorter molecules

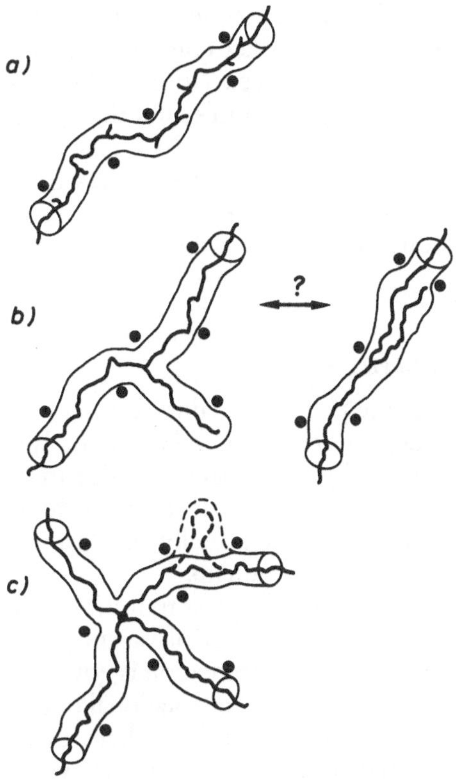

Fig. 12. Schematic representation of lateral constraints of branched molecules (only part of the molecules shown) (a) Tube model for short chain branched molecules; (b) Ramification of the tube due to long chain branches and alternative arrangement of the side chain parallel to the backbone in one tube; (c) Modified reptation of the long arms in star-shaped molecules via a tube leakage

orientation by reptation is slow. After unloading, the oriented long molecule, tending to re-establish its equilibrium conformation, will cause a partially reversed displacement of volume elements that are still connected to its chain. In other words, the orientation of the long molecule gives rise to an increased elasticity of the blend. The experimentally observed drastic increase in recoverable strains due to low concentrations of a high molar mass component in binary blends of monodisperse components [14, 40] is also found, but less pronounced, for broadly distributed commercial polymer melts (see below).

c) *Influence of branching*

At a given molar mass, branched molecules have a shorter main chain compared to a linear molecule. For branches of equal size the extent of the main chain is reduced, not only by a growing number of branches but also by an increase in their length at a given degree of branching. Also the radius of gyration is reduced. Thus a reduction in the reptation time and viscosity is expected for short chain branches if the latter do not essentially modify the tube geometry (cf. Fig. 12a).

In the case of long chain branches a ramification of the tube may occur (Fig. 12b) which will act as a hindrance to reptation. Here, the long side chain can move, to some degree, independently of the backbone. However, there might be a tendency for the branches to arrange parallel to the main chain in one tube such that the reptation motion is less impeded than by a tube ramification. This parallel arrangement seems more likely for high deformation rates.

It is easy to imagine that the simple reptation motion of the centre of mass is blocked in the case of star-shaped molecules. Here, a modified diffusion of the branches via a lateral tube leakage is discussed [39, 44, 67] (Fig. 12c). This kind of motion presumably plays also some role in long chain branched melts.

The impact of the branching structure on the elasticity of the melt is difficult to predict. Short chain branches should, on one hand, as a first approximation, only influence the reptation time and thus the viscosity level but not the steady-state compliance J_e^o. On the other hand, if long branches gave rise to additional entanglements of two adjacent molecules this would increase the network point density. As a result, the level of the relaxation modulus would be increased and the steady-state compliance decreased, since a given elastic deformation would only be achieved at a higher stress level.

d) Influence of temperature

The mobility of the polymer chains necessary for reptation mainly depends on two conditions: firstly, a sufficiently high flexibility of the backbone due to C-C-bonds that can rotate and, secondly, the presence of enough vacancies in the melt to actually enable a spacial rearrangement of chain segments.

At temperatures far above the glass transition temperature a sufficiently high number of vacancies are available, such that the speed of reptation is governed solely by the jump probability over rotational potentials. This yields an Arrhenius-type dependence of the relaxation times τ_i on temperature. In that case a variation in temperature from T_o to T causes all relaxation times of melts that behave 'thermorheologically simple' to change by the same shift factor a_T

$$a_T = \exp\left[\frac{E_o}{R}\left(\frac{1}{T} - \frac{1}{T_o}\right)\right], \tag{53}$$

E_o being the flow activation energy and R the gas constant.

The value of E_o is dependent on the chemical nature of the monomer unit and will be increased by a steric impediment of the C-C-rotation. Thus the spacial extension of side groups will influence the temperature dependence of a_T. Since branches also impede the free rotation of the backbone, we expect the flow activation energy to increase with the number of branches and also to depend on the side chain length. In addition, as the rotational potentials of chain units with and without branches are different, the shift factor may no longer be the same for all relaxation times in melts with non-uniform long chain branches.

The steady-state compliance J_e^o is more or less temperature invariant. Rubber-like liquid theory yields a temperature dependence proportional to $1/\varrho T$. Since the changes of ϱT are rather small in the molten state, the relaxation strengths in Equation (12) may be regarded as temperature invariant. It follows from Equation (22) that the temperature dependence of the zero shear viscosity is proportional to the shift factor:

$$a_T = \eta_o(T)/\eta_o(T_o). \tag{54}$$

This is why the flow activation energy in Equation (43) can be directly determined from the temperature dependence of η_o.

4. Experimental results on melts of different molecular structure

a) Influence of the molar mass distribution on the behaviour in simple shear

Figure 13 shows the steady-state viscosities η_s of two polypropylene melts versus shear rate. The two materials have different molar mass distributions. This is demonstrated by the GPC curves inserted into the Figure. PP1 (solid line) contains higher molar mass fractions compared to PP2 (broken line). As a result, both the average molar mass M_W and the ratio M_w/M_n have higher values in the case of PP1.

The data points of the viscosity functions in Figure 13 have been determined by quite different experimental techniques: shear creep tests using a sandwich-type rheometer (△) [45] and a cone-plate rotational rheometer (□) [46] (Rheometrics Stress Rheometer); capillary rheometry using a nitrogen gas driven low pressure viscometer (▽●) [47] and a plunger driven high pressure viscometer (◇■) (Goettfert Rheograph 2000); and finally oscillatrory shear experiments (see below) by means of a modified Weissenberg rheogoniometer (○◀) [27]. In the latter case the absolute values of the complex dynamic viscosity $|\eta^*(\omega)|$ have been converted to steady-state viscosities, as proposed by Cox-Merz [48]:

$$\eta_s(\dot{\gamma}) = |\eta^*(\omega)| = \frac{1}{\omega}[G'(\omega)^2 + G''(\omega)^2]^{1/2}$$
$$\text{for } \dot{\gamma} = \omega. \tag{55}$$

There is a nice overlap of the results from the different rheometers, and obviously, the Cox-Merz relation is valid for polypropylene melts (cf. also [49]).

The investigated shear rate range covers both the linear (Newtonian) range of constant viscosity ($\eta_s = \eta_o$) and the range of pronounced shear-thining ($\eta_s \ll \eta_o$). PP1 (unfilled symbols) and PP2 (full symbols) exhibit nearly coinciding viscosity functions in the shear

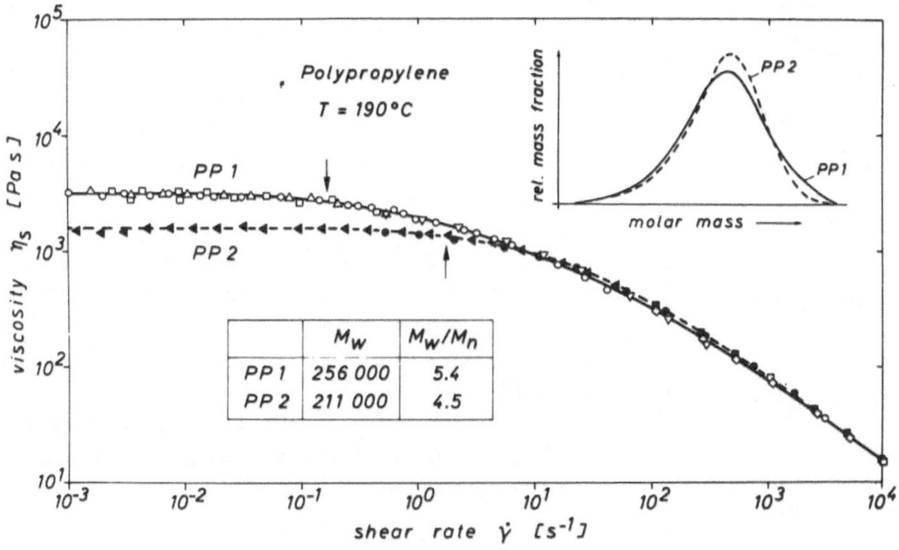

Fig. 13. Viscosity functions of two polypropylene melts having different molar mass distributions. The data points were obtained using different experimental techniques (see text)

rate range of extrusion and injection moulding (10 $s^{-1} < \dot{\gamma} < 10^4$ s^{-1}). In fact, both melts have the same melt flow index MFI [50] of 7 g/10′ at 190 °C.

The differences between the molar mass distributions of PP1 and PP2 mainly show up in the viscosity behaviour at low shear rates. Here PP1 has a higher zero shear rate viscosity due to its higher M_w, in agreement with Equation (46). Also it is clearly seen that the onset of significant shear-thinning is observed at smaller shear rates in the case of PP1. This is due to the differences in the characteristic retardation times of the two melts ($\tau_{R1} = 6.1$ s for PP1 and $\tau_{R2} = 0.37$ s for PP2,

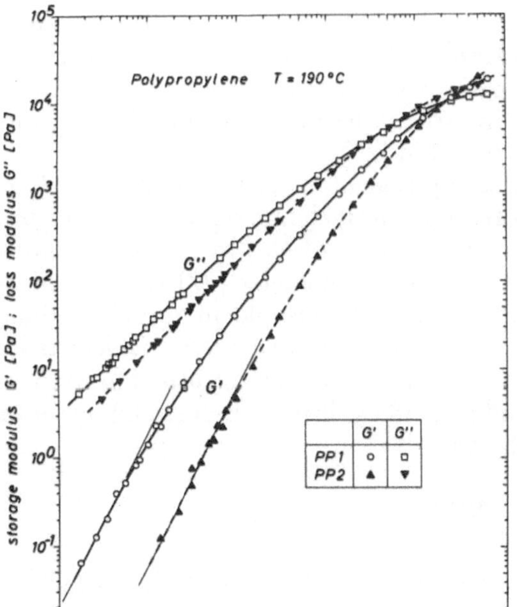

Fig. 14 a). Loss and storage moduli of the two polypropylene melts from Figure 13 versus angular frequency

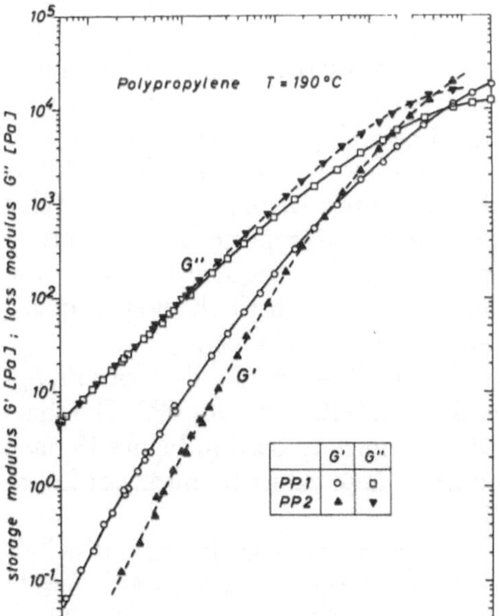

Fig. 14 b). Molar mass invariant representation of the dynamic moduli versus reduced angular frequency

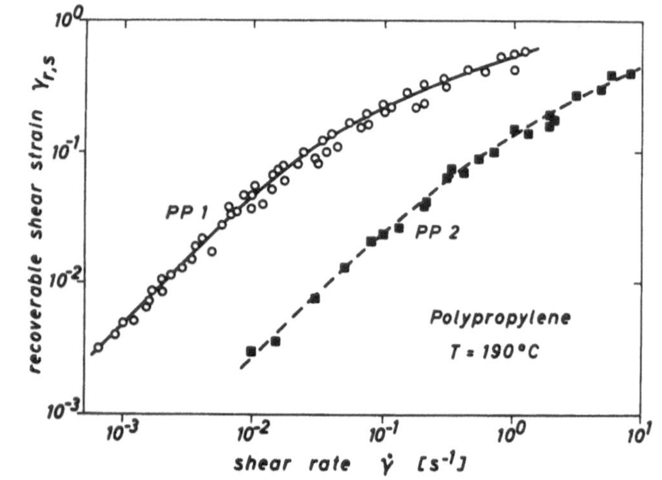

Fig. 15. Recoverable shear strains of the two polypropylene melts in steady-state shear flow versus shear rate. Recoil measurements were determined using the modified Weissenberg rheogoniometer (■), or different types of rheometers (O) (see [25])

see below). The arrows in Figure 13, marking the shear rates $\dot{\gamma} = 1/\tau_R$ for both materials, demonstrate that a pronounced shear-thinning is observed if the shear rate exceeds the reciprocal of the characteristic retardation time.

Measurements of the storage modulus G' and the loss modulus G'' versus angular frequency are shown in Figure 14a. As discussed in section 2 the loss moduli increase proportional to ω at the limit of small frequencies. The limiting course of the storage moduli (Eq. (34b)) is a straight line of slope 2 in the double-log plot (depicted by the thin lines in Fig. 14a). PP1 exhibits considerably higher values of the storage modulus than PP2 and reaches the asymptote $G' \sim \omega^2$ only at significantly smaller angular frequencies. From the dynamic moduli the following basic rheological quantities are obtained (cf. Eqs. (35 a, b)):

$\eta_{01} = 3.2 \cdot 10^3$ Pas and $J_e^o = 1.9 \cdot 10^{-3}$ Pa^{-1} for PP1, $\eta_{02} = 1.6 \cdot 10^3$ Pas and $J_e^o = 2.3 \cdot 10^{-4}$ Pa^{-1} for PP2. The characteristic retardation times τ_R used in Figure 13 have been calculated from these values by means of Equations (33), (48).

For melts having comparable molar mass distributions, a change in the average molar mass M_w merely causes a horizontal shift of $G'(\omega)$ and $G''(\omega)$ [51]. Thus, to eliminate the influence of different M_w on the viscoelasticity, a reduced representation of the moduli versus $\eta_o\omega$ is of advantage (Fig. 14b). This procedure yields an M_w-invariant asymptote $G''(\eta_o\omega) = \eta_o\omega$ for the loss moduli versus reduced angular frequency.

In the normalized plot it is clearly seen that PP1 exhibits considerably higher G'-values than PP2 at a given level of G'' at the limit of small frequencies. This corresponds to a higher degree of elastic deformations at a given shear rate. However, the differences become smaller with increasing reduced frequency and finally a crossover of the G'-curves is found. The curvature of the moduli versus reduced frequency is more pronounced for the melt with the broader molar mass distribution (PP1). As a consequence, the intersection point of the dynamic moduli ($G' = G''$) is found at higher values of $\eta_o\omega$ and at a lower moduli level for PP1 than for PP2.

The higher elasticity of PP1 due to its high molar mass shoulder clearly shows up, too, in measurements of the steady-state recoverable shear strain $\gamma_{r,s}$ versus shear rate (Fig. 15). For details of the experiments the reader is referred to [25]. The differences in $\gamma_{r,s}$ are most pronounced in the limit of small shear rates. Here the recoverable strains increase proportional to the shear rate (compare Eq. (24)). In the range of slope 1 in the double-log plot we obtain from the ratio $\gamma_{r,s}/\dot{\gamma}$ values of $\tau_R = 4.8$ s for PP1 and $\tau_R = 0.27$ s for PP2. Due to experimental difficulties in the γ_r-measurement of quickly recoiling melts these values are somewhat smaller than the characteristic times τ_R from oscillatory shear measurements (see above). When the shear rate is increased, the recoil increases less than proportional to $\dot{\gamma}$ and the differences due to differing shapes of the molar mass distributions become less pronounced. A comparison between the prediction of Equation (42) and the measured shear rate dependence of $\gamma_{r,s}$ for PP1 can be found in [25].

The steady-state normal stress coefficients θ_s of the two polypropylene melts are represented in Figure 16. Direct measurements of $N_{1,s}$ by means of a modified Weissenberg rheogoniometer [27] (circles and solid squares) were only possible in the shear-thinning range. Values of the normal stress coefficients in the Newtonian range have been calculated from the storage moduli G' in oscillatory shear (stars) by means of the relation

$$\theta_s = 2\,G'/\omega^2 \quad \text{for} \quad \omega = \dot{\gamma} \tag{56}$$

and from the recoverable shear strains $\gamma_{r,s}$ according to

$$\theta_s = 2\,\eta\,\gamma_{r,s}/\dot{\gamma}\,. \tag{57}$$

Strictly speaking, these relations are only valid in the range of a constant normal stress coefficient (see Eqs.

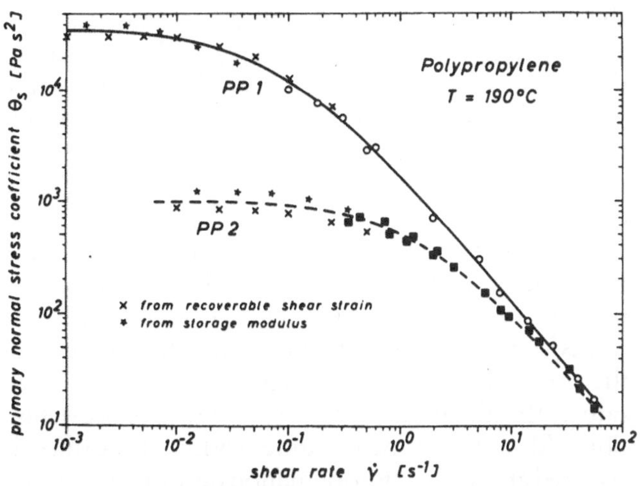

Fig. 16. Steady-state primary normal stress coefficients of the polypropylene melts versus shear rate. Direct normal stress measurements in a cone-plate rheometer (○, ■); measurements calculated from $\gamma_{r,s}$ using Equation (57), (×); and from G' using Equation (56), (★)

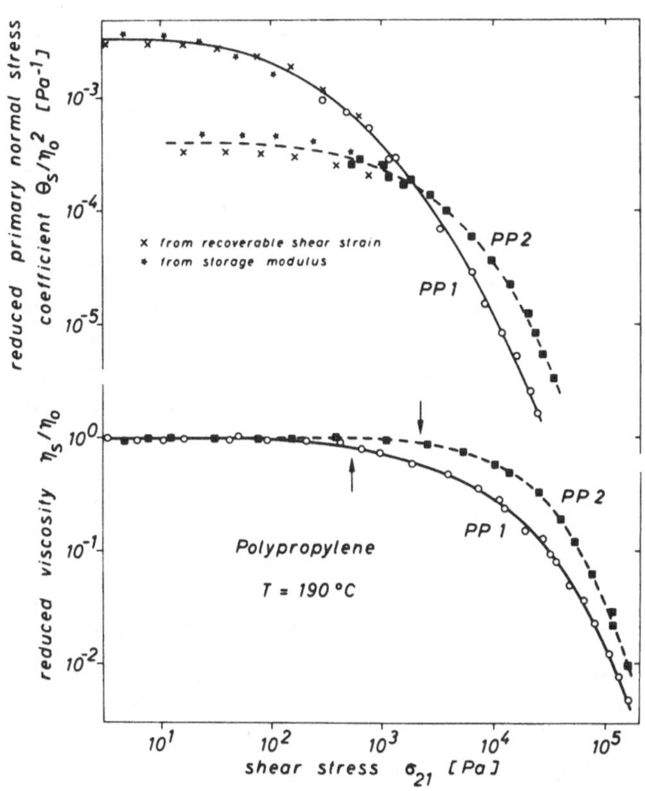

Fig. 17. Reduced representations of the steady-state viscosities (Fig. 13) and normal stress coefficients (Fig. 16) versus shear stress

(24), (34a)) but they they may also be used as approximations in the shear rate range close to the onset of shear-thinning. Generalizations of Equations (56), (57) for non-Newtonian melts are discussed in [25].

The zero shear rate primary normal stress coefficient θ_o of PP1 has a much higher level than that of PP2. This is due to the fact that θ_o is proportional both to η_o^2 and J_e^o (Eq. (25)). These two quantities have higher values in the case of the PP1 with the high molar mass shoulder.

The $\theta_s(\dot{\gamma})$-curves of the two melts converge when the shear rate is increased. However, for a comparison of normal stress coefficients of various melts, differences caused by non-equal values of M_w should be eliminated, by plotting the reduced normal stress coefficient θ_s/η_o^2 versus shear stress (Fig. 17, upper part). Such a plot in general gives molar mass invariant curves if the molar mass distributions are similar.

In the reduced representation, the constant value θ_o/η_o^2 in the limit of small shear stresses corresponds, except by a factor of two, to the steady-state compliance J_e^o. When the shear stress is increased a decrease of the reduced normal stress coefficients is already observed at smaller stress levels, in the case of the more broadly distributed PP1 melt. This material also shows the steeper decrease of the reduced normal stress coefficient. Finally, a crossover of the curves is found such that at high shear stresses the reduced normal stress coefficients of PP1 are significantly smaller than those of PP2.

The lower part of Figure 17 shows the reduced viscosities η_s/η_o versus σ_{21}. It is clear that the onset of shear-thinning occurs at lower shear stresses in the case of PP1. An estimate of the necessary shear stress level is possible by the characteristic shear stress σ_R

$$\sigma_R = \eta_o/\tau_R = 1/J_e^o . \tag{58}$$

The σ_R-values are indicated by the arrows at the reduced viscosity functions in Figure 17. The differences in η_s/η_o between PP1 and PP2 show up most clearly in the shear stress range of 10^4 Pa $< \sigma_{21} < 10^5$ Pa. At still higher stress levels, the viscosity functions tend to come more close to each other with increasing σ_{21}. A comparison of the reduced normal stress coefficients and reduced viscosities clearly shows that a shear rate dependence of the quantity θ_s already becomes significant at considerably smaller values of the shear stress than σ_R.

b) Influence of the molar mass distribution on extrusion properties

A rheological characterization of melts in the well defined simple shear flow has the advantage of an experimentally simple realizability. However, for processing more complex flow situations are relevant. As an example, the extrusion through a capillary, as the most simple step of processing, implies both a superposition of shear and elongation and spatially inhomogeneous deformation rates. Capillary rheometry can supply important information for the practical application of melts, such as entrance pressure losses, extrudate swell, and critical flow rates for the occurrence of flow instabilities.

Extrudate swell is due to recoverable elongational strains induced by the converging flow at the die entrance as well as by recoverable shear strains originating within the capillary [7]. Part of the elastic strains decay during the passage of the melt through the die. The recoverable strains still remaining in the material, when it leaves the die, give rise to a shrinkage of the extrudate in the direction of flow. Since the melt is incompressible this causes a radial swell.

The magnitude of swelling is the highest for an orifice (capillary with negligible length). Here the impact of the converging entrance flow is most pronounced, since only a short time elapses until the deformed volume elements of the melt are unloaded at the orifice exit. As a result, the decay of molecule orientations remains the smallest for the orifice compared to a die of non-zero length [52].

Figure 18 shows the extrudate swell of the two polypropylene melts versus apparent shear rate D (wall shear rate of a Newtonian liquid at the same flow rate) as measured from an orifice (schematically depicted at the bottom of Fig. 19). It is obvious that, in spite of the very similar shear viscosities of the two melts in the investigated shear rate range (cf. Fig. 13), the high molar mass shoulder of PP1 gives rise to a considerably higher extrudate swell compared to PP2. The reason for this is a higher degree of recoil due to the orientations of the high molar mass components (cf. Fig. 10)). The increase of extrudate swell with growing wall shear rate is in agreement with Figure 8. In fact, an increase of the flow rate gives rise to higher elongational strain rates at the die entrance, viz. a higher degree of orientation. Also the residence time within the orifice is reduced and with it the orientation decay by reptation.

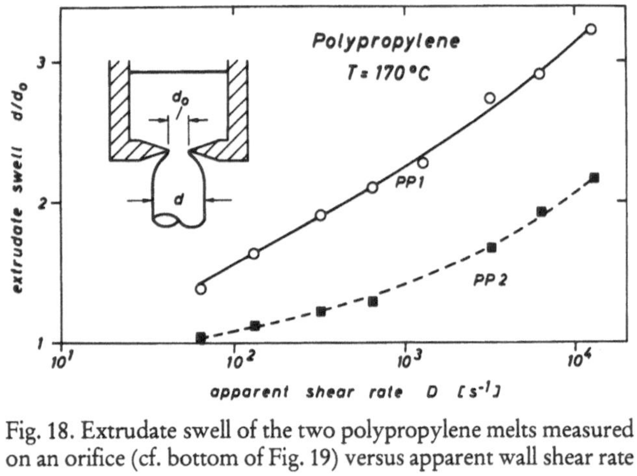

Fig. 18. Extrudate swell of the two polypropylene melts measured on an orifice (cf. bottom of Fig. 19) versus apparent wall shear rate

Another important quantity in extrusion is the entrance pressure loss p_c due to the elongational flow and, in some cases, secondary flows such as vortices. They can be experimentally determined by the Bagley method. If the measured extrusion pressure p for a constant apparent wall shear rate D is plotted versus the length L to radius R ratio of the capillary, straight lines are obtained (Fig. 19). The non-zero extrusion pressure p_c obtained by an extrapolation of the Bagley straight line to $L/R = 0$ is equal to the sum of the entrance pressure loss and an exit pressure loss. Since for normal thermoplastics the exit pressure loss is small compared to the entrance pressure loss, the Bagley correction p_c essentially represents the latter quantity.

Figure 19, for example, shows a detail of Bagley plots for a high density polyethylene (HDPE) melt. The arrows indicate that the straight lines actually meet the data points for $L/R = 60$ which are not contained in the diagram for scaling reasons! It is interesting to note that the p_c-values extrapolated from the extrusion pressures measured on dies of different L/R (open symbols) fully agree with the extrusion pressures determined at an orifice (solid symbols). Thus the Bagley correction can easily be determined by a single measurement! The geometry of the orifice used is depicted at the bottom of Figure 19. A is a replacable steel plate of 0.1 mm thickness with a hole diameter 0.2 mm less than that of the conical supports B and C.

The Bagley corrections of the polypropylene melts PP1 and PP2 are plotted in Figure 20a versus wall shear. Such a diagram, in general, gives stress molar mass invariant curves for melts with similar molar mass distributions (compare [42] and Figs. 20b, 21). PP1 with more high molar mass components compared to PP2 shows markedly higher p_c-values. This may be

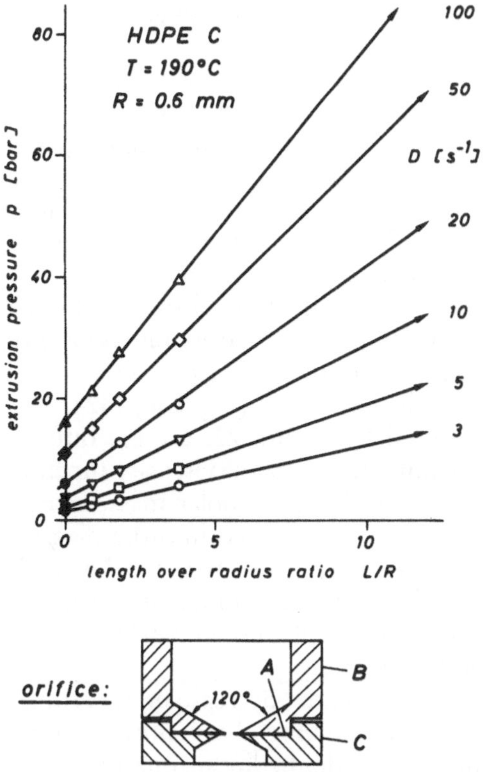

Fig. 19. Bagley plots of the extrusion pressures measured at contant shear rates versus L/R of the die (open symbols). The solid symbols represent the Bagley corrections directly determined using the orifice schematically depicted below

explained by a higher level of transient elongational viscosities at a given wall shear stress, if one assumes the flow kinetics to be rather similar [1].

A third characteristic quantity for extrusion is the onset of flow instabilities. For that purpose, Figure 21 shows viscosity functions versus wall shear stress of high density polyethylenes with similar molar mass distributions (produced by means of the same catalyst) but very different average lengths of the molecules. The average molar mass is characterized by the intrinsic viscosity $[\eta]$ (dimension *dl/g*) measured in decalin at 130 °C. The extrusion measurements were performed using a plunger driven high pressure capillary rheometer.

At the wall shear stresses marked by the vertical bars with tics a pulsation of the extrusion pressure, caused by periodic stickslip transitions [53], is observed. In that range also a periodic pulsation of the flow rate occurs, and the surface of the extrudate is no longer smooth but shows periodic irregularities. It is clearly seen that the critical value of the shear stress for the onset of wall slip is independent of te viscosity level of the four HDPE melts. (The slightly higher value of the lowest molar mass melt can be explained by some dissipative heating due to the high wall shear rates of that sample).

A possible explanation for the molar mass invariant critical shear stress is based on the assumption that wall slip can only occur if a critical level of recoverable strain, and with it molecular orientation, is achieved [54]. For similar molar mass distributions, the same level of elastic strains is reached at a given shear stress level. However, the necessary shear rate varies with the average molar mass, proportional to the reciprocal of the characteristic retardation time. This can be proved by means of Equations (40), (42), if all relaxa-

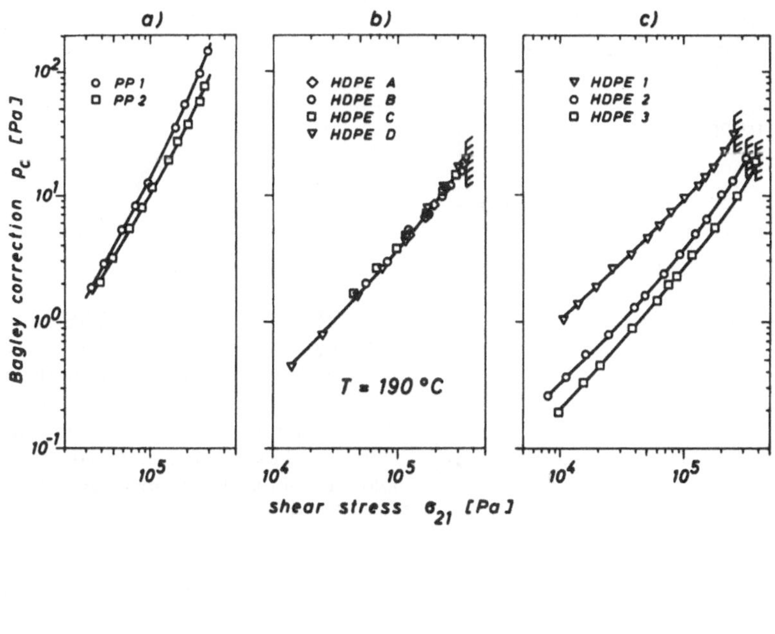

Fig. 20. Plot of Bagley corrections versus wall shear stress. Vertical bars mark onset of wall slip; a) polypropylenes PP1 and PP2 from Figure 13; b) linear HDPE melts with similar molar mass distributions (see Fig. 21); c) HDPE melts with different shapes of the GPC curves (see Fig. 22)

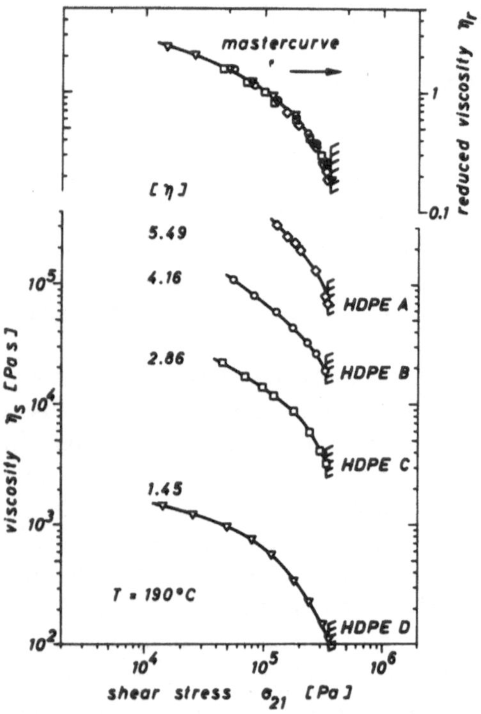

Fig. 21. Viscosity functions of HDPE melts with similar molar mass distributions but very different average molar masses (characterized by the intrinsic viscosity [η]). Vertical bars with tics mark the onset of flow instabilities due to wall slip. A mastercurve is obtained by relating the measured viscosities to the η_s-value at $\sigma_{21} = 10^5$ Pa

tion times of the spectrum simply change by the same factor.

Due to the similar molar mass distributions of the four HDPE melts, the shape of their viscosity functions is the same. A master curve $\eta_r(\sigma_{21})$ (upper part of Fig. 21) can be constructed by relating the viscosities to the η_s-value measured at a given shear stress. In Figure 21 the value at 10^5 Pa has been used. This is in agreement with the results for another series of HDPE melts given in [55]. In addition, the same shape of the molar mass distributions and of the viscosity functions gives rise to coinciding values of the Bagley corrections, if plotted versus wall shear stress (Fig. 20b) [42].

As demonstrated by Figure 22, a quite different result is obtained if one compares viscosity functions of HDPE melts with disimilar molar mass distributions. The GPC data are inserted into the diagram. HDPE3 has the most narrow distribution, whereas the GPC curve of HDPE2 shows a comparable shape but a broader distribution. Finally, HDPE3 exhibits a very broad bimodal distribution.

In the same sequence, the decrease of the reduced viscosity η_s/η_o with growing shear stress becomes more pronounced outside the linear (quasi Newtonian) range of constant viscosity. For instance, at a shear stress of 10^5 Pa the viscosity of HDPE1 is only about $\eta_o/1000$ compared to a viscosity reduction by a factor of five in the case of HDPE3. Such an extreme shear-thinning behaviour can be of great importance for

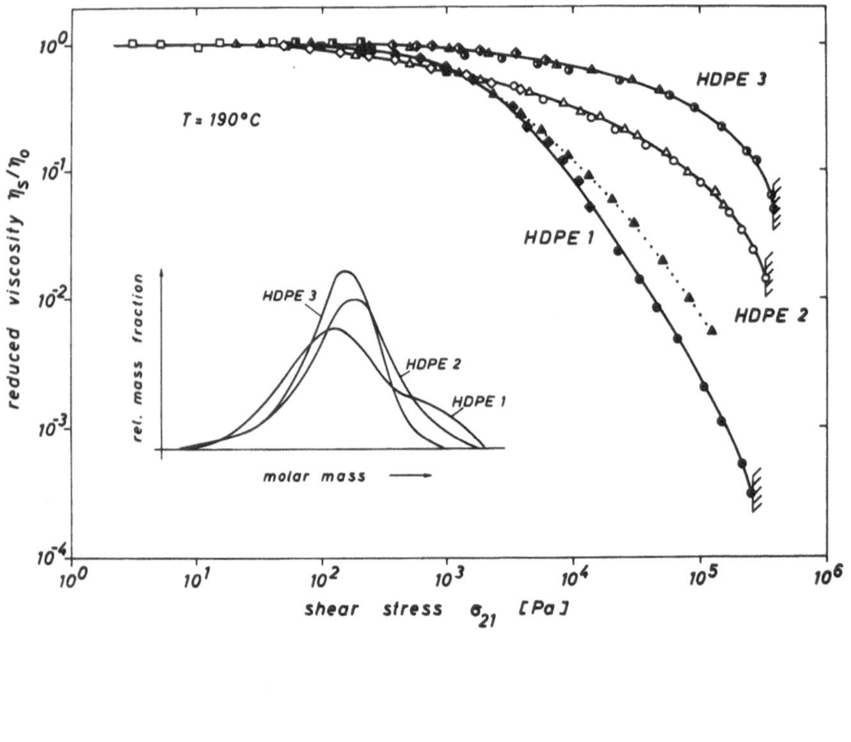

Fig. 22. Reduced viscosities of HDPE melts with quite different shapes of the molar mass distributions (cf. Table 1) versus shear stress. Steady-state viscosities (——); $|\eta^*|$ of HDPE1 for $\sigma_{21} = |G^*|$ (·····). It demonstrates that the Cox-Merz relation does not hold for that material in the range of high shear stresses

Table 1. Comparison of rheological quantities (at 190 °C) and molecular properties of the three HDPE melts having different molar mass distributions

quantity	dimension	HDPE1	HDPE2	HDPE3
$[\eta]$	dl/g	3.42	2.54	1.75
M_w	g/mol	409000	207000	93000
M_w/M_n		22	8.8	3.9
η_0	Pa s	$1.6 \cdot 10^6$	$5.0 \cdot 10^4$	$4.0 \cdot 10^3$
η_s [a])	Pa s	$5.1 \cdot 10^5$	$2.3 \cdot 10^4$	$3.2 \cdot 10^3$
$N_{1,s}$ [a])	Pa	$4.3 \cdot 10^3$	$2.7 \cdot 10^3$	$1.0 \cdot 10^3$
$\lvert\eta^*\rvert$ [b])	Pa s	$5.2 \cdot 10^5$	$2.2 \cdot 10^4$	$3.0 \cdot 10^3$
G'/G''		0.83	0.49	0.19

[a]) for $\sigma_{21} = 3000$ Pa; [b]) $\lvert G^* \rvert = 3000$ Pa

extrusion, since it actually enables the processing of high molar mass materials at a reasonable extrusion rate.

Going from HDPE3 to HDPE1 the onset of wall slip is markedly shifted to smaller wall shear stresses. This effect is seen more clearly in the Bagley correction plot in Figure 20c. Also the p_c-values increase when the molar mass distribution becomes broader.

Table 1 gives additional experimental results for the HDPE melts. Both the average molar masses M_w and the ratios M_w/M_n, determined by GPC, decrease from HDPE1 to HDPE3. A comparison of the steady-state normal stress differences $N_{1,s}$ at a shear stress of 3000 Pa shows an increase with the breadth of the molar mass distribution (line 6). Line 8 contains the ratios of storage and loss moduli G'/G'' measured at $\lvert G^* \rvert =$ 3000 Pa. These ratios, being correlated to recoverable shear strains of the melt, increase from HDPE3 to HDPE1. Finally, a comparison of the steady-state shear viscosities η_s in line 4 with the viscosities $\lvert\eta^*(\omega)\rvert$ in line 7 (the comparison is made for $\lvert G^* \rvert = \sigma_{21}$) shows that the Cox-Merz relation Equation (55) seems to be valid for HDPE2 and HDPE3 over the whole shear stress or $\lvert G^* \rvert$ range, respectively, investigated here. For HDPE1, however, the Cox-Merz relation only holds for small shear stresses (cf. broken line in Fig. 22). In the range of high stresses $\lvert\eta^*(\omega)\rvert$ clearly overestimates the true steady shear viscosity.

The experimentally observed decrease of the critical wall shear stress for wall slip of broader distributed melts can be related to the differences in the elastic compliances. As is demonstrated by Table 1, HDPE1 shows both higher values of the primary normal stress difference at a given shear stress and higher values of the ratio G'/G'' at a given level of the absolute value of the dynamic modulus $\lvert G^* \rvert$. This indicates that the recoverable shear strains of HDPE1 are higher than those of HDPE2 and HDPE3. Therefore, a critical level of the recoverable shear strain to allow wall slip, will be reached for HDPE1 at a smaller shear stress.

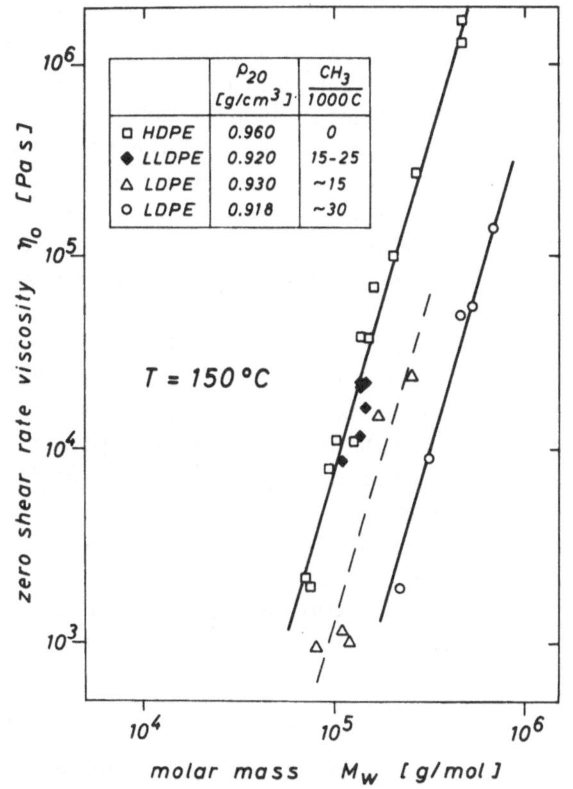

Fig. 23. Zero shear rate viscosities at 150 °C of polyethylene melts with different branching structures (see text) versus average molar mass. The straight lines have a slope of 3.4

c) *Influence of sidegroups and branching structure*

In Figure 23, the zero shear rate viscosities of polyethylene melts having different branching structures are plotted versus the weight average molar masses determined by GPC. Within the accuracy of the M_w-measurements, Equation (46) is found to be valid for the linear HDPE melts (squares) having a density of $\varrho = 0.960$ g/cm³ at 20 °C (the slopes of the straight lines in Figure 23 are 3.4). These materials have been polymerized in low pressure processes.

The validity of Equation (46) is also found for a group of long and short chain branched LDPE products (circles) that have been produced by the same type of high pressure process. Their density is 0.918 g/cm³ and they all have the same degree of branching, characterized by 30 CH_3-end groups per 1000 C-atoms. Compared at a constant value of M_w, however, the viscosities of these samples are by about two powers of ten lower than those of the unbranched HDPE.

When the degree of branching of the LDPE is reduced to 15 CH_3/1000 C in the high pressure process, such that the density is increased to 0.930 g/cm^3 (triangles), the viscosity level is almost in the middle, between those of the linear HDPE and the LDPE with 30 CH_3/1000 C.

Linear low density polyethylenes (LLDPE) have a reduced density solely compared to HDPE soley due to short chain branches mostly resulting from copolymerization with butene or hexene in a low pressure process. Depending on the type and number of the short side chains their zero shear rate viscosities are comparable to those of the linear HDPEs or slightly lower at a given average molar mass (full rhombs). It is noteworthy here that, according to [56], the viscosity of poly-α-olefins decreases exponentially with the length of the side chains.

The influence of regular sidegroups on η_o is shown in Figure 24. The highest viscosity level for a given M_w is found for HDPE (data points of Fig. 23 converted to 190 °C). All straight lines again have a slope of 3.4. In polypropylene one H-atom at each alternate C-atom is replaced by a CH_3-group (cf. Table 2). As a consequence, the zero shear viscosities of PP are lower by about a factor of ten. The second CH_3-group at the same C-atom in polyisobutylene (PIB) gives rise to an increased viscosity level compared to polypropylene. This is presumably due to a reduced rotational mobility of the chain units. Compared to PP the CH_3-group is replaced by the more bulky benzene ring in polystyrene (PS). This yields a viscosity level that is only slightly lower than that of HDPE at 190 °C. However, one should take into account here that the temperature dependencies of the viscosities of the melts in Figure 24 are not identical (cf. Table 2). Therefore, the differences in the viscosity levels are somewhat dependent on the arbitrarily chosen refrence temperature.

The η_o-values of the PIB samples in Figure 24 cover more than six powers of ten. Taking into account that the steady-state compliances of the PIB melts lie arround $J_e^o = 5 \cdot 10^{-5}$ Pa^{-1} (cf. e.g. [43]), since they all have similar molar mass distributions with $M_w/M_n \approx 2$, we can calculate characteristic retardation times using Equation (48) of $1.5 \cdot 10^{-3}$ s $< \tau_R < 3 \cdot 10^3$ s for 190 °C. At room temperature these values will be increased proportional to the rise of η_o by another 3 powers of ten [57]. The resulting time scale range of

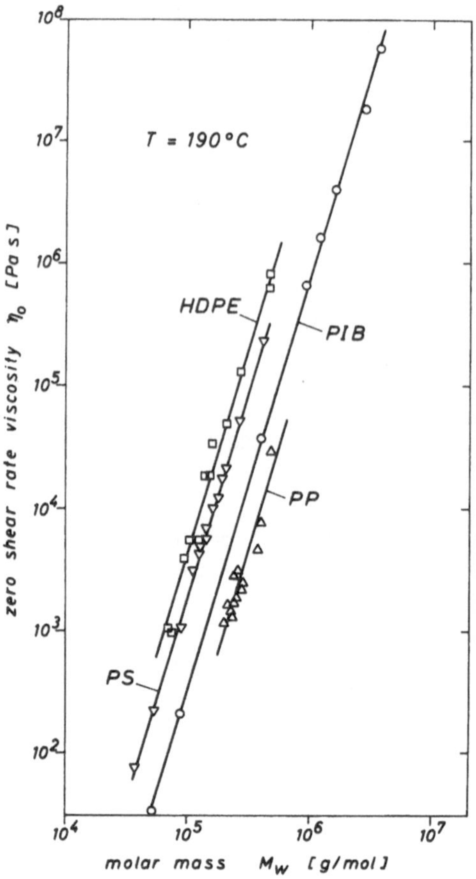

Fig. 24. Zero shear rate viscosities at 190 °C of melts with different regular side groups (cf. Table 2) versus average molar mass. The polystyrene data are reproduced from reference [64].

about 1 s to 10^6 s makes it clear that at room temperature the low molar mass PIB melts appear as fluids, whereas the high molar mass samples behave like elastomers.

For the semicrystalline polyethylenes the molten state is reached only at temperatures of more than 100 °C above the glass transition temperature. This is why the temperature dependence of the zero shear rate viscosity is of the Arrhenius-type (Eq. (53)). Figure 25 shows semi-log plots of η_o versus the reciprocal of the absolute temperature for polyethylene melts having different branching structures. Straight lines are obtained, the slopes of which give the flow activation energies E_o listed in the table inserted into the figure.

Whereas the lowest flow activation energy of 28 kJ/mol is found for the linear HDPE we get a value as high as 60 kJ/mol for a long chain branched LDPE melt with 30 CH_3-end groups for each 1000 C-atoms. When the degree of branching is reduced to 15 CH_3/

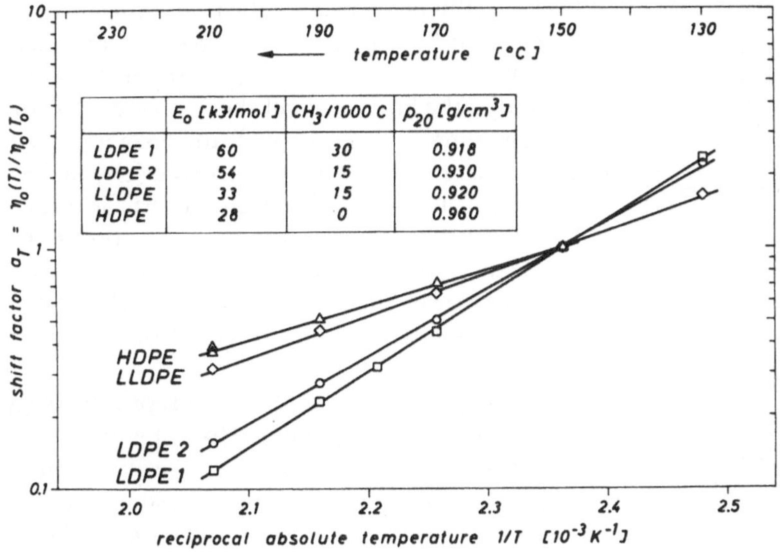

Fig. 25. Arrhenius plots for the shift factors a_T determined from the zero shear rate viscosities of polyethylene melts having different branching structures. The inserted table gives the flow activation energy E_o, the degree of branching, and the density of the materials at 20 °C

1000 C, still in the same high pressure process, the flow activation energy is also reduced to 54 kJ/mol. However, when a comparatively low density of the material at 20 °C is achieved only by short chain branches, as in the case of linear low density polyethylenes (LLDPE), a flow activation energy of only 33 kJ/mol is found, which is only slightly higher than that of HDPE. If follows that long chain branches are responsible for the high E_o-values of LDPE compared to HDPE, and that a growing number of the branches increases the flow activation energy.

Table 2 gives a comparison of the flow activation energies of melts with different regular side groups. The temperature dependence of the shift factor a_T over a wide range of $1/T$ for the non-crystalline PIB and PS samples is better described by a WLF-equation [14]. The values of E_o listed in Table 2 are thus only valid in a narrow temperature range around 190 °C. Going from HDPE to P(1-butene) the flow activation energy is increased with growing length of the side groups (compare [57]). When the ethyl-group in P(1-butene) is replaced by two adjacent methylgroups, as in PIB, E_o is reduced. However, compared to PP the bulky benzene ring of PS causes a distinct increase in the flow activation energy, which is to be expected from an increase of the height of the rotational potential.

d) *Peculiarities of long chain branched LDPE melts*

Long chain branches in LDPE not only increase the temperature dependence of the viscosity but also cause

Table 2. Chemical structures and flow activation energies E_o for the temperature range around 190 °C of melts having different regular sidegroups

material	monomer unit	E_o [kJ/mol]
HDPE	H H | | – C – C – | | H H	28
PP	H H | | – C – C – | | H CH_3	44
P(1-butene)	H H | | – C – C – | | H CH_2 | CH_3	66
PIB	H CH_3 | | – C – C – | | H CH_3	56 [43, 57]
PS	H H | | – C – C – | | H C_6H_5	125 [63]

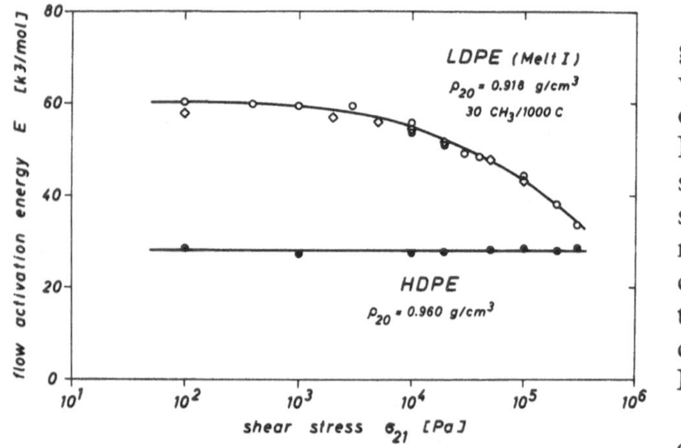

Fig. 26. Flow activation energies as determined from Arrhenius plots if the temperature dependence of viscosity is evaluated at different constant shear stresses. Whereas the flow activation energy is constant shear stresses. Whereas the flow activation energy is constant for the linear HDPE (●) a distinct stress dependence of E is found for the investigated LDPE melt (○◇) (see text)

deviations from a thermorheologically simple behaviour [58]. This is clearly seen by comparing the flow activation energies at different constant shear stresses.

Figure 26 shows plots of the flow activation energies versus shear stress. The data are based on careful measurements of the viscosity functions of a HDPE melt (150 °C < T < 210 °C) and an LDPE melt (115 °C < T < 210 °C). By using stabilized samples thermally induced changes of the melts at low shear stresses and thus relatively long durations of the experiments could be kept negligible. In the high shear stress range dissipative heating of the melts was controlled by using a capillary rheometer and different radii of the dies. Arrhenius plots (cf. Fig. 25) of the steady-state viscosities evaluated at a given shear stress yielded straight lines in all cases. The activation energies E shown in Figure 26 were determined by

$$E = R \frac{d \ln \eta_s}{d\, 1/T}\bigg|_{\sigma_{21}} . \qquad (59)$$

The index o of the symbol E is omitted here to indicate that the evaluation is no longer restricted to the zero shear rate viscosity. For HDPE (full circles) we find $E = E_o$ over the whole shear stress range investigated. This proves that HDPE is thermorheologically simple. Therefore, the viscosity functions measured at different T can be shifted in a log-log plot to coincide on a temperature invariant master curve.

However, this is not true for the LDPE melt investigated here. For Melt I [12] a shear stress independent value of the flow activation energy E (open circles) is only found in the limit of small shear stresses, viz. the Newtonian range of viscosity. With growing shear stress a distinct reduction of E is observed (cf. [58, 59]) such that finally the E_o-level of HDPE is nearly reached at the highest shear stresses. The rombs also contained in Figure 26 represent activation energies that have been evaluated by the same method from viscosity functions of the same material published by Meissner [59].

They clearly demonstrate the good reproducibility of the effect. It has to be concluded from Figure 26 that LDPE Melt I is not thermorheologically simple. Therefore, strictly speaking, the viscosity functions measured at different temperatures cannot be shifted to coincide on a mastercurve in a log-log plot of η_s versus shear rate or shear stress! This is obvious for plots versus shear stress. In the representation versus shear rate, however, the change of the shape of $\eta_s(\dot{\gamma})$ with temperature is not seen so clearly, since, at high shear rates, the viscosity function runs nearly parallel to lines of constant shear stress.

A possible explanation for the observed effect in LDPE takes into account that ramifications of the tube (cf. Fig. 12b) will probably increase the barrier of the rotational potential of C–C-bonds near the branching. At low shear rates all relaxation times contribute fully to the viscosity level, as given in Equation (22). The longest relaxation time of the spectrum corresponds to a co-operative motion of the whole chain. The temperature dependence of that relaxation time will be distinctly influenced by the reduced mobility of the tube ramifications, giving rise to a high value of the flow activation energy. However, at high shear rates and shear stresses, according to Equation (40), the contribution of longer relaxation times to the viscosity is reduced by the factor $1/(1 + n\dot{\gamma}\tau_i)^2$. The short relaxation times correspond to motions of only parts of the molecule, most of them having no branches, and thus are governed by the low activation energy of a linear chain. In conclusion, we can say that the high activation energy of the long relaxation times only shows up clearly at small shear rates in the Newtonian range of viscosity. With increasing shear stress the short relaxation times, with an activation energy close to that of HDPE, become more and more dominant.

In addition, we expect the long chain branches to partially arrange in parallel to the mainchain at high shear rates (Fig. 12b, right side). This conformational

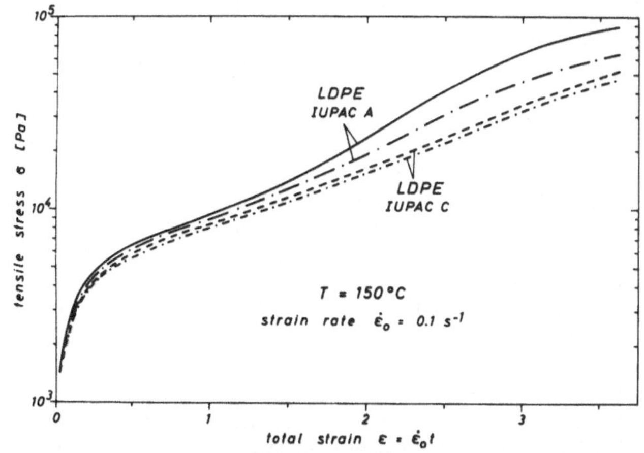

Fig. 27. Stress-strain curves of an LDPE melt measured after different mechanical pretreatments. (------), (— — —) IUPAC A and C, respectively, after extrusion on a laboratory extruder; (--·--) and (-·--·-) same melts extruded from a capillary rheometer (see text)

change is presumably responsible for the observed influence of a mechanically pretreatment on elongational properties of LDPE melts. For more details the reader is referred to [61].

As an example, Figure 27 shows stress-strain curves measured on LDPE samples at constant strain rates of $\dot{\varepsilon}_0 = 0.1\ s^{-1}$. Sample IUPAC C is the same batch as IUPAC A but has been submitted to an additional extrusion process before pelletizing. This treatment does not affect the molar mass distribution [62]. When rod-like samples are prepared from the pellets by a gentle extrusion in a capillary rheometer a distinct lower level of the tensile stress is found for sample IUPAC C (broken line) compared to IUPAC A (full line). However, the preparation of rod-like specimens on a laboratory extruder, which imposes high shear rates and squeezing flows, yields a markedly reduced tensile stress only for IUPAC A (—·—·—), such that the stress strain curve is now closer to that of IUPAC (-·-·-·-). The changes of the rheological properties of IUPAC A and C due to different mechanical pretreatments are reversible, as could be experimentally verified by dissolving and subsequently reprecipitating the samples [61].

The observed behaviour may be interpreted by Figure 12b. We assume an arrangement of the long chain branches in LDPE parallel to the mainchain to occur at high shear and stretch rates, e. g. on a laboratory extruder. After unloading of the mechanically pretreated melt, the long chain branches tend to re-establish their equilibrium conformation, viz. to find a tube independent of the mainchain. If the energy levels of the conformations with and without ramification of the tube are very close, the time scale for this conformational rearrangement will be rather long compared to the characteristic retardation time. In other words, a full decay of the overall orientation of the molecule and a complete stress relaxation can occur, although this does not mean that the long chain branches have found their equilibrium conformation. The time necessary for the latter process may be so long that during the residence times needed, a thermally induced degradation can occur.

In solution, however, the molecules are more or less separated such that the mobility of the long arms is increased. After reprecipitation the conformation in the bulk material corresponds more to the equilibrium state (Fig. 12b, left side). This conformation of the long chain branches obviously gives rise to higher tensile stresses in stretching tests.

5. Final remarks

In this paper we have made use of the idea that the decay of orientation of a molecule laterally trapped by adjacent molecules is essentially achieved by a reptation motion. This rough model is very helpful in understanding some fundamental features of the rheological behaviour. However, it is important to note that the model, as it is discussed here, is rather general. In fact, neither details of the mutual arrangements of the molecules or of the real nature of the entanglements have been defined. This means that the qualitative and semiquantitative relations between molecular structure and rheological properties, like viscosity, elasticity, and orientation, given here, are not at all appropriate for making a distinction between more detailed molecular models. So far, both coil and bundle models are consistent with the experimental results.

The analytical description of linear and nonlinear material functions given in this paper is based on the relaxation time spectra of the melts. It has been assumed that these are known for a given material. Methods to determine the spectra from measurements are described elsewhere (e. g. [25]). This phenomenological approach appears to be the only possibility at the moment for practically used systems. Mostly details about the molar mass distribution and the branching structure are not avialable, nor are well established molecular theories that enable us to calculate the spectra directly from the molecular data.

Rubber-like liquid theory is a pertinent starting point to analytically described the time-dependent behaviour of melts in the range of a strain-independent time average of the entanglement density. An extension to shear-thinning polymer melts is possible simply by introducing a global strain dependence of the relaxation modulus. This strain dependence actually describes the decrease of the entanglement density with increasing deformation. Again, the attenuation function must be determined experimentally. This approach has been found to be very successful for simple shear and uniaxial elongation.

Acknowledgements

The author wants to express his thanks to Mr. M. Reuther for preparing the diagrams, and to Messrs. Ch. Kaduk, F. Landmesser, A. Schmidt, P. Schweizer, and J. Ulmerich for the carefull experimental work. Thanks are also due to my colleague Dr. W. Ball for supplying the GPC data.

References

1. Cogswell FN (1981) Polymer Melt Rheology, Wiley, New York
2. Han CD (1976) Rheology in Polymer Processing, Academic Press, New York
3. Janeschitz-Kriegl H (1969) Adv Polym Sci 6:170
4. Janeschitz-Kriegl H (1983) Polymer Melt Rheology and Flow Birefringence, Springer, Heidelberg
5. Retting W (1975) Coll & Polym Sci 253:852
6. Jones TT (1976) Pure & Appl Chem 45:39
7. Lodge AS (1964) Elastic Liquids Academic Press, New York
8. Meißner J (1971) Kunststoffe 61:576
9. Treolar LRG (1956) In: Stuart HA (ed) Die Physik der Hochpolymeren, Bd IV, Springer, Heidelberg
10. Kuhn W, Grün F (1942) Kolloid-Z Z Polymere 101:248
11. Wagner HM (1979) Rheol Acta 18:33
12. Laun HM (1978) Rheol Acta 17:1
13. Laun HM (1981) Coll & Polym Sci 259:97
14. Ferry JD (1980) Viscoelastic Properties of Polymers 3rd Ed, Wiley, New York
15. Pfandl W, Schwarzl FR (1985) Coll & Polym Sci 263:328
16. Schausberger A et al. (1983) Rheol Acta 22:550
17. Montfort JP et al. (1979) Rheol Acta 18:623
18. Berstedt BH (1979) J Appl Polym Sci 23:1279
19. Heron H, Pedersen S, Chapoy LL (1976) Rheol Acta 15:379
20. Laun HM, Münstedt H (1978) Rheol Acta 17:415
21. Münstedt H, Laun HM (1979) Rheol Acta 18:492
22. Wagner MH (1976) Rheol Acta 15:136
23. Wagner MH (1977) Rheol Acta 16:43
24. Wagner HM, Meißner J (1980) Makromol Chem 181:1533
25. Laun HM (1986) J Rheol 30:459
26. Laun HM, Münstedt H (1976) Rheol Acta 15:517
27. Meißner J (1975) Rheol Acta 14:201
28. Gortemaker FH et al (1976) Rheol Acta 15:256
29. Graessley WW (1974) Adv Polym Sci 16:1
30. Pechhold W (1980) Coll & Polym Sci 258:269
31. Genannt R, Pechhold W, Großmann HP (1977) Coll & Polym Sci 255:285
32. Pechhold W (1984) Makromol Chem Suppl 6:163
33. De Gennes PG (1971) J Chem Phys 55:572
34. Doi M, Edwards SF (1978) J S C Faraday II 74:1789, 1802, 1818
35. Doi M (1981) J Polym Sci, Letters 19:265
36. Doi M (1983) J Polym Sci, Physics 21:667
37. Wendel H (1981) Colloid Polym Sci 259:908
38. Goldbach G, Retting W (1978) In: Ullmanns Encyklopädie der technischen Chemie, Verlag Chemie, Weinheim 15:219
39. Graessley WW (1982) Adv Polym Sci 47:47
40. Pechhold W, von Soden W, Stoll B (1981) Makromol Chem 182:573
41. p 249 in [14]
42. Laun HM (1979) Rheol Acta 18:478
43. Zosel A (1971) Kolloid-Z u Z Polymere 246:657
44. Klein J et al. (1984) Rheol Acta 23:277
45. Laun HM, Meißner J (1980) Rheol Acta 19:60
46. Franck AP (1984) J Rheol 28:492
47. Laun HM (1986) IN: Proceedings of the 2nd Conference of European Rheologists, Prague, to appear in Rheol Acta
48. Cox WP, Merz EH (1958) J Polym Sci 28:619
49. Pfandl W et al. (1984) Rheol Acta 23:277
50. DIN 53753 (1983) ASTM 12/83
51. Zosel A (1971) Rheol Acta 10:215
52. Vlachopolus J (1981) Rev Def Behav Mat 3:219
53. Uhland E (1979) Rheol Acta 18:1
54. Laun HM (1982) Rheol Acta 21:464
55. Fleißner M (1981) Angew Makromol Chem 94:197
56. Wang J-S et al (1978) J Polym Sci, Physics 16:1709
57. Wang J-S et al (1970) J Polym Sci, Letters 8:671
58. Rokudai M, Fujiki T (1981) J Appl Polym Sci 26:1343
59. Meißner J (1965) In: Proc IVth Int Congress on Rheology, Part 3, page 437, Interscience Publishers, New York
60. Figure 6 on page 579 in [8]
61. Münstedt H (1981) Coll & Polym Sci 259:966
62. Meißner J (1975) Pure & Appl Chem 42:553
63. Münstedt H (1978) Kunststoffe 68:92
64. Montfort JP et al. (1978) Polymer 19:277
65. Graessley WW, Struglinski MJ (1986) Macromolecules 19:1754
66. Montfort JP, Marin G, Monge Ph (1986) Macromolecules 19:1979
67. Graessley WW (1982) Macromolecules 15:1164

Received September 18, 1986;
accepted March 5, 1987

Nonlinear rheological properties of polymer melts and their prediction based on the relaxation time spectrum

H. M. Laun

Kunststofflaboratorium, BASF Aktiengesellschaft, Ludwigshafen, Rhein, F.R.G.

During processing, melts are submitted to high deformation rates and high total deformations. Under these conditions their rheological behaviour, in general, is nonlinear. For instance, the viscosities, normal stress coefficients, and elastic compliances will be deformation rate dependent.

Yet the starting point of a rheological characterization should be the determination of the linear viscoelastic behaviour. This can be done by measuring any linear viscoelastic material function, e.g. the shear relaxation modulus $\mathring{G}(t)$. It may be convenient to represent the material function by a relaxation time spectrum [1]. For simplicity we use a discrete spectrum:

$$\mathring{G}(t) = \sum g_i \exp(-t/\tau_i). \tag{1}$$

The pertinent relaxation strengths g_i and relaxation times τ_i must be determined by a fit to experimental data, since mostly, no appropriate molecular theories nor precise molecular data are available.

It is important to note that a considerably large amount of information on nonlinear material functions, too, can be drawn from linear viscoelasticity. Firstly, the linear material functions are found to be asymptotes to the real material behaviour. Secondly, there exists a variety of (empirical) relations that interrelate linear and nonlinear material functions (e.g. Cox-Merz-relation [2], Gleissle's Mirror-law [3], etc. [4]).

Interestingly enough, only a little attention has been paid to the consequences of linear viscoelasticity on the prediction of recoverable strains. In the case of simple shear

$$\gamma_r(t) = \gamma(t) - \frac{1}{\eta_o} \int_{-\infty}^{t} \sigma_{21}(t')\, dt' \tag{2}$$

relates the recoverable strain γ_r to the total strain γ and the history of the shear stress σ_{21}, η_o being the zero shear rate viscosity [5]. Analogous equations are valid for recoverable elongational strains. From Eq. (2) it follows that recoverable strains and flow birefringence (validity of the stress-optical law is verified) are different measures of molecular orientation.

As a starting point for a constitutive equation of melts hodge's rubberlike-liquid theory [6] can be used, in which the melt is regarded as a temporary network:

$$\sigma(t) = -p\,1 + \int_{-\infty}^{t} \mathring{m}(t-t')\, C_t^{-1}(t')\, dt'. \tag{3}$$

Here σ is the stress tensor at current time t, p the hydrostatic pressure and $C_t^{-1}(t')$ the Finger relative strain tensor with respect to the past time t'. The only material dependent quantity is the linear viscoelastic memory function $\mathring{m}(t-t')$,

$$\mathring{m}(t) = -\frac{\partial \mathring{G}(t)}{\partial t}, \tag{4}$$

which is unaffected by the flow. In other words, the creation and loss of entanglements is independent of the magnitude and the rate of deformation. As a result, both the viscosity and the normal stress coefficient in simple shear are independent of the shear rate.

By introducing a factorized strain dependent memory function

$$m(t-t', I_1, I_2) = m(t-t')\, h(I_1, I_2) \tag{5}$$

into Equation (3), Wagner [7-9] could considerably improve the description of the real material behaviour. The attenuation function is dependent on the invariants I_1 and I_2 of the Finger relative strain tensor and describes the additional loss of entanglements due to increasing strains. An experimental verification of the factorized memory function can be given by step strain tests [1, 4].

The time and shear strain dependent modulus $G(t, \gamma)$ in a step shear experiment of magnitude γ is now given by

$$G(t, \gamma) = \mathring{G}(t)\, h(\gamma) \tag{6}$$

since the two invariants are identical in simple shear and can be expressed by the shear strain.

For many practical applicatons, the attenuation function $h(\gamma)$ may be approximated by a single exponential

$$h(\gamma) = \exp(-n\gamma). \tag{7}$$

Equation (7) has the advantage that only one additional parameter governs the nonlinearity and that analytical expressions for the nonlinear material functions can be obtained [1, 4, 10].

In spite of the success of the Wagner-type constitutive equation, which is formally equivalent to a BKZ theory [11], as long as reverse deformations are excluded, examples for the limitations of the validity of Equation (6) can be given both for simple shear and uniaxial elongation.

Using the same relaxation modulus $\mathring{G}(t)$ as in Equation (6), the attenuation function can be independently evaluated from the time dependent viscosity in constant shear rate tests [8]. For polystyrene, for example, a distinct difference of the $h(\gamma)$ functions determined by step shear and step shear rate experiments was observed [4]. When applied to constant strain rate tests in uniaxial elongation of a LDPE melt, the attenuation function was found to partially depend on the strain rate [12], in contradiction to the basic assumption, viz. separability of time and strain.

In addition, the irreversibility of entanglement destruction has to be taken into account in the case of decreasing strains [11]. This is of importance for high amplitude oscillatory shear deformations and the prediction of reversible strains.

The irreversibility also complicates the calculation of recoverable strains in the nonlinear region. As a consequence, the inversion of the integral in the constitutive Equation (3) can only be done numerically. It has been found, however, that recoverable strains after steady-state flow can be obtained without integral inversion by using the concept of deformation rate-dependent effective relaxation strengths [4].

References

1. Laun HM (1978) Rheol Acta 17:1–15
2. Cox WP, Merz EH (1958) J Polym Sci 28:619
3. Gleissle W (1980) In: Astarita G, Marrucci G, Nicolais L (eds) Rheology, Vol 2, Plenum Press, New York, pp 457–462
4. Laun HM (1986) J Rheol 30:459–501
5. Laun HM (1981) Coll & Polym Sci 259:97–110
6. Lodge AS (1964) Elastic Liquids, Academic Press, New York
7. Wagner MH (1976) Rheol Acta 15:136–142
8. Wagner MH (1979) Rheol Acta 18:33–50
9. Wagner MH, Meissner J (1980) Makromol Chem 181:1533–1550
10. Laun HM, Wagner MH, Janeschitz-Kriegl H (1979) Rheol Acta 18:615–622
11. Wagner MH, Stephenson SE (1979) J Rheol 23:489–504
12. Laun HM (1980) In: Astarita G, Marrucci G, Nicolais L (eds) Rheol, Vol 2, Plenum, New York, pp 419–424

Author's address:

H. M. Laun
BASF Aktiengesellschaft
Kunststofflaboratorium - G 201
D-6700 Ludwigshafen, F.R.G.

Discussion

WEYMANS:

The time-dependent rheological functions you measured not only depended on time, shear rate and mean molecular weight, but also on the molecular weight distribution, as Grassley and others have shown. How do you take these effects into account in your theory?

LAUN:

I start with measuring the linear viscoelastic behaviour which is strongly dependent on the molecular mass distribution. I'm not interested in the spectrum itself, but in the analytic representation of the modulus $G^0(t)$ over a very wide time scale range. There is also some influence of the molecular mass distribution on non-linearity, but in my experience the influence on non-linearity is minor compared to that on the linear viscoelasticity.

ILAVSKY:

At the end of your talk, you showed that there are some problems in describing experimental data. Generally one can ask to what extent of shear rates the dependences of the memory function can be separated into time dependences and deformation dependence. We tried to do this for rubbers and we found, for chemically cross-linked networks, that in the region, say from elongation 1 to 2.5, this can be separated into deformation and time dependence. But if we then go to higher extensions, the stress relaxation, which depends on time and on extension, is no longer separable into these two parts. How it is in the melts? Are there some limits of separation?

LAUN:

There are some limits. Separability of time and deformation in a step shear test is found in the range of long relaxation times which are relevant for viscous flow. Experimental data by Osaki on polymer solutions clearly show that in the range of short times, say milliseconds, the modulus becomes independent of the shear strain. In that range, separability does not work. The behaviour of melts is most probably similar to that observed in concentrated solutions.

In addition, our experimental results on long-chain branched LDPE melts clearly show that even at relatively low Hencky strain rates, the attenuation functions, evaluated from the time dependent stressing viscosities, are different for the various strain rates. It must be said, however, that the failure of separability is distinct only at relatively high total deformations.

WINTER:

Is there a limit to high rates of deformation? Can you apply it to high shear rates?

LAUN:

The validity of the factorized memory function can be examined by comparing time-dependent shear stresses in different types of deformation. Here, we find that it obviously makes a difference whether we do a step shear experiment, where we have very high shear rates only during the "step", or if we do a constant shear rate experiment, where we have a low deformation rate throughout the whole test. For polystyrene we get quite different attenuation functions. It is difficult, however, to give numbers for the strain rates at which separability begins to fail.

WINTER:

I think you want to make a correlation that the short-time behaviour in step strain might also be seen in a steady shear flow at very high rates, where you also look at a short-time scale.

MEISSNER:

I think what you told us here is new that you found that for step shear and step shear rate experiments the attenuation functions are not exactly the same. Is the reason for this that you used another material? Your old findings and the findings of Dr. Wagner on LDPE were that they both agreed. The time scale on which this factorization is possible is, of course, also a matter of the material chosen?

LAUN:

Part of the data have already been presented at the International Congress of Rheology in Naples in 1980 [12]. The simple concept of a factorized memory function is of great practical value for engineering applications. Since it is an empirical approximation, we tried to examine its limitations to know the range of validity. Originally, we were developing equipment to perform step shear tests as the most simple method to directly determine the attenuation function. Later we realized that the $h(\gamma)$-functions evaluated in this way were not appropriate to predict constant shear rate tests of linear polymers. It was, in fact, by accident that the single exponential, which has been introduced by Wagner, nicely coincided with the attenuation function from step shear tests for the LDPE melt we have thoroughly investigated as a model melt.

MEISSNER:

You showed us the recoverable deformations which are really extremely high. You have much more recoverable elastic deformations than you have in natural rubber, for instance. So this again supports the concept of a rubber-like liquid. I would like to ask you just to give us again the maximum values which you had in elongation and in shear, so that we get a feeling of how large the recovery of this material is.

LAUN:

In elongation we have steady-state recoverable strains up to 2.2 in the Hencky strain measure. This corresponds to a recoverable stretch ratio of $\lambda_r = e^{2.2} \approx 9.0$. In shear we measured recoverable shear strains up $\gamma_r \approx 9$.

BOUÉ:

Did you make any comparison with the Doi-Edwards model, for example, if you did step shear?

LAUN:

We have measured the attenuation functions for linear materials but these were not narrowly distributed. These damping functions are decreasing much more steeply with shear strain than those of highly branched material like LDPE.

Comparisons with the literature from other investigators show that for narrowly distributed materials, the Doi-Edwards prediction comes close to the experimental results.

KILIAN:

I would like to ask you whether you can derive the mean degree of orientation from the birefringence experiments?

LAUN:

We did not do birefringence measurements on our own, but have used data of Janeschitz-Kriegl. From his data, we determined spectra and so on, using the stress optical law. We have not used birefringence to evaluate the degree of orientation.

Progress in Colloid & Polymer Science Progr Colloid & Polymer Sci 75:140–145 (1987)

Transient networks by hydrogen bond interactions in polybutadiene-melts

R. Stadler

Institut für Makromolekulare Chemie – Hermann-Staudinger-Haus, Freiburg, F.R.G.

Abstract: Transient networks are intermediate between covalently crosslinked networks and entangled polymer melts. The linkages between polymer chains are formed by secondary valence interactions. As a consequence, transient networks may have properties of both extreme cases. The questions of equilibrium network properties and dynamic properties in relation to the network structure are of special interest for the understanding of these systems. To study structure-property relationships in transient networks, polybutadienes with few functional groups, that form hydrogen bond complexes, are used as model system.

Key words: Transient networks, rheological properties, polymer modification, hydrogen bonding.

Introduction

Polymer networks are three-dimensional infinitely large arrangements of long chain molecules. Polymer networks may be classified by the strength of the network forming linkages between linear polymer chains. In the case of covalently crosslinked systems, the energy to break the linkage is in the order of 100–400 kJ/mol. On the other hand, the "bond energy" of entanglement networks is zero[1]). Transient networks are characterized by crosslink stabilities between these two extreme cases. Secondary valence interactions cover this range of bond energies. Thus the type of networks to be discussed is based on crosslinks by dipolar, coulombic or hydrogen bond interactions.

The energy required to break such linkages may be obtained either by heating the material to higher temperatures (below the temperatur at which chain scission occurs) or by adding a convenient solvent. At lower temperatures or after evaporation of the solvent the network may form again, i. e. transient networks are reversible in many cases. Typical examples of reversible networks are block copolymers, ionomers and many biopolymer gels.

To give a complete description of a polymer network, information about formation, structure, dynamics and mechanical properties is required. In covalently crosslinked systems, structure information may be obtained from the analysis of the gelation process (see Ilavsky and Dusek, this volume) and/or from neutron scattering experiments of endlinked "model" networks as well as of statistically crosslinked networks. This structure information may be correlated to the elastic and orientational properties of these networks. Much less information about structure-property relationships is available for transient networks.

Before some results on a transient "model" network are presented, some basic differences between permanent and transient networks will be firstly discussed:

[1]) The analysis of the temperature dependence of the viscoelastic functions according to WLF [1] (viscosity, logarithmic shift factors a_T) shows, that the local activation energy of flow E_a is independent of the molecular weight (the constants C_1 and C_2 in the WLF-equation are independent of the molecular weight). Thus the topological constraints that are responsible for the molecular weight dependence of rheological quantities (viscosity, width of the rubbery plateau) are of different nature in comparison to the localized interactions in transient networks, where the linkages are based on dipole, coulombic or hydrogen bond interactions. On the other hand, the activation energy of local flow in covalently crosslinked systems of low crosslink density is similar to the melt, i. e. the few fixed linkages do not influence the local chain dynamics.

Formation

In covalently crosslinked systems, the introduction of the chemical unit that forms the linkage and the formation of the three-dimensional structure occur simultaneously. If a crosslink once is formed, it is fixed (changes in the network structure of sulfur cured systems are neglected for simplification). In most of the transient networks based on secondary valence interactions, the introduction of the chemical linking unit and the network formation are separate or can be separated easily. Due to the lower free energy of formation, a considerable amount of chemical linking units, that do not form crosslinks may be present in the thermodynamic equilibrium.

Structure

In covalently crosslinked systems any type of network defects, like loose ends, loops, inhomogeneities by improper crosslink conditions, will be fixed and not change with time and by the application of an external force. The structure can be controlled by the crosslink conditions (dilution, stochiometric ratio, curing temperature). Transient networks may change their structure with temperature (change of the equilibrium constant), time (healing), by the action of an external force or by adding a solvent, depending on the kinetics of formation and breaking of the secondary valence forces.

Properties

Covalently crosslinked networks are viscoelastic solids, i. e. they show an equilibrium retractive force after the application of a load, even after a very long time. Molecular theories of rubber elasticity can be used to describe the stress-strain and the optical behavior of covalently crosslinked networks. On the other hand the entanglement networks are viscoelastic liquids, i. e. no equilibrium modulus is present in these systems. Roughly speaking, the different properties of the networks reflect the fact that in entangled polymer melts no discrete localized junctions are present. Consequently, transient networks take an intermediate position. The existence of a pseudo-equilibrium state, or the occurrence of viscous flow, may depend on the conditions of network formation and testing (i. e. temperature) and the type of secondary valence interaction.

According to the theories of rubber elasticity, the mechanical properties of networks are influenced by topological constraints imposed on the network chains by the action of the crosslinks. In transient networks, the topological constraints can be varied quite easily. Thus transient networks – though no model is present to understand theoretically the systems in detail – are convenient model systems to investigate the relations between entanglement networks and covalently crosslinked systems.

To investigate the relations between formation and properties of transient networks a special type of model system has been prepared: by polymer analogous reaction a certain number of specifically interacting groups are attached to a polybutadiene backbone (Fig. 1 a). As functional groups 1, 2, 4-triazolidine-3, 5-diones (urazoles) are used [2–5].

The urazole groups form hydrogen bond complexes (Fig. 1 b). In Figure 2 the infrared spectra of the carbonyl-streching vibrations of the urazole ring system are shown for different temperatures. The vibration of complexed (1701 cm^{-1}) and free (1723 cm^{-1}) carbonyl groups are clearly distinguished [6]. The association between two urazole groups is given by a simple thermodynamic equilibrium reaction

$$U + U \xrightarrow{K} U_2$$

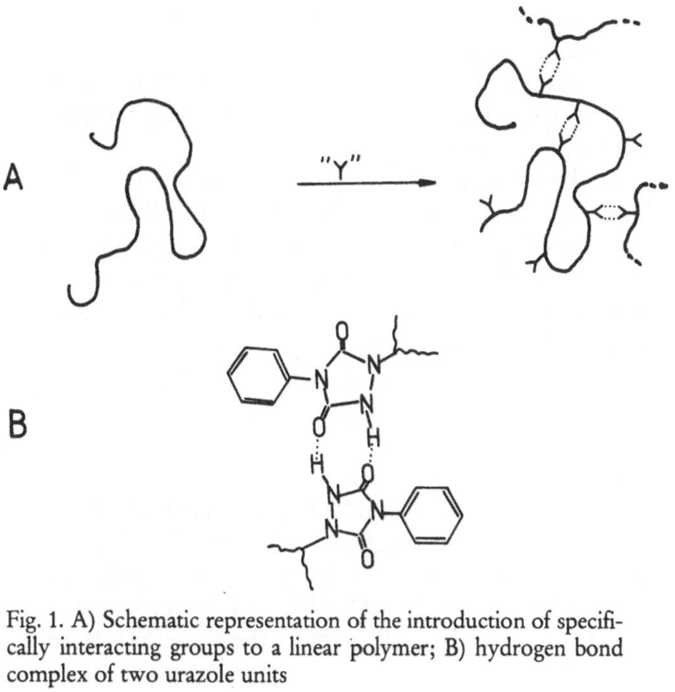

Fig. 1. A) Schematic representation of the introduction of specifically interacting groups to a linear polymer; B) hydrogen bond complex of two urazole units

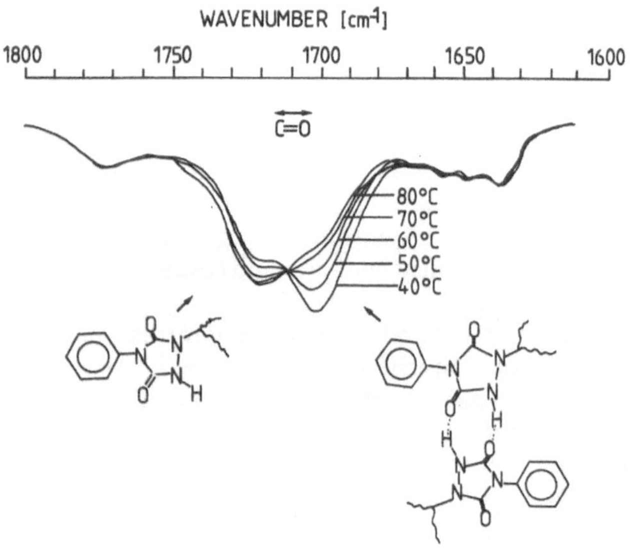

Fig. 2. Temperature dependent infrared spectra of the $C{=}O$-streching vibration; 1% modification, temperatures as indicated

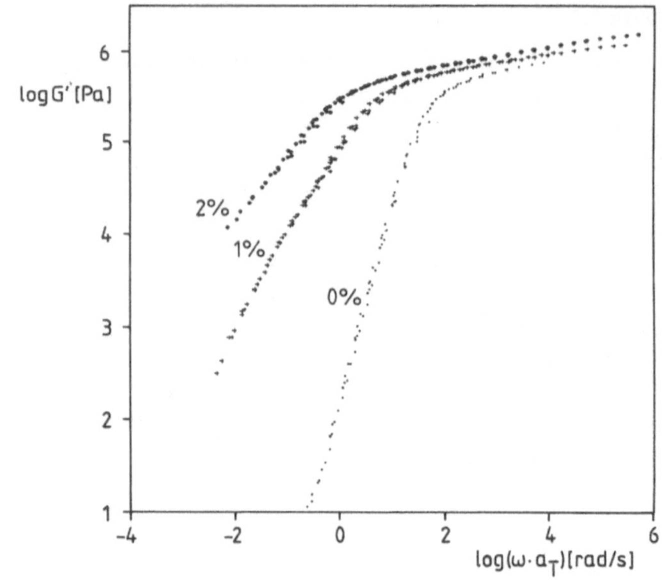

Fig. 3. Storage modulus G' as a function of the reduced frequency $\omega \cdot a_T$ for various urazole contents: (●) 0%; (+) 1%; (*) 2%

characterized by the equilibrium constant K. Thus this system should offer the possibility of giving a quantitative correlation between the number of associating functional groups and the mechanical properties of the resulting transient elastomeric networks. The network structure should be a unique function of this thermodynamic equilibrium state, which may be varied by changing temperature, number of urazole groups per chain and concentration.

In addition, by chemical modification of the urazole moiety, associating structures of increasing complexity may be realized [7].

In the following, some experimental results will be given to visualize how hydrogen bonds influence the rheological properties of an unpolar polymer melt matrix. Finally, some open questions concerning the formation and structure of these reversible networks will be discussed briefly.

Results

Polybutadienes of different molecular weights and narrow molecular weight distribution (MWD) have been used. The results are given here for a polybutadiene of $M_n = 48\,500$ $M_w/M_n = 1.06$. For a more detailed experimental description the reader is referred to [4, 5].

In Figures 3 and 4, the storage and the loss modulus master curves are given for samples where 0%, 1%, and 2% of the chain repeating units carry urazole groups. The functions were obtained by simple horizontal shifting of reduced isothermal data [4]. Smooth

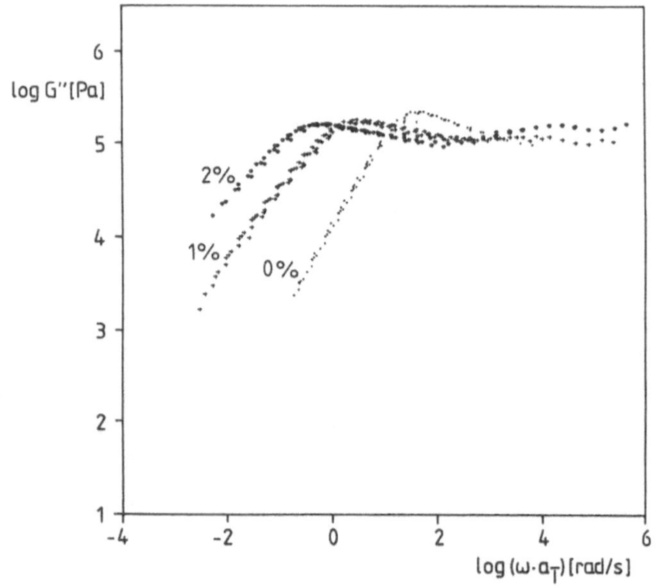

Fig. 4. Loss modulus G'' as a function of the reduced frequency $\omega \cdot a_T$; symbols as in Figure 3

viscoelastic master curves indicate a thermorheological simple material, at least in the experimental frequency scale (0.04–7.5 Hz). It is evident that the transition from the rubbery plateau to the flow region is shifted to lower frequencies with increasing number of urazole groups. The loss modulus maximum – a measure for the longest relaxation times in monodisperse polymer melts [8] – is shifted to lower frequencies. In addition the pronounced sharp maximum of the unmodified sample is lowered and broadened, indicating changes in the relaxation mechanisms.

In Figure 5, the dynamic viscosities η' are given. In the non-Newtonian high frequency region the viscosity of the unmodified and the modified samples are the same, i. e. the influence of the hydrogen bonds on the properties is of minor importance, if the rubbery plateau is reached. The basic difference between modified and unmodified samples is the increase of the zero shear viscosity η_0 with increasing modification, and the onset of the non-Newtonian regime at much lower frequencies $\omega \cdot a_T$. The same is observed, if polymers of increasing molecular weights are compared [1]. Thus the first conclusions are:

– by the association of urazole groups the zero shear viscosity increases, corresponding to an apparent increase of the molecular weight. This increase is related to the number of urazole groups per chain [5]

– the observed transition to flow indicates, that the hydrogen bonds form no network structure, that shows an equilibrium network modulus. This can be shown in detail by the quantitative analysis of the relaxation time spectrum [5, 9].

In Figure 6 the storage compliance J' is plotted versus the frequency. For narrow distributed polymers, the limiting value at low frequencies J_e^0 is independent from the molecular weight [1, 8, 10], but increases strongly with polydispersity [1, 11]. The same behavior is observed for the modified samples, i. e. the hydrogen bond interaction has a similar effect on the storage compliance J' as an increase in the molecular weight distribution. This result can be interpreted along the same line as the viscosity data:

By the reversible association between urazole groups, large macromolecular clusters are formed. Due to the statistical distribution of the functional groups along the chain contour, the association process leads to highly branched clusters of broad apparent MWD. Of course the comparison of the highly branched dynamic clusters in the hydrogen bond networks with linear chains of varying molecular weight and molecular weight distribution is somewhat artificial. For each temperature and testing frequency a new equilibrium network structure will form. Consequently no equilibrium network modulus is detecta-

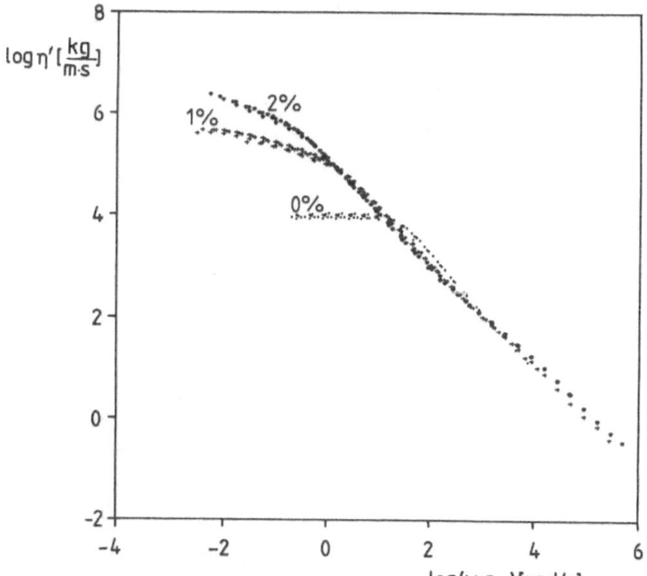

Fig. 5. Dynamic viscosity η' as a function of the reduced frequency $\omega \cdot a_T$; symbols as in Figure 3

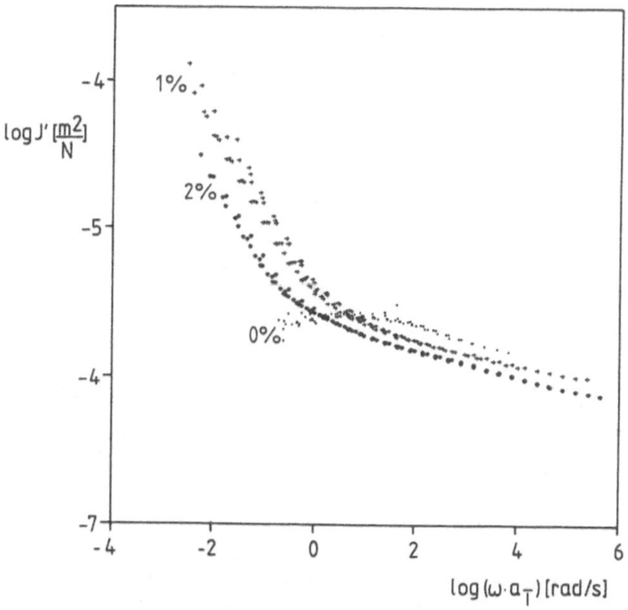

Fig. 6. Storage compliance J_e' as a function of the reduced frequency $\omega \cdot a_T$; symbols as in Figure 3

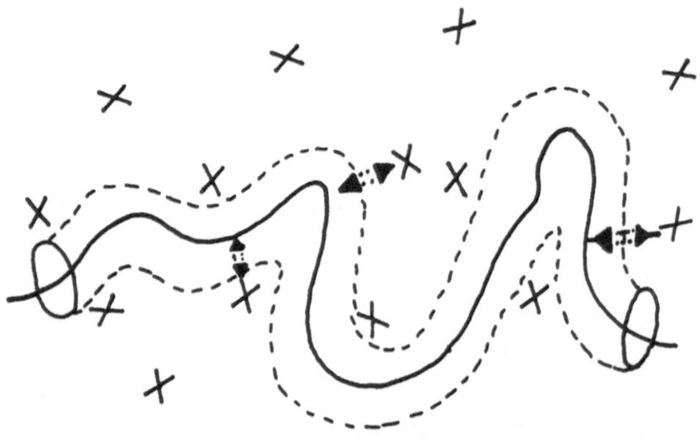

Fig. 7. Schematic representation of the hindered reptational motion of a chain, carrying associated functional groups

ble. As discussed above, this disadvantage should be overcome by the fact that the equilibrium constant that determines the network formation is available by independent spectroscopic techniques.

In the first attempt the experimental findings can be modelled by a hindered reptational motion (Fig. 7). A corresponding model, obtained for ionomeric systems in dilute solution [12], is presently extended to the hydrogen bond networks [13].

A first analysis shows, that the equilibrium constant K is not only dependent on the temperature, but also on the number of functional groups per chain: each hydrogen bond complex induces additional restrictions on the polymer chain. Thus these topological restrictions reduce the number of formed complexes, as they would be expected to from the concentration and the equilibrium constant of a low molecular weight model [14], or of a system with a low degree of modification.

Acknowledgement

The author is indebted to L. de Lucca Freitas, J. Burgert, and Dr. M. Möller for helpful discussions. Financial support from the Bundesministerium für Forschung und Technologie (Project BASF-Freiburg) and Stiftung Volkswagenwerk (Joint-Project Freiburg-Porto Alegre) is gratefully acknowledged.

References

1. Ferry JD (ed) (1980) Viscoelastic properties of polymers, John Wiley & Sons, New York
2. Leong KW, Butler GB (1980) J Macromol Sci A-14:287
3. Stadler R, Burgert J (1986) Makromol Chem 187:1681
4. Stadler R, de Lucca Freitas L (1986) Coll & Polym Sci 264:773
5. de Lucca Freitas L, Stadler R (1987) Macromolecules, in press
6. Stadler R, de Lucca Freitas L (1986) Polym Bull 15:173
7. Burgert J, Stadler R (1987) Chem Ber 120:691
8. Graessley W (1982) Adv Polym Sci Vol 47
9. de Lucca Freitas L, Stadler R, Rheol Acta, to be published
10. Carella JM, Graessley WW, Fetters LJ (1984) Macromolecules 17:2775
11. Struglinski MJ, Graessley WW (1985) Macromolecules 17:2630
12. Gonzalez AE (1983) Polymer 24:77
13. Stadler R, Macromolecules, in press
14. Auschra C, de Lucca Freitas L, Abetz V, Stadler R, Ber Bunsenges Phys Chem, submitted

Received January 16, 1987;
accepted March 5, 1987

Author's address:

R. Stadler
Institut für Makromoleculare Chemie
Universität Freiburg
Stefan-Meier-Straße 31
D-7800 Freiburg, F.R.G.

Discussion

PETER:

At the beginning, you had a very interesting question. Can thermoplastic elastomers substitute chemical crosslinked rubbers? What is your opinion?

STADLER:

Not this type of thermoplastic material. I think all types of thermoplastic elastomers that are not phase-separated like the sulphonated systems, like the type presented here show, at long-times, flow behaviour. For applications where long-time properties are required, these are not adequate materials.

PETER:

Do you think that it is possible to make soft particles, maybe in the shore range of 50, 55 by this type of modification?

STADLER:

Yes, I think there would be a possibility of making a suspension of rubber latex particles and fixing the hydrogen bond-forming group on the surface of this suspension, in order to stabilize the colloidal particle, so that the polybutadiene can be processed more easily.

LAUN:

I would like to know whether you have any idea of the life-time of your hydrogen bonds compared to the reptation time of a molecule in the melt?

STADLER:

The dynamics of the hydrogen bond formation are known, from low-molecular weight chemistry, to be in a frequency range 10^{11}–10^{12} s^{-1}. These rates are very high, compared to the frequency range we look at in the mechanical measurement. So, I think the important quantity is not the actual life-time, i. e. the dynamics of the hydrogen bonds, but the fraction of time that a hydrogen bond is formed or is open. This time fraction is just the inverse of the equilibrium constant [1]. The reciprocal of the equilibrium constant describes the fraction of groups that is involved in hydrogen bonds or, in other words, the fraction of time that one group is not involved in a hydrogen bond. We just work on the application of a modified Doi-Edwards theory, where we assume that only the chain can reptate or move for a certain time span. The time span when a chain is free should be proportional to the equilibrium constant of the hydrogen bond to the number of effective bonds. It seems that this works quite reasonably [1].

[1] Stadler R, Freitas L: Macromolecules, submitted.

WINTER:

The mechanism might be quite uniform in your case but with a normal regular polymer we have a distribution of relaxation times. Your interference is in a very specific manner (that makes it little simpler) but it is distributed.

VILGIS:

Can you say something about the glass transition temperature as the function of the number of these units?

STADLER:

The glass transition temperature for the butadiene is about − 100 °C, and if we only go up to 5 % of modification, the glass transition only increases to about 4–5 °C, but then it increases very strongly for higher degrees of modification. So if we go to 20 % hydrogen bonds, then the glass transition is very hard to detect in the range of 280 °C and 300 °C. At very low levels of modification, we have a small influence.

Leading Contribution, published in Colloid & Polymer Sci 264:829-846 (1986)

Multiaxial elongations of polyisobutylene and the predictions of several network theories

A. Demarmels[1]) and J. Meissner

Institut für Polymere, Eidgenössische Technische Hochschule Zürich, Zürich, Switzerland

Abstract: The experimental investigation of the rheological behaviour of polymer melts is mostly performed by simple shear and elongational tests. In former communications we described the performance of more general flows, viz. multiaxial elongations with different ratios of the principal components of the strain rate tensor. In this paper the results obtained with polyisobutylene at 23 °C are compared with the predictions from different constitutive equations for these types of flows in order to examine whether these equations can describe the general nonlinear deformation behaviour. Network theories are known to do so for simple shear and simple elongation. Therefore, four types of network theories are taken into consideration. Their predictions agree only qualitatively with the measured data, and this means that large differences may exist. It is astonishing that for tests with a variation of the test mode during the deformation period, out of all equations taken into consideration here the linear viscoelastic equation predicts the best description of the experimental data. The conclusion is that the generalization of the material behaviour known from simple test programs to the behaviour in more general deformations may lead to very wrong predictions.

Key words: Polymer melt rheology, multiaxial elongation, network theory, polyisobutylene, linear viscoelaticity

[1]) Present address: Brown, Boveri & Cie., Zentrallabor Kunststoffe CH-8050 Zürich, Switzerland.

Authors' address:

Prof. Dr. J. Meissner
Institut für Polymere
Eidgenössische Technische Hochschule Zürich
CH-8092 Zürich, Switzerland

Discussion

LAUN:
We have performed measurements in uniaxial elongation and simple shear on the same material. And we have compared our results with yours, and our conclusion is that, considering the non-linearity in the rheological behaviour or the deviation from the linear viscoelastic case of simple elongation, uniaxial elongation and simple shear represent the extreme cases.

MEISSNER:
Correct.

LAUN:
If you want to say whether non-linearity, or how non-linearity, is influenced by the structure of the material, you should do these tests. Do you agree?

MEISSNER:
Yes, with one exception, namely planar elongation. To me it is fascinating that we found the opposite behaviour in two perpendicular directions. In the direction of stretch, the stress increases above the linear viscoelastic case, whereas in the perpendicular direction of constant dimensions the stresses are dramatically reduced with respect to the linear viscoelastic prediction. I do not know which structural parameter is responsible for this deviation, because until now we have worked with one material only.

LAUN:
But the stresses are still time dependent. The two planar elongational viscosities are within the limit given by simple shear and uniaxial elongation. This is at least our conclusion.

But let me come to my second question:

Based on your viscosity results of planar elongation, you come to the conclusion that the factorized time and strain-dependent memory function, Wagner's approach for planar elongation, does not work. I'm sorry to say, I'm not fully convinced because at the moment it is difficult to judge the reliability of the experiments. Of course, we do not have other experiments. We are glad we have these experiments, but we cannot judge the reliability range at the moment. So we do not see what part of the discrepancy may be due to experimental uncertainties and what is a failure of the theory. In that respect, I think we have the question of temperature constancy, because polyisobutylene is extremely temperature sensitive (around room temperature) and you do not use a temperature bath. So I think temperature control makes some problems.

MEISSNER:

I agree where the last part of your comment is concerned. We had to learn that room temperature is not room temperature, but when we found that out we reduced all the data to 23 °C. As I mentioned already we cannot argue about the precision of our data. But the stresses which we measure in planar elongation for the perpendicular direction, are so much smaller than the linear viscoelastic predictions that this is really out of the inaccuracy which you mention. We do not intend to improve the present rheometer but to build a new one instead, with much smaller dimensions and an extended temperature range up to 200 °C.

STADLER:

Don't you have troubles with stress-induced crystallization of the polyisobutylene?

MEISSNER:

No, these stresses are not so high and I should mention that our material was not an extremely high molecular weight polyisobutylene. We work with Oppanol B 15 and its crystallization occurs at very large stresses only.

TOMKA:

To clarify this point, I could tell you an experimental result we worked out with the same polyisobutylene in simple elongation. The maximum elongations which were reached in the multi-axial experiments by Demarmels were 25-30. We achieved, in simple elongation, values of 500, 600, and up to even these large values, the orientation of the material is so low that we don't achieve a birefringence of more than 10 % of the inherent birefringence. So the second moment of the orientation distribution function of the segments is less than 0.1.

STADLER:

But just polyisobutylene is a very complicated system in which to detect the orientation from birefringence, because the calculations on the base of the isomeric state models show that it doesn't work in the case of polyisobutylene.

TOMKA:

You don't need this. You know the birefringence, you can compare X-ray diffraction data of the amorphous halo distribution (e. g. R. S. Stein et al.). From that you can well estimate the intrinsic birefringence of this material. If you reach 10 % of this value, the material is oriented very little, and this corresponds to $\lambda \sim 400, 500$. In these experiments, we had so low a birefringence that we couldn't even measure it and we really can exclude crystallisation.

WINTER:

A model which is very popular in America is the so called Giesekus model. Did you ever try to compare your data with predictions from that model?

MEISSNER:

No, we did not. We checked only the equations which I showed.

WINTER:

Then you realized that they were not able to describe the experiments. People think the Giesekus equation is a very suitable equation to describe experimental observations.

GIESEKUS:

A certain handicap for the use of the model under discussion [1] is that it was elaborated and discussed in detail only as a one-mode model, which does not allow fitting to a presupposed linear-viscoelastic behaviour.

Originally it was, however, designed as a multi-mode model, and recently G. Eder, a doctoral student of H. Janeschitz-Kriegl, has applied it in this form, although with some simplifying suppositions, to multiaxial extensional and to interrupted shear flows [2]. Fairly good predictions of experimental results are observed, which are not too different from those of the Marrucci [3] and Phan Thien-Tanner [4] models. Advantages of our model are, however, that it predicts – in contrast to the Marrucci model – a non-disappearing second normal-stress difference and does not require, in contrast to the Phan Thien-Tanner model, a generalized Oldroyd (or Gordon-Schowalter) derivative which is known to lead to strange predictions in some more complicated types of flow, cf. Giesekus [5].

1. Giesekus H (1966) Rheol Acta 5:29-35; ibid (1982) 21:366-375; (1982) J Non-Newtonian Fluid Mech 11:69-109; ibid (1983) 12:367-374; ibid (1985) 17:349:372
2. Eder G (1986) Auswertung rheologischer Zustandsgleichungen für Polymerschmelzen, Dissertation Johannes-Kepler-Universität Linz
3. Marrucci G (1972) Trans Soc Rheology 16:321-330; et al. (1973) Rheol Acta 12:269-275; (1976) J Non-Newtonian Fluid Mech 1:125-146
4. Phan Thien N, Tanner RI (1977) J Non-Newtonian Fluid Mech 2:353-365
5. Giesekus H (1984) J Non-Newtonian Fluid Mech 14:47-65

MEISSNER:

I am very happy that I can spread the message that there are more modes of deformation now available than just simple elongation and simple shear. And if the people who work on the models, have those data, they should keep in mind not only to compare simple shear and simple elongation, but to determine the model predictions for a variety of additional modes of deformation.

WINTER:

What may happen on molecular terms? I would just like to put forward a few things by thinking a loud. First, à question: What was your Polyisobutylene?

MEISSNER:

It was Opponal B 15 of BASF AG, Ludwigshafen, Rhein, F.R.G.

KELLER:

Was it monodisperse?

MEISSNER:

Definitely not. It was just a commercial polyisobutylene.

KELLER:

After this, you see that the extensional flow field of an individual molecule, now depends on two factors: the strain rate and the maximum strain achieved. Now, for molecular weight distribution, you could consider that as a solution of the longer molecules in the shorter ones, which would tie up. And in such a multi-component solution, the longer molecules will respond to lower strain rates to overcome the conformation relaxation time, that is the resistance to extension. So they may start to extend while the shorter molecules are still relaxing. Now what final orientation you get will then depend on the relaxation time, that is on the strain rate appropriate to each molecular weight component and the final strain, that is the maximum extension you reach. So this I want to raise, for consideration of what final orientation you get, or whether crystallization could set in ultimately.

MEISSNER:

I only want to repeat that we can exclude this type of influence because we have relatively low stresses or small orientations. From a practical point of view, I can only make the following statement. From a theoretical point of view, you want data from melts with a small molecular weight distribution. If you really have such a material, the melt is brittle, it breaks. If you have a broad molecular weight distribution, the small molecules act as grease or lubricant, and so you can have large deformations; the very large molecules give the strength of the melt. Dr. Laun can confirm that we were very fortunate to pick up a material with a rather broad molecular weight distribution. With linear molecules and a narrow molecular weight distribution, you have a lot of trouble from the experimental point of view.

In polymer processing, the emphasis is also on materials with broad molecular weight distribution.

KELLER:

That is probably an expression of the fact, as I see it, that it is the longest molecule that we start extending in a relaxing short-molecular solvent.

Viscoelasticity and thermal equilibrium

Z. Rigbi

Technion – Israel Institute of Technology, Haifa, Israel

Abstract: The coupling between the force support by a viscoelastic rod, its length, changes in its temperature and the time elapsed from the beginning of loading is examined. General conclusions are established which show that an isothermal relaxation at constant length is a contradiction in terms. Similar conclusions are drawn relating to creep.

Key words: Viscoelastic, creep, relaxation, coupling.

This contribution deals with the behaviour of a rod made out of a viscoelastic material, for instance, polyisobutylene, under the influence of a suddenly applied uniaxial deformation or load. For the purpose of this study, several imaginary experiments are undertaken and described in the following, and the resulting behaviour of the rod is analysed.

In the first experiment, an extension to a uniform rod of a given length l_0 and volume V is applied. It is assumed that the rod has any axisymmetrical cross-section, and that the change in volume as a consequence of small changes in the length or temperature of the rod is zero or negligible. Simultaneously with the applied extension, as a result of which the length becomes l, a force is brought into being in the rod which, in accordance with its viscoelasticity, changes with time. Energy being lost, its temperature will also undergo a change. Normally, however, the temperature distribution in the cross-section will not be uniform because of the thermal transport properties of the material, and in order to limit this effect, an additional constraint may be imposed, namely, that the section is uniformly very thin, so that the temperature at any instant is spatially constant and depends upon the ambient conditions and the emissivity of the surface of the surface of the rod.

From the point of view of the person conducting the experiment, then:

1. The force, F, developed by the rod is a function of the instantaneous length, l, imposed and the time elapsed since the beginning of the experiment, t. This is actually the usual relaxation test, which is conducted without the precautions stated above.

2. The temperature T will change as a function of the same variables.

It is preferable to deal with this problem in infinitesimal terms in order to avoid considerations of changing geometry; the broad conclusions will, however, remain the same.

In general, these relations may be described by the expression $f(F, l, t, T) = 0$, the particular form of the unknown function depending upon material properties and ambient conditions. For the purposes of this analysis, we may parameterize this relation as follows:

$$F = F(l, t); \quad T = T(l, t). \tag{1}$$

In this situation, the total differentials of the force and temperature are

$$dF = (\partial F/\partial l)_t \, dl + (\partial F/\partial t)_l \, dt \tag{2}$$

and

$$dT = (\partial T/\partial l)_t \, dl + (\partial T/\partial t)_l \, dt. \tag{3}$$

Be definition, $(\partial F/\partial l)_t$ is similar in nature to the elastic modulus of the material measured at time t, the difference lying in the neglect of the initial dimensions of the rod. In order not to use the customary symbols, we shall name this derivative ε_t or ε, which is a material constant for a fixed geometry. $(\partial F/\partial t)_l$ is the rate of relaxation of the force, also a material property, symbolised by r. $(\partial T/\partial t)_l$ is the rate of cooling, (θ) which is influenced by ambient conditions, the geometry and the nature of the surface of the rod, and generally, by the diffusivity of the material, although this last condition has been removed by the foregoing considerations.

If the length of the rod is monitored to keep the load constant, Equation (2) becomes

$$(\partial l/\partial t)_F = r/\varepsilon \tag{4}$$

which is the usual creep experiment for a viscoelastic material.

Rewriting Equation (3) as follows:

$$\left(\frac{\partial T}{\partial l}\right)_F = \left(\frac{\partial T}{\partial l}\right)_t + \left(\frac{\partial T}{\partial t}\right)_l \Big/ \left(\frac{\partial l}{\partial t}\right)_F \tag{5}$$

and observing that $(\partial l/\partial T)_F$ is the expansion of the rod due to a unit change of temperature, $\alpha\, l$, we obtain the expression for the change of temperature of the rod after it had been extended for a fixed period t:

$$\left(\frac{\partial T}{\partial l}\right)_t = \frac{1}{\alpha\, l} - \frac{\varepsilon\, \theta}{r}\,. \tag{6}$$

Operating in a similar fashion, we have

$$\left(\frac{\partial l}{\partial t}\right)_T = \frac{(\partial T/\partial t)_l}{(\partial T/\partial l)_t} = \frac{r}{\varepsilon + r/\alpha\, \theta} \tag{7}$$

and

$$\left(\frac{\partial F}{\partial t}\right)_T = -\frac{r}{1 + \varepsilon\, \alpha\, \theta/r}\,. \tag{8}$$

If we compare Equation (7) with the definition of the creep rate, Equation (4), we find that the creep rate at constant force, and the creep rate at constant temperature are different, the latter being smaller. Similarly, the relaxation rate at constant temperature is smaller than that at constant length. We are therefore forced to the conclusion that creep cannot be determined at one and the same time under a constant load and a constant temperature, and that there is no such thing as isothermal relaxation at constant length. In fact, this is a contradiction in terms.

From Equations (7) and (8) we can derive

$$\varepsilon_T = \left(\frac{\partial F}{\partial l}\right)_T = -\frac{r}{\alpha\, \theta}\,. \tag{9}$$

This modulus differs from $\varepsilon_t = -(\partial F/\partial l)_t$ and depends on ambient conditions through θ. Experimentally, it can only be determined by infinitely slow changes of l so that the heat generated is dissipated before it can cause a temperature rise.

We now specify that this material has no elastic component due to entropy, and further assume that it behaves irreversibly; this implies that its internal energy remains unchanged, or that in an adiabatic experiment

$$V\,dQ = F\,dl - V\,\varrho\,dT = 0 \tag{10}$$

in which ϱ is the specific heat per unit *volume* of the rod.

Equation (10) is not an exact differential but a Pfaffian differential form, which can be converted into an exact differential by means of an integrating factor $\mu = \mu\,(l, T)$. We thus obtain the thermodynamic function ϕ as

$$d(V\,\phi) = V\,d\phi = \mu\,F\,dl - V\,\mu\,\varrho\,dT = 0\,. \tag{11}$$

The following relations can be written down directly from Equation (4) and Equation (11):

$$\left(\frac{\partial \varepsilon}{\partial t}\right)_l = -\left(\frac{\partial r}{\partial l}\right)_t \tag{12}$$

$$\left(\frac{\partial [\mu\, F]}{\partial T}\right)_l = -V\,\varrho\left(\frac{\partial \mu}{\partial l}\right)_T\,. \tag{13}$$

In our present state of knowledge, we do not know the meaning of the function μ, or what physical behaviour it represents.

I have made attempts to locate reports in the literature of experiments which would give substance to these speculations. Unfortunately, very little accurate work has been performed, almost all referring to elastic materials with small dissipative components or none at all. Most of this has been carried out by Müller [1] using his very accurate extension calorimeter. On the other hand, there is a considerable body of theoretical work [2–7], including the present, which is unsubstantiated by experiment. In accordance with the statement [8] made earlier in these proceedings, I appeal for and experimental program to support this and other speculations.

I will not be deviating very far from the subject if I state my belief that, of necessity, calorimetric experiments, even of elastic materials, must be carried out on thin films of fibres in order to avoid the non-uniform thermal fields mentioned above. Apparently, Godovsky's calorimeter [9] is so designed, although his original description [10, 11] in Russian are not available to me.

References

1. Muller FH (1965) In: Eirich FR (ed) Rheology, Vol 5, Academic Press, New York, p 457 et seq
2. Astarita G (1974) Polym Eng Sci 14:730
3. Coleman BD (1964) Arch Rat Mech Anal 17:230
4. Stastna J, Vodak F (1976) Int J Eng Sci 14:143
5. Lyon RE, Farris RJ (1984) Polym Eng Sci 24:908
6. Kilian H-G, Vilgis Th (1984) Coll & Polym Sci 262:691
7. Enderle H-F, Kilian H-G, Vilgis Th (1984) Coll & Polym Sci 262:696
8. Rigbi Z (1975) this volume also, J Appl Polym Sci 19:1611
9. Godovsky YuK (1986) Adv Polym Sci 76:31
10. Godovsky YuK (ed) (1982) Thermal Physics of Polymers, Khimia, Moscow
11. Godovsky YuK (ed) (1976) Thermophysical Methods of Polymer Characterization, Khimia, Moscow

Received October 6, 1986;
accepted April 14, 1987

Author's address:

Zvi Rigbi
Technion – Israel Institute of Technology
Haifa, Israel

Discussion

ILAVSKY:

I don't understand your first figure. Why do you suppose that T is a function of the length of the sample? Because from what you were trying to tell us, we cannot make a creep experiment.

RIGBI:

You have a rod, you pull the rod, it's a viscoelastic material, some of the force vanishes, right? Therefore, some of the work has gone somewhere into heat. So the temperature of the rod must change. I don't control it. It changes and it changes irrespective of the temperature of the walls or of the air or of the water. You cannot help that. If you take any sample of viscoelastic material and give it a quick pull, you see the rise in temperature.

KILIAN:

I have problems with the formulations presented, since you are using time like an "extensive variable". When applying the formalism of the irreversible thermodynamics, as given for example by Meixner, time-dependent internal variables (conjugated to their affinities) allow the description of coupled processes in terms of an Onsager approach.

RIGBI:

I agree with you.

KILIAN:

It appears to me as if you have used the condition of integrability so as to arrive at the results presented. In terms of the Meixner approach, one finds a well-defined series of relaxation times; the relative magnitudes of which depend on the set of variables that is taken to be constant.

RIGBI:

Yes, I agree with you. I agree with this slightly different formalization of the same problem. But it is actually a very simple formalization, and I think it gives us at least a possibility of arguing with Dr. Ilavsky, on phenomenological grounds, without going into much mathematics, and that I think is an advantage.

Dynamics of permanent and temporary networks: Small angle neutron scattering measurements and related remarks on the classical models of rubber deformation

F. Boué[1]), J. Bastide[1])[2]), M. Buzier[1]), C. Collette[3])[4]), A. Lapp[1])[2]), and J. Herz[2])

[1]) Laboratoire Léon Brillouin, CEN-Saclay, Gif-sur-Yvette, France
[2]) Institut Charles Sadron, C.R.M., Strasbourg, France
[3]) Laboratoire de Physico-Chimie Macromoléculaire, Ecole Supérieure de Physique et Chimie, Paris, France
[4]) Present address: Atochem, Levallois-Perret, France

Abstract: We observed the neutron scattering of mixtures of deuterated and normal polystyrene in bulk for samples (strips) undergoing a stress relaxation after a step-strain. Some samples were just melt (i.e. molten at $T > Tg = 100\,°C$), others were crosslinked and thus some were rubbers at $T > Tg$. The scattering measured at different times elapsed after the step-strain gives access to the dynamics at the submolecular scale (300–10 A). It is compared with the theoretical predictions that one may extract from the classical models. Observed discrepancies are tentatively interpreted following several "remarks" on the topology of the network and the possible mechanisms of deformation at the corresponding scales.

Key words: Polymer networks, rubber de formation, small angle neutron scattering, dynamics, relaxation, polystyrene, crosslinking.

1. Introduction

This paper has two aims. The first is to report our experimental work on polymer networks, which is presently the study of their dynamics and their equilibrium state under permanent deformation, through the scattering of neutrons by long labelled paths in the network. This experiment pertains to only one branch of the whole set of the neutron scattering investigations; some other branches are described in these proceedings by Richter and Picot. The second aim is to report some remarks, which arised during this work, on the physical structure of rubbers and on their deformation. We will not present them as explanations for our results, but we think that the open-minded nature of this meeting allows such a method.

These remarks are grouped in section 4. Some experimental details are given in section 2, but only a few: we will refer the reader to more detailed descriptions [1]. Concerning experimental results described in section 3, we have also chosen a presentation which is not completely systematic. One reason for this is that the treatment of our data is still incomplete.

2. Description of the experiment

Inside the network, we want to observe as large a part of it as possible, by labelling it. We then want to observe its dynamics by deforming the network suddenly, and measuring the elastic neutron scattering during the transient relaxation of the network towards equilibrium. We call equilibrium the stationary response to the given deformation, measurable a very long time after applying the strain.

The large labelled part of the network will be what we call a *labelled path.* At this stage we have to cope with problems in crosslinking the chains and characterizing the final rubber by physicochemical methods. After the crosslinking is achieved, we have to perform the sudden deformation and the SANS measurements during the transient relaxation, by means of quenching the sample at different times after the deformation.

Preparation of the network before deformation

In order to obtain large labelled parts inside the network, first of all, long deuterated polystyrene chains ($M_{WD} = 1.5\ 10^6 - 2.10^6$) in solution were mixed with non-deuterated chains of as close a molecular weight as possible (see Table 1). Then as many crosslinking tetrafunctional molecules as necessary were added to bridge each chain to other chains by many junctions (between 10 and 100). Thus, a labelled chain was transformed into a succession of labelled meshes – the part of chain between two crosslinks – one following the other, which we call a labelled path. This object is quite large: we have checked in this work that it has a Gaussian configuration inside an unnstrained bulk network; the radius of gyration of a polystyrene chain of $M \sim 2.10^6$ is then about 500 Å. Information may be recorded at all scales between 10 Å and 500 Å by SANS, if one is able to cover, say, 2.10^{-3} Å^{-1} – 10^{-1} Å^{-1} in the range of the scattering vector.

The genuine chains were produced in CRM, Strasbourg, by some of us, A. Lapp and J. Herz, and characterized after fractionation by careful light scattering and GPC measurements. Mixing the H chains with the D chains was done in solution in good solvent (toluene, chloroform). The solution was then cast. This leads to dry solid samples (in the glassy state as T_g for polystyrene is 100 °C). A few of these samples were kept; the remainder were crosslinked.

Gamma-irradiation crosslinking

In this method of preparation, the dry samples were slightly swollen, under a very low air pressure (10^{-5} Torr), by cyclopentane. The container was sealed and exposed to the irradiation of a Co60 source. Cyclopentane forms radicals under irradiation, which can be transmitted to the chains. This noticeably enhances the efficiency of the chain bridging with respect to a direct irradiation of the chains in bulk. The samples thus contained a small fraction (20 %) of cyclopentane. Because this is a theta solvent at room temperature and because the concentration in polymer is high, the chains are initially in the Gaussian state. We have to keep in mind that they are probably very entangled. After exposures of about 1 month (corresponding to 60 Mrad) the containers were broken and the samples washed; the sol part was extracted. The swelling ratio at swelling equilibrium, Q, was measured (Q is of the order of 20). Then the gels were dried back to the bulk glassy state, in which they appeared in the same shape as before crosslinking. However, they had actually undergone, from their state of preparation (20 % of solvent), a slight 3-dimensional compression of volume ratio 1.2; in one dimension, this corresponds to a compression of a factor of $(1.2)^{1/3} \sim 1.06$.

We encountered difficulties in measuring the sol part. The solid residue of the washing solvent was altogether, i.e. including solid residue, contained in about 2 solvent, about 30–40 mg, ~ 3 % of the sample weight. Absolute GPC measurements allow an estimation of 1–2 % of PS chains of molecular weight around 50 000. We believe that these small chains are residues of the original distribution, being too small to have a high probability of bridging. In a way, the crosslinking is here achieving a fractionation.

Table 1. Molecular weight of the polymers before preparation of the sample. An additional thermal degradation is not taken into account

Sample	Deuterated polymer	Non-deuterated polymer
γ-rubber and "parent" melt	PSD B1 F1 $M_{wD} = 2.6\ 10^6$ (light scattering) $M_{wD}/M_{nD} = 1.33$	PSH 507 $M_w = 1.6\ 10^6$ $M_{wD}/M_n = 1.28$
Friedel-Craft rubbers and "parent" melt	PSD 436 F1 $M_{peak} \sim 1.5\ 10^6$ $M_{wD} = 1.8\ 10^6$ (light scattering)	PSH 2441 $M_{peak} = 1.1\ 10^6$ $M_{wD}/M_n \sim 1.3$

Friedel Craft crosslinking

The dry uncrosslinked samples were swollen by dichloroethane at 20 % or 30 % of concentration, with a crosslinking agent (dichloromethylbenzene) in a fraction of x moles per mole of polystyrene monomer, and a coreactive, $SnCl_4$. Heating was at 60 ° or 80 °C for 2–3 days. Then the same operations – washing, sol extraction, swelling ratio measurement, drying – performed on gels were performed for these samples as well. Data are presented here for $x =$ 0.02 % and 0.05 % mole/mole, but we could not extract precise information on the crosslinking density because of the presence of side reactions. (The latter appear to perform the crosslinking in the given solvent-catalyst system, even with no crosslinking agent). However when x increases, the swelling ratio decreases (15–16 for 0.02 % and 11–12 for 0.05 %) as would happen if the crosslinking density increased. In all cases we think that crosslinking would benefit much by many improvements, towards a truly quantitative method. Our samples need more characterization. We will, however, use the data within this uncertainty.

Mechanical part of the experiment

If the samples are heated above T_g, the chains will be in the molten state. If uncrosslinked, the samples will be melts, entangled on a large time scale, flowing for very long periods [2, 3]. If crosslinked, the samples correspond strictly to the definition of a rubber; over very long periods, they will not flow, but keep a constant shape under the effect of both chemical permanent junctions and, probably also, trapped entanglements [3].

The stretching, from an initial length L_0 to a final length $L = \lambda L_0$ was performed at a relatively low temperature, held constant at $T=$ 117 °C, i. e. $T_g + 17$ °C, for a time of 5 s ($\lambda = 1.46$), 10 s ($\lambda = 2.14$), 20 s ($\lambda = 4.6$). The sample was then maintained at constant shape and temperature for 1 min, then quenched. Heating was achieved by dipping the sample into a temperature controlled oil bath. Quenching was achieved by taking the sample out of the bath. The samples were observed by SANS in this quenched state. Then they were dipped again in the bath, but now at a temperature of $T = 150$ °C, left for 1 min, quenched, and observed again. Finally they were relaxed for a third time, for 30 min at 150 °C, quenched, and observed. Basically, we performed measurements for a set of three values of λ, and three relaxation times at each λ, for melts and different rubbers. Given data here are for γ-rubbers irradiated at 60 Mrad ($Q = 18$–20) and Friedel Craft rubbers with 0.05 % ($Q = 11$–12) and 0.02 % ($Q =$ 15–16).

Neutron scattering

The technique of neutron measurements of stretched samples, using a bidimensional detector for anisotropic scattering, has been detailed formerly [1] and is summarized in Appendix 1. Here data are given for the scattering vector q between 10^{-2} and 10^{-1} Å^{-1} only, although other measurements were performed.

3. Some results

We will first present the data and present some results which, while scientifically new and meaningful, seem "normal" from a "common-sense" understanding of the physics of rubber: the "good" results. Then we will present more puzzling behaviours, called the "striking" results. At this stage we will make a comparison with the form factors predicted by the classical models of rubber elasticity. The latter are recalled in section 4.

3.1 Some "good" results

General behaviour

In Figure 1 the form factor is given for two values of the angle between $\vec{q}$ and the stretching axis, θ, 0 deg (||) and 90 deg ($\perp$), for different elongations λ. The general behaviour can be seen, in the two directions θ, choosing any given λ. Note that Figure 1 is a Kratzky plot, namely $q^2S(q)$ against q. An isotropic Gaussian chain will give a curve increasing from zero at $q=0$ towards a plateau at $q>1/R_g$ (R_g, radius of gyration). This is the case for the chain in the melt (Fig. 1); we can also check that this is the case for the labelled path in the rubber (Fig. 2), which basically overlaps the melt data, in spite of the slight deswelling (1.06) detailed above. Joining the plateau value, $q^2S(q)=1$ occurs at a higher q than probably expected from the first estimate for $1/R_g$ we have given above, $R_g \sim 500$ Å: the plateau should be reached within 1% for $q>5/R_g$, i.e. $q>10^{-2}$ Å^{-1}. This may be the characteristic of some thermal degradation

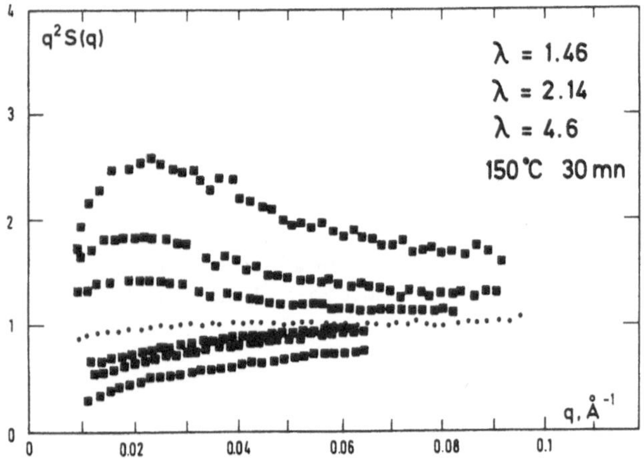

Fig. 1. $q^2S(q)$ versus q for data for γ-rubber unstretched (•), and stretched (■) at three elongation ratios for a duration of relaxation of 30 min. at 150 °C. Above the curve in dots: perpendicular direction; below: parallel

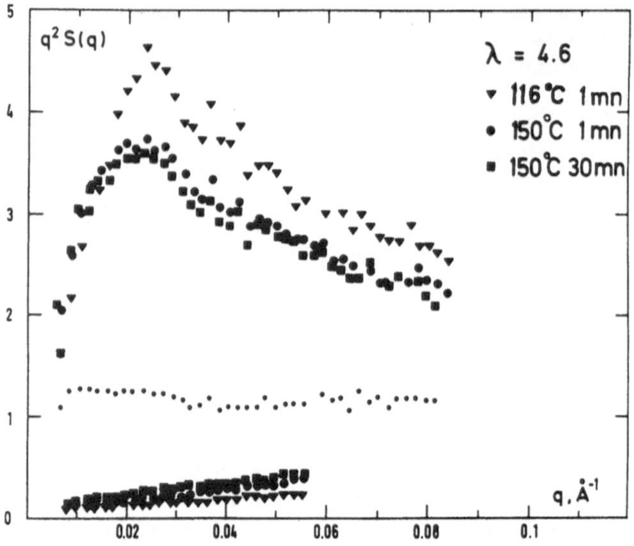

Fig. 2. Effect of the duration of relaxation for a Friedel Craft rubber (Q = 11–12 in toluene). Above the dots perpendicular direction; below: parallel

of these long chains during the drying of samples in a vacuum oven [1].

If we look now at one curve in the parallel direction, it is always lower than for the isotropic samples, and continuously increasing with q. This means that the density of monomers is smaller in the || direction, as the chains are stretched, and it returns to the isotropic value when decreasing the scale of observation. This is a well known feature, present in any model of rubber elasticity [4, 5] (see section 4): it is this progressive loss of affinitiy from the macroscopic scale down to that of the undeformed chemical bond, which allows such high elastic deformation in rubbers. On a macroscopic scale, the rubber is a solid, elastic; on the monomer scale, it is a liquid, locally much less sensitive to the macroscopic strain.

We will now consider a curve in the perpendicular direction: it is always *above* the curve for the isotropic sample. Concerning the curve on the right hand side of the maximum that it exhibits, towards the large q values, one records a continuous *decreasing*. These two features are the congrous with what was observed in the parallel direction. On the left hand side of the maximum, there is an increasing $q^2S(q)$ when q increases. This occurs whatever the direction, if $q \ll 1/R_g$ or $q \sim 1/R_g$, because in this regime $S(q) \sim S(0)(1-q^2R_g^2/3)$ varies slowly. This regime is covered here only for the perpendicular direction, $R_{g\perp}$ being small enough; the balance of the decrease at large q produces a maximum. In the parallel direction, it would not produce a maximum, but the range $q < 1/R_{g\parallel}$ would not be covered because $R_{g\parallel}$ is too large.

If the deformation was *totally* affine, i. e. if

$$\forall |\vec{r}|,\ \vec{r} \rightarrow (\bar{\bar{\lambda}}) \cdot \vec{r}, \quad \text{where } (\bar{\bar{\lambda}}) = \begin{Bmatrix} \lambda_x & 0 & 0 \\ 0 & \lambda_y & 0 \\ 0 & 0 & \lambda_z \end{Bmatrix} \tag{1}$$

in appropriate coordinates, $\bar{\bar{\lambda}}$ being the same at any scale of the sample, even the smallest, we would have, at large q, $S(q) \sim \frac{1}{(q\sqrt{\lambda})^2} \sim \frac{\lambda}{q^2}$ in the perpendicular direction, and $S(q) \sim \frac{1}{\lambda^2 q^2}$ in the parallel direction. $q^2 S(q)$ would then reach a plateau of λ in the perpendicular direction and $1/\lambda^2$ in the parallel. Experimental data show that this plateau is never reached: in the q range investigated, the form factor is very sensitive, to the loss of affinity characteristic of the rubber.

Effect of the elongation ratio

Figure 1 shows, in a simple way, how sensitive the anisotropy of the data is to the elongation ratio for any sample. It is coarsely following the predictions of classical models. However, as seen in more detail in section 3.2, the discrepancies with model curves are enhanced by increasing λ.

Effect of relaxation

Figure 2 shows the form factor after three relaxations of increasing duration for the same sample; in both directions, it approaches more and more the isotropic form factor when t increases. In other words, the distance at which the loss of affineness starts increases with time. In a liquid, the mean square distance at which a point after a time t has diffused is proportional to t if the motion is Brownian. In the Rouse model of the dynamics of a free chain [6], it is $\sim t^{1/2}$. This model can be applied to Brownian dynamics of rubbers. When looking at the dynamics of the melt [1b] we were led to calculate the form factor for the Rouse model. Using this calculation, and estimating the parameters, we arrive at an agreement with that model at short time (e. g., $t = 1$ min at 117 °C). At long periods, for rubbers, this "liquid" diffusive behaviour is expected to cease when t is large enough. Then one should reach the form factor of chains in a deformed rubber at equilibrium under deformation. We will discuss in section 3.2 whether the times involved at $T = 150$ °C here are large enough for that or not.

Again the behaviour shown in Figure 2 is general for all the samples that we have observed in addition to the data presented here.

Effect of temperature

A time temperature superposition principle is generally accepted for glass-forming polymers. We have found elsewhere [1.b] for melts, a general agreement with the WLF formula [3] using the coefficients of Tobolsky.

Effect of crosslinking

When using the Friedel Craft reaction described above, we varied the concentration x of the crosslinking agent. As a result, the main features exhibited by the obtained networks (swelling degree and velocity of swelling in good solvent, elastic and viscoelastic behaviour) were consistant with an increase in the crosslinking density with increasing x. The only quantitative information that we may dispose of at the moment is the variation, with x, of the swelling degree, Q, of the samples in pure toluene. We obtained $Q \simeq 15$–16 for $x = 2.10^{-4}$ mole/mole, $Q \simeq 11$–12 for $x = 5.10^{-4}$ mole/mole and $Q \simeq 8$–9 for $x = 10^{-3}$ mole/mole. As mentioned before, because of the presence of side reactions, there is probably no simple proportionality between x and the actual crosslinking density. We are currently working on a more complete characteriza-

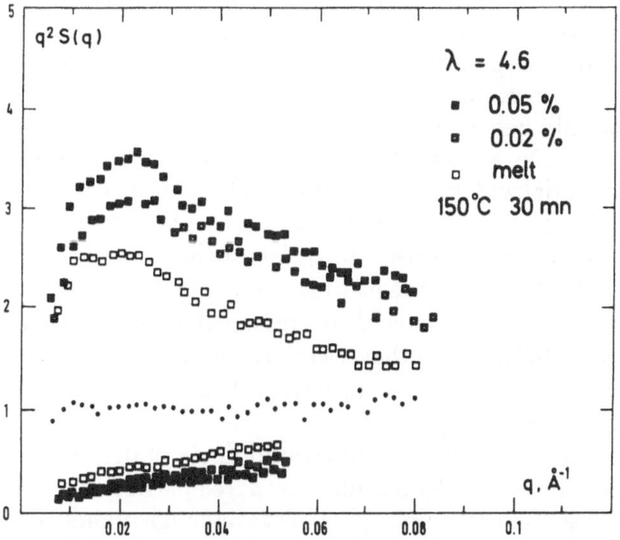

Fig. 3. Effect of the fraction of crosslinking agent at given λ and duration of relaxation: (■) $Q = 11$–12 in toluene; (■) $Q = 15$–16 in toluene; (□) parent melt. Above: perpendicular; below: parallel.

tion of the samples and also on achieving a more quantitative crosslinking reaction. We will simply present the trends of the results, which again follow trivial rules. In figure 3, we show two form factors corresponding to labelled paths in networks ($x = 5.10^{-4}$ m/m and $x = 2.10^{-4}$ m/m) and one corresponding to the same polymer chains dispersed in a deformed melt ($x = 0$). It can be seen that the larger x, the further the form factor from the isotropic one, i. e. the larger the deformation in both the perpendicular and parallel directions (at all moles probed here). In other words, the value of q at which $S(q)$ departs from affinity is larger, or the corresponding scale is smaller. This is in agreement with the idea that this scale is related to the size of the average elementary mesh: the latter is expected to decrease, as does the molecular weight of the mesh when one increases the number of crosslinks in a chain of a given length. However, it may be that the relation between the threshold of loss of affineness and the crosslinking density will not be the one predicted by the models in a more precise analysis. Figure 3 also shows a result for an uncrosslinked melt: we will discuss this in the following section.

3.2 Some more "striking" results

Comparison of melts and rubbers

As far as the melt is concerned, the current theoretical background [2, 3, 6, 7] leads to three regimes:

1. In very short periods, the material does not even "know" whether crosslinks are present or not, these being marginal in respect of other interactions. Rubber and uncrosslinked material of the same polymer should behave in the same way (i).

2. On a very large time scale, the uncrosslinked material will flow and relax completely towards a state of zero stress. It follows that the form factor will relax down to the isotropic form factor (iii).

3. On an intermediate time scale, if the chains of the uncrosslinked material are entangled, they are expected to behave *as if* they were crosslinked temporarily. The "entanglements" play the role of non-permanent crosslinks (regime 2).

Regime 1 should end around times of the order of the maximum Rouse time for a free part of chain between two crosslinks (M_c) or two entanglements (M_e). Taking $M_e = 2.10^4$ for PS melt, $M_c = 5.10^3 - 5.10^4$ for our crosslinked samples gives 10–300 at $T = 117$ °C, or less than a few seconds at $T = 150$ °C [1].

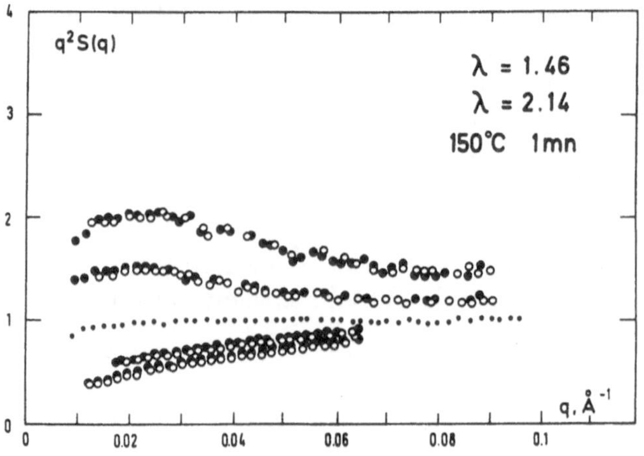

Fig. 4. Comparison of the form factors of a γ-rubber ($Q = 18$–20) (●) with its parent melt (○) for same relaxation parameters. Above: perpendicular; below: parallel

Regime 3 should start around the maximum relaxation time for relaxation of the stress, which we estimate [1], for $M \sim 1.5\ 10^6 - 2.10^6$, as 10^4–10^5 s at 150 °C.

In practice, we have to compare a given rubber with its "parent" melt, i. e. a mixture of the same H and D polymers which have been used to make the rubber.

In Figure 4, the rubber is compared with its "parent" melt for two elongation ratios and a time $t = 1$ min at 150 °C. This time should correspond to regime 2. The overlap is striking. The simplest conclusion is that the melt is well within the plateau regime and behaves as the rubber. However, the molecular weight of the apparent mesh of the melt should not be exactly the same as that of the rubber, and this difference should appear in the form factor. We will postpone detailed discussion and simply recall that our rubbers probably contain many trapped entanglements. As far as the chemical mesh is a part of a chain of the same length as the one between two entanglements, the one would be mixed with the other. If we now increase the chemical crosslinking density, as for the Friedel Craft rubbers, we should notice an effect: indeed, data in Figure 3 show a difference between the rubbers and their parent melt. It should be noted, however, that the duration of relaxation is larger, 30 min at 150 °C.

Overrelaxation at large times

In Figure 5 the form factor of the irradiated rubber for 1 min and 30 min at 150 °C is displaid for $\lambda = 2.14$. In principle, both times are far beyond the maximum Rouse time of a free part of chain between two crosslinks: ~ 100 s at 117 °C i. e. much less than one second

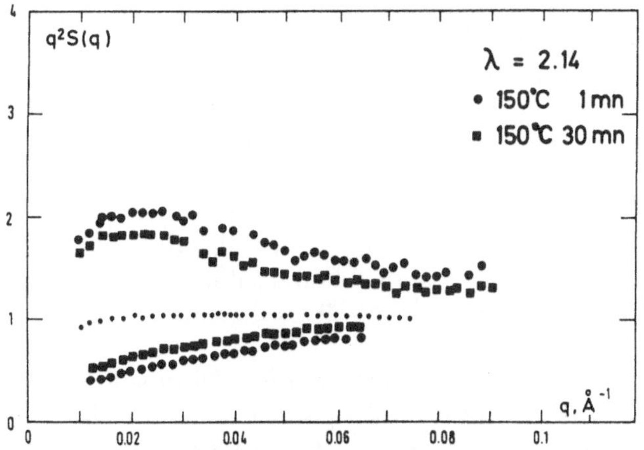

Fig. 5. Relaxation of the form factor of a y-rubber stretched 2.14 times between 1 min (●) and 30 min (■) at 150 °C

at 150 °C. It follows that equilibrium under a constant strain should be obtained. However, we see still an important relaxation at these long times.

This is also the case for one chemically crosslinked sample (Fig. 2): the relaxation (between 1 min and 30 min at 150 °C) in the perpendicular direction is weak this time, but the one in the parallel direction is quite large[1]).

A comparison between the gel relaxed for 30 min at 150 °C and the "parent" melt relaxed for the same time, is not currently available, but we can compare the gel with the "parent" melt of F.C. rubbers melt (the molecular weights are slightly different) relaxed for the same time at the same temperature. The data are roughly overlapping.

In summary, we are confronted with several anomalies:

– A melt sample is not expected to relax much within this time range, unless the estimates for time and fraction of small molecular weights are seriously wrong.

– A rubber is not expected to relax at all within this time range, even if our estimated values for the crosslinking are wrong, unless the estimates for time are seriously wrong.

– However, one rubber and its "parent" melt behave very closely. When the crosslinked samples display a somewhat smaller swelling ratio, Q, rubbers and melt behave much less closely; the additional relaxation remains, however.

We cannot exclude a number of errors in our experiment, such as the polydispersity of the melt after the moulding - deformation-relaxation process, or the uncertainties in crosslinking, such that other characterizations are necessary. However, we think that the uncertainty is not so large that it would not be interesting to display these behaviours.

Comparison with form factors calculated from models

The calculation itself of such form factors is detailed elsewhere [1] for the two models that we want to use here. Some of the features of the models are recalled in section 4. Let us recall here that in the junction-affine model [4], the mean positions in time of the crosslinks are displaced affinely to the macroscopic deformation and the fluctuations around this position are nought, or, equivalently, because SANS averages in both time and the different chains, the fluctuations also are affinely deformed. In the phantom network model [5], the mean positions are displaced in the same way, but the fluctuations remain isotropic, and equal to the one inside the network, which is calculated by the model, and proportional to the radius of gyration of an elementary mesh. Roughly, the phantom network is equivalent to the junction-affine model with an elementary mesh twice as long.

In Figure 6a, for $\lambda = 1.46$, $t = 30$ mm at 150 °C, it is possible to fit the data rather well to calculated curves. But it should be noted that the curves in parallel and perpendicular *do not correspond to the same value of molecular weight of the mesh, and are not calculated following the same model.* Namely, the calculated curve in the perpendicular direction is for the junction – affine model with $M_{mesh} = 10\,000$, and the calculated curve in the parallel direction is for the phantom network model with $M_{mesh} = 50\,000$. This can be summarized by saying that there is much less deformation appearing in the parallel direction than in the perpendicular direction.

In Figure 6b the comparison with dàta for $\lambda = 4.6$, $t = 30$ min at 150 °C leads to the observation of a similar but enhanced tendency. In the parallel direction, the molecular weight necessary for the mesh to fit data is now larger than 50 000, apparently 100 000, using the phantom network model. In the perpendicular direc-

[1]) An error in normalisation by scattering volume, a priori avoided carefully, could not explain this effect because it would affect both directions by the same factor. It follows that an effort to make data for 1 min and 30 min at 150 °C overlap in the perpendicular direction would increase the difference in the parallel direction, and that an effort to reduce the present difference in parallel, which is very large – remember that this representation magnifies the variation in the perpendicular direction and reduces the one in the parallel – would lead to a noticeable difference in the perpendicular direction.

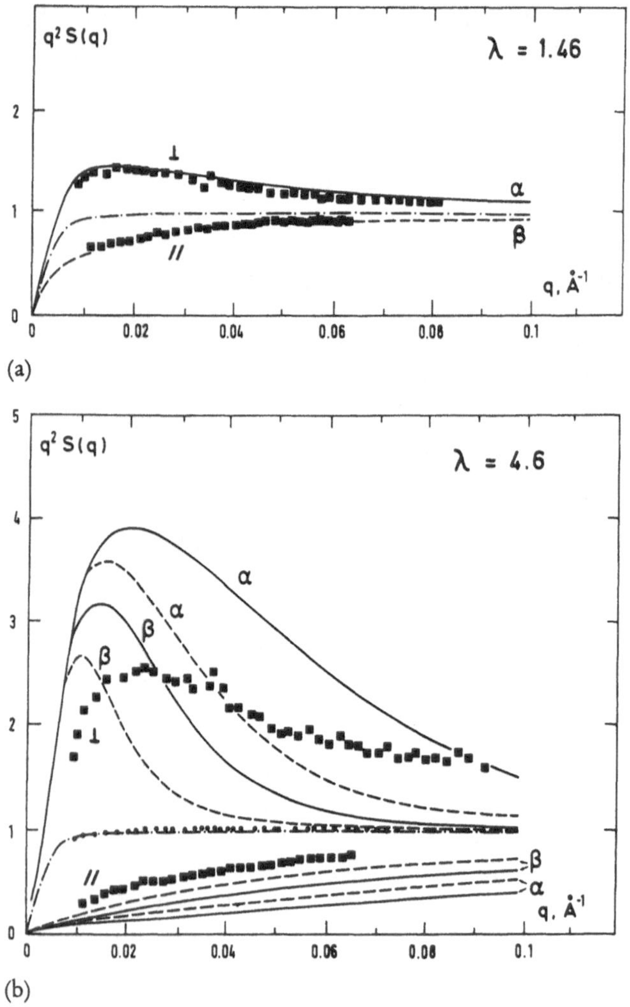

Fig. 6. Comparisons of data for γ-rubber (irradiated), $Q \sim 20$ and calculated form factors. (——) junction affine, (- - -) phantom network for elementary chain of molecular weight 10 000 (α) and 50 000 (β). (a) $\lambda = 1.46$. (b) $\lambda = 4.6$; (-·-·-) isotropic form factor

tion, it would be the same at very low q; meanwhile, $M = 10\,000$ with a phantom network is enough at around $q = 5.10^{-2}$ Å^{-1}. Actually, no good fit for the data is possible here.

We will briefly recall two classical arguments about such a discrepancy: the first one concerns the pendant chains, which would remain isotropic. It is contradicted here by two facts: first, the agreement in the perpendicular direction, with a molecular weight 10 000 for the mesh, is very good, and pendant chains could only make the measured form factor less deformed than the calculated one. Second, one would need a different fraction of pendant chains to explain the data in the perpendicular and parallel direction. Such a difficulty has been encountered in other cases [6], in attempts to fit the variation with λ of the radius of gyration of labelled meshes (see section 4).

Isointensity curves

In the former sections we have only discussed data for the perpendicular and the parallel directions. It is not mandatory that this should suffice to characterize the anisotropy, and we will see that it does not indeed suffice.

A very simple assumption for deformation is the affineness to the macroscopic deformation at all scales, that is Equation (1). This leads to $S(\vec{q}) = S_{\mathrm{iso}}\,(\lambda_q\,\vec{q})$, where $\lambda_q = (\vec{q}\;\tilde{\tilde{\lambda}}\;\vec{q})/q^2$ and it is straightforward that isointensity levels are ellipses. If the material is observed at a very short time after deformation, we know that it will be close to the affine deformation in our q range, or, equivalently, that the departure from affineness will show up at q larger than 10^{-1} Å^{-1}. It follows that isointensity levels will also be as close to ellipses for $q < 10^{-1}$ Å^{-1}; we recall that at very large q they should return to a circular shape, as no important deformation should show up at the scale of the chemical bond. The actual data indeed show some very elliptical shapes for isointensities when t is short, as in Figure 7a, b. When the material relaxes, there is no real reason why this elliptical shape should last. An exception is the very small q regime ($qR_g \ll 1$), for which the form factor depends only on the inertia tensor, which may always be diagonalized. Then, the dependence on three numbers only, R^2_{gx}, R^2_{gy}, R^2_{gz}, leads to ellipses at $qR_{gx\,(y,z)} \ll 1$ (see Appendix 2). However, the characteristic dependence on the deformation of $S(q)$ following the classical models – mainly the assumption of affine displacement of the junctions for both models [4, 5] – leads in practice to curves close to ellipses, at least ovoïd.

If we now include into the models the presence of some chemical defects — pendant chains, intrachain crosslinking — leading to the existence of an undeformed or weakly deformed part of chain, the shape will evolve towards a combination of circle and ellipses: in practice, calculations lead to some kind of "lemon" shape.

We have observed a very different behaviour in many situations: Figures 7 b, c, d, e are only representative cases. The shape of the isointensity levels is much closer to that of a lozenge. It is not a detector effect, nor a crosslinking effect, as this is also observed for the melt. The relation with quenching can be still

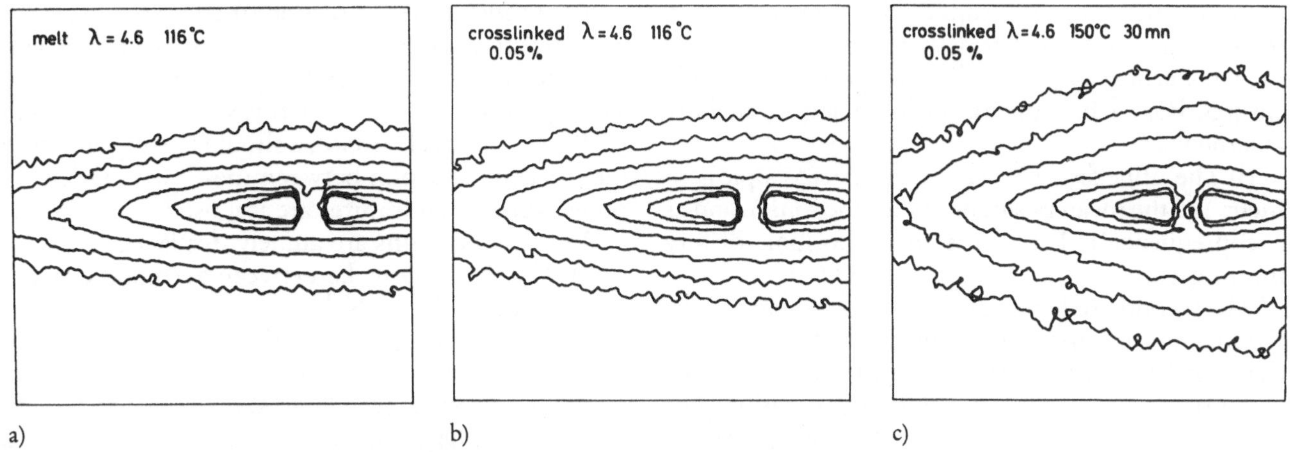

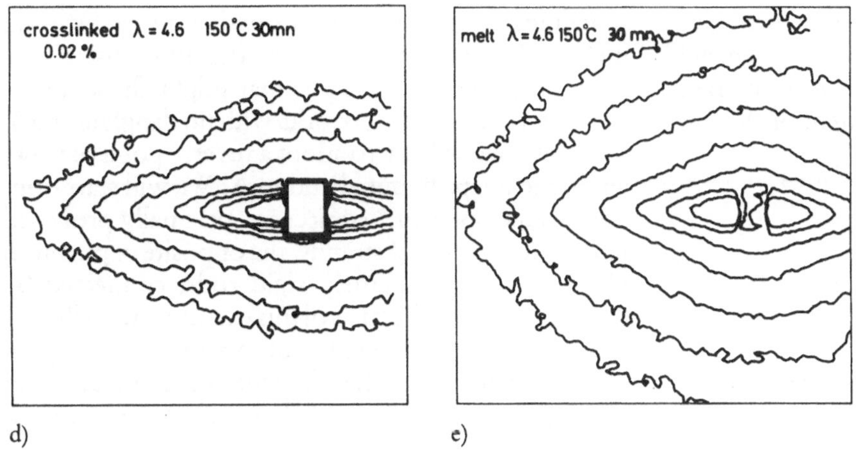

Fig. 7. Shapes of the isointensity curves on the bidimensional detector

questionned. In a recent experiment, the sample was quenched below T_g only once, then reheated and observed successively at different t while the temperature is always kept above T_g; we saw the lozenge effect also. This time, only the first quenching could have been effective. The formation of crazes or of fibers, related or not to the quenching, is a possible origin, but it remains to be explained why crosslinked and uncrosslinked samples yield the same data, while crazes and fibers are known to weaken when the material is crosslinked. The last caution must be for the polymer itself, polystyrene: equivalent measurements for another polymer are not available.

It should be noted, however, that it is possible to enhance the lozenge shape by actions which are deeply related to deformation and relaxation. After a long relaxation at a large λ, e.g. 4.6, or 3, one can let the sample shrink under its own restoring force, as fast as possible, and then quench it. The signal is now very "lozenge-like", even more, namely the curvature of the levels changes in sign around angles lying around $\pi/4$ (larger than 0 and smaller than $\pi/2$) and symmetric values. Further more, this behaviour is a bit smoothed after an additional relaxation at the new length, which leads to lozeanges closer to ellipses.

4. Miscellaneous

The classical models describing the response of a rubber to a deformation and, also the dynamics of this rubber, consider only two kinds of objects:

– The elementary chains, which are at least submitted to the constraint of having their two ends localized at the junctions.

– The junctions, submitted to the restoring force of the chains.

In general, no further object is considered in the system, no other possible collective fluctuation. Quite often, moreover, the chains are assumed to be "phantom". This means that their volume is not taken into account and they are supposed to interact only through the junctions. The network is then considered as a collection of essentially independent entropic springs. For such a system, the following is generally assumed:

(i) The affine displacement (in the macroscopic geometry) of the mean positions of the junctions[2]), and as a consequence[3]) that of the mean position of any monomer.

The second prediction of these models concerns the fluctuations around the mean positions and their dynamics, first in an unstrained network, and then the variations of these quantities upon deformation. For the network of phantom chains, it was first proposed by James and Guth [5] that:

(ii) Fluctuations are not affected by the deformation. This arises also from a more elaborated mean field model of the junctions [6]. Four main works come to very similar conclusions about the extent of the fluctuations [5–8]:

(iii) The elastic force f is given by

$$f/A_o = \frac{1}{2}\,\nu\,kT\left(\lambda - \frac{1}{\lambda^2}\right) \tag{2}$$

A_o is the surface area perpendicular to the force, in the unstrained conformation, and ν the number of elementary chains per unit of volume.

λ is the extension (or compression) ratio, for an isochore uniaxial deformation.

From References [5] and [6], this is related to a fluctuation $\langle \Delta R^2 \rangle = (1/2)\, N_{mesh}\, a^2$ of the end-to-end distance of the mesh, where N_{mesh} is the number of monomers in one elementary unit.

An alternative to (ii) and (iii) is to propose

(iv) To reduce the fluctuations or to make them depend upon the deformation ratio.

This is actually the proposal of the oldest model, where there was no fluctuation at all [4]. A completely affine deformation of the fluctuations leads to the same result, as far as both time and chain averages are concerned (e. g. for the form factor of some labelled chains in the net).

[2]) The consequence of assuming the affine deformation of at least two points of the network of linear springs has been assumed in earlier and in later models.

[3]) In an end-to-end pulled chain (of end-to-end vector $R_o - R$) any monomer fluctuates around a mean position $R_i = R_o + (R_o - R_N) \times i/N$; the fluctuation in time being the same as for an unconstrained chain, such that $\langle (R_i - R_j)^2 \rangle \sim (i - j)\, a^2$.

The models tested in section III correspond, for the junction affine model, to assumptions (i) and (iv) with zero fluctuation and, for the phantom network, to assumptions (i), (ii) and (iii). It is very often proposed [7–9] that the "real" network should correspond to an intermediate situation between these two models: in other words, two features are imposed:

A) Mean positions are affinely displaced

B) The fluctuations lie between zero and a maximum value $\langle \Delta R^2 \rangle = Na^2/2$.

Because of discrepancies with data, we would like to know if it is possible to escape these commitments. Firstly, we propose accepting A, but trying to modify B by allowing larger fluctuations. This can be achieved in a very particular topology for the network, the latter remaining a net of phantom Gaussian springs. We will include this argument in a first attempt to discuss the different possible topologies that might show up for terminated nets (i. e. networks with no dangling ends). Secondly, we will try to adopt a reverse point of view, aiming to adapt B, but abandon A. We will start from the "disinterspersion" conjecture, originally proposed [10] for a net form of strands, the opposite of phantom Gaussian springs, namely rigid rods, connected by flexible links. On the basis of this example, we will propose a different combination of rotation and extension of the elementary meshes than the affine displacement A. We can obtain different behaviour, closer at least to some experimental data.

Finally, we will come back to the lozenge effect, in order to produce a comparable trend by calculation, using a particular distribution of rotation and extension, in spite of its oddness.

4.1 Effects of the structure of the networks[4])

Drawing a network on a blackboard, for example, often leads firstly to a set of unbricated lines, secondly to a simple 2-D square lattice. On the other hand, it is possible to draw most particular arrangements of chains, and even easier nowadays, after the efforts in theories of branching, gelation and aggregation developed over the years. It is, however, generally assumed that they could apply to an incomplete network, as one obtains just after the onset of gelation, but not when the network is, as in the spirit of this section, completed (as

[4]) Ideas presented in References [12, 13, 15 and 2 p. 147–152] are the basis of this paragraph, even if here their different arrangement makes their explicit quotation sometimes difficult.

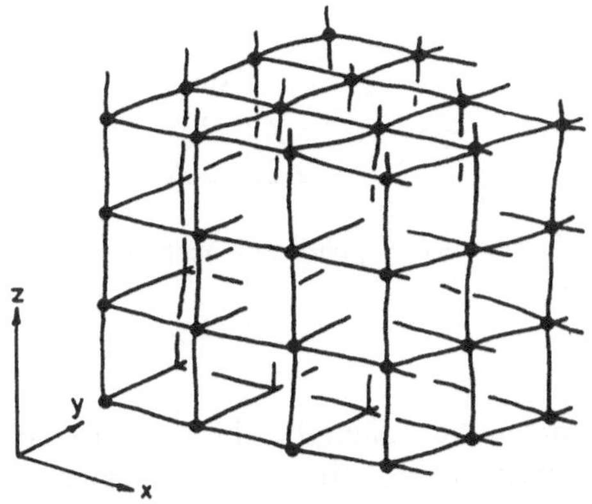

Fig. 8. A "standard" network, i. e. a crystallographic net

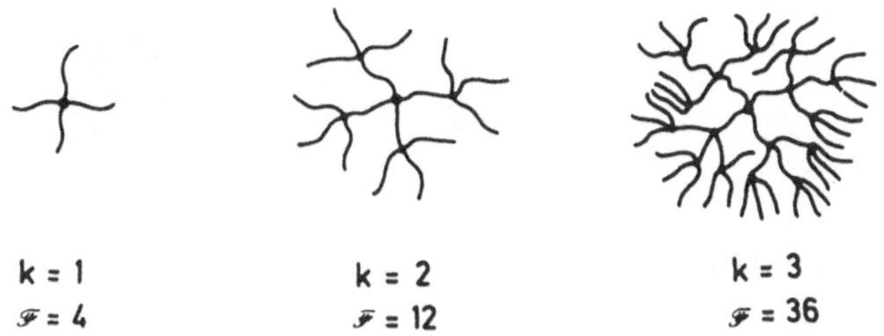

Fig. 9. Trees of increasing generation rank k

would be the result of a percolation process when the fraction of active bonds is $p=1$). Actually, special connectivities may exist, even in a completed network or a terminated net, but we cannot preclude their relative importance. Therefore we shall examine some of them.

Unfolded structure

Once the junctions are chemically formed, the network of chains may change shape noticeably by pulling on the chains, but also simply by displacing and rotating them: the net of a fisherman, or a 3-D net, can be made compact or folded to obey some density constraint, for example. We will consider here the unfolded network, which makes the connectivity of the chains more visible: this indeed matters only for the properties of a network of phantom chains which are the burden of this section.

Comparison with crystal lattices

A category of networks comes naturally to mind, the *"crystallographic nets""*, which regroups all networks, the connectivity of which can be mapped on classic cristalline lattices (Fig. 8). Usual percolation procedures on lattices would produce nets of this kind for $p=1$.

Trying now to imagine a *"non-crystallographic network"*, that cannot be mapped on a crystalline lattice, at once raises the question of the "shorter closed path". If we imagine a walker along the chains, starting from a given node, and passing from one node to another, what are the shortest paths allowing him/her to come back to the starting point? On a cubic lattice, these paths are made of four steps each (one step corresponds to an elementary mesh in the network); in a diamond lattice they consist of six steps each. But we can imagine networks of the same functionality as in a crystal lattice (four in the latter case) for which the "shortest closed paths" are not distributed in such a regular way.

Networks of trees

A first attempt to increase, at least on average, the length of the shorter loops, can be made using objects with no loops, i. e. trees. It is only the connection of two, or several, trees by the extremities of their branches, which are still active, that will produce loops. However the number $\mathcal{F}$ of active sites increases quite fast with the rank k of the tree (Fig. 9). This makes impossible to make all the sites react together, for $k > 2$ or 3, by connecting the trees only by their outside, i. e. without allowing the interspersion of the different objects: a lot of dangling chains are formed. On the contrary, two or several trees may grow at the same place, become interspersed, then at a given rank interconnect by their extremities, and form a cluster of reduced external functionality $f_{ext} = (f-1)\,\mathcal{F}$. The clusters can then be connected one with another, in order to form a terminated net with no dangling end. However, their fractal dimension 4 [11, 12] soon makes trees more dense that the polymer bulk, except if the branches – the elementary chains – are very long. This means that such a net can be formed only if the crosslinking density is weak. Moreover, interspersion of the trees requires a high concentration during the preparation [13]. Even under these conditions, one notices that k can never be very large.

Finally, it should be noted that trees were used by Graessley [8] to calculate the elastic modulus of a phantom network, by using the assumption of affinely displaced fixed extremities only. This gives the phantom network result (iii) when the rank k is $\gg 1$.

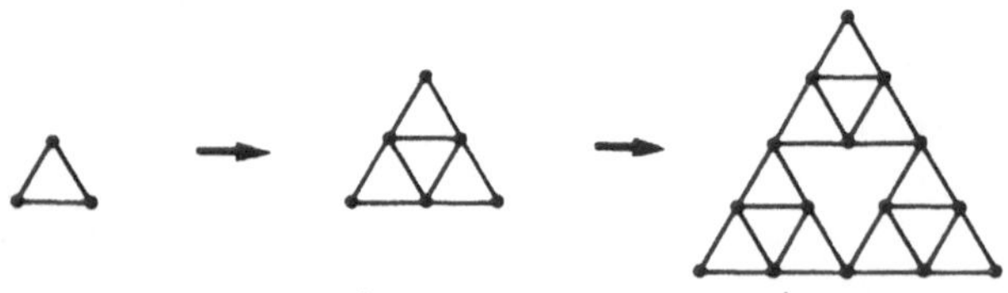

Fig. 10. Chain clusters having a Sierpinski gasket topology

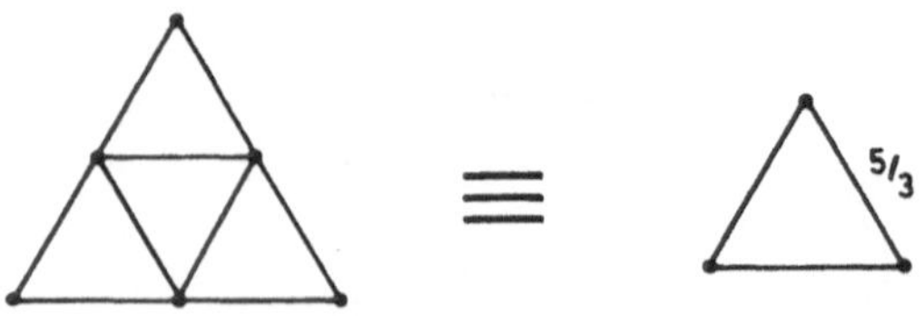

Fig. 12. Star triangle equivalence applied to a Sierpinski element formed of ideal chains

Decreasing k leads to some increase of the elastic modulus, which becomes closer to the Kuhn-Flory prediction [4, 7].

Networks formed of objects of low dimensionality. Self-similar terminated nets

We have seen that the lack of small loops produces objects of high density and high external functionality. It would be only by imposing a lot of dangling ends at different ranks that we could escape the two problems. We will now consider an object containing even smaller loops than in the corresponding crystal lattices of the same functionality. Together with the enforcement of very short loops some "holes", will appear, thus a much weaker density and no increase in the external functionality. A special way to satisfy this requirement is to map a net on self-similar objects. This is the case for the celebrated Sierpinsky gasket [14, 15] for which $f = 4$, as in the diamond lattice. (We could also use a tri-dimensional gasket made of tetrahedras instead of triangles, with $f = 6$). Figure 10 shows the initial stages; the segments represent elementary chains, which can all be of the same chemical length. There are no dangling ends, except of course at the summits: it is a completed network, perfect from the chemistry point of view. The shorter loops are of three branches, compared to four for a 2-D square lattice of six in a diamond lattice, which have the functionality $f = 4$. It also appears as containing many "holes". In contrast to the trees, the external functionality of these objects does not increase with k, the generation rank. It remains constant, equal to $(f-2)\mathscr{F} = 6$ $(f = 4, \mathscr{F} = 3)$ allowing for a connection of the objects between themselves at any k, thus at any scale. Such an interconnection will provoke the self-similarity to stop, as illustrated in Figure 11, again with the important presence of holes. Again, in contrast to the trees, one expects a decrease in density with k inside a given object, since the fractal dimension is inferior to 3. The growth will not be limited by reaching the bulk density. At this point, it is tempting to reverse the argument and to wonder if a low concentration during the crosslinking would favour high ks, by limiting the interconnection without limiting the growth.

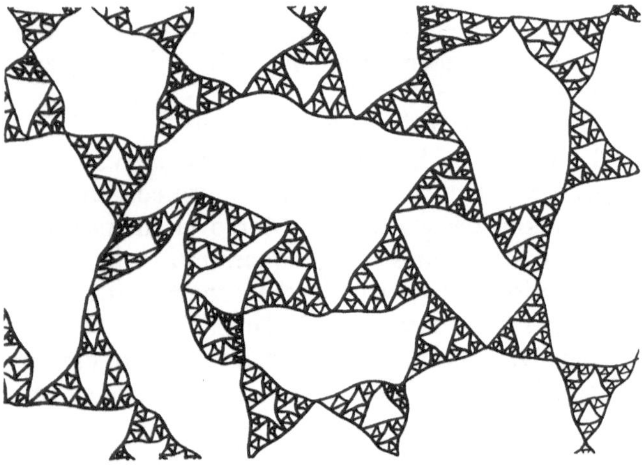

Fig. 11. Network formed by interconnection of Sierpinski elements

Ultralow modulus and ultralarge fluctuations

We will recall the basis of the derivation [16], as detailed [17] and some conclusions.

Similarly to the "micronetwork" model of Graessley [8] one can insulate a set of f chains connected to a crosslink by one end, forming a star. The f other ends are assumed to be fixed in space, the centre being free. Via a "star-triangle" equivalence, this is equivalent to the f fixed ends connected two by two by equivalent chains of modified length. The centre has vanished. Doing that three times on a gasket of rank 2 leads back to a gasket of rank 1, with a 5/3 larger length of chain (Fig. 12). An object of rank k will correspond to length $(5/3)^k$ larger. For a given k, we may assume interconnection. We then can take the most constraining conditions on the summits of the gaskets of rank k, namely the Kuhn affine displacement: even in that case we can obtain an overall modulus smaller than the phantom network one, for $k > 1$, or $k > 4$, depending on additional assumptions (the one leading to $k > 1$ actually being the most sensible). Thus, even within the framework of the assumptions of the phantom network model, a particular topology may lead to overcome the phantom network "absolute limit" of the elastic modulus. This is the same for the fluctuations of the different crosslinks of the gasket, some of them being very large (this allows for the small modulus): this would produce a noticeably less deformed form factor.

Structure and ability to folding

The possibility of changing the shape of a network by folding-unfolding has been proposed [10] for swelling-deswelling, and could be considered for stretching; this is briefly mentioned in the next section. Here we wish to look only at how much a network can be folded depending on its topological structure. The "natural intuition" is that a very regular network, such as a crystallographic one, is less able to be folded than an irregular one, containing larger loops and holes. Some practical tests on fabricating different species by hand did not allow us to confirm this intuition: a reasonably large piece of freely jointed rods with a diamond lattice topology appeared as flexible as an irregular net built for the purpose of creating very large return paths.

To conclude, for this argument we acknowledge that topologies as described above can be rare: a real rubber is performing an average over many possible topologies. This is the basis of the mean field theory [6]. However, we are not completely convinced that there could be no dependence at all on the crosslinking methods. For example, our two opposite examples, tree and self-similar gasket, which could both arise from an end-linking synthesis, are not fitted to the same range of concentration of chains[5]). More generally, it could be that the conditions of preparation (concentration, kinetics) favour the growth of certain kinds of topologies, reduce some others to nought, and then displace the theoretical averages. An outside experimental result, obtained by Mayen [18], could confirm this belief: applying mechanical or electrical low frequency vibrations in the course of gelation, does not modify the kinetics of the chemical reaction (as checked by UV-, IR-measurements) nor the final degree of advancement (quasi-total), but noticeably modifies the swelling degree at swelling equilibrium in a good solvent. This could be due to a change of topology.

4.2 Orientation versus elongation

All the models of deformation of a network actually represent a combination of *orientation* and *elongation* (or contraction) of the end-to-end vectors of the elementary chains. The simplest example is the junction affine model [4] in which the ratio of rotation over elongation depends on the initial orientation of the end-to-end vector, as sketched for one vector in Figure 13 a, b. We now want to try to abandon the assumption of affine displacement of the junctions in order to get a smaller anisotropy in the form factor: it appears below that this is possible by privileging the rotation. For the likelihood of this process, we can lean on a simple hardware model: a network of rigid rods connected by free universal joints can be swollen (deswollen) by unfolding (folding) as a kind of 3-dimensional accordion [10] but it may also, less easily, be stretched, provided that it was not completely unfolded before. Here deformation means only rotation, and this process is extremely non-affine.

We choose here to start from the junction affine model but to weaken the elongation-contraction. This could be done for each end-to-end vector by allowing it to rotate and elongate first, and then obliging it, without changing its orientation, to recover its initial length (see Fig. 13 c). For the sake of simplicity we can simply assume that the length averaged over all orientations is equal to the undeformed one. In the junction affine process, Equation (1) can be rewritten:

$$\vec{r} = \begin{pmatrix} x \\ y \\ z \end{pmatrix} \rightarrow \vec{r}' = \begin{pmatrix} \lambda_x & x \\ \lambda_y & y \\ \lambda_z & z \end{pmatrix}$$

and the average is

$$\langle r^2 \rangle_\theta = \left(\Sigma_\alpha \frac{\lambda_\alpha^2}{3} \right) \langle r^2 \rangle_{\text{iso}}, \; \alpha = x, y, z \,. \tag{3}$$

[5]) In Reference [2] p 147–152, it is shown that already the gelation process may depend significantly on the crosslinking conditions (e. g. concentration).

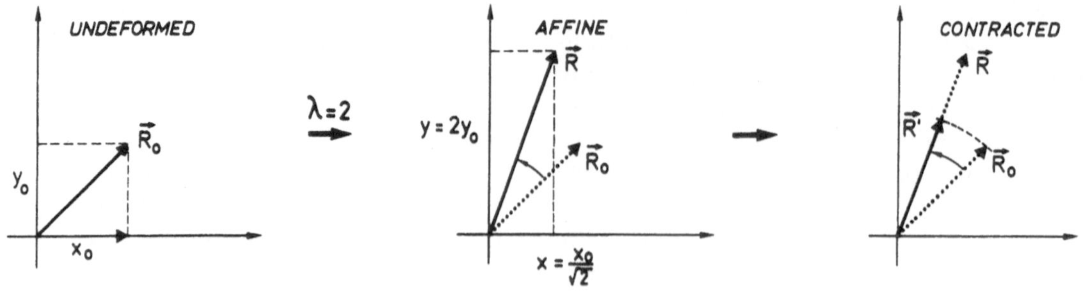

Fig. 13. Affine deformation ($a \rightarrow b$) and contracted affine deformation ($b \rightarrow c$)

We impose then

$$r' \rightarrow r'' = \begin{pmatrix} \lambda_x/(\Sigma_\alpha \lambda_\alpha^2/3)^{1/3} & x \\ \lambda_y/(\Sigma_\alpha \lambda_\alpha^2/3)^{1/2} & y \\ \lambda_z/(\Sigma_\alpha \lambda_\alpha^2/3)^{1/2} & z \end{pmatrix}$$

Using the expression of the squared radius of gyration of ideal chains of fixed end-to-end vector $R = (R_x, R_y, R_z)\rangle$

$$\langle R_g^2 \rangle_\alpha = \frac{1}{6}\left(\frac{Na^2}{2} + \frac{3}{2}\langle R_\alpha^2 \rangle\right) \alpha = x, y, z \qquad (5)$$

(Eq. A 7. F Appendix 2).

We obtain a different variation of $\langle R_g^2 \rangle_\alpha$ than for the junction affine model: it is smaller in the direction parallel to the stretching, but *larger* in the perpendicular one, although there is *no* elongation of the chains.

It is simple to extend the contraction to the phantom network, for which everything happens as if λ_α^2 was replaced by:

$$(\lambda_\alpha^*)^2 = \lambda_\alpha^2 \frac{(f-2)}{f} + \frac{2}{f} \qquad (6)$$

for a functionality f of the junctions. Using Equation (6) leads to Equation (A8) of Appendix 2.

Figure 14 shows comparison with data from our colleagues [19] for the radius of gyration of labelled meshes, well suitable to this comparison (the use of data for long labelled paths needs heavier calculations). For the largest mesh size, the agreement is better than for the junction affine and phantom network model. Note that the "disinterspersion" process was proposed to exist [10] essentially for widely interpenetrated meshes, i. e. for large molecular weights of the latter. Furthermore, two features are satisfying. Firstly, the

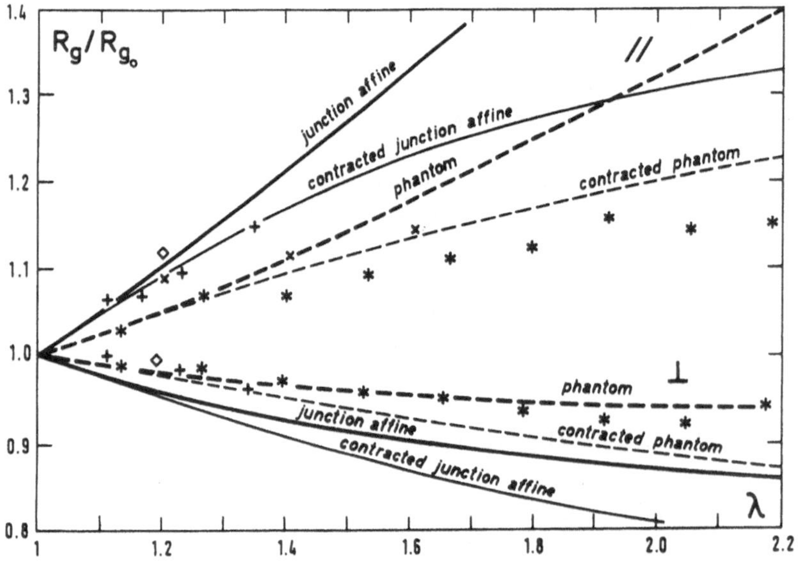

Fig. 14. Variation with the elongatin ratio λ of the radius of gyration in parallel and perpendicular direction for labelled meshes: data (Ref. [19]) and models (junction affine, phantom network, and their "contracked" versions, Equations (A7) and (A8) of Appendix 2)

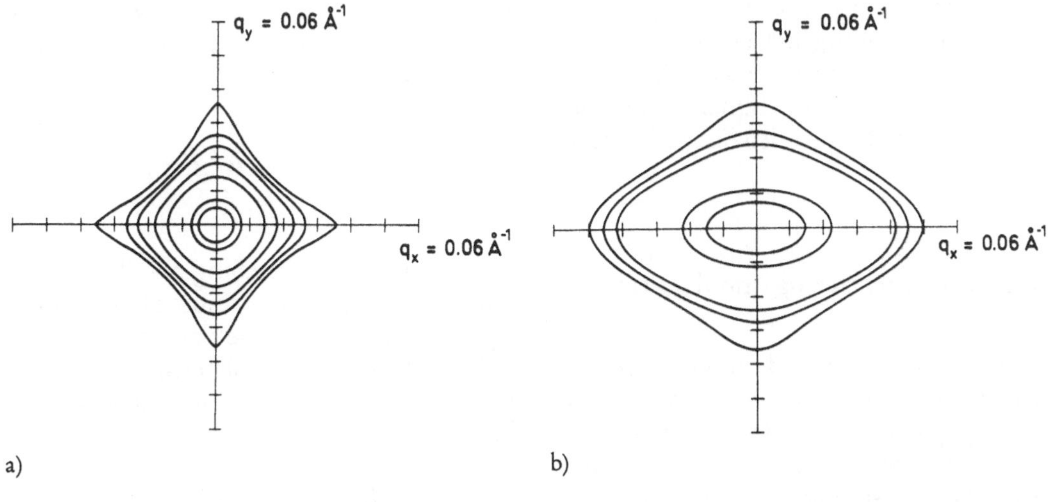

Fig. 15. Isointensities for (a) crossed chains of same end-to-end vector ($R^2 = 2Na^2$). (b) crossed chains of end-to-end vectors affinely deformed

curvature direction of the calculated line is the same as for the data, which is not the case for the classical model. Secondly, the contracted phantom network model is slightly more oriented than the data in the *two* directions: if we were now taking into account a fraction of dangling chains (which are very likely), a good fit could be obtained for two directions at the same time (always for the larger mesh size). This is not the case with the classical models, for which such a fit needs a very different fraction of dangling chains for the two directions. We should recall, however, that we consider this model as an example essentially showing that a change in the radius of gyration is not univocally bound to an extension of the chains.

4.3 Lozenge effect from special orientation

Calculations of the form factor based on the classical models [4–6] do not lead to lozenge shapes for the isointensities in the geometry of our experiment. The adjunction of "classical" adjuvents, dangling ends, closed loops, produces a "lemon" shape from the superposition of a quasiellipse and a circle. We cannot certify that lozenges are impossible to obtain, in particular we have not investigated entanglement effects. But we were, however, struck by the fact that a very simple model, unrealistic indeed, reproduces the experimental result to some extent.

Basically, we suppose only two possible orientations (orthogonal) of the end-to-end vectors of the elementary chains. Practically we can also suppose a given value of the length of the vector, say twice its natural length, $(Na^2)^{1/2}$ and consider only a labelled mesh, not a labelled path. The calculation is a simple addition of form factors of one chain with an supposed end-to-end vector [20]. Figure 15a shows at a q value of the order of the experimental one, a square shape is obtained. If we now deform our end-to-end vector distribution by keeping in one direction an extension equal to $2(Na^2)^{1/2}$, but contracting the orthogonal end-to-end distances to $(1/\sqrt{2})(Na^2)^{1/2}$ we obtain Figure 15b. At small q, the curves are still ellipses, as they must be from the features of the Guinier range (see section 3). At intermediate q we cross over to lozenge shapes, and at even larger q the curvature is inverted, as it appears faintly in some experimental measurements.

5. Comments

Our opinion is that the experimental process must be refined, in particular with respect to the crosslinking procedure, which must become quantitative and to several checks about possible artefacts in the processing of the sample, mechanical and thermal (quenching, crazes). A part of the behaviour can be related to the polymer used, polystyrene. However, a lot of the behaviours described here are close to the one observed with SANS for different polymers, different crosslinking procedures and different labelling (e. g. labelled mesh in an end-linked polydimethylsiloxane).

Concerning comparison with models, the *effect of the entanglements*, must be investigated; also the rigidity of the chain which is actually closer to a wormlike chain. It should be remembered that the remarks of section 4 were orignally included for an open discussion at the meeting.

Appendix 1: Neutron scattering and data treatment

The procedure was the standard one developed for uniaxially deformed samples [1]. The incident wavelength was 12 Å ± 10% (mechanical selector). The collimation basis was 3 m, the iris 10 mm; thus delta-theta incident ~ 3.10^{-3} rd. The sample detector distance was 2.82 m; the distance between two cells was 0.5 cm. The detector was bidimensional and we regrouped the cells of the same q within a sector of mean angle ϕ with the stretching axis. Here we present data for $\phi = 0°$ (∥) and 90° (perp.); but we also looked at the shape of the curves joining all the cells of equal counting (isointensity). After regrouping, the data were treated from transmission, scattering volume, detector efficiency, leading to an absolute spectrum ($S(q)$ in barns). The incoherent background obtained from an absolute spectrum of a non-deuterated sample of the same volume was then subtracted. All spectra may then be compared directly.

Appendix 2: Expressions for apparent radius of gyration in anisotropic scattering

When an object is anisotropic, it is no longer possible to define the radius of gyration. The inertia tensor can always be defined, but it is not this quantity which is relevant to scattering. It is actually an apparent, direction-dependent, radius of gyration, $R_g(\theta)$. We can write, for small q's

$$\begin{aligned} S(\vec{q}) &= \sum_i \sum_j \langle e^{i\,\vec{q}\cdot\vec{r}_{ij}}\rangle \\ &\simeq \sum_i \sum_j \langle 1 - i\,\vec{q}\,\vec{r}_{ij}\rangle \\ &\simeq N^2\left(1 - \frac{1}{N^2}\left\langle \sum_{i,j} (\vec{q}\cdot\vec{r}_{ij})^2\right\rangle\right) \\ &\simeq N^2\left(1 - \frac{1}{N^2}\sum_{i,j} \cdot\, \Sigma_\alpha\, q_\alpha^2\, \langle r_{ij\alpha}^2\rangle\right) \end{aligned} \quad \text{(A1)}$$

i and j being the indices of two monomers of the labelled chain, and $\alpha = x, y, z$. $\langle\quad\rangle$ is an average over chains and time. We can then define [21]

$$S(q) \simeq N^2\left(1 - \Sigma_\alpha \frac{q_\alpha \cdot R_{g\alpha}}{3}\right). \quad \text{(A2)}$$

For a uniaxial stretching parallel to the x axis, R_{gx} will be the "parallel radius of gyration", $R_{g\parallel}$, and $R_{gy} = R_{gz}$ the transverse one, $R_{g\perp}$.

For a Gaussian flexible chain of end-to-end vector fixed at a value $\vec{R}$ of coordinates (R_x, R_y, R_z), it is known [4, 20] that

$$\langle r_{ij\alpha}^2\rangle = \frac{(i-j)\,(N-(i-j))}{3N}\,a^2 + \frac{(|i-j|)^2}{N}\,R_\alpha^2. \quad \text{(A3)}$$

It follows that

$$\begin{aligned} R_{g\alpha}^2 &= \frac{1}{N^2}\Bigg[\sum_i\sum_j \frac{|i-j|\,(N-|i-j|)}{3N}\,a^2 \\ &\quad + \sum_i\sum_j \frac{(i-j)^2}{N}\,R_\alpha^2\Bigg] \\ &\sim \frac{1}{6}\left(\frac{Na^2}{2} + \frac{3}{2}\,R_\alpha^2\right). \end{aligned} \quad \text{(A4)}$$

Now, if R varies for the different chains

$$R_{g\alpha}^2 = \frac{1}{6}\frac{Na^2}{2} + \frac{3}{2}\,\langle R_\alpha^2\rangle_{\text{chains}} \quad \text{(A5)}$$

$\langle\rangle$ chains is an average over the chains.

For the contracted junction affine model, we have $R_\alpha \to \frac{\lambda_\alpha}{(\Sigma_\alpha \lambda_\alpha^2/3)} \cdot R\alpha$ for any end-to-end vector $\vec{R}$. The distribution of the $\vec{R}$s before deformation is isotropic and Gaussian with $\langle R_\alpha^2\rangle = \mathrm{Na}^2$, for any α. Thus after deformation contraction

$$R_{g\alpha}^2 = \frac{1}{6}\frac{Na^2}{2} + \frac{3}{2}\frac{\lambda_\alpha\, Na^2}{(\Sigma_\alpha \lambda_\alpha^2/3)} \quad \text{(A6)}$$

giving in uniaxial deformation ($\lambda_x = \lambda$, $\lambda_y = \lambda_z = 1/\sqrt{\lambda}$)

$$R^2_{g\parallel} = \frac{Na^2}{6}\,\frac{2\lambda^2 + 1/\lambda}{\lambda^2 + 2/\lambda}$$

$$R^2_{g\perp} = \frac{Na^2}{6}\,\frac{\lambda^2 + 5/\lambda}{2\lambda^2 + 4/\lambda} \qquad \text{(A7)}$$

(contracted junction affine).

For the contracted phantom network model, we replace λ_a^2 by $\lambda_a^{*2} = \frac{\lambda_a^2 + 1}{2}$ (see Eq. 6 in the text), for a functionality $f = 4$ of the crosslinks, which leads to

$$R^2_{g\parallel} = \frac{Na^2}{6}\,\frac{2\lambda^2 + 1/\lambda + 3}{\lambda^2 + 2/\lambda + 3}$$

$$R^2_{g\perp} = \frac{Na^2}{6}\,\frac{\lambda^2 + 5/\lambda + 6}{2\lambda^2 + 4/\lambda + 6} \qquad \text{(A8)}$$

(contracted phantom network)

Acknowledgements

Neutron experiments were run at I.L.L., Grenoble, and L.L.B., Saclay. We wish to thank M. Cruz, R. Oberthür, A. Rennie (I.L.L.), B. Farnoux, A. Brulet, F. Gibert, T. Krebs (L.L.B.). We wish to thank Z. Mankovsky for his very kind help for the use of the Co60 chamber (CNRS, Labs of Vitry-Thiais). Deuterated and undeuterated chains were made and characterized in I.C.S. (C.R.M.), Strasbourg by some of us and P. Rempp. We have benefitted from the help and the advice of G. Beinert, F. Isel, and C. Strazielle. We are indebted for discussions about networks to H. Benoit, S. F. Edwards, L. Leibler, C. Picot, P. Rempp, G. Weill, R. Oeser, and T. Vilgis (MPI Mainz). L.L.B. and C.R.M. are "laboratoires propres" of CNRS.

References

1. a) Bastide J, Herz J, Boué F (1985) J Physique 46:1967
1. b) Boué F (July 1987) Transient relaxation mechanisms of polymer melts and rubbers investigated by SANS, Adv Pol Sci, Vol 82
2. de Gennes PG (1977) Scaling Concepts in Polymer Physics, Cornell University Press, Ithaca, New York
3. Ferry JD (1983) Viscoelastic properties of polymers, J Wiley
4. Kuhn W (1936) 76:258; (1939) 87:3
5. James H, Guth E (1947) J Chem Phys 15:669; (1947) 15:651
6. Deam RT, Edwards SF (1976) Phil Trans R Soc Lond A, 280:1296
6. b) Doï M, Edwards SF (1986) The Theory of Polymer dynamics, Oxford Univ Press, Walter Street, Oxford
7. Doï M, Edwards SF (1978) J Chem Soc Faraday Trans 2, 74:1802
7. b) Flory PJ (1976) Proc Soc 351:351,for a review of the major original models
8. Graessley (1975) Macromolecules 8:186 (1975) Macromolecules 8:865
9. Erman B, Flory PJ (1982) Macromolecules 15:800; (1978) J Chem Phys 68:5363
10. Bastide J, Candau S, Picot C (1980) J Macromol Phys B19:13
11. Zimm BH, Stockmayer WH (1949) J Chem Phys 17:1301; Stockmayer WH (1943) J Chem Phys 11:45; Flory PJ, Principles of polymer chemistry, Cornell University Press, Ithaca, New York
12. Daoud M, Bouchaud E, Jannink G (1986) Macromolecules 19:1955
13. de Gennes PG (1977) J Phys Lett 38:355
14. Mandelbrot B (1977) Fractals: form, chance and dimensions, Freeman, San Francisco
15. Alexander S (1985) In: Boccara N, Daoud M (eds) Physics of Finely Divided Matter, Springer Proc Phys 5, Springer Verlag, Heidelberg; Alexander S (1984) J Physique, Paris 45
16. Bastide J, Boué F (1986) Physica 140 A, 251
17. Bastide J, Boué F, to be submitted to Macromolecules
18. Mayen M (1985) Europ Poly J 21:903
19. Beltzung M, Picot C, Herz J (1984) Macromolecules 17:663
20. Kuhn W, Künzle O, Katchalsky A (1948) Helvet Chimica Acta Vol XXXI:VII
21. Benoit H (1964) C.R.M. lectures, S. Levy Thesis, Strasbourg; Ullman R (1979) J Chem Phys 71:436

Received March 19, 1987;
accepted July 13, 1987

Authors' address:

F. Boué
Laboratoire Léon Brillouin
CEN-Saclay
F-91191 Gif-sur-Yvette, France

Discussion

D. RICHTER:

I have a question with respect to the time waited for the relaxation of the sample. Are you sure that you really reach an equilibrium, when you now compare with the theories? I mean, you stretch and let it relax, but in order to compare it with the phantom network or whatever, it must be in equilibrium after that.

BOUÉ:

Yes, first the stretching is done at low temperature, at something like 110 °C, and this is done in about 15 s. So when you now look at something which is at parity and multiply the time by 10^5 or more, you see that you have completely forgotten about what occurred in those 7 or 15 s at the beginning.

Also, if you now compare this to all the data on rubbers, you feel quite safe. If we compare with the end of the transition regime, I said it was something like 1 min at 110 °C, so we look at 10^5 min compared to that one. To say that you are under equilibrium is very difficult. For a melt, if you wait too long, you will go beyond the terminal time, and your sample will flow. For the rubber samples, as all the people working with rubber know perfectly well, you never reach equilibrium in a rubber. But this is not in the classical theories. I think if you rely a bit on theories, and you must always do that in physics, you are fully within the rubbery plateau, as classically defined for melt and for rubber. A more detailed discussion of this is in Reference [1a] of our paper in these proceedings.

ILAVSKY:
If you irradiate your sample, how many crosslinks do you introduce into the sample?

BOUÉ:
I estimated the mesh at 30 000.

ILAVSKY:
But you have the same response from the melt and from the rubbers. I would expect that if you introduce crosslinks into the melt, something should happen with neutron scattering.

BOUÉ:
I would expect that, too. Actually, a clear difference between melt and Friedel Craft rubbers is observed. But for our irradiated rubbers, melt and rubber behave in the same way. The problem is quite tricky because the molecular weight between two crosslinks is, in this case, about the same as that for the melt.

ILAVSKY:
Between entanglements?

BOUÉ:
Yes, because between entanglement it is 20 000 and I estimate for the irradiated crosslinked material it is 30 000. In this latter estimate we include the entanglements.

ILAVSKY:
Did you try to determine something like the equilibrium modulus of your crosslinked sample or the crosslinking density, say from the maximum swelling degree of the sample?

BOUÉ:
Yes. We measured the maximum swelling degree Q, but not the modulus.

ILAVSKY:
So that you had the finite value to the Q?

BOUÉ:
Yes, $Q = 18$ to 20. We had some other samples with the same or other irradiation rates, which were more carefully measured on other properties. So we think we are more or less sure.

ILAVSKY:
Then I cannot believe the small angle neutron scattering results. You should have the difference.
For swelling I asked only whether some crosslinking has really proceeded. And you are saying that you could not deswell the crosslinked samples, so there are some crosslinks. But then small angle neutron scattering simply does not see them, because you have the same response from the melt and from the crosslinked sample.

BOUÉ:
Yes, that was the experiment.

ILAVSKY:
You should first, I think, increase the crosslink density.

BOUÉ:
Yes, we have done that. If we increase the crosslinking density a lot, then we start to see a difference. These data are not complete, I cannot present everything here, but these will be in the written version.

MEISSNER:
If you think in terms of rubber-like elasticity, the deformation of polystyrene at 110 °C is very dangerous. We found that, at about this temperature, a large portion of energy is converted into internal energy, because of the energy difference between gauche and trans conformations. So, especially just above Tg, the deformation of polystyrene melts is not exclusively entropy-elastic, there is a large contribution from energy elasticity.

BOUÉ:
I was not discussing those data at 110 °C, it was just for a kind of pedagogic presentation. I have investigated this effect that you are talking about in studies of stretched polystyrene melt in my Thèse d'Etat, and I agree with you. We remarked that the enthalpic part of the stress was relaxing much faster than our quenching, anyway.

SAUTTER:
You agree that this result means that a chain in perpendicular direction to the strain will be very much shorter after relaxation: compressed, so to speak. Is that correct?

BOUÉ:
I am not sure I understand the question. What appears is that after the relaxation, in perpendicular direction, the chain does not seem more relaxed than predicted by the models, i. e. you can fit even with the junction affine model, which is considered as the upper limit of rubber elasticity. But in parallel, it is the opposite. It is less deformed than the phantom network after deformation.

KILIAN:
Have you compared the high q-range with X-ray measurements by which the orientation is observed up to 20 Å? I could not figure out how you would interpret the orientation pattern (i. e. the lozenges).

BOUÉ:
I will propose something in the second part of my talk, but it is just a proposal. If you imagine that you have only some chains with their end-to-end vector parallel to stretching and some chains with their end-to-end vector perpendicular to stretching, then you can get a lozenge pattern.

KILIAN:
The problem for me is that you can only derive a characterization of the orientation under the assumption of having stretching invariant elements. It could therefore be helpful to compare the X-ray results with those in the high q-range.

BOUÉ:
But in X-rays you will see just the halo.

KILIAN:

Yes, but it is possible to draw from these experiments, the orientational distribution function up to 20 Å.

BOUÉ:

I remember that a long time ago, we did X-ray on some samples which were less carefully done than this. It gave a halo but the same in the two directions. It was not a very accurate measurement.

PICOT:

I just wanted to mention that this deformation, which is more accentuated as seen by neutron scattering in the transverse direction compared to the parallel direction, seems to be quite a general phenomenon.

BOUÉ:

Yes, you could see that, for example, from the work of Beltzung-Picot on PDMS for the radius of gyration as a function of λ. In perpendicular, *Rg* is not very far from the junction affine model and essentially more deformed than predicted by the phantom network. In parallel, it is always close to the phantom network. In the experiments presented here you see it on a large q-regime and you see how it occurs. The idea was to follow this relaxation until reaching the equilibrium under strain. It is better than looking at it at the end only. You clearly see how and when data depart from model predictions, as the time passes. One illustration of the failure of the models at long times is that even between 1 min at 150 °C and 30 min at 150 °C you still see a relaxation. If it agreed with the model predictions on rubber for PS, you would not observe it. It should be finished at 1 min. We were surprised to see an additional relaxation at this long time.

KILIAN:

How do you calculate the intensity distribution function? In the Debye-region you may consider the molecular coil as an object with a constant mean density. The form and size of this coil determine the constructive interference in the Debye-region. If there is no coil-to-coil interference, the scattered intensity should then correspond to that of an oriented coil-gas. One has to know the orientational distribution of the anisotropic coils. In the large q-range, the orientation pattern is determined by small, probably stretching, invariant segments. This can be described by means of correlation functions. In the intermediate range serious problems to a solid understanding might arise. What is your opinion about this?

BOUÉ:

You consider the labelled chain as a path passing through N_c crosslinks. The distribution of the positions of the crosslinks is given by the junction affine model. It is possible to calculate the form factor of a chain with some monomers fixed at the fixed positions of the crosslinks, and then to average over the distribution of positions (see Ref. [1 a] in my paper in these Proceedings). In the phantom network, the crosslinks are not independent. This is more difficult and we have used the calculation of Warner and Edwards.

VILGIS:

There is, of course, a process which does allow for such structures, if you allow for entanglements, because they will destroy the structure at this alteration point completely. There is no reason why the distance between the crosslinks is a relevant distance in the system.

BOUÉ:

It is an interesting proposal. I have not spoken at all about trying to fit using an entanglement model. We have tried to calculate something in which we included the entanglements but we did not finish the calculation to compare them.

KILIAN:

If there are chains of different lengths in the network, they were stretched to different degrees. If these chains of different lengths are fixed to the same crosslinkage, I find your idea that rotation occurs not so bad. This might lead to complicated relaxation processes. These effects should, to my understanding, induce lower anisotropies in respect of the direction of strain.

BOUÉ:

I agree: for example, bi-modal networks should give affine displacements of the junctions only if you accept the theorem of Sam Edwards, but if you put any interaction in addition to the chain, then it destroys that.

WINTER:

Did any of the properties suggest self-similarity? At the end you proposed the self-similar model and that should be expressed somewhere in it.

BOUÉ:

No, this model is just a remark to show how a giant fluctuation of the crosslinks could occur, but different topologies may coexist and I do not think that in this experiment we can find any self-similarity at the moment.

WINTER:

You don't look for the solution of the problem in that?

BOUÉ:

It may be that such an object exists but there is no sign of fractals in the experiment – directly in the data. We are thinking of other experiments to show it in swollen gels, eventually.

PIETRALLA:

At first sight, I would presume that the fluctuations should also become anisotropic at the end. Are there arguments against this? There would still be another point of view, looking at the deformation of a network or melt of long flexible molecules. We are all speaking of fluctuating network junctions deforming affinely. But the simplest model would be a volume element deforming affinely. The chain can move within that volume element, and you exert a force as well, because you now have an anisotropic time-averaged form of the molecule. In the melt this molecule can relax to the normal isotropic conditions, whereas in the rubber, it is forbidden by the junction points. But at the first status of deformation with high speed, we have a similar problem for both. At long times, you have a different behaviour because of the junction points forming the fluctuating "walls" of the constraint volume. I think this would be the simplest way of looking at the deformation of long molecules.

BOUÉ:

This is a good approach. Looking at what gives the chain, if, for example, you put it in a cube and you deform affinely, in this cube you have even two contributions: a kind of extension contribution and a kind of confinement contribution. So you take a textbook and you can start to calculate that. I'm not sure what it would give; we thought of looking at something such as the fact that the chain has got a finite volume so that it can be compact to a certain distance. This gives a new cut-off, a new crossover distance, different from the mesh size. This was proposed by Bastide and Leibler. I must add that tube models are not far from your idea, in a way. The Edwards description first, and also the work of Gaylord and Gotlieb, may be quoted.

Progress in Colloid & Polymer Science Progr Colloid & Polymer Sci 75:171–178 (1987)

A Model of cooperative motions in dense polymer systems by means of closed dynamic loops*)

T. Pakula

Max-Planck-Institut für Polymerforschung, Mainz, F.R.G.

Abstract: A microscopic model of cooperative motions in dense polymer systems is described. The mechanism suggested consists of collective replacements of molecular subsegments within closed loops of motion. Application of the model to the description of various effects related to mobility in dense polymer systems are presented. These include: (1) computer simulated "melting" of ordered chain structures as a method of generation of amorphous dense systems, (2) computer simulation of liquid-glass transition in polymer systems and (3) a qualitative analysis of chain motions in dense polymer systems.

Key words: Cooperative motion, self diffusion, glass transition, model of dense polymer system, computer simulation.

Introduction

Recent understanding of molecular motions in bulk polymers is based on two independent concepts related to two different size scales. To justify the local mobility of molecular segments above the glass transition, the "free volume" concept has been introduced [1–3], whereas the diffusive motions of polymer chains are commonly modelled as "reptation" processes [4, 5]. Both these concepts, however succesful in describing particular phenomena, do not constitute an uniform picture which would consider a natural interdependence between motions on various size scales.

Recently, a mechanism of relaxation in dense polymer systems has been suggested [6] which involves coupling between localized short-range segmental motions and motions of whole molecules over long-range distances. In this mechanism collective replacements of chain segments within closed loops of motion are assumed as a natural consequence of close packing of the chains. The movement consists of two contributing mechanisms. The first is based on position exchange between chain elements belonging to two different chains in intimate contact along some short portion of the chains. The second consists of translational movement of chain segments along the chain contour between subsequent position exchanging areas. Through each position exchange unit the motion can be transfered from one chain to another. Within the loop of motion each chain element is shifted to a new position occupied previously by another adjacent element also belonging to the loop. A simple local mode of this type of motion is illustrated in Figure 1 for the case of linear chains on a simple cubic lattice. In this example a kink on a chain and an element parallel to the kink top are assumed as the position exchanging unit. The model involving the above types of motion has been used for computer simulations which demonstrate that overall mobility in condensed systems involving, for example, long range displacements of linear chains, can be observed [7]. In this paper we present some initial results illustrating the applicability of the above model to the description or simulation of various effects related to mobility in condensed state of macromolecular systems.

"Melting" of ordered structures as a method of generation of amorphous dense systems

A system filled completely with chains each of N freely joined monomers is considered. It is assumed

*) Paper presented in part at the "Discussion Meeting on Networks", Gomadingen 1986.

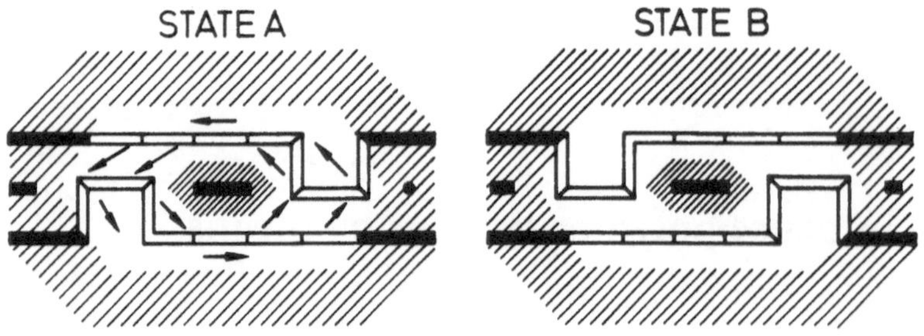

Fig. 1. An example of a simple case of motion within the loop for chains on the lattice. Arrows denote directions of element displacements in rearrangement from state A to state B

that the cooperative relaxation mechanism is the only one operating in such a system. Each cooperative relaxation is the closed loop motion of short chain segments belonging to one or a few chains. Within loops the motion is transfered from one chain to another by position exchanging areas for which special local conformational arrangement is necessary. The local conformation at which the loop of motion can enter the chain will be called here "motion acceptor" and a place at which the loop can leave the chain will be regarded as "motion donor". In the simplest case one can consider kinks and chain ends as "motion acceptors" and chain ends as well as all chain segments parallel to kink tops or to chain ends as "motion donors". This is illustrated in Figure 2 for the case of chains on cubic lattice. A "motion acceptor" and a "motion donor" constitute the position exchange unit. Each position exchanging unit can be regarded as an element transfering the motion from one chain to another.

The simplest case of the initial state of a system with chains which fill the space completely is the perfectly oriented state in which all chains are parallel as shown in Figure 3. The motion introduced in such a system involves a kind of "melting" leading finally to the isotropic system. A series of cross-sections through a three dimensional system with chains of the length $N = 32$ undergoing such a process is shown in Figure 3.

Computer simulations have been performed with periodic boundary conditions. As an indication of the transition of the system from the oriented state to an isotropic state the change of the chain segment orientation distribution within the model can be given. The number of chain segments remaining parallel to the initial orientation after various times is plotted in Figure 4a versus the distance from the bottom face of the model. These dependences indicate that the process starts at model edges containing chain ends but finally the state is reached in which 1/3 of the chain segments are oriented in all directions homogeneously throughout the system. Chain ends concentrated initially at the model faces become homogeneously distributed in the system final state as shown in Figure 4b. The overall orientation changes can be characterised by the time dependence of the order parameter defined as

$$\langle f \rangle = \left\langle \frac{3}{2}\left[\left(\frac{\bar{n}-n_x}{n}\right)^2 + \left(\frac{\bar{n}-n_y}{n}\right)^2 + \left(\frac{\bar{n}-n_z}{n}\right)^2\right]\right\rangle \quad (1)$$

where n is the number of chain segments in the system, $\bar{n}$ is the mean number of chain segments oriented in each direction in a completely isotropic system ($\bar{n} = n/3$) and n_x, n_y, n_z are the numbers of segments oriented in the x, y and z directions, respectively. The order

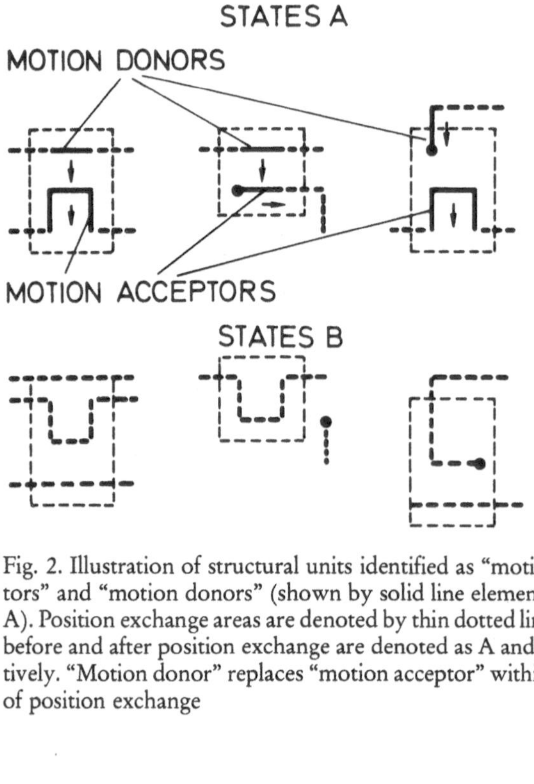

Fig. 2. Illustration of structural units identified as "motion acceptors" and "motion donors" (shown by solid line elements in state A). Position exchange areas are denoted by thin dotted lines. States before and after position exchange are denoted as A and B respectively. "Motion donor" replaces "motion acceptor" within the area of position exchange

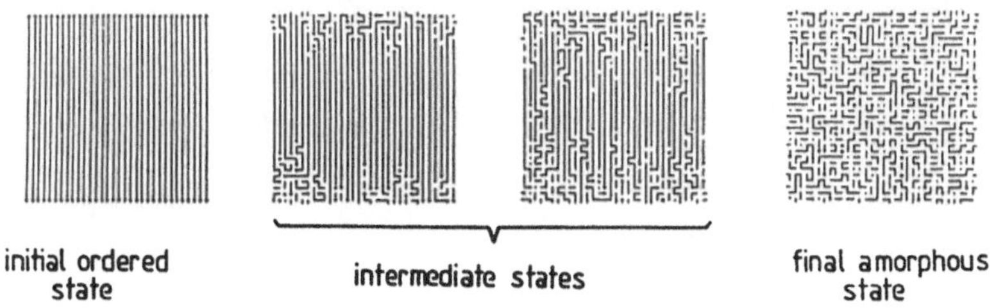

Fig. 3. Illustration of structural changes during "melting" process in a system with chain length $N = 32$. Cross-sections through the three-dimensional model are shown

parameter so defined assumes a value of 1 for the system with perfect uniaxial orientation and is 0 for a completely isotropic system. The results are shown in Figure 5. As a measure of time, the overall number of steps taken in searching mobile loops is used. This dependence shows that the initial perfectly oriented system of chains transforms to the final stable isotropic state ($\langle f \rangle = 0$).

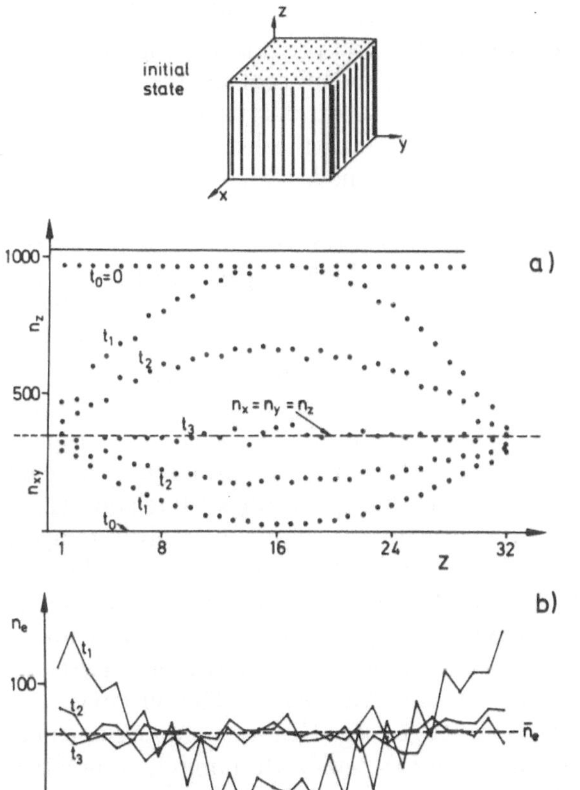

Fig. 4. (a) Distribution of numbers of chain segments n_z and $n_{xy} = (n_x + n_y)/2$ oriented parallel and perpendicular to the initial orientation direction respectively. Various distributions are recorded for various times $t_3 > t_2 > t_1 > t_0 = 0$. (b) Distribution of numbers of chain ends along the initial chain orientation direction for various times

An analysis of the final chain conformations for various chain lengths, reported elswehre [6], has indicated that Gaussian chains can be achieved in such equilibrated systems.

This means that the process observed provides a method for generating isotropic dense systems with randomly coiled chains. The main advantage of this method is that in the initial ordered state various parameters of the system can be easily introduced as, for example, chain length, chain shape (linear or cyclic chains) or composition if various chains are considered.

Overall mobility scheme in amorphous state. Temperature dependence

When the amorphous isotropic state is achieved the traces of loops can be regarded as randomly coiled rings because they consist of parts of random chains. Each position exchanging unit can be regarded as a virtual "bridge" between chain elements being locally in close contact and through which the motion can be

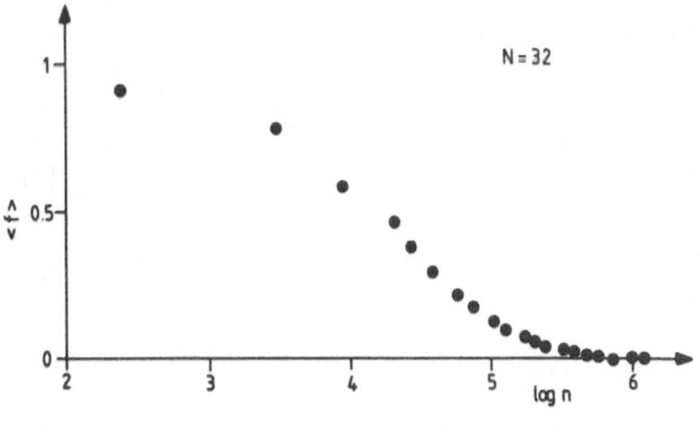

Fig. 5. Time dependence of the overall orientation factor during the "melting" of the ordered structure

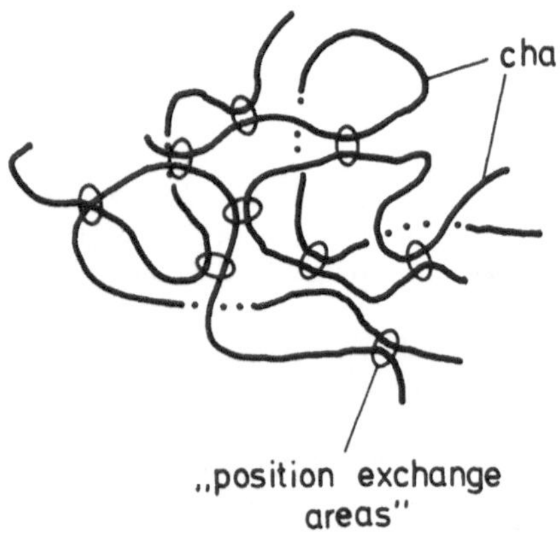

Fig. 6. Schematic illustration of the virtual mobility network (Rings denote position exchange areas)

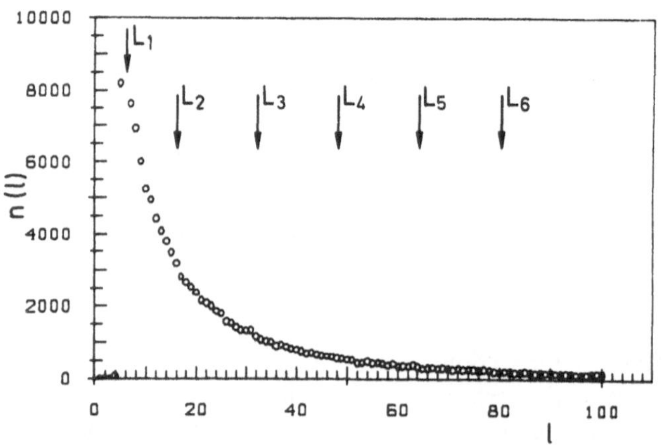

Fig. 7. Size distribution of mobile loops in a system with chains of the length $N = 512$

transfered from one chain to another. The whole system can be then treated as a virtual network of potentially mobile loops. The density of "bridges" in such a system will determine structural conditions for motion possibilities. The virtual network is illustrated schematically in Figure 6. Small rings in the figure denote position exchange units consisting of a "motion acceptor" on one chain and a "motion donor" on another chain or another part of the same chain. Each closed trace in such a random network can be regarded as a potentially mobile loop. Loop sizes can be broadly distributed in such a system. An exemplary distribution of loop length determined for a simulated system with chains of length $N = 512$ is shown in Figure 7. This distribution can be related to the distribution of probabilities that the closed trace of length l will be found if one moves along paths in the virtual network consisting of all chain contours and of "bridges" through which the motion can be transfered between chains.

The temperature can be introduced into such system if we assume that the cooperative rearrangement within a closed loop requires a certain sufficient fluctuation in energy in order to move the local system (the loop) into another configuration. According to the theoretical considerations of Adam and Gibbs [8] the probability of cooperative rearrangement at a given temperature is related to the size of rearranging region as follows

$$p(z) = A \exp(-z\,\Delta\mu/kT) \qquad (2)$$

where z is the size of rearranging region (loop length in our case) and $\Delta\mu$ is the potential energy hindering the cooperative rearrangement per monomer segment. In this way the motion probability in our system will be determined by a structurally conditioned loop length distribution such as that presented in Figure 7 and by the temperature dependent probabilities of rearrangements within loops of various sizes.

In the system for which the distribution presented in Figure 7 was determined, we have assumed that all structurally possible rearrangements are equally probable (independent of loop size). This can be regarded as the situation at very high temperature. Lowering the temperature will involve a modification of the size distribution of rearrangable regions, making rearrangements less probable, especially in longer loops.

We report here a very simplified case of the temperature dependence of mobility in a computer simulated system. In this system, instead of modifying the length distribution of rearrangable loops by relation (2) we have assumed that there exists, for a given temperature, an upper limit of the loop size at which the motion is performed. Under such conditions the motion has been observed in a dense system with randomly coiled chains generated according to the procedure of the "domino" model [9]. The system generated in this way can be regarded as a volume element cut out from a larger system with infinitely long chains, which means that in the volume considered the chains are cut at model faces. These ends of chains have been regarded as fixed when the motion in the system has been generated. Because of these fixed chain ends at model faces, only the central part of the model, consisting of 16^3 elements, has been examined, while motion has been generated in the whole system consisting of 32^3 elements. An element in the central part

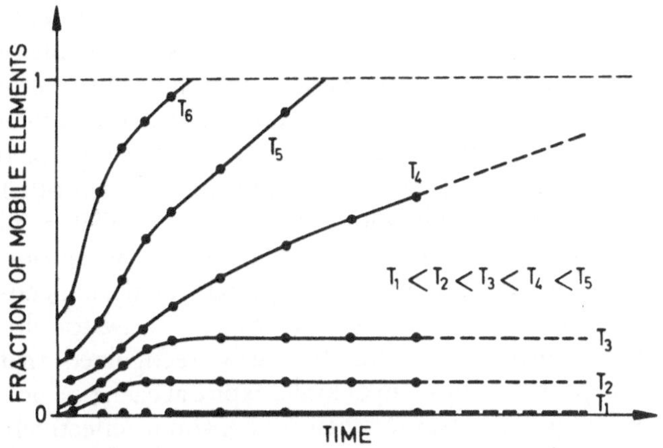

Fig. 8. Time dependence of fractions of mobile elements in systems with various limits for allowed lengths of mobile loops ($L_1 = 6, L_2 = 16, L_3 = 32, L_4 = 48, L_5 = 64, L_6 = 80$)

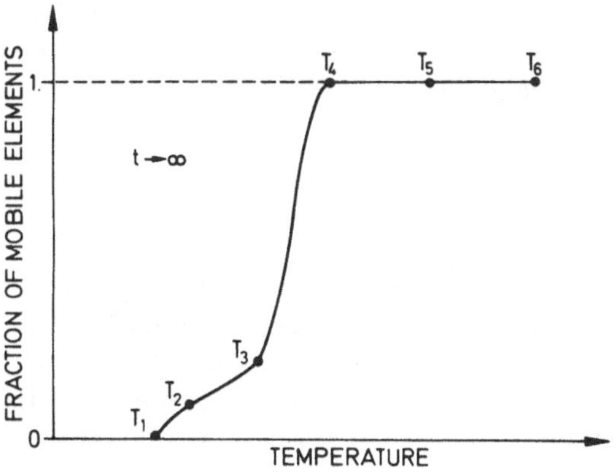

Fig. 10. Temperature dependence of final mobile fraction in simulated systems. Temperature scale is $T = C\,L$ with an arbitrary chosen constant C

of the model has been regarded as mobile when it has contributed to motion as a position exchanging element at least once. The time dependence of the fraction of such mobile elements for various temperatures related to the upper limit of loop length L by an arbitrary constant ($T = C\,L$) are shown in Figure 8. It is seen that at low temperature only a small fraction of the system can be regarded as mobile. Because of the low limit of loop length only local motion modes are possible which, when not influencing each other, can not lead to the overall mobility of the system. An example of such local type of rearrangement is shown in Figure 9. At higher temperature, when motion within longer loops is possible, the overall mobility of the system is achieved so that each element can change its position. In Figure 10 the temperature dependence of the final fraction (at $t \to \infty$) of mobile elements is plotted. This dependence shows that at a certain temperature, a transition betwen a nonmobile and a mobile state of the system takes place. However, localised small scale motions are possible below this transition.

The simulated temperature dependence presented here, though very simplified, shows that the model of motion considered is able to describe at least qualitatively the glass-liquid transition in dense polymer systems.

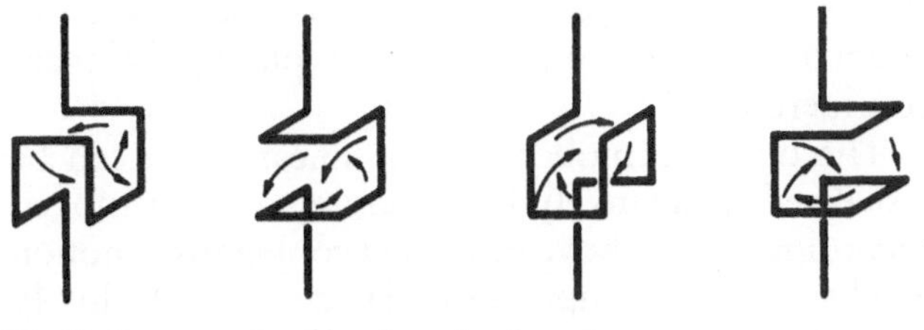

Fig. 9. An example of local mode of motion

Motion of a chain

A chain in the system moves locally when a mobile loop enters his contour at the "motion acceptor". The loop leaves the chain at another chain element, the „motion donor", at some curvlinear distance d from the first element. A pass of the motion loop through the chain causes a shift of the "motion acceptor" (a kink) along the chain by a distance d which generally depends on the concentration of position exchanging units in the system. Each shift of a kink along the chain by distance d effectively moves the kink mass by vec-

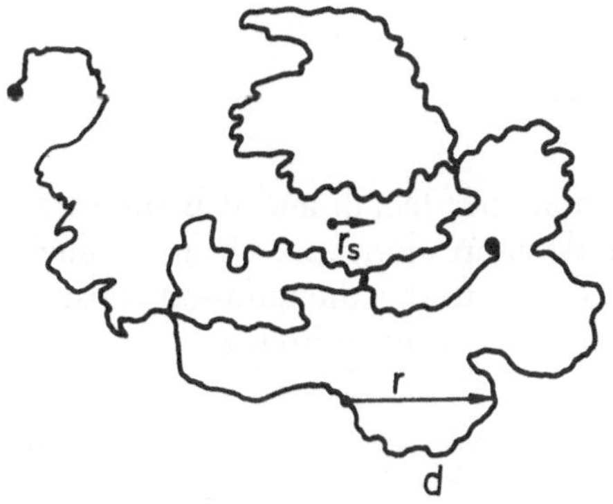

Fig. 11. Schematic illustration of an alemental displacement of the center-of-mass of a chain caused by the shift of a kink along the chain by a curvlinear distance d

tor r in the space as illustrated in Figure 11. This involves a shift of the chain's centre-of-mass by a parallel vector of the length

$$r_s = \frac{r}{N} \tag{3}$$

where N is the chain length expressed in the number of monomers in the chain. This describes the elemental displacement of the center-of-mass of the chain caused by a single loop of motion. The diffusive motion of the chain can then be considered as a sum of such elemental steps

$$\langle R_s^2 \rangle = \sum_{i=1}^{n} \langle r_{si}^2 \rangle + 2 \sum\sum_{1 \leq i < j \leq n} \langle r_{si} r_{sj} \rangle . \tag{4}$$

For randomly coiled chains without correlations between bonds we have $\langle r_{si} r_{sj} \rangle = 0$ for $i \neq j$. Under the simplifying assumption that each shift has a constant length r_s we can express the mean square displacement of the center-of-mass in time t as follows:

$$\langle R_s^2 \rangle = r_s^2\, f\, t \tag{5}$$

were f is the frequency with which a chain is entered by mobile loops. This frequency will depend on the number of "motion acceptors" on the chain and on the frequency f_o with which a single "acceptor" is moved along the chain. This can be expressed as

$$f = f_o\, N\, c_a \tag{6}$$

where c_a is the concentration of motion acceptors on the chain per monomer unit. Introducing Equations (3) and (6) into Equation (5) and considering that $r^2 = a^2\, d$ we get

$$\langle R_s^2 \rangle = \frac{1}{N}\, a^2\, d\, f_o\, c_a\, t \tag{7}$$

where a is the monomer length and d is the mean length of the kink jump along the chain in single motion step. According to this the diffusion constant for a chain of length N can be written as

$$D_o = \frac{1}{N}\, a^2\, d\, f_o\, c_a . \tag{8}$$

Such a result implies, however, that for any chain considered, the whole space is available for movement, as it would be in the dilute solution regime. In fact, however, in dense systems the displacements of kinks are restricted to a certain complicated space of freedom formed in the surrounding system moving with the particular chain cooperatively. Through each position exchange the chain can slide between other chains. The possibility for a chain to move will depend in this way on the concentration of motion "acceptors" in the surrounding space and especially on the concentration of these "acceptors" which can release topological constraints. It can be easily recognized that "acceptors" in the form of kinks expire at each position exchange and in this way can not contribute effectively to constraint release. On the other hand, the chain ends can be considered as permanent motion "acceptors". It seems therefore to be reasonable to postulate that the effective freedom for a chain for long distance displacements will be determined by the concentration of chain ends in the surrounding space. This can be described as a friction effect with the coefficient of friction $\varphi \sim 1/c_4$ where c_e is the concentration of chain ends. The diffusion constant D_o related to the nonconstrained motion of a chain (as described by Eq. (8)) can then be modified by the effect of such a friction

$$D = D_o/\varphi . \tag{9}$$

If the surrounding chains are of uniform length N' the friction coefficient will be proportional to N' and consequently

$$D = \frac{A}{N}\, \frac{a^2}{N'}\, d\, f_o\, c_a \tag{10}$$

where A is a constant. In the case of self diffusion when $N = N'$ this describes the chain length dependence

$$D \sim N^{-2} \tag{11}$$

for sufficiently long chains ($N >> d$). This result agrees well with the computer simulated dependence [7] and coincides with the prediction of the reptation model [5]. It has to be noticed, however, that the picture of topological constraints considered here is different from that usually used when reptating chains are considered.

The diffusive motions of the linear chain can be visualized qualitatively by observing a chain moved according to the above described cooperative motion mechanism in a computer simulated system of closely packed chains. In Figure 12 various states are shown of

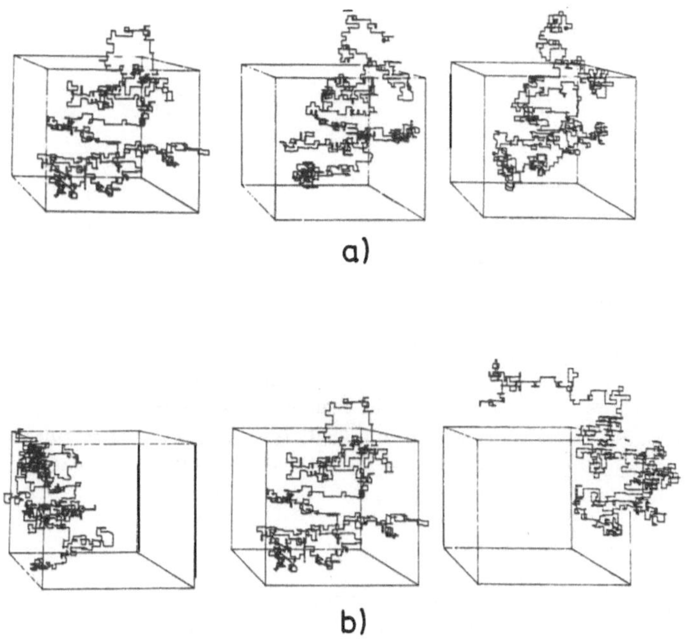

Fig. 12. Computer simulated motion of the chain of length $N = 512$. The time intervals between various states in case (a) are 100 times smaller than in case (b)

a single chain with the length $N = 512$ diffusing in the system filled completely with chains of the same length. The time intervals between various states have been chosen so that changes of chain conformation and position taking place over a short time scale (Fig. 12a) and over a longer time scale (Fig. 12b) can be observed. It can be easily noticed that various types of motion contribute to the observed chain diffusion. The chain changes locally its internal form within relatively short time intervals and moves translationaly and rotationaly when observed over a longer time scale. It would be difficult to distinguish in the above observed motion that it is restricted to a kind of tube surrounding the chain contour. The chain observed has rather a large degree of freedom for internal shape changes.

Some special cases of restricted motion of a chain

From the above analysis comes out that the chain motion is caused by a random walk of kinks along chains. Each step of this walk is part of a closed loop cooperative rearrangement in which one or few chains can take part. The kinks can enter or leave chains through "motion acceptors" (chain ends or kinks) and they can be generated by local accumulation of conformational changes related to walks of various kinks. This constitutes a peculiar case of random walk of kinks taking place between fixed absorbing and generating barriers (chain ends) in which the walking element can also be generated or absorbed at any point between fixed barriers. If the later possibility is forbidden the problem is equivalent to the case considered by De Genes in the solution of the reptation model [4]. We would like to demonstrate qualitatively here that the internal mobility of a chain which moves according to the introduced mechanism is not necessarily related to the mobility of chain ends as one expects in the reptation model. Computer simulated chain motions have been performed in which the nature of barriers at chain ends has been changed from absorbing and generating to reflecting ones. In Figures 13 examples are shown of chain motion with length $N = 128$ for which one chain end is assumed to be a reflecting barrier in the first case (Fig. 13a) and this condition is extended to both chain ends in the second case (Fig. 13b). The chain ends regarded as reflecting barriers become fixed in the space, which certainly excludes the translational motion of the whole chain when observed over a long time. In both cases, however, the chain remains mobile and its centre-of-mass can fluctuate or rotate around the fixed end in the first case or around the end-to-end axis in the second case. These effects can only be achieved when kinks can be generated and absorbed at any place between fixed barriers related to chain ends.

The above examples demonstrate that the motion mechanism and the computer simulation based on it are suitable for the analysis of some special cases of

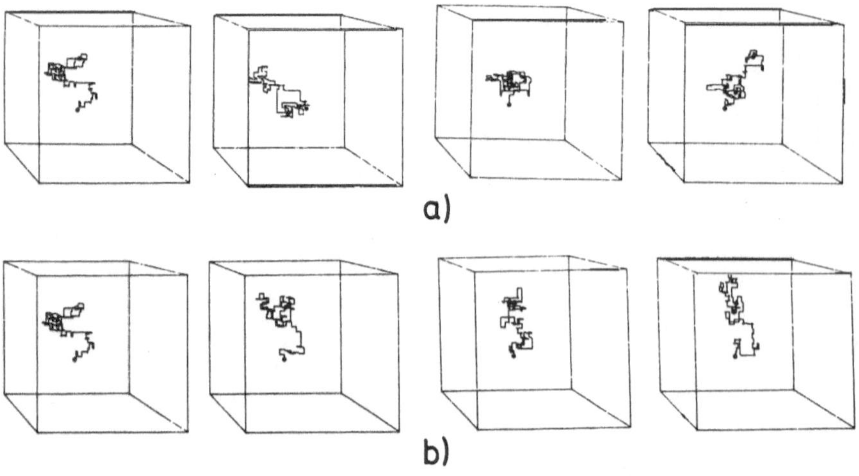

Fig. 13. Examples of restricted motion of the chain in a dense system with chains of the length $N = 128$: (a) chain with one end fixed, (b) chain with both ends fixed

chain motions such as star molecules (the case shown in Figure 13a can be regarded as an arm of a star), polymer networks (the case in Figure 13b can be regarded as a motion of a chain between cross-links) or ring molecules for which the translational motion has been demonstrated [6] and can only be observed when kinks are generated and absorbed on the whole ring contour.

References

1. Fox GT, Flory PJ (1950) J Appl Phys 21:581 (1951) J Phys Chem 55:221; (1954) J Polym Sci 14:315
2. William ML, Landel RF, Ferry JD (1955) J Am Chem Soc 77:3701
3. Cohen MH, Grest GS (1979) Phys Rev B 20:1077
4. de Gennes PG (1971) J Chem Phys 55:592
5. Doi M, Edwards SF (1982) J Chem Soc Faraday Trans 2 74:1789
6. Pakula T (1987) Macromolecules 20:679
7. Pakula T, Geyler S, Macromolecules, in press
8. Adams G, Gibbs JM (1965) J Chem Phys 43:139
9. Pakula T (1987) Polymer 28:1293

Received August 17, 1987;
accepted August 20, 1987

Author's address:

T. Pakula
Max-Planck-Institut für Polymerforschung
Postfach 3148
D-6500 Mainz, F.R.G.

Discussion

TOMKA:

Would it be possible to make similar calculations, such what you showed in your last slide, but with the confinement that both ends of the molecule should not leave a space? Measurements of light scattering from entrapped chain molecules in porous silica exhibit a very large change of the translational diffusion coefficient because now the molecule can freely move only in a confined space.

It would be a very good test for your model to see whether you get this enormous change of the translation and diffusion coefficient.

PAKULA:

I don't know what the result would be on the mathematical model, but with computer simulation we could certainly do so.

D. RICHTER:

If you have this collective type of motion, then there should be quite the difference in the coherent or incoherent scattering function from such a system, because in the coherent function, the collective response should come up, while in the incoherent one, you would just see the mean square displacement of the particles, that means the $t^{1/2}$ law. So it would be very interesting to calculate the mean square displacement, as well as the coherent response. Then you could compare with the experiments which are already there.

PAKULA:

We would like to do this. And in relation to the diffusion of free chains in networks it is also possible to do this.

Progress in Colloid & Polymer Science Progr Colloid & Polymer Sci 75:179–200 (1987)

Entanglements in semi-dilute solutions as revealed by elongational flow studies

A. Keller, A. J. Müller and J. A. Odell

H. H. Wills Physics Laboratory, University of Bristol, Bristol, G.B.R.

Abstract: Our methodology of chain stretching using elongational flow enables identification of circumstances in which macromolecules in solution extend cooperatively, in contrast to their extending in isolation. This opens a new window for the detection and study of entanglements. In addition to defining the concentration for the required degree of coil overlap, the present studies also define the time scale on which the geometric entanglement becomes mechanically effective, thus introducing a dynamic element to the identification and classification of entanglements. The entanglements develop through a sequence of patterns, periodic both in time and space with increasing strain rate. The strain patterns thus arising modify the flow field locally, assessed by velocimetry. Further, the macroscopic flow resistance ("elongational viscosity") is determined and correlated with the various stages of chain stretching and network formation. These findings link molecular behaviour and macrorheology. Examples from engineering applications indicate how entanglements may help to account for the various flow modifying actions of polymeric additives, and conversely, how experience gained in engineering applications can potentially further the study of entanglements.

Key words: Elongational-flow, semi-dilute solution, elongational-viscosity, entanglements, birefringence.

1. Introduction

This article describes a new approach to entanglements, their identification and characterization, based upon the response of semi-dilute solutions to elongational flow fields. It will relate some works published recently [1] with results from ongoing work still awaiting publication [2].

The approach is based on our recently acquired ability to extend isolated macromolecules virtually fully, and to register this chain extension (see Refs. [3–4] and review [5]). If the chains do not extend as isolated entities, but in unison with others within their close environment, this can be readily diagnosed and interpreted as the consequence of mechanically active entanglements, which then provides a sensitive tool to explore this fundamental, but hitherto elusive macromolecular phenomenon. The way in which this is achieved will be described, together with some of its impact on solution rheology and consequent applications. Firstly, however, the preliminaries relating to the isolated molecule will be presented, as this is required for the appreciation of the entanglement studies.

2. Extending isolated molecules; principle and realization

2.1 Principle

The behaviour of an isolated molecule is best approached by dilute solutions. The molecule itself will generally be considered as highly flexible and non-free draining for the present article; the approach is readily extendable to more rigid and permeable chains.

In order to extend a molecule by flow, the flow field has to be of extensional character, which means that the fluid must be persistently extending. The molecule contained by the fluid will follow this extension to the degree determined by its frictional interaction with the fluid. However, it will only be able to do so when the elongational strain rate $\dot{\varepsilon}$ characterizing the flow exceeds a critical value $\dot{\varepsilon}_c$ of the order of the reciprocal of the conformational relaxation time τ (where $\tau = f/K$; f and K being the friction coefficient and the entropic restoring force constant, respectively) of the coil. In this case, the extensional influence of the flow, as transmitted by the friction, will overcome the entropic resistance of the coil tending to retain its random coil conformation. Once this stage is reached and coil extension starts, the coil will become increasingly more permeable to the fluid, hence the frictional interaciton, and the extensional forces will increase. In consequence, the extension becomes a run-away process until it is halted by the increasing restoring forces. It can be shown [6] that the latter will only occur very close to full chain extension, hence, for all practical considerations, virtually full chain extension should be achievable. It follows from the above, as is indeed predicted by preceding theory [7–8], that chain extension is critical in strain rate ($\dot{\varepsilon}$); below ($\dot{\varepsilon}_c$) the coil remains virtually unextended while above it extends almost fully, where this critical coil stretch transition is defined by

$$\dot{\varepsilon}_c \tau \approx 1 . \quad (1)$$

In view of the fact that for a usual long flexible polymer, τ is of the order of milliseconds the required $\dot{\varepsilon}$ needs to exceed $10^3\ s^{-1}$ for most systems used in practice.

However the attainment of $\dot{\varepsilon} > \dot{\varepsilon}_c$ is necessary, yet not sufficient for full chain extension, because a further criterion, that of sufficient strain (ε), also needs to be satisfied. The latter is a rather demanding criterion in view of the fact that for a long molecule the coil stretch transition may require strains of 500 x or more. In experimental terms, this means that the molecule must have a long enough residence time within a flow field of $\dot{\varepsilon} > \dot{\varepsilon}_c$ for the required strains to be attained.

2.2 Method

We succeeded in realizing the above dual condition for full chain extension, namely, sufficient strain rate and its maintenance for a long enough time for a given fluid element to attain the required strain. In the process, we also tested and verified the predictions inherent in Equation (1). This was achieved experimentally by generating controlled elongational flow fields either through double suck jets creating uniaxial stretching flows, or through a cross-slot device, or alternately through a four roller mill producing simple shear flows. Of these, the double jet case, most widely used

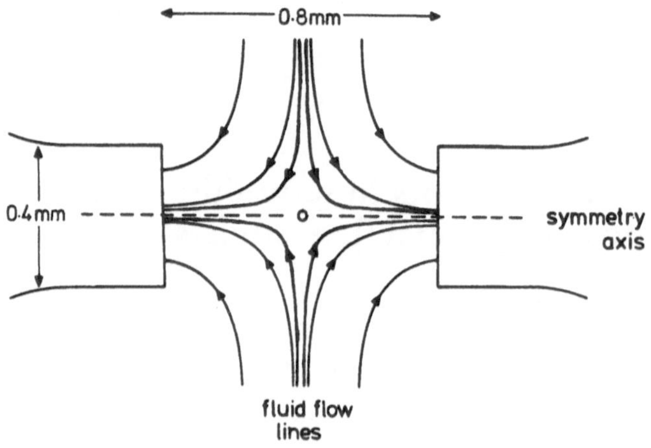

Fig. 1. The opposed jets. Polymer solution is sucked into the jets along the symmetry axis. O represents the stagnation point in the centre of the flow-field

for the purposes of the present paper, will be briefly described.

Figure 1 illustrates the principle involved, and Figure 2 the method of realization. With suction applied, extensional flow between the jets is generated. The chain extension, when it sets in, is then apparent through birefringence, which for dilute solutions is seen as a birefringent, hence a bright line against a dark background between crossed polars (Fig. 3). The birefringence can be measured, or as most usually done for continuous monitoring, the intensity (I) of transmitted light is recorded as a function of $\dot{\varepsilon}$ (for low values the retardation is proportional to $\sqrt{I}$).

The two requirements raised above are readily achieved: the high strain rates through appropriate dimensions of the orifices and their separation, coupled with appropriate suction speeds with due attention to stay within the laminar flow limits. The high strains themselves are automatically achieved thanks to the intrinsic nature of the flow field. Namely, the flow field contains a stagnation point (O in Fig. 1) which means that fluid elements along streamlines passing through it will have, in principle at least, infinite residence time, hence will become correspondingly extended. This has the further important consequence that chain extension will be confined to those streamlines which pass through or close to the stagnation point. The result of the latter is a high degree of *localization* of chain extension within the flow field, even when the strain rate itself would be sufficient for chain extension to occur over the whole volume between the

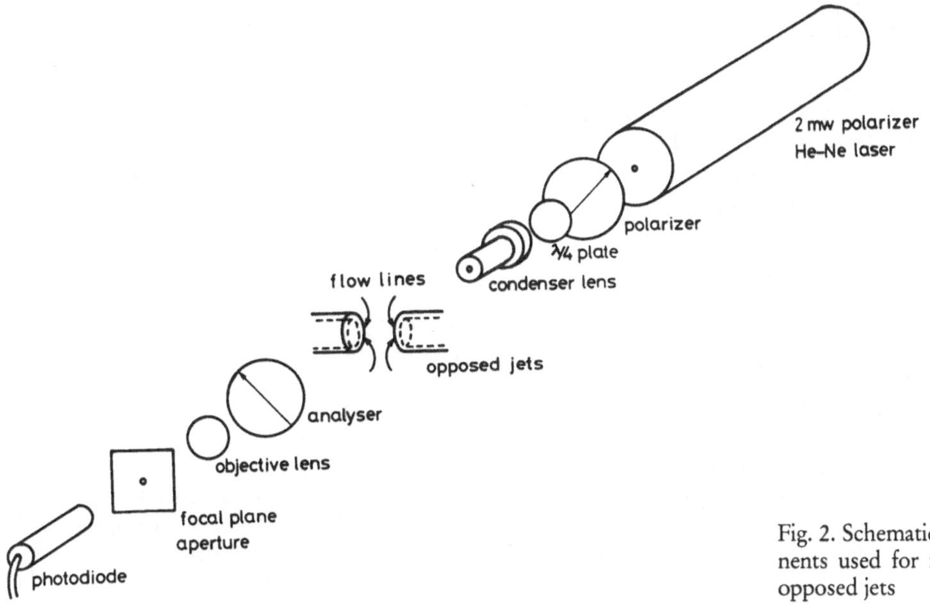

Fig. 2. Schematic representation of the optical components used for measurement of birefringence in the opposed jets

jets. It follows that the localization defines the regions of adequate strain. The narrow birefringent line in Figure 3 (as compared to the full jet orifice) corresponds to this localization. The appreciation of this localization is paramount for what is to follow, where departures from the sharply defined localization will acquire special significance.

2.3 Establishment of criticality

Using the methodology just outlined, the criticality inherent in Equation (1), as predicted by theory, could be verified, as far as we know, for the first time. In view of the molecular weight dependence of τ (see later) this verification required ideally monodisperse material, which amongst readily available polymers is best approximated by anionically polymerised polystyrene, a material used in much that is to follow.

It was found that the birefringent line in Figure 3 indeed appeared suddenly as $\dot{\varepsilon}$ was increased, as did the birefringence in a corresponding trace (Fig. 4). This verifies the criticality of chain extension and also provides a value for $\dot{\varepsilon}_c$, and through Equation (1) for τ. This proved to be a most valuable parameter for characterizing the isolated macromolecule. A few instances will be listed; for further information see Reference [5].

3. Determination of τ in the service of molecular characterization

3.1 Draining characteristics of the random coil; hysteresis

The numerical values of τ, determined as above, were found to be close to expectations from the non-free draining Zimm model within any uncertainty arising from choice of input parameters. This numerical value is a well defined function of molecular weight M (see below), and for a given M of the viscosity (η). In the dilute solutions considered here, η is that pertain-

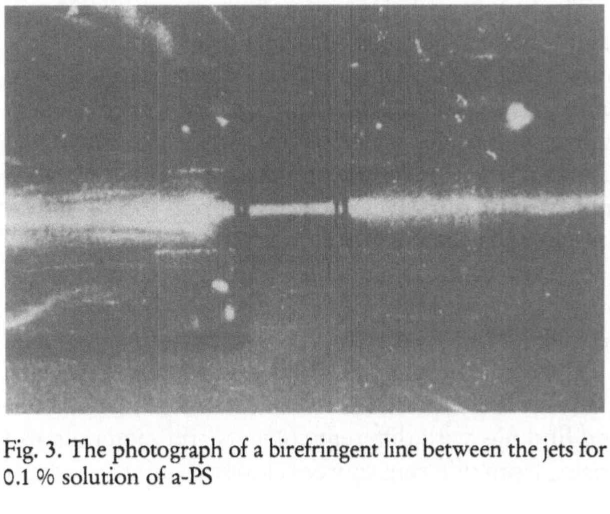

Fig. 3. The photograph of a birefringent line between the jets for a 0.1 % solution of a-PS

ing to the solvent in line with expectations from solution theory. It is important to realise that this characterizes the response of the chain in its initial undeformed state, and is different from the relaxation time(s) corresponding to the extension process itself, and is significantly different from the relaxation time associated with the retraction from the extended state. Preliminary experiments indicate that the latter is very much longer, by 1–2 magnitudes. Such a hysteresis between extension and retraction has been predicted by de Gennes [8] which, thanks to our new methodology, could now be made accessible to experimental verification. In physical terms, this longer time scale of contraction is due to the increased frictional contact of the chain with the solvent when in its fully extended form, in which case it will approach that in a free draining situation.

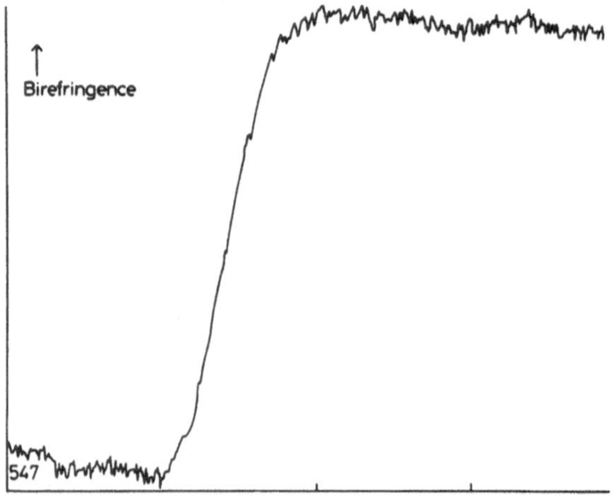

Fig. 4. Experimentally determined flow-induced birefringence versus strain-rate for a-PS. The maximum value of birefringence is 1.15 × 10^{-5} which corresponds to an optical retardation of 27 nm for a cell depth of 1.8 mm (0.1 % solution, $\bar{M}_w = 5.5 \times 10^6$ in Dekalin)

3.2 Molecular weight dependence, and a method for molecular weight determination

It is anticipated, even by qualitative considerations, that the longer molecules will be increasingly readily extendable, and more quantitatively, that

$$\tau \propto M^{\alpha} \quad (2)$$

where α should be in the range of 1.5–2.0. The value of 1.5 is expected for the non-free draining chain in the θ state, and the larger values correspond to increasingly expanded and increasingly non-hydrodynamically screened or more free draining situations. In the rather limited number of cases where narrow molecular weight samples of sufficiently high molecular weight were available (see below) this molecular weight dependence could be tested.

The validity of the power law, Equation (2) was established in all cases. For a-PS and polyethylene oxide (PEO) α was found to be 1.5 and, rather remarkably, irrespective of solvent quality in the few instances where this was varied. A value of 1.7 for α has just been reported for the good solvent bromobenzene (not used by ourselves) from a different source [9]. In the case of the bacterial polysaccharide Pullulan, which happened to be obtainable in comparatively sharp fractions, α was reported as 1.8 in what is believed to be good solvent [10].

The actual α values as obtained in this way, lead to many weighty questions relevant to solution theories, and findings with different solvents and materials and arising from different sources clearly require coordination. We shall not be concerned with these issues here; we shall merely avail ourselves of the situation that a single α value exists for a given polymer, or polymer-solvent system, which, if identified by suitable calibrants, can be used for determining molecular weights, not only an average value but the actual distribution. The situation for a single molecular weight is approximated by Figure 4. In case of a broad molecular weight spectrum the I v. $\dot{\varepsilon}$ curve will be a superposition of a continuous spectrum of such curves, appropriately weighted and displaced along the $\dot{\varepsilon}$ axis. Such a full curve, when expressed in terms of Equation (2) with the appropriate α, in fact provides the cumulative molecular weight distribution, from which the actual differential molecular weight can be readily constructed.

The inverse relation between $\dot{\varepsilon}_c$ and M [from Eqs. (1) and (2)] is to be noted. This has two important consequences. Firstly, that this method of molecular weight determination will be most readily applicable to the longest molecules, i. e. to those least accessible by conventional methods. Secondly, that there will be a cut off for chain extension at a low molecular weight limit determined by the maximum $\dot{\varepsilon}$ that can be realized in a given experiment.

3.3 Flow induced chain scission

If the chains are overstretched, i. e. $\dot{\varepsilon}$ increases along the plateau in Figure 4, they will eventually break [11,

12]. The molecular weight of the fracture products can then be determined by the method outlined in the preceding section. Thus, through our technique we can both break chains in a controlled manner, and analyse the resulting fracture products within the same apparatus enabling a systematic study of flow induced chain fracture. The principal results of consequence later on, are as follows:

a) The chains break into two closely equal halves, hence accurately at their centre.

b) The critical fracture strain rate ($\dot{\varepsilon}_f$) obeys the relation

$$\dot{\varepsilon}_f \propto \frac{1}{M^2} \,. \tag{3}$$

c) It follows from a combination of Equation (1) (with $\alpha < 2$) and (3) that there is a limiting upper M beyond which the chains cannot be stretched out, unavoidably breaking in the process. This has been confirmed experimentally [12].

Results a, b and c, can be simply interpreted by applying Stokes' law to a stretched out string of beads; in fact, the applicability of Stokes' law in this way is possibly the clearest evidence (for others see Refs. [5, 11, 12]) that the chains are indeed close to their stretched out state. The fracture stress itself can be calculated, and was found to correspond to that of the breaking strength of a C-C backbone bond, well within an order of magnitude. The fact that chain breakage is almost exactly (and not only statistically) central can be accounted for by considering the fracture as an activated process, where the exponential nature of the transition probability from the intact to the fractured state confines the probable fracture sites close to the exact geometrical centre of the stretched out chain [12, 13].

3.4 Chain stiffness – polyelectrolytes

The extensibility of the chains is greatly affected by their stiffness, and vice versa; the stiffer the chain, the more readily it is extended. In the limit of a fully rigid rod there is no extension, only orientation. More explicitly, for stiffer chains, both the strain rate and the strain required for full chain extension are lower. In terms of our experiment, this lower strain rate requirement means that $\dot{\varepsilon}_c$ will be reduced as far as the extension remains critical in $\dot{\varepsilon}$. However, with increasing chain stiffness the criticality in $\dot{\varepsilon}$, which is associated with entropic retraction within the chain itself, becomes less pronounced, until for a closely rigid rod system the chain (rod) alignment itself becomes a smooth function of strain rate determined by the orientability (the rotational diffusion coefficient D_r) of the rigid entity. The requirement of lower total strain has the additional consequence that there will be a lower degree of localization, i. e. the birefringent line will become broader with increasing stiffness, because for a given $\dot{\varepsilon}$ chain extension can be achieved for shorter residence times, hence along more peripherally situated streamlines. It follows from the above that our elongational-flow-based approach provides a new method for assessing chain stiffness or flexibility.

Based on these principles, we have studied extensively the orientability of a rigid rod system, (PBT, polybenzthiazole) in terms of its rotational diffusion coefficients and the mutual interference ("entanglement") of the rods as the concentration is increased [14, 15]. Similar studies are also being conducted here on biomolecules forming rigid or worm-like systems such as collagen (where denaturation can be followed) and xanthan gum [16]. Closer to the present subject of flexible molecules are our studies on the polyelectrolyte polystyrene sulphonate (PSS), where the flexibility (or rigidity) of the chains can be sensitively influenced by the ionic environment [17]. For the stochiometric PSS salt, the chain is in a highly expanded conformation due to mutual repulsion of incompletely screened charged groups along the chain. Such an expanded chain collapses into what would characterise a flexible neutral molecule on addition of excess ions in the form of salts as a consequence of the more complete charge screening which results. In a previous work, these changes in flexibility/rigidity have been followed in terms of Equation (1), in the flexibility regime where criticality in $\dot{\varepsilon}$ still pertains. Here, τ values, which were larger by 2–3 magnitudes than in the corresponding collapsed chain, could be observed in the most expanded salt free system. The simultaneously occurring substantial delocalization (broadening of the line) has been subsequently analysed in terms of chain stiffness [6]. Such environmentally influenced chain expansions are particularly relevant to the issue of the onset of entanglements, as will be referred to in section 7.2.1.

4. Identifying entanglements

4.1 Disentanglement and network formation

It is to be expected that as the concentration is increased, and as the chains begin to overlap and hence

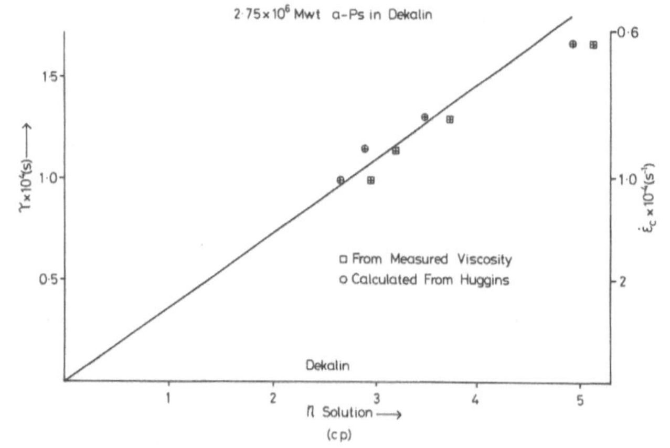

Fig. 5. The coil-stretch relaxation time (τ) as a function of solution viscosity

a)

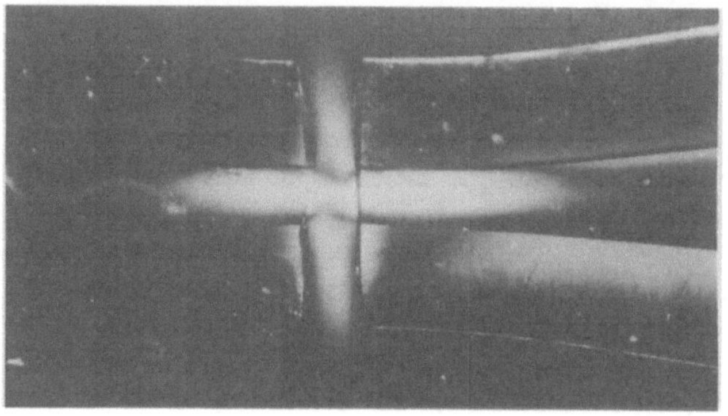

b)

Fig. 6. a) The narrow birefringent line observed between opposed jets in a semidilute solution with $\dot{\varepsilon}_c < \dot{\varepsilon} < \dot{\varepsilon}_p$ (0.2 %, 7.2 × 10^6 $\bar{M}_w$, a-Ps). Photograph, crossed polaroid; b) Flaring of birefringence in the same solution with $\dot{\varepsilon} > \dot{\varepsilon}_n$

interact, the extension behaviour, as diagnosed by our method, will be affected. As stated in the Introduction, this should provide a sensitive tool to identify the stage at which the initially isolated entities will begin to stretch out cooperatively.

With solution concentration increasing, the first effect observed remained, as before, the onset of a narrow birefringent line along the jet axis setting in at a critical strain rate $\dot{\varepsilon}_c$ (Fig. 6a). At this stage, no discontinuity has so far been observed regarding this effect with increasing concentration (c): the only influence of concentration at this point was a steady linear increase in τ ($\equiv \dot{\varepsilon}_c^{-1}$), and thus in the viscosity η, where η is now the solution viscosity (which itself increases with concentration, in the first approximation linearly) (Fig. 5). This, together with the persistance of extremely high localization (Fig. 6a) means that even for concentrations at which chain overlap would otherwise be expected (from what is to follow), the chains continue to extend as they have done in the most dilute solutions where their isolated state can be regarded as assured. In fact, the observed concentration, hence viscosity dependence, is consistent with the same trend that was found for dilute solutions when the solvent was varied, for a constant concentration. It therefore appears that τ is proportional to the viscosity of the medium in which the molecule extends, irrespective of how far this is constituted by molecules of the same or of different species. In other words, on the time scale of $\dot{\varepsilon}_c$, a given chain continues to extend as if in isolation, irrespective of coil overlap, and is "seeing" other molecules of its own kind only as a source of viscous energy dissipation and not as a source of physical entanglement.

Nevertheless, when $\dot{\varepsilon}$ is increased beyond $\dot{\varepsilon}_c$ new effects set in beyond a sharply defined concentration to be denoted c^+. At $c \geq c^+$ a series of occurrences are observable for $\dot{\varepsilon} > \dot{\varepsilon}_c$ even by visual inspection, of which the final and most dramatic is illustrated by Figure 6b. In what follows, we shall present and discuss this dominant effect first; the individual stages leading up to it will receive attention subsequently.

Figure 6b shows one particular variant of a whole range of appearances, which we collectively term 'flare'. Their common features are: (a) delocalization of the birefringence, which here can occupy the whole volume between the jets, can spread into the jets and spill over and beyond the entry side of the jet system; (b) the flare is unstable, the bright area, like a flame, is seen in continual motion. The flare effect sets in suddenly at a critical $\dot{\varepsilon}$ to be termed $\dot{\varepsilon}_n$ (n stands for network – see below). $\dot{\varepsilon}_n$ is a decreasing function of concentration; this decrease is very rapid shortly beyond c^+, gradually levelling out for higher concentrations (Figs. 11, 12, upper curves, see section 5).

We interpret the flare effect as a cooperative extension of molecules over the volume seen as birefringent. This would arise if the chains were behaving as if connected, i. e. like a transient network. The fact that this connectedness arises only beyond a given concentration signifies that a cooperative effect is involved, implicit in the above network model. The fact that it only appears at a critical strain rate, $\dot{\varepsilon}_n$, however, introduces a new and important factor: the time dependence of the connectedness. Namely, the connectedness can only exist over a very short period with a long time limit defined by $\tau_n = \frac{1}{\dot{\varepsilon}_n}$. Thus for times of τ_n, or shorter, and only for these times, an assembly of mutually overlapping chains can act as a mechanically connected network.

At this point an essential interjection needs to be made. For the given experimental circumstances, the passage time of the molecule within the gap between the jets is very short compared to the time scale of the change in strain rate. Thus, the coil stretch and flare effects (in fact, the whole range of effects in between the two, to be described in section 5 to follow) are *not* consecutive responses of the same molecule to the increasing strain rate, as a given chain will have long left the system before the strain rate could have increased substantially enough to evoke a noticeably different effect. Each consecutive event therefore pertains to a *different* group of molecules, such as successively enter (and leave) the 'active' portion of the flow field.

The overall message of the above experiments is now obvious. For a given molecular weight, the chains will stretch out virtually completely once a critical strain rate $\dot{\varepsilon}_c$, corresponding to the coil stretch transition, is attained. This remains true irrespective of the existence or degree of chain overlap, and is only affected by the *mean* viscosity of the medium in which the chains are located (i. e. the solution viscosity). The fact that chains extend, even in a molecularly overlapping environment, means that on the time scale of $\tau = 1/\dot{\varepsilon}_c$ they can still disentangle. This ability of the chains to disentangle will persist up to the shorter time scale of τ_n, when the overlapping chains will begin to "grip each other", giving rise to the network effect.

It follows that two criticalities are involved in the creation of the transient network: a critical minimum concentration (c^+), and for each concentration $c \geq c^+$ a critical minimum time (τ_n). The time effect will be considered first.

4.2 *Criticality in time scale*

The quantity τ_n readily determinable through the flare effect, represents the shortest time during which a given molecule can disengage itself from its neighbours in the course of chain extension, and will be termed as the 'disentanglement time' accordingly. Its dependence on concentration for a particular a-PS sample is given in Figure 7. As can be seen, it is an increasing function of concentration. The actual times are of the order of 10^{-5} s, a plausible time scale in terms of molecular dynamics for chains in that length range [1, 18].

We may examine this time scale more specifically in terms of other available information. Possibly the most relevant is the translational diffusion coefficient D measured exhaustively in (non-flowing) systems by Leger et al. [19]. We can interpolate their data points to obtain D for our molecular weights and concentrations. In the knowledge of D we can readily calculate the distance (x) through which a molecule would diffuse during a given time (t) from the relation $x = Dt^{1/2}$. For the times represented by τ_n in Figure 7, the values range from 7 to 38 nm. We may usefully compare these with the radius of gyration of the same molecule under θ conditions, known with some confidence to be 70 nm, a value which should be closely applicable to our solvent system. Thus our τ_n values correspond to times during which, within a stationary system, our molecule would diffuse over a fraction (0.1 to 0.5) of a coil radius, as expressed by the radius of gyration. This

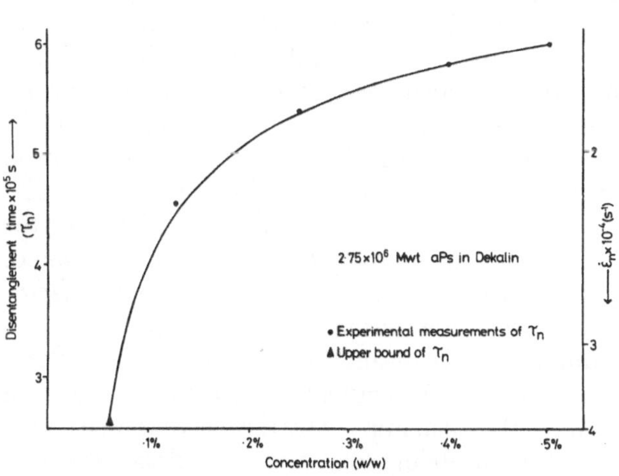

Fig. 7. Disentanglement time (τ_n) as a function of solution concentration

is at least consistent with the picture that the coils overlap through a fraction of their diameter and will behave as isolated chains on extension if they are given the time to diffuse out of each other's environment, a time which we may identify as the disentanglement time. Admittedly, the invoking of values for D which pertain to a non-flowing system may well be questionable. Even so, the mere fact that the values thus arrived at seem reasonable is noteworthy in itself and, if taken at face value, could even suggest that the flow itself may have little effect on the disentanglement as such, at least at strain rates not exceeding $\dot{\varepsilon}_n$.

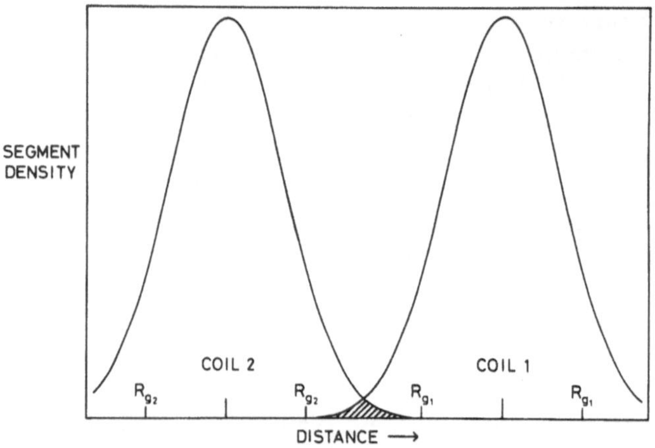

Fig. 8. A schematic representation of the overlap of two normal distributions of segment density. R_{g1} and R_{g2} define the respective radii of gyration. The overlap area is shown as shaded

4.3 Criticality in concentration

The sharp down-turn of the disentanglement time vs. concentration curve for the flare effect (Fig. 7) indicates that the flare sets in within a narrow concentration range as c is increased. This is the critical concentration already indentified as c^+. The postulation of the existence of a critical lower concentration is consistent with the concept of a network, as for a network to arise there needs to be a pathway of mutual contact across the whole macroscopic system, the establishment of which, in common with all such percolation problems, is a critical phenomenon. In our case, the minimum contact criterion is clearly the existence of coil overlap.

The criterion of coil overlap is familiar in solution theory, where it represents the transition from dilute to semi-dilute solutions, the corresponding concentration being denoted as c^*. This value is calculated on the basis that the chains can be represented as spheres of dimensions defined by the radius of gyration (R_g), and corresponds to the concentration at which contact between spheres is established. This criterion for c^* has been calculated in slightly different ways by the different authors. Depending on the criterion used, for our system of a-PS with $M = 2.7 \times 10^6$, c^* is in the range of 1–5 %. The experimental determination using neutron scattering yields a c^* of 5 % for the same system [20].

As seen from Figure 7, c^+, the lower limit for the flare effect, is somewhat below 0.1 % (upper bound of τ_n in Fig. 7), i.e. it is at a concentration lower by a factor of 10–50 than the calculated c^*. Without any further assumption, this means that the elongational flow-induced extension registers a much lower degree of chain overlap than implied by the conventional c^* criterion for semi-dilute solutions. Considering the matter in another way, our methodology provides a considerably more sensitive test for chain overlap, hence entanglements, than the traditional criterion for semi-dilute solutions.

The answer to this is likely to reside in the fact that a single parameter, such as R_g, is inadequate for defining conditions of interchain contact, and the segment density in the overlap region as a function of centre-to-centre distance, which in turn is a function of concentration, needs considering instead. The approach to this is as follows:

The distribution of the segments of flexible coils in the θ state (in the absence of excluded volume effects) is well described by a normal distribution ($\sigma(r)$) [21]

$$\sigma(r) \propto \frac{M}{R_g^3} \exp - \left[\frac{3r^2}{2R_g^2}\right]. \tag{4}$$

Two such coils can presumably interpenetrate freely. Figure 8 shows two coils overlapping; it is clear that interactions can occur with separations much greater than $2 \times R_g$, and thus at concentrations considerably below the conventional c^*.

Considering two coils (1) and (2) l apart, the probability of coil 1 interacting with coil 2 is the integral of $\sigma_1\sigma_2 dV$ which leads to:

$$P_{12} \propto \frac{M^2}{R_g^3} \exp\left[\frac{-3l^2}{4R_g^2}\right]. \tag{5}$$

This yields a number which characterizes the degree to which two coils in the θ state "see" each other. The effect is overlayed by any specific interactions between coils. In the simplest case it describes the opportunity for topological entanglements. Thus entanglement effects which show up on a macroscopic time-scale are characterized by a particular degree of interaction P_{12} (e.g. neutron scattering [20], viscometry). At short times, the degree of interaction P_{12}, required for connectivity, is much reduced since even tenuous entanglements on the periphery of the coils cannot disentangle quickly enough.

Setting P_{12} to a constant to describe a given degree of interaction, or the onset of connectivity, or even gelation, gives different values of c^* which depend upon the nature of the interaction concerned and particularly on the time-scale of observation. Indeed, from Equation (5), including logarithmic terms, and utilising $M \propto R_g^2$ approximates to:

$$c^* \propto M^{-0.6} \tag{6}$$

in contrast to the −0.5 exponent generally assumed. Indeed, based upon gelation criteria, there is some evidence that favours the higher exponent [22,23].

5. The evolution of the network

In the preceding experimentation, the flare was observed to set in suddenly at $\dot{\varepsilon}_n$. However, on more detailed scrutiny it could be seen to develop through a well defined sequence of stages as $\dot{\varepsilon}_n$ was being closely approached. The existence of these stages does not detract from the decisive role of the flare, and of issues based upon it, which remain central to our argument; yet the individual steps leading up to the flare have much more, and rather subtle, information to convey. In particular, this information relates to the gradual development of the 'infinite' connected network, both in terms of spatial spread and in terms of the time scale of connectedness. In the course of this development, the two criticalities, that is strain rate (time scale) and localization (space), became apparent in a rather instructive way.

Two different systems will be referred to: closely monodisperse a-PS (2.75 × 10^6 $\bar{M}_w$) and polydisperse high molecular weight PEO ($\bar{M}_w$ > 6 × 10^5, 4 × 10^6 and 5 × 10^6). The former permitted quantitative deductions about relaxation times and overlap concentrations, as already discussed. The polydisperse PEO, while less suitable for quantitative assessment, proved convenient to display qualitatively the sequence of events involved. This is because, for each $\dot{\varepsilon}$ a given species within the broad molecular weight spread available becomes extended, leading to a ready display of the full sequence within an accessible range of $\dot{\varepsilon}$ for one and the same material.

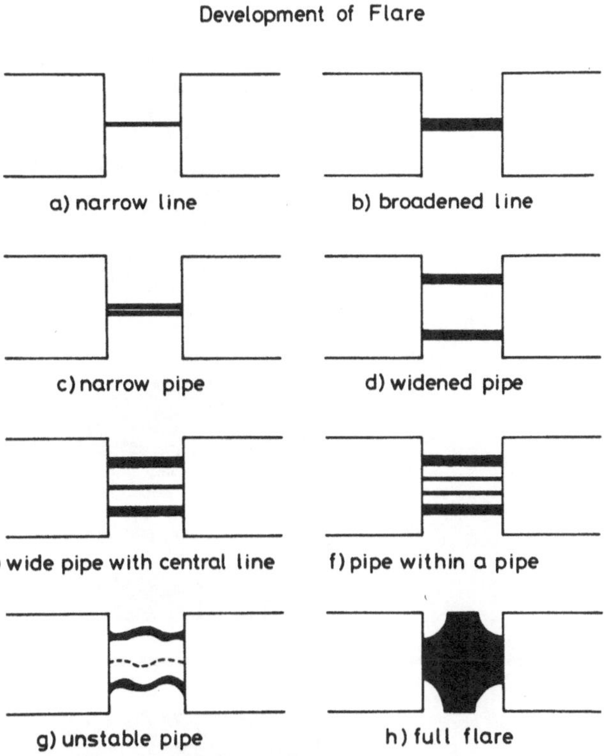

Fig. 9. Diagrams showing the sequence of characteristic images as seen between the jets, representating the stages in the development of full connectivity as the strain rate is increased towards $\dot{\varepsilon}_n$

5.1 Observations

As $\dot{\varepsilon}$ is increased beyond $\dot{\varepsilon}_c$ a sequence of new effects is observed visually. This sequence is depicted graphically in Figure 9 with photographic examples of individual stages, selected from a given experimental series shown in Figure 10.

Firstly, the sharply localized birefringent line (Figs. 9a, 10a) gradually broadens (Figs. 9b, 10b). This broadening is seen at all concentrations, even in the highest dilutions. However, all the further following steps require a critical lower concentration threshold. Below this threshold (c_p^+ – see below) the broadening of the line tends to a limit, while above it, a rather remarkable series of events develop which we shall describe.

With $\dot{\varepsilon}$ approaching $\dot{\varepsilon}_n$, a dark central line appears along the widened birefringent region (Figs. 9c, 10c). On increasing $\dot{\varepsilon}$, this dark line broadens progressively, both in absolute terms and relative to the overall width of the whole birefringent zone which itself keeps on widening with $\dot{\varepsilon}$ (Figs. 9d, 10d). As a result of the latter, the birefringence may become confined to two thin peripheral bright stripes. It was ascertained that the

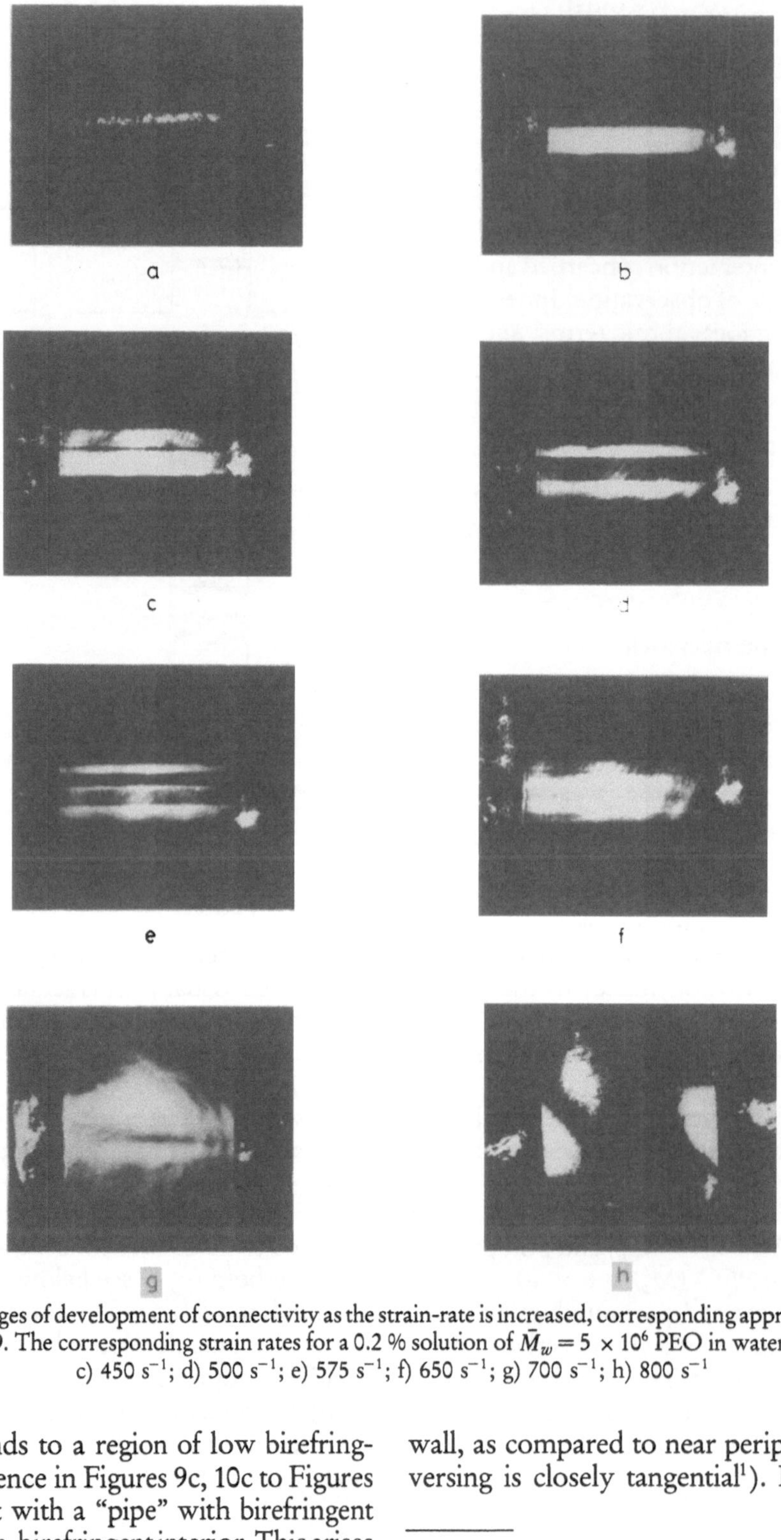

Fig. 10. Photographs of stages of development of connectivity as the strain-rate is increased, corresponding approximately to the schematic representations in Figure 9. The corresponding strain rates for a 0.2 % solution of $\bar{M}_w = 5 \times 10^6$ PEO in water are: a) 150 s^{-1}; b) 200 s^{-1}; c) 450 s^{-1}; d) 500 s^{-1}; e) 575 s^{-1}; f) 650 s^{-1}; g) 700 s^{-1}; h) 800 s^{-1}

dark zone corresponds to a region of low birefringence. Thus, the sequence in Figures 9c, 10c to Figures 9d, 10d, is consistent with a "pipe" with birefringent walls containing a non-birefringent interior. This arises because the optical path is much shorter when traversing the pipe centrally, i.e perpendicular to the pipe wall, as compared to near peripherally when the traversing is closely tangential[1]). Beyond this widened

[1]) An alternate explanation for the dark line effect has been reported by us [12] (and also by Farinato [24]) and is an issue discussed elsewhere [2].

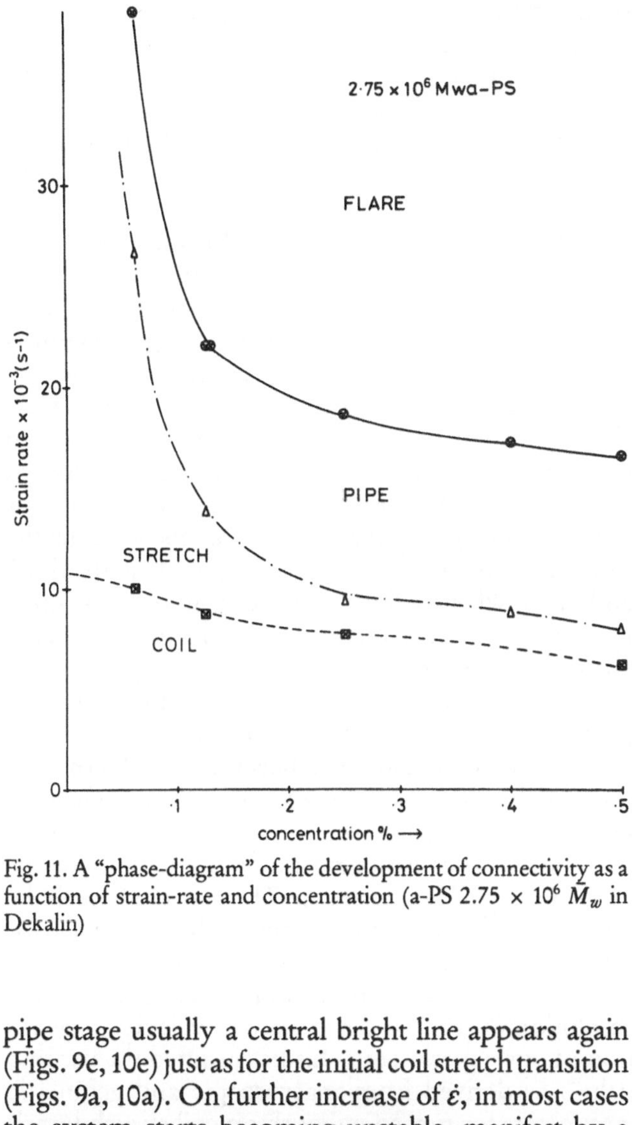

Fig. 11. A "phase-diagram" of the development of connectivity as a function of strain-rate and concentration (a-PS 2.75 × 10^6 $\bar{M}_w$ in Dekalin)

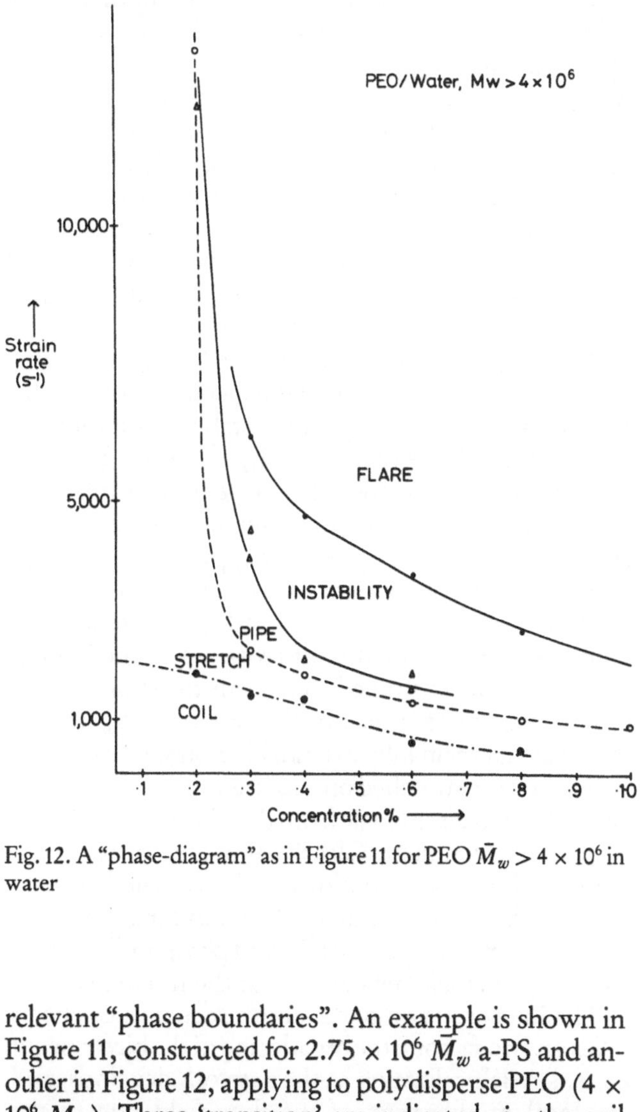

Fig. 12. A "phase-diagram" as in Figure 11 for PEO $\bar{M}_w > 4 \times 10^6$ in water

pipe stage usually a central bright line appears again (Figs. 9e, 10e) just as for the initial coil stretch transition (Figs. 9a, 10a). On further increase of $\dot{\varepsilon}$, in most cases the system starts becoming unstable, manifest by a 'wobbling' of the whole pipe system (Fig. 9g, 10g). At slighly higher $\dot{\varepsilon}$ this pipe system bursts and the previously identified flare, the dominant feature of the whole sequence, is reached. (Figs. 9h, 10h). Not unexpectedly, the appearance close to the instability is highly variable. In some cases, it is possible to observe yet a further stage between e) and g) (in Figs. 9, and 10): the central bright line within the pipe developing a further dark central line within, thus creating a pipe within a pipe (Figs. 9f, 10f), before the whole system becomes unstable, giving rise to the flare.

The strain rates for the whole sequence of effects were found to be strongly concentration dependent, starting at the lower limiting concentration c_p^+ (where the subscript is for pipe) already mentioned. A "phase diagram" can thus be constructed to embrace the relevant "phase boundaries". An example is shown in Figure 11, constructed for 2.75 × 10^6 $\bar{M}_w$ a-PS and another in Figure 12, applying to polydisperse PEO (4 × 10^6 $\bar{M}_w$). Three 'transition' are indicated: $\dot{\varepsilon}_c$, the coil stretch, $\dot{\varepsilon}_p$ the first sign of the pipe and $\dot{\varepsilon}_n$, the flare formation. All three are decreasing functions of concentration, yet with some differences. These will now be commented on in turn.

In contrast to $\dot{\varepsilon}_p$ and $\dot{\varepsilon}_n$ there is no limiting lower c for $\dot{\varepsilon}_c$. While not apparent from the c ranges in Figures 11, 12, $\dot{\varepsilon}_c$ extends to and levels off at infinite dilutions. The slow decrease with c within the c range of the "phase diagrams", is due to the increase in solution viscosity, as already discussed.

Both $\dot{\varepsilon}_p$ and $\dot{\varepsilon}_n$, after setting in at c_p^+ and c_n^+, respectively, first decrease very rapidly and then more slowly with c, the $\dot{\varepsilon}_p$ vs. c curve becoming roughly parallel to $\dot{\varepsilon}_n$ but displaced to lower strain rates.

The low-concentration cut off signifies that both $\dot{\varepsilon}_p$ and $\dot{\varepsilon}_n$ are manifestations of molecular interactions. As a broad overall statement, to be expanded upon further below, we interpret $\dot{\varepsilon}_p$ as signalling the first stages of chain entanglement in terms of connectedness and range, and $\dot{\varepsilon}_n$, as the strain rate where the entanglement network becomes fully developed, spanning the whole region between the jets and extending beyond. We are inclined to apply the same distinction to the respective lower critical concentrations c_p^+ and c_n^+. As seen from the "phase diagrams" the two are very close to each other. Whether they would actually coincide at higher $\dot{\varepsilon}$ would be difficult to say with certainty.

The whole sequence shifts both to lower concentration and to lower strain rate with increasing molecular weight. This follows from the established molecular weight dependence of $\dot{\varepsilon}_c$ (Eq. [2]) and, even qualitatively, from the expectation that longer chains will overlap at lower concentrations and that lower $\dot{\varepsilon}$s are required to extend a more overlapping system.

The "phase diagram" has a further notable feature. Although the curves for $\dot{\varepsilon}_c$, $\dot{\varepsilon}_p$ and $\dot{\varepsilon}_n$ at first seem to converge with c they do not cross, up to the highest concentrations examined, but seem to retain a finite near constant difference. In the case of $\dot{\varepsilon}_c$ this means that chains remain fully extendable as individuals, provided they are stretched on a long enough time scale to allow them to disentangle from their increasingly overlapping environment. Clearly, the issue of what the upper limiting concentration may be, or whether there is one at all, is of great potential interest for attempts to stretch out chains in the condensed phase and, in particular, for entanglement studies in the molten state.

The interpretation of the above sequence of effects relies on two factors: the broadening of the birefringent zone, i.e. delocalization as the network develops, and the local modification of the flow field due to the extending network itself. In what follows, evidence for the latter will be first presented as a preliminary to develop an explanation for the full sequence of observations. The rheological effects to be reported in the later sections, while primarily of interest for their own sake, will then provide strong additional support for the model proposed.

5.2 Origin of the "pipe"; local flow velocities

For an understanding of the origin of the pipe effect, the influence of the chain extension on the local flow field, which has initially created the chain extension, needs invoking. Local flow modification arises through a combination of two factors: (i) the energy requirement for chain extension, (ii) the localized nature of the chain extension. It follows from (i) and (ii) that the flow velocity is expected to become modified at strain rates when the chains start to extend, i.e. at $\dot{\varepsilon}_c$ and beyond, but only within the localized region of the chain extension. Establishment of this feature calls for local flow velocimetry.

Within our studies, such flow velocimetry work has been undertaken in two stages with initially different objectives. In the first stage, we were concerned with the establishment of the new methodology and with its application to the stretching of the isolated molecule. In this context we had to know the correct strain rates for I vs. $\dot{\varepsilon}$ plots such as in Figure 4. In particular, we had to ascertain whether it was justified to use $\dot{\varepsilon}$ values as obtained from the measured overall flow velocity for the given flow geometry between the jets or within the slots (the latter were used more frequently at the time) and, if so, whether this remained valid over the full $\dot{\varepsilon}$ range employed extending to $\dot{\varepsilon}_c$ and beyond when the chains become stretched out.

Laser-Doppler velocimetry was used for this purpose, carried out in cooperation with Gardner and Pike, at the Royal Radar Establishment, Malvern, [25] then employing the slot system where the exit channel, rather than the more desirable stagnation point region, was chosen because of positional limitations of the available experimental set up. The expected parabolic velocity profile was observed across the slot, which for dilute solutions remained unaffected by chain extension at the coil stretch transition, to within the spatial resolution (5 μm^3) and accuracy of velocity registration of our experiment. This has justified the use of $\dot{\varepsilon}$ values as derived from macroscopic and geometric parameters for preceding work on coil stretch transitions. Nevertheless, significant flow modifications were observed for higher concentration in the form of very pronounced dips at the apex of the parabolic velocity profile at high strain rates, as also observed previously by Lyazid et al. [26]. However, at the time, these striking findings could not be paralleled with birefringence observations between the slots, and their relation to the flare image could not thus be ascertained, and the origin of the effect has remained unexplained. The more recent flow velocimetry work was undertaken to correlate strain patterns with local velocity variations, also taking note of the newly recognized pre-flare birefringence effects.

In this recent, second stage of our Laser-Doppler velocimetry work, we were scanning across the flow lines spanning the jet orifices in defined stages of the

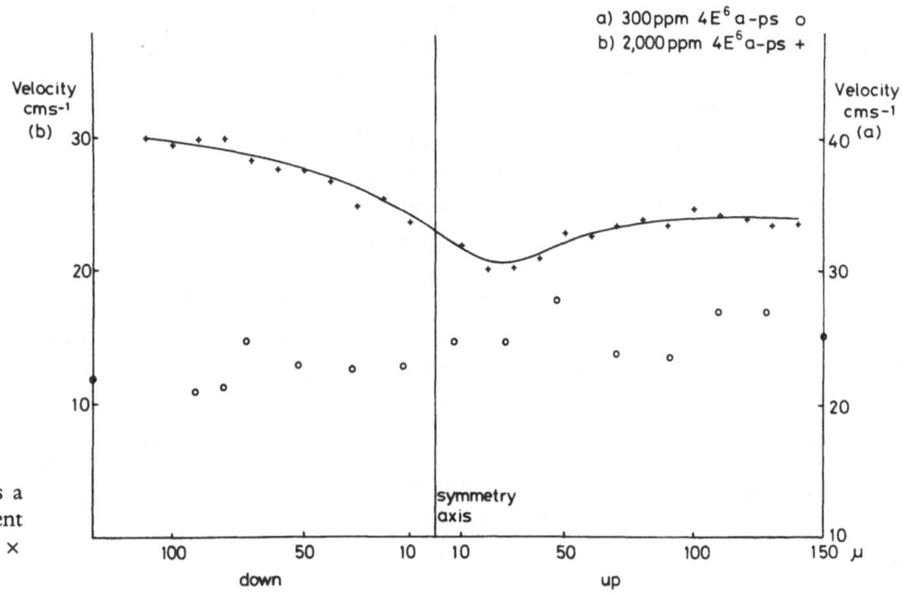

Fig. 13. Velocimetry scans. Velocity as a function of position across the birefringent line. (a) dilute, (b) semidilute solution (4 × 10^6 a-PS in Dekalin)

development of the birefringent image with a spatial volume resolution of 10 μm^3. The most significant result for the present purpose is typified by Figures 13, where Figure 13a represents a scan across a uniform birefringent line of 30 µm width obtained from a dilute (0.03 %) solution of a-PS above $\dot{\varepsilon}_c$, and Figure 13b a scan across a pipe of 100 µm width obtained from a 0.2 % solution of the same polymer at $\dot{\varepsilon}_p$. The latter displays a clear 'dip' in the velocity corresponding to the axis of the jet system where the dark line is seen in the image simultaneously.

The above result is quite general. If follows that the appearance of the pipe is associated with a corresponding local reduction in flow velocity, and conversely, that the extending chains produce observable flow modificatioin only beyond the concentration (c_p^+) and strain rate ($\dot{\varepsilon}_p$) where the pipe effect appears.

The result that the central dark line of the pipe is associated with a zone of lower flow velocity offers a ready explanation for the pipe effect. Namely, that the broadened chain extended regions 'screen' the flow within, so that the flow velocity, and hence the extensional strain rate, drops within the central zone. When $\dot{\varepsilon}$ drops below criticality, $\dot{\varepsilon}_c$, chain extension and consequent birefringence ceases along the central axis, giving rise to the effect observed.

5.3 Interpretation

We are now in a position to attempt an interpretation of the sequence of events taking place with increasing strain rate in a solution of geometrically overlapping molecules.

At $\dot{\varepsilon}_c$, as already discussed, the chains stretch out individually giving rise to the birefringent line, which at first is confined to the central axis passing through the stagnation point, where residence time, hence the achievable strain is a maximum (localization of birefringence). For strain rates beyond $\dot{\varepsilon}_c$ the chains also become extended along stream lines, increasingly further removed from the central axis corresponding to shorter residence times causing the birefringent line to broaden. This "delocalization" is the consequence of the fact that with increasing strain rate, virtually full chain extension can be achieved during shorter times, a point which, together with the existence of a limiting birefringent line width, can be demonstrated quantitatively [6].

It is an important but subtle point to note that at $\dot{\varepsilon}_c$ the molecular strain is much lower than the fluid strain, but will increase towards the value of the fluid strain as the strain increases beyond $\dot{\varepsilon}_c$ (treated quantitatively in Ref. [6]). This has the consequence that with increasing $\dot{\varepsilon}$ the chains become extended along progressively more peripheral stream lines, in spite of the fact that the residence time is reduced due to the faster passage of fluid, with an ensuing broadening of the birefringent line. However, this broadening will reach a limit when the shortened residence time due to faster flow at increased $\dot{\varepsilon}$ is not compensated any longer by an enhancement in molecular strain (because the latter approaches the limiting strain which is that of the fluid): for further increase in strain rate, the birefringent line width will thus remain invariant. It is this limiting zone width which is exceeded when the transient network formation sets in – see main text.

Simultaneously chain extension extracts energy from the flow, which must mean increasing flow resistance (i.e. viscosity, see later), and thus, for a given driving force, a reduced flow velocity within the localized regions. It is further expected that this screening effect will become larger towards the centre of the birefringent zone. It is now a fact of observation that such a velocity reduction becomes significant (hence registrable by our present technique) only beyond a certain concentration (i.e. c_p^+), and for a given concentration only beyond a certain stage of broadening (not yet assessed quantitatively). Accordingly, the central flow velocity, hence the associated $\dot{\varepsilon}$, can then drop below the critical coil stretch transition ($\dot{\varepsilon}_c$) which will 'switch off' the birefringence there, producing the central dark line, hence the pipe.

At this point some comments are appropriate about the role of concentration in screening the flow field. Beyond the self evident reason that with more chains to extend there is more energy extracted, there are several further interconnected factors which lead to an enhancement of screening with increasing concentration, in particular with the onset of network effects. In the first place, it is likely that the transient network state itself provides more effective screening than the same molecules stretched out but in a non-interacting manner, as judged from the criterion of a critical concentration for the pipe effect. A more subtle and fundamental effect arises from the continuing outward spread of the pipe walls which proceeds beyond the broadening limit pertaining to the birefringent zone for the isolated molecule. Thus for a transient network, which arises beyond c_p^+, much less strain is required for full extension than for isolated molecules. It follows, that network chain extension will spread out to correspondingly more peripheral stream lines. In addition, we need to consider that network formation itself is enhanced as $\dot{\varepsilon}$ increases, because new mechanically effective junctions will be generated on the shorter time scale. This means that progressively less strain is required for stretching out a network element on increasing $\dot{\varepsilon}$ with the consequence that the network, hence the birefringent zone, by this stage in the form of a pipe, will broaden without bounds.

The reality of the above situation is supported by considering the strains involved in the actual experiments. For the broadest pipes which could be observed, the outermost diameter was ~ 300 μm. From the knowledge of the flow field, the fluid strain there was found to be only 10 ×. This is 1–2 orders of magnitude lower than needed for extending the isolated molecule, a fact which in itself requires the invoking of a network in which the extending element, the mesh length, is correspondingly shorter.

Returning to the sequence of events taking place *within* the expanding pipe, one can now envisage how the process, which has originally *created* the pipe, can repeat itself in the following manner. The screening of the pipe interior is not expected to be complete: the velocity will rise there as well, on further increase of the mean $\dot{\varepsilon}$, and progressively so when the outwards expanding pipe walls become relatively thinner and the isotropic interior, where there is no energy dissipation due to chain extension, broader. Thus the central strain rate could reattain the value $\dot{\varepsilon}_c$, when a new localized central birefringent line should arise, as is, in fact, observed. It will be apparent that the process can repeat itself on further increase of $\dot{\varepsilon}$. The new central line itself (together with the primary pipe) will again begin to broaden, where the newly arising screening will again start reducing the central velocity. When the latter drops below $\dot{\varepsilon}_c$ a new pipe, a "pipe within a pipe" will appear as is, in fact, sometimes observed (Figs. 9 f, 10 f).

Each stage of the above sequence will correspond to as yet another phase in the network development, albeit of a curiously structured network. Accordingly, the network is spreading increasingly outward, yet developing an internal periodicity concentric with the jet axis, corresponding to the system going periodically on and off "critical". The complex orientational morphology contrives to minimise the strain rate seen by the extending molecules, by storing elastic energy and locally reducing the flow velocity. In this way the strain rate is controlled just below that which gives rise to a continuous network. With further increasing externally imposed flow rate, nevertheless, a continuous network is eventually produced. Here, the deforming connnected entity will not be able to comply with the macroscopic flow geometry: we have reached $\dot{\varepsilon}_n$, the transition to flare and associated flow instability. Or looked upon otherwise, the entangled system is really a gel when considered on a time scale $\leq \tau_n \equiv (\dot{\varepsilon}_n)^{-1}$ and a gel cannot flow without it being continually broken.

Strictly speaking, all of the above relates to one particular type of flow field. Special, and perhaps artificial as it may appear, it is possibly the simplest realization of elongational, hence necessarily spatially inhomogeneous laminar flow field. While recognizing this fact, we maintain that the conclusions reached on this system, should be pertinent to more complex inhomogeneous flow fields with predominantly extensional components, such as may arise in practical applications.

The present jet system is not quite uniquely defined, even for the present purposes. Thus, the thickness of the jet walls (in addition to the orifice diameter and separation) has an effect on the stability-instability criterion. For all these reasons, exact quantifications of the analysis of the events here presented, beyond giving order of magnitudes, would be of no general significance, except as pertinent to the particular experiment involved. What we believe is of wider significance, is the newly recognized general trend of progressive network formation, its particular manifestations through criticalities, both time-wise and spatial, and the introduction of a new methodology for the study of either entanglements or of flow phenomena produced by polymeric additives. Having hopefully made an inroad in the entanglement subject it is to the second, to the flow aspect, we shall be turning in what follows.

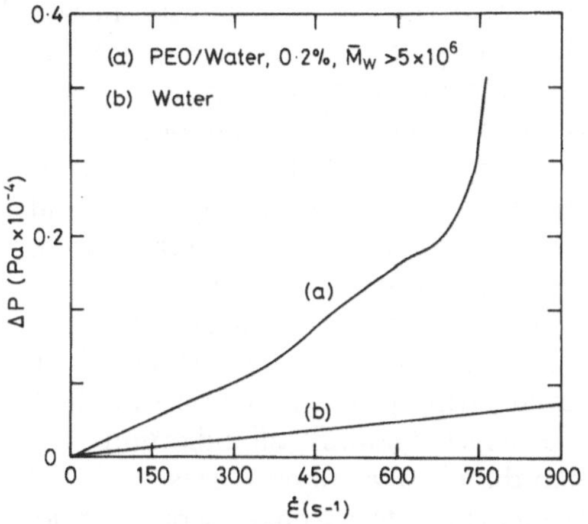

Fig. 14. Pressure drop across the jets ΔP versus $\dot{\varepsilon}$ (Bernoulli corrected curves) (a) semidilute PEO and (b) water

6. Macroscopic flow behaviour

It is to be expected that the dramatic effects associated with the formation of transient networks will also have a correspondingly pronounced influence on the macroscopic flow behaviour, and in particular, that the visually observed events will be reflected by changes in the flow resistance presented by the jet system as a whole. The latter, besides being of considerable practical significance, could then create a link between molecular behaviour and hydrodynamics involving polymeric additives. In addition, the results which emerge add further to the molecular picture.

6.1 Flow resistance – elongational viscosity

In order to quantify flow resistance, we have measured the pressure drop across the jets (ΔP) as the strain rate ($\dot{\varepsilon}$) is increased [2]. The ΔP vs. $\dot{\varepsilon}$ traces thus obtained showed a curvature, even with pure solvent, which originated from the Bernoulli effect due to convergent flow, an effect which could then be corrected for analytically. Figure 14 shows such a corrected trace. The slope of the corrected ΔP vs. $\dot{\varepsilon}$ curve then provides an expression of an effective elongational viscosity η_e'. This extensional viscosity is a well defined function of the macroscopic strain rate, but, because of the presence of the stagnation point, it must correspond to a combination of a wide range of fluid strains. Because of the way ΔP is measured, there will also be a contribution to η_e' from Poiseuille flow inside the jets. We shall present our arguments in terms of η_e' (as just defined) vs. $\dot{\varepsilon}$ plots in what follows, which is not only an important parameter per se but also accentuates features of interest for the present.

Figure 15 shows η_e' as a function of $\dot{\varepsilon}$ over a $\dot{\varepsilon}$ range spanning $\dot{\varepsilon}_c$, $\dot{\varepsilon}_p$ and $\dot{\varepsilon}_n$, and typifies the simplest sequence of events. The initial constant value corresponds to a pseudo Newtonian viscosity. At the point at which the pipe is formed there is a pronounced peak, in the present case at $\dot{\varepsilon} = 465\ s^{-1}$. At still higher $\dot{\varepsilon}$ when the birefringence flares the flow becomes irregular,

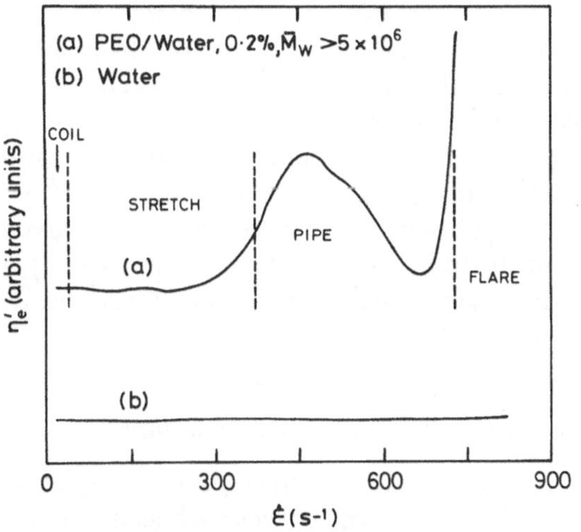

Fig. 15. Elongational viscosity η_e' versus $\dot{\varepsilon}$ curves derived from Figure 14, (a) semidilute PEO and (b) water

with the extensional viscosity enormously increased (with only the initial portion of the rise shown); (for a trace embracing more of the flare see Figure 18 in section 7.2.1 to follow). The reality of all these effects is borne out by comparison with the totally invariant η_e' value for the pure solvent (water) recorded under otherwise identical circumstances. Closer inspection of Figure 15 reveals further details. Thus, there is a shoulder on the high end of the peak associated with the pipe effect. Other traces may in fact show a double (Fig. 16) or even a triple peak (some traces in Fig. 17) within the pipe effect region. By all indications, these periodicities in the η_e' vs. $\dot{\varepsilon}$ curves correspond to the periodic sequences in the visual image of pipe development described previously.

In summary, we thus have a correspondence between the visually registered development of transient entanglement formation and the macroscopic flow characteristics. Accordingly, entanglement formation is associated with an increase in flow resistance and in η_e' derived therefrom, leading up to a massive increase when the network becomes 'infinite', as signalled by the flare. The overall direction of the effect, i.e. an increase in η_e' with $\dot{\varepsilon}$, follows a trend such as might be anticipated by general considerations. However, the drop (or drops) of η_e' along the way, as represented by the minima, i.e. the intermittant decreases with increasing $\dot{\varepsilon}$, are unexpected. Nevertheless, even these effects, contrary to intuition per se, seem to fall into a consistent scheme in the light of the sequence of visual images observed.

The above findings should have wider implications for the study of inhomogeneous but still laminar flows. Namely, any macroscopically measured flow parameter can only be an aggregate expression of the wide spatial variations within the flow field resulting from inhomogeneities in molecular strain. Conversely, it follows that molecular behaviour cannot be extracted from macroscopic flow measurements alone without local probing of the "morphology" of both strain and flow fields. The above statements apply even for a given constant mean strain rate, to which the intricate, yet tractable effects associated with the strain rate dependence add as yet a further dimension.

6.2 *Reversibility – flow induced degradation*

A further important aspect emerges from the reversibility characteristics of the curves such as in Figures 16, 17. The ΔP (and the derived η_e' vs. $\dot{\varepsilon}$) plots are strictly reproducible for reruns up to the onset of flare, including the pipe effect (Fig. 16). However, once the flare stage has been reached, the same plots for subsequent consecutive reruns show a reduction in the initial value of η_e', and the flare and associated effects are shifted to higher strain rates (Fig. 17). These effects are consistent with a progressive reduction in molecular weight as chains are broken by stretching beyond $\dot{\varepsilon}_n$. (Such chain scission was indeed confirmed by separate molecular weight measurements). Clearly, a continuous network cannot conform to the elongational flow,

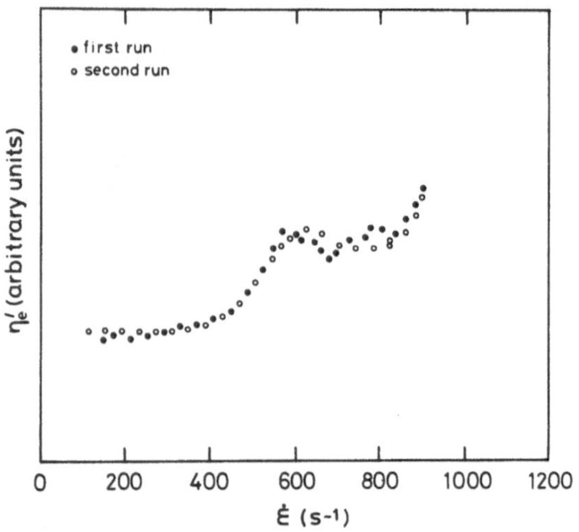

Fig. 16. Elongational viscosity η_e' versus $\dot{\varepsilon}$ for two successive runs of a 0.2 % PEO/water solution ($\bar{M}_w > 5 \times 10^6$) where the experiment was stopped at $\dot{\varepsilon}_n$

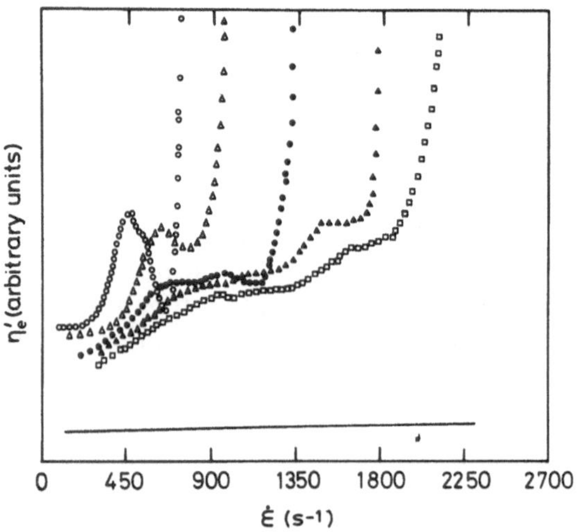

Fig. 17. Elongational viscosity η_e' versus $\dot{\varepsilon}$ for successive runs of a 0.2 % PEO/water solution ($\bar{M}_w > 5 \times 10^6$): (○) first run, (△) second run, (●) third run, (▲) fourth run, (□) fifth run, (–) pure water

but must tear in the process, which is indeed reflected both by the observed chain scission and associated flow instability.

We thus see that in addition to its intrinsic significance, the rheological behaviour provides strong additional support for our picture of network formation, in particular the interpretation of the flare effect. Furthermore, and rather unexpectedly, it conveys information on flow-induced chain scission. We have now identified two kinds of flow induced degradation. The first, explored extensively previously [11–13], applies to the isolated chain, when the stretched out entity breaks virtually exactly in the centre, yielding two closely equal chain halves as the elongational flow rate is increased far enough beyond the coil → stretch transition. The second mode of scission is found presently and applies to overlapping chains which break as the extended transient network attempts to comply with the flow field. Although not yet fully confirmed, here, fracture is not expected to be closely central to the molecule. While both fracture modes can occur, the second, involving entanglements, takes place at lower strain rates.

7. Applications

7.1 On polymeric additives in general

It is well known that soluble high molecular weight polymeric additives can influence macroscopic flow behaviour with diverse practical applications [27, 28]. Understanding the behaviour of the often minute amount of polymeric additive is essential for the design and control of the flow system. There is growing awareness that elongational flow-induced chain extension must be ultimately responsible for the flow modifying action of many polymeric additives; however, in real applications and even in engineering modelling, the situation is far too complex for identification of the basic phenomena involved. We believe that our type of flow experimentation has much to offer in this applied sphere, and conversely, that interpretation of practical flow effects from our new point of view can further basic understanding. The latter holds in particular for entanglements, the reason for inclusion of this section here.

The extension of chains must extract energy from the flow. For practically relevant situations this can have two diametrally opposed consequences.

The first is the intriguing Thoms effect, also known as 'drag reduction'. In macroscopic terms, this consists of turbulence suppression and consequent energy saving in fluid transport, through the addition of even minute amounts of very high molecular weight polymer. Through presently shaping views [29] this effect is likely to be due to chain extension within the localized elongational flow field arising between incipient vortices. Here the energy expended on localized chain extension suppresses further vortex development and consequently much larger energy dissipation. In view of the extreme dilutions involved, these effects are attributable to the extension of truly isolated molecules [29, 30], and as such will be of no further concern for the present article.

In the second family of applications the energy dissipation associated with chain stretching is utilised directly in the form of enhanced viscosity, in cases where this is desirable. However, this is not the trivial matter of increasing fluid viscosity all round, as this would lead to uneconomically high expenditure of energy when driving a viscous fluid along its full path. Rather, it is often desirable to create high flow resistance within an otherwise low viscosity fluid at a particular locality or state of flow. Such a localized or transient viscosity enhancement will arise in elongational flow when favourable circumstances for chain stretching are created. Here two examples will be quoted.

In enhanced oil recovery, fluid is pumped into porous rock to expel the oil. For this the viscosity has to be sufficiently high so as to displace the oil and not to bypass it. The flow through a complex, open pore system will be repeatedly and consecutively convergent and divergent, hence locally of elongational character, with stagnation points behind the obstacles and at bifurcations, all presented or created in abundance by the porous structure. Clearly, conditions for chain stretching and corresponding viscosity enhancement, here confined to within the pore bed, will be met. The fluid being used in engineering practice is water with polyacrylamide as the most frequent additive. Much research (e.g. [31, 32]) on the subject is centred on flow through model pore systems (e.g. a bed of beads). One common result is the registering of a sudden rise in flow resistance at a particular macroscopic flow velocity (most commonly as expressed by the Reynolds number). This rise may or may not level off, and can occur even at very high dilutions (0.01 %) [31]. The effect is interpreted, by some sources at least [31, 32], as arising through the stretching of *individual* chains in the elongationally effective portions of the flow field.

There are numerous applications when a fluid is to be finely dispersed (say, spraying in agriculture), or spray formation is to be prevented, as for ignition suppression in escaping aerofuel in case of accidents. Focussing on the latter, which is a topical issue [33], the formation and further subdivision of droplets is a clear case of stretching flow where an occluded macromolecule could become stretched out. Conversely, in the presence of such macromolecules, energy is absorbed by the stretching process which in turn causes suppression of droplet formation, hence the demisting effect.

7.2 Relevance of our elongational flow experiments with implications for entanglements

We have applied our elongational flow technique using the double jet (in one case the four-roll mill) on fluids, such as have potential outlets in oil recovery and demisting. We conclude that the most effective viscosity enhancement is created by transient entanglements, the subject of this publication. While, in view of the foregoings, this finding may not appear surprising, we draw attention to its potential relevance to the class of effects which feature in the hydrodynamic and engineering literature. To be brief, here we shall restrict ourselves to a few broad comments taking the two above mentioned topics, viscosity enhancement with relevance to oil recovery and demisting, in turn.

7.2.1 Viscosity enhancement

We experimented with polyacrylamides [34], the polymer most widely referred to in the oil recovery related literature. With partially hydrolysed polyacrylamides, we could identify large increases in flow resistance as corresponding to the flare stage between the jets, i.e. to the stage which we associate with the full development of the transient entanglement network. Effects due to the stretching out of individual chains (at $\dot{\varepsilon}_c$) and to that of restricted networks (the pipe effect at $\dot{\varepsilon}_p$ and immediately beyond) were also identificable but were much smaller.

To convey an impression of scale, in Figure 18 we show an example of a ΔP vs. $\dot{\varepsilon}$ curve comprising the full flare. Here the rise in ΔP is the dominant effect, in contrast to Figure 14 (and Fig. 15 expressed in terms of η_e') which concentrated on the comparatively minor initial features associated with the pipe effect. The flow resistance may increase apparently indefinitely, as in Figure 18, or may level off on repeated circulation, both of which have counterparts in the engineering literature. We associate the continuing rise with always

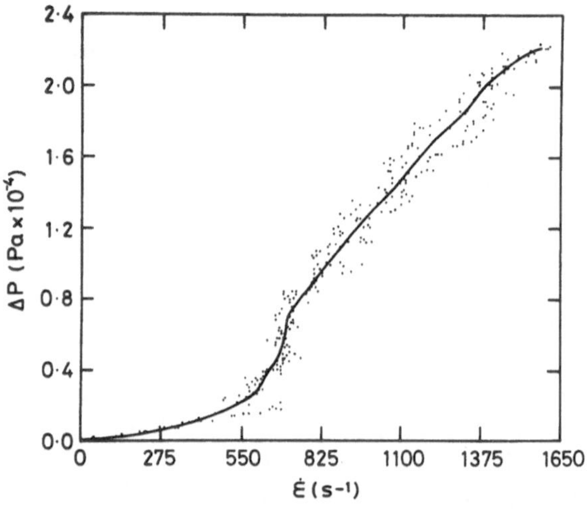

Fig. 18. Pressure drop across the jets ΔP versus $\dot{\varepsilon}$ for a 0.2 % PEO/water solution ($M_w > 5 \times 10^6$): (·) experimental data, (—) smoothed data. The large rise in ΔP at high strain-rate corresponds to the full development of the flare (compare with Fig. 14 which shows ΔP on an expanded scale and up to the initial stage of the flare only). The scatter of experimental points around the smoothed curve is indicative of flow osciallations which set in beyond the pipe stage.

new, previously unstretched molecules entering the flow field, while the levelling off with molecules which have been previously stretched while entangled and have correspondingly undergone scission. It should be noted that to some extent the latter will have always occurred by the time the fluid has advanced deeply into the porous bed along its consecutively and repeatedly converging and diverging flow path, provided that entanglements are involved.

If the concentration of the partially hydrolysed polyacrylamide may appear too low for entanglement formation (we have observed flares and corresponding rises in flow resistance down to 0.005 % concentration) it must be remembered that available characterization of these materials is incomplete. This is particularly so at the high molecular weight end, which matters most, both for chain extension and for chain overlap (see sections 3.2, 4.3). The same applies to the molecular conformation of what, in the hydrolysed state, is essentially a polyelectrolyte. In fact, we found that by adding salt we were shifting the entanglement effect, as identified by our criterion, to much higher $\dot{\varepsilon}$ and could even move it out of our $\dot{\varepsilon}$ range altogether. This is consistent with the expectation that in the presence of excess salt, the initially expanded polyelectrolyte coil should collapse, as is indeed identified in our own elongational flow work in the case of isolated chains within dilute systems [17] (see section 3.4).

The last experiments have two important implications. Firstly, that entanglements can indeed be involved even at extremely low (down to 0.005 %) concentrations provided that the chains are sufficiently expanded, as in the absence of excess salt. Secondly, that entanglement formation depends sensitively on the state of coil expansion (expansion of coils promotes entanglements) which, in the case of polyelectrolytes, we are in a position to vary and explore systematically.

Finally a comment on the influence of flow-induced degradation, a basic limitation to the use of polymeric additives. We have demonstrated earlier (see section 3.3) that the fully developed entanglement network (beyond $\dot{\varepsilon}_n$, i.e. in the flare stage) cannot flow without chain breakage. All our experiments in the range of large rises of flow resistance are in fact within that regime, as ascertained by the reversibility test pertaining to Figure 17. According to our work this scission would be primarily of the more random entanglement type, as compared to the exact halving of the isolated chain which occurs in more dilute solutions at appropriately higher strain rates. It follows that the corresponding distinction, namely as to whether entangled or isolated molecules are being broken is equally pertinent in applications, oil recovery in particular. It is then an important corollary that for entanglements, chain breakage is inevitable, while for the isolated chain it need not be so.

7.2.2 Demisting agent; transient bonds

The material of our experimentation [1] was FM9 (ICI trade name). Even though composition was undisclosed, it is known to form hydrogen bonds in non-aqueous solutions. For this, and further reasons which will become apparent below, this material has provided us with pointers towards broader issues.

Three salient experimental observations will be invoked. First, as received, the material did *not* show a localized coil stretch transition. The first birefringence effect on increasing $\dot{\varepsilon}$ was the flare, setting in at a rather low, yet non-zero, $\dot{\varepsilon}$ value (i.e. at $\dot{\varepsilon}_n$), most conveniently observed in a four-roll mill. Secondly, at the flare stage, the flow resistance (here measured by the driving torque of the rollers), which up to that point had stayed about constant, started to increase dramatically (see Fig. 12 in Ref. [1]). Thirdly, on addition of acetic acid, which here acts as a hydrogen bond breaker, the system was converted to that of a usual dilute high molecular weight solution: namely, a localized birefringent line appeared at a critical strain rate between the jets (used here because of the much higher strain rates needed to see any affect).

The above three effects allow the following preliminary conclusions to be drawn. As before, the onset of flare signals network formation. In view of the fact that this relies on the presence of hydrogen bonding, the network junctions here must be the intermolecular hydrogen bonds themselves, in contrast to the merely geometrical (presumably van der Waals force based) contact in the more unspecific molecules quoted so far. The fact that the network effect has a criticality, i.e. it sets in only at a specific strain rate $\dot{\varepsilon}_n$, signifies that the network is temporary; below $\dot{\varepsilon}_n$ the molecules flow as disconnected entities, but at and beyond $\dot{\varepsilon}_n$ they become linked in a mechanically effective way. As before, $\dot{\varepsilon}_n$ defines the time scale at which the system behaves as if connected, which, in this case, acquires added significance because it must relate to the life-time of transient hydrogen bonds. The observed sudden upswing of the driving force needed to maintain the flow at $\dot{\varepsilon}_n$ should, in line with all the foregoings, be a reflection of network formation, but in this instance in the specific context of the life-time of transient hydrogen bonds. We may generalise this situation to any other kind of bonding which can create interchain contact on a time scale characterized by the appropriate $\dot{\varepsilon}_n$. When $\dot{\varepsilon}_n$ becomes very low, comparable to the time of carrying out a conventional laboratory test, then the system will acquire the characteristics of a permanent network, such as conventionally classified as a gel. Thus, conventionally defined 'physical' gelation represents the long time end of the spectrum of transient network. We see this possibility of probing the linking effect of transient chemical bonds, together with their respective life-times and their influence on the resulting networks, as a new horizon created by these experiments on the demisting agent.

Finally, we note from the above that we now have a situation in which $\dot{\varepsilon}_n < \dot{\varepsilon}_c$. That is when transient hydrogen bonds constitute the junctions, single chains cannot disentangle, even on the longest time scales applied so far. It should follow that under these conditions, chains cannot be stretched out as individuals within their mutually overlapping environment any longer, a point of potential practical significance, say in melt or solution spinning or fibre drawing. Nevertheless, the hydrogen bonds are removed, as shown by our experiment using acetic acid. This demonstrates that through the present approach, it is possible to control and diagnose the mechanically effective, yet transient, entangling and disentangling of chains within a molecularly overlapping system, including the time scale of the events. It is on this note that we conclude the present article.

Acknowledgement

We wish to acknowledge contributions by Dr. Andrea Chow (Stanford) whose still unpublished works, while at Bristol, are quoted throughout. The work covered by this review has been supported by the Science and Engineering Research Council and by BP Venture Research Unit in its earlier and later stages respectively, both of which we gratefully acknowledge.

References

1. Odell JA, Keller A, Miles MJ (1985) Polymer 26:1219
2. Chow A, Keller A, Müller AJ, Odell JA (1987) Macromolecules, in press
3. Pope DP, Keller A (1978) Coll & Polym Sci 256:751
4. Farrell CJ, Keller A, Miles MJ, Pope DP (1980) Polym 21:129
5. Keller A, Odell JA (1985) Coll & Polym Sci 263:181
6. Odell JA (1988) J Polym Sci, Polym Phys Ed, in press
7. Peterlin A (1966) J Polym Sci (B) 4:287
8. De Gennes PG (1974) J Chem Phys 60:5030
9. Brestkin YuV, Saddikov IS, Agranova SA, Baranov VG, Frenkel S (1986) Polym Bull 15:147
10. Atkins EDT, Attwool PT, Miles MJ (1988) in preparation
11. Odell JA, Keller A, Miles MJ (1983) Polym Communications 24:7
12. Odell JA, Keller A (1986) J Polym Sci, Polym Phys 24:1889
13. Odell JA, Keller A, Rabin Y (1988) J Chem Phys, in press
14. Odell JA, Atkins EDT, Keller A (1983) J Polym Sci Lett 21:289
15. Odell JA, Keller A, Atkins EDT (1985) Macromolecules 18:1443
16. Atkins EDT, Miyamato Y (1988) in preparation
17. Miles MJ, Tanaka K, Keller A (1983) Polymer 24:1081
18. Martin JE (1984) Macromolecules 17:1279
19. Leger L, Hervet H, Rondelez F (1981) Macromolecules 14:1732
20. Cotton JP, Nierlich M, Bove F, Daoud M, Farnoux B, Janninck G, Dupplesix R, Picot CJ (1976) J Chem Phys 65:1101
21. Flory PJ (ed) (1966) Principles of Polymer Chemistry, 5th Edition, Cornell University, Ithaca
22. Tan H, Moet A, Hiltner A, Baer E (1983) Macromolecules 16:28
23. Odell JA (1988) in preparation
24. Farinato RS (1986) Abstract and Lecture, Bristol Conference, Flexibility of Macromolecules in Solution, Inst Physics, London
25. Gardner K, Pike ER, Miles MJ, Keller A, Tanaka K (1982) Polymer 23:1432
26. Lyazid A, Scrivener O, Teitgen R (1980) In: Asterita G, Marruci G, Nicolais L (eds) Rheology, Plenum Pub Corp, New York, V2:141
27. Gampert B (ed) (1985) Proc IUTAM-Symposium, The influence of polymer additives on velocity and temperature fields, Essen, FRG, 26–28th June, 1984, Springer Berlin
28. Bird RB, Armstrong RC, Hassager O (eds) (1977) Dynamics of polymeric liquids, Vol 1 and 2, John Wiley and Sons Inc, New York
29. Lumley JL (1969) Ann Rev Fluid Mech 1:367-384
30. Odell JA, Tucker IM, Ferry M, Müller AJ (1988) in preparation
31. Haas R, Durst F (1982) Rheol Acta 21:150
32. Durst F, Haas R, Kaczmar BU (1981) J Appl Polymer Sci 26:3125
33. Chao KK, Child CA, Grens EA, Williams MC (1984) AIChE J 30:111
34. Keller A, Müller AJ, Odell JA (1988) Polymer, to be published

Received February 2, 1987;
accepted April 16, 1987

Authors' address:

A. Keller
H. H. Wills Physics Laboratory
University of Bristol
Tyndall Avenue
Bristol BS8 1TL, England

Discussion

WEYMANS:
Just a very simple question: What are the dimensions of the pipe which you observed?

KELLER:
The central orifice about 0.5 mm. The wall thickness incidentally makes a difference to achievement to ultimate strain rates. It is a secondary effect, but it practically makes a difference of how thick the walls are. The birefringent "pipes", when they appear, have diameters of 0.02 to 0.2 mm according to jet geometry and strain rate.

WEYMANS:
Are the sizes of the jets polymer structure dependent?

KELLER:
Certainly they are dependent on molecular weight. For higher molecular weight, they develop at lower strain rate and become larger. I wouldn't dare to say systematically. We used two kinds of material here, which I collapsed into one; monodisperse polystyrene and polydisperse polyethyleneoxide. If we have a polydisperse system, the system chooses its own molecular weights to create the extension. Below that molecular weight, the effect of the polymer is to only enhance the viscosity. In fact, in polydisperse polymer systems, the experiment is easier to control.

WINTER:
How do you define the rate of extension in your experiment?

KELLER:

In the first instance from the geometry: by knowing the flow rate and the geometry, and knowing that it is zero in the middle. Then we check this, of course, by means of laser velocimetry.

RIGBI:

Obviously, when you break the chains you form free ions. Do you have any in the liquid to stop it from reforming again?

KELLER:

Well, this is a very pertinent question. I don't think that they will rejoin again because they are in a dilute solution. But they are bound to react with something and whether to make use of them to do chemistry, is a very tempting question.

MEISSNER:

It is fascinating that you really can correlate your measureable qualities to the molecular weight in the central part. You had higher concentrations, you say, and you convinced us that you noticed there were entanglements. Do you have any qualitative measure of the junction density?

KELLER:

Not at this stage.

WINTER:

In the region of the flare, when you compare the birefringence to the maximum possible birefringence, what ratio do you get? Do you get full extension?

KELLER:

We do not know because we don't know the path length.

GIESEKUS:

Recently we made some observations on the discharge of a dilute polymer solution through a small hole after preshearing in a Couette rheometer. The "orifice viscosity" which has a good approximation to the character of an extensional viscosity, was observed to increase moderately with increasing shear rate, but to grow much more intensively after Taylor vortices had occured [1]. This last observation gives an indication of flow-induced associations, built-up in those regions of the Taylor vortices where extensional flow predominates, with a not-too-short dissociation time.

In a much earlier investigation, we observed that in the entrance, as well as in the discharge region of the flow of a dilute polymer solution through an orifice, the liquid became turbid, indicating associations with the magnitude of the light wave length. With some kinds of polymers this turbidity disappeared soon after the throughflow, but with some others it became permanent [2].

1. Vißmann K. Extensional flow behaviour of dilute polymer solutions after simple shear flow (to be published in the Proceedings of the Second Conference of European Rheologists, Prague, June 17-20; supplementary volume of Rheol Acta).
2. Giesekus H (1969) Rheol Acta 8:411-421

KELLER:

There are various factors which play a part here. One is what I already said: the hysteresis. Once the chain is extended, its retraction takes much longer than the time scale on which it extends. That may give an impression as if it had a prehistory, which it did have but of a specific kind.

The other thing on the phase transformation: yes, if one is close to a phase boundary of the liquid-solid type, one, not surprisingly, would expect to promote phase separation. An extended chain will promote the crystallization or transformation into an ordered structure. But we are somewhat at a loss a priori to predict what to expect in a liquid-liquid phase segregation, whether chain extension promoted it or not. Here, I refer to works by Wolff.

The final effect I wish to mention is not specifically due to the elongational flow but occurs when very high molecular weight, and I emphasize, very high molecular weight solutions pass along solid walls. They adsorb, which of course is not new. What is new is that this absorption can lead to the building up of absorbed layers which can become so thick that they can even block the flow. In fact, in the last few issues of Colloid Polym Sci we had a series of papers on it. This adsorption has many consequences. One of them is that it affects measurements of intrinsic viscosity, by constricting the capillary diameter.

The adsorption layer can form under any kind of flow. An interesting point is that effect which, for example, Metzner and Wisbrun now interprete as phase transformation. We interpret it as due to the adsorption-entanglement layers. Amongst these there can be visible turbidity which is initiated by gel particles at the walls.

GIESEKUS:

The turbidity which you mentioned is generated at the walls. What I spoke of happens in the free fluid. In the middle region of a solution – perhaps a distance away of a hundred times the capillary diameter or more – you see a zone which is turbid. Hence, in this case, it cannot be explained in this way but of course you may be right under other conditions.

PAKULA:

Can you estimate how many fully extended chains you need to be able to observe the onset of the jet?

KELLER:

There is virtually no limit because it just depends on the sensitivity of the photon counter. In fact we are engaged in an exercise to actually identify individual chains by autocorrelation. We think that 1000 in a given volume element which is 3 or 4 μm^3 should be identifiable.

ILAVSKY:

You found that the critical concentration for the appearance of the effect is much lower than c^*?

KELLER:

Yes.

ILAVSKY:

Then I would like to know: if you use higher concentrations than c^*, how far can you go down with the molecular weight? Can you then arrive at the number which is usually known from mechanical measurements, like the critical molecular weight for entanglements for the appearance of the networks in your experiments?

KELLER:

We haven't done it. But that's a very pertinent question. Our ongoing rheology work, measuring flow resistence parallel to observing birefringence (i. e. strain) effects will hopefully lead to it.

TOMKA:

I would like to add a historical fact to your experience with gel formation in Coutte-flow of very large molecules. Signer was able to purify DNA, which is a very large molecule, with this technique, so far so that Rosalind Franklin could get the first very good X-ray diffraction patterns. These diffraction patterns could be properly interpreted by Watson and Crick. This was the clue: the purification of the DNA-sodium salt, and this was done by the Couette flow gel formation. But now I would like to ask you a question on the behaviour of polymer melts. Could you also observe this thin line formation at a critical strain rate or some other phenomena?

KELLER:

On polymer melts we could see the crystallizing line appearing, a very sharp line exactly at the centre of the jet. Incidentally, not only in the centre, but if we put an obstacle here (and this is of general significance), a crystalline line formed behind each obstacle. The obstacle is the source of elongational flow, creating a stagnation point. Now, the birefringence itself does not show a sharp line against a dark background, because the entire polymer becomes birefringent. Although away from the central axis the polymer is only very weakly oriented, we have a big path length, hence large retardation and intensity. This is in contrast to solutions where the solvent is not birefringent, hence, we see a bright line in a dark background. Even so in the melt, we see isochromatic curves (curves of equal retardation) and these bunch, having a sharp cusp at the centre-line between the jets defining a highly localized region of high chain extension.

TOMKA:

But have you done similar experiments in non-crystallizable melts?

KELLER:

Not so far.

LAUN:

I would like to know what your opinion is on the ductless siphon method. In the ductless siphon test – I think it should be quite similar to your opposite jet method – you also measure some kind of elongational viscosity. Do you think that you can get your cold stretched transition in the ductless siphon test, too?

KELLER:

Well, I know about it, but we have no experience with it. First, we were not setting out to measure elongational viscosity. We wanted to supplement our visual observations with viscosity measurements or something else equivalent to viscosity.

So we needed a method which could be used in conjunction with the jets, while with the ductless siphon I don't know what we would see. I ought to add that in our method we do not rely on a self-supprting fluid thread, as in all other methods of measuring elongational viscosity.

WINTER:

You have to have a concentrated system. The network limit published could be reproduced.

KELLER:

But in this system we can measure irrespective of whether it's viscous, or highly fluid.

Progress in Colloid & Polymer Science Progr Colloid & Polymer Sci 75:201–212 (1987)

Filled Networks

Surface properties of fillers and interactions with elastomers

A. Vidal and J. B. Donnet

Centre de Recherches sur la Phyisco-Chimie des Surfaces Solides – CNRS, Mulhouse, France

Abstract: In recent years, an increase in the use of composite materials consisting of a polymeric matrix and a filler, such as carbon black or silica, has been observed. It is associated with the outstanding mechanical properties exhibited by such materials, which are dependent on the nature of the polymer-filler interface. In this respect, the physical or chemical interactions the solid surface can exchange with the matrix play an important role.

After a short review of the parameters used for the characterization of fillers, and of some of the methods allowing an estimation of their reinforcing ability, the nature of the interactions able to take place between the elastomer and the filler will be examined. For this purpose, two examples will be provided. In the first, the kind of interaction the surface active sites of a filler can exchange with an elastomer will be considered, in the second, the consequences of a selective deactivation of the filler surface on its reinforcing ability will be reported.

Key words: Elastomers, fillers, reinforcement, surface activity, surface free energy.

Introduction

Natural and synthetic elastomers exhibit such a number of homogeneous physical and chemical properties that they currently form a large family, able to meet a wide range of requirements. It follows that rubbers rank among the basic wares of the modern world. It is, however, obvious that their numerous domains of use result not only from their vulcanization, but above all from their blending with fillers such as carbon black or silica.

Originally, these fillers were intended to play the role of extenders, in order to cut down manufacturing costs. However, it was soon realized that some of them imparted unexpected properties (hardness, strength, etc.) to the processed materials. This improvement of properties was called reinforcement.

This phenomenon covers of course, many mechanisms [1], and Figure 1 provides a qualitative illustration for interpreting modulus change of an elastomer upon filler blending [2]. A hydrodynamic or strain amplification effect, the existence of filler-elastomer bonds and the structure of carbon black [3], all play a part in this modulus increase.

Characterization of reinforcing fillers

First of all, it is obvious that the morphological characteristics of the filler play a significant part in reinforcement. Indeed, large-size particles dispersed in a gum only increase the stress applied to the elastomer, thus contributing to its weakening. Conversely, reinforcement effects are observed when the particle diameter

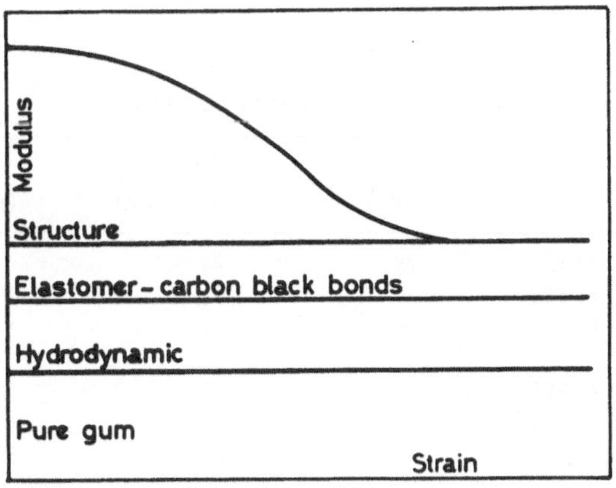

Fig. 1. Contributions to the stiffening effect of carbon black (with permission from Ref. [9])

becomes sufficiently small, much lower than 10 μm anyway.

The single selection of particle diameter for the characterization of a reinforcing filler is, however, not appropriate because, on the one hand, only fillers exhibiting a very poor reinforcing effect consist of independent spherical particles, and, on the other hand, gum-filler interactions take place at the elastomer-filler interface and are thus conditioned by the accessibility of the surface. A knowledge of the specific surface area of the filler is thus a prerequisite. It can be determined either by gas adsorption, thus taking into account its microporosity, or by adsorption of larger molecules, which will provide an evaluation of its accessible surface area [4, 5]. Other parameters should also be known, such as the degree of particle aggregation, which is called the structure and is measured by DBP absorption, the bulkiness, the shape and the anisometry of the aggregates [6].

Some of the above-mentioned parameters are tabulated, for a few carbon blacks of very different morphologies, in Table 1. It appears that in spite of its complex morphology, the structure of a black can generally be fairly well evaluated by means of its specific surface area and its DBP absorption number [7].

From a technological standpoint, numerous methods have been developed to appreciate the reinforcing ability of a filler, once incorporated in an elastomer. Two of them will be discussed: the α_F parameter [9] and the analysis of the empirical constants of the Mooney-Rivlin relationship [10].

The α_F parameter

This is deduced from the rheological properties of filler elastomer-blends. Its evaluation requires knowledge of the filler loading level and the value of the shear torque applied at the start of the vulcanization, D_a, and at full cure, D_∞; D_a^o and D_∞^o corresponding to the same values for the pure gum. If the curing mechanism of the elastomer is assumed to be unaffected by the filler, the increase of shear torque observed upon its blending can conceivably be attributed to the contribution to the overall crosslinking process of the polymer-filler interactions. The latter can therefore be estimated and the α_F factor will be defined by the following relation:

$$\alpha_F = \frac{\left[\dfrac{D_\infty - D_a}{D_\infty^o - D_a^o}\right] - 1}{m_f/m_p}$$

where m_f and m_p correspond to the quantities of black and elastomer, respectively. From the relations:

$$D_{spe} = \frac{D_\infty - D_a}{D_\infty^o - D_a^o} \quad \text{and} \quad C_g = m_f/m_p$$

we arrive at the equation,

$$D_{spe} = 1 + \alpha_F \cdot C_g$$

which is quite similar to that of Einstein. From α_F one can directly estimate the part played by the morphological characteristics of the filler in the dispersion polymeric medium. α_F is, of course, dependent, on the nature of the polymer and its evolution can be attributed to conformational particularities of the polymer chains in the vicinity of the surface. These conformations are of course in relation to the nature and extent of polymer-filler interactions. α_F is thus able to provide valuable information about the behavior of a filler

Table 1. Structural parameters for various furnace blacks

ASTM code		d (nm)	DBP (cm³/100 g)	$A \times 10^2$ (μm²)	N_p	Anis.	Surf. area (m²/g)
HAF type	N326	28	74	3.62	136	1.75	82
	N330	29	101	7.22	278	1.78	83
	N347	26	132	8.52	331	1.88	86
ISAF type	N219	–	73	2.38	103	1.67	116
	N220	22	116	5.45	261	1.82	112
	N242	–	141	8.44	380	1.78	–

d = diameter of elementary particles; DBP = DBP absorption number; A = projected area of an aggregate; N_p number of particles contained in an aggregate; Anis. = aggregate anisometry. From Reference [8]

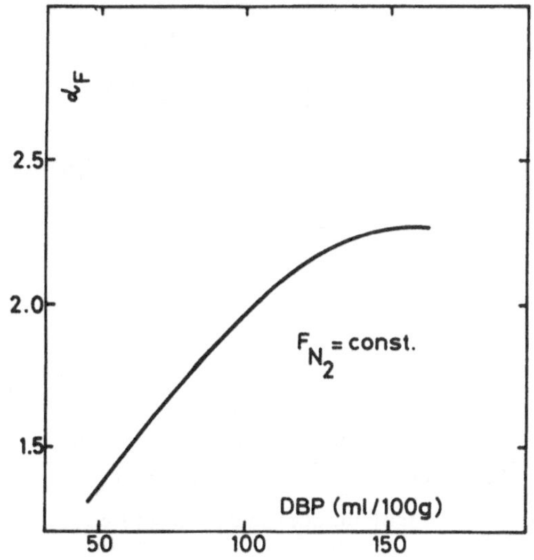

Fig. 2. Evolution of α_F with DBP number (from Ref. [10])

blended with a given polymer (in situ characterization). In Figure 2 one can check the relation which exists between α_F and the DBP absorption number, i. e. the structure of the black. It is particularly evident that, beyond given limits, an increase in structure does not bring any benefit from a technological point of view.

The empirical constants of the Mooney-Rivlin relationship

Another method of characterization of the reinforcing ability of a filler is to proceed through an analysis of the meaning of the constants C_1 and C_2 of the Mooney-Rivlin relationship [10]. It is known that for small deformations in elongation, the stress, σ, to which a sample is subjected can be related to its elongation, λ, by the semiempirical equation of Mooney and Rivlin:

$$\sigma = 2(C_1 + C_2\lambda^{-1})(\lambda - \lambda^{-2})$$

in which C_1 and C_2 are two constants independent of elongation. According to such a relation, after a first cycle of deformation, the plotting of the stress-strain properties of some samples filled with different types of carbon black is provided in Figure 3.

Upon a second cycle of deformation, i. e. vulcanizates having passed the so-called Mullins effect, the interpretation of the stress-strain curves according to the Mooney-Rivlin equation will provide C_1 and C_2 values which are different from the previous ones. The results obtained for several carbon blacks after both cycles of deformation are reported in Table 2.

From these results, it appears that after a first deformation cycle, the values of C_1 do not deviate from one

Table 2. Constants C_1 and C_2 (MPa) of the Mooney-Rivlin equation for carbon black filled vulcanizates (filler loading lever: 20%)

Filler	Deformation cycle (First)		Deformation cycle (Second)	
	C_1	C_2	C_1	C_2
N110	0.26	1.04	0.23	0.43
N242	0.20	0.96	0.19	0.46
N330	0.19	0.89	0.19	0.45
N550	0.25	0.79	0.25	0.50
N660	0.21	0.69	0.19	0.43
N765	0.24	0.60	0.24	0.38
Unfilled	0.13	0.17	0.12	0.16

From Reference [10]

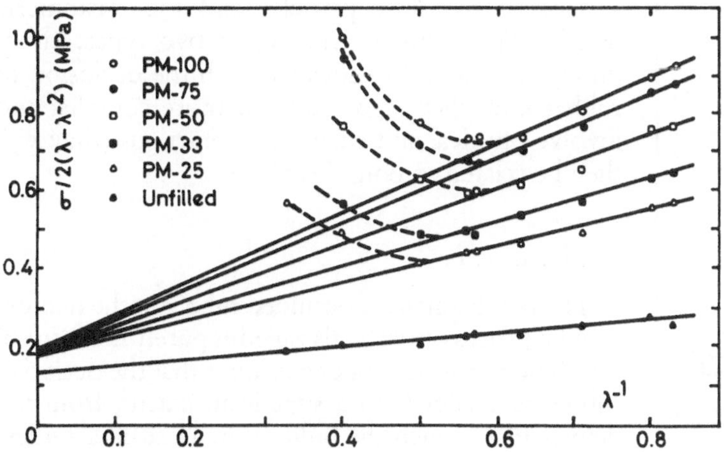

Fig. 3. The Mooney-Rivlin equation applied to vulcanizates based on SBR loaded with various types of Soviet carbon blacks (from Ref. [10])

another to a considerable extent (the unfilled vulcanizate, however, yields a sligthly lower value); those of C_2 in contrast, are widely different. They increase when moving from inactive to active blacks. Upon a second deformation, while the values of C_1 are in good agreement with the previous ones, a considerable change in those of C_2 is observed. At this stage, there is no longer very much difference between the various carbon black types. These results can be interpreted [10] by considering that in a filled elastomeric network, the total number of active network chain segments, v_{eff}, can be expressed by the sum of three components:

$$v_{\text{eff}} = v_p + v_k + v_t$$

where v_p = the number of active network strands developed during curing;
v_k = the number of active network strands due to entanglements;
v_t = the number of active network strands due to filler-elastomer interactions.

Since C_1 is independent of the black, its physical meaning is decisively determined by the number of active network chain segments developing in the elastomer upon curing, thus C_1 is dependent on v_p. As for C_2, its increase with the structure of the black and its decrease on Mullins softening suggest that it is proportional to the changing (destroying and newly developing) number of active network strands, i. e. to v_k and particularly v_t. C_2 can thus be used to appreciate the intensity of rubber-filler interactions. In this respect, one can check in Figures 4 and 5 the relationship which

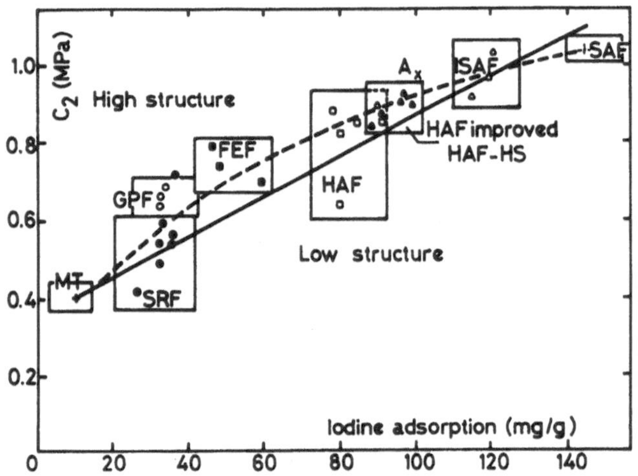

Fig. 4. Relation between specific surface area and C_2 (from Ref. [10])

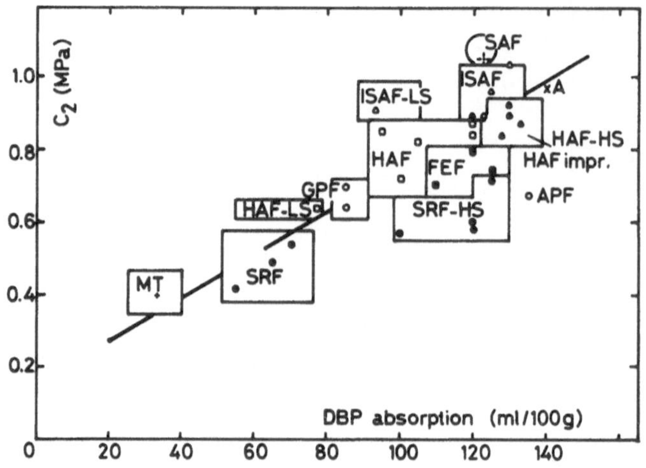

Fig. 5. Relation between DBP number and C_2 (from Ref. [10])

exists between C_2 and the conventional filler characterization parameters, such as the active surface area of the filler and its structure.

To sum up, it can be stated that the plotting of stress-strain measurements according to the Mooney-Rivlin representation can provide a characteristic constant C_2, which includes the effect of the accessible surface area of the filler and its structure and is thus suitable for the characterization of its reinforcing ability.

Having defined the parameters allowing a characterization of the reinforcing ability of a filler, what are the effects exerted by these aggregates on the properties of filled elastomers?

Nature of filler-elastomer interactions

All investigators concerned with rubber reinforcement agree that this effect is strongly dependent on the interfacial interactions able to take place between the surface of the filler particles and the elastomeric matrix. These interactions are of two types, either purely mechanical (associated with the occlusion of rubber into filler aggregates), or more complex and involve physical and chemical interactions (they will then be related to bound rubber).

Occluded rubber

The involvement of occluded rubber in the reinforcement process must be classified as purely mechanical [11]. The simple fact of occlusion is that the occluded rubber is shielded, to a significant extent, from the deformation which the bulk of the elastomer undergoes when subjected to stress. As a consequence, occluded rubber is trapped by the carbon black aggregates, without necessarily restricting the segmental motion of the chains located in the vicinity of the filler. The concept of molecular occlusion leads to an increase in the effective volume of the filler in rubber due to the filling of interparticulate voids by the elastomer. Such an effect has been interpreted, according to Kraus [12, 13] and Medalia [6], by a structure-concentration equivalence principle. This is examplified in Figure 6 for an SBR rubber, reinforced with various amounts of furnace blacks of identical specific surface area, but exhibiting widely different structures [1–3]. It appears that the stress at 300% extension is a function of the effective carbon black concentration "aV_2".

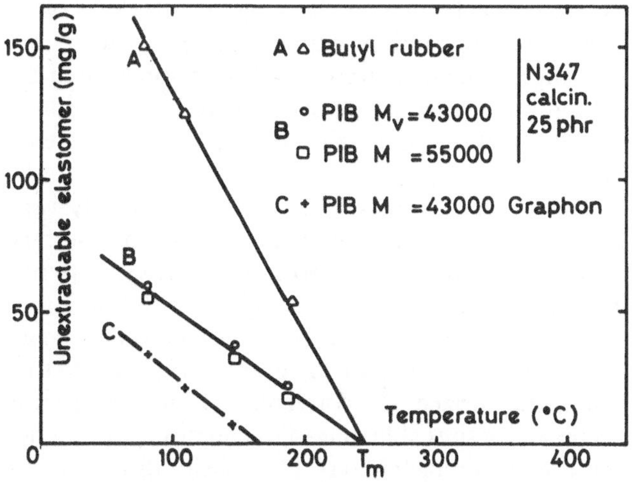

Fig. 7. Unextracted elastomer as a function of extraction temperature (with permission from Ref. [14])

Bound rubber

While the effects of occluded rubber are purely mechanical in nature, bound rubber is associated with more complex processes involving physical as well as chemical interactions. When an uncured filler-elastomer blend is submitted for an extended period of time to a solvent extraction, only part of the elastomer can be recovered, even with a very good solvent for the rubber. This quantity of unextractable elastomer adhering to the filler is known as "bound rubber". The data in Figure 7 indicate that the amount of unextractable rubber is unaffected by the molecular weight of the polymer [14], they lead to a recognition of the role of chemisorption in the bonding of elastomers to fillers. A practicle means of assessing the stability of the created bond has been suggested by Sircar and Voet [14] who plotted the unextractable amount of rubber from the filler as a function of the temperature of extraction, for various carbon blacks and elastomers. They observed the experimental points related to a given polymer-filler blend falling on straight lines, the extrapolation of which to zero grafting ratio provided the so-called solvolysis temperature, T_m. T_m, which appears to be independent of the solvent used, represents the temperature theoretically required to eliminate all bonds between carbon black and elastomer and is therefore indicative of the bond strength. It appears that a graphitized carbon black, which exhibits only a very small amount of bound rubber, leads to a much lower T_m. This result confirms the role played by chemisorption in the formation of bound rubber. It is, however, necessary to discriminate between bound rubber and the effects it exerts on the neighbouring elastomer molecules which make up a rubber shell of finite thickness surrounding the carbon black particles [15]. These effects associated with a reduced mobility of the rubber chains affect the physical properties of the sample.

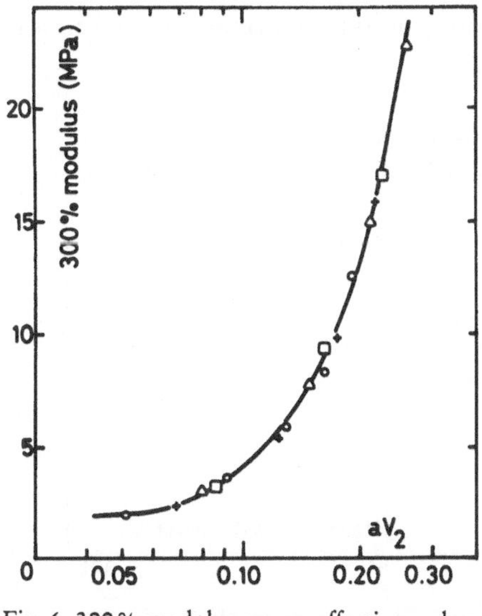

Fig. 6. 300% modulus versus effective carbon black concentration (with permission from Ref. [1] and [12])

What is the part now played by the surface activity of the filler in the reinforcement process?

Filler surface activity and reinforcement

The interactions able to take place between the elastomer and the surface of a solid can be of either a chemical or a physical nature. In this respect, two examples will be discussed. In the first, the kind of interactions the surface active sites of a filler can exchange with an elastomer will be considered; in the second, the consequences of a selective deactivation of the filler surface on its reinforcing ability will be examined.

The surface of carbon blacks is energetically very heterogeneous. This filler exhibits not only surface chemical functions, but also free valencies, network imperfections, etc. However, only a small fraction of an accessible carbon black surface corresponds to high energy sites. For instance Rivin et al. [16] showed that 5% of the surface of a conventional furnace black is occupied by sites able to lead to chemisorption, whereas the remaining 95% are available for dispersive interactions. Of course, the latter play a role in polymer-filler adhesion but do not contribute significantly to reinforcement.

The chemical aspect of reinforcement involves reactions taking place between the elastomer and the surface functional groups of the filler. What are these surface chemical functions? Table 3 shows that for a carbon black, the surface area of which is 120 m^2/g, elemental analysis reveals the presence of carbon (96%) but also 3.2% oxygen and 0.55% hydrogen. These heteroatoms are part of several reactive functional groups on the solid surface, which have been identified as phenolic, hydroxyl, quinone, carboxyl, lactone and peroxide groups, among others.

Rather than exhaustively investigating the different types of reactions that can take place between the surface functional groups of the black and the elastomer [16–31], we shall focus our attention on the effect of the hydrogen atoms, which are located on the surface of the filler, at the edges of the polyaromatic structures which make up the black. The role played by these surface hydrogens can be outlined by answering the following questions [32, 33]:

1. Are surface hydrogens of carbon black reactive with SBR or, in other words, is there any hydrogen transfer?

Table 3. Elementary analysis of a carbon black

Particle diameter (nm)		24.6
Specific surface area (m^2/g)		120
Elementary analysis (%)	C	96
	O	3.2
	H	0.55
Surface chemical functions (μEq/g)	–COOH	50
	–OH	100

2. Is hydrogen reactivity related to reinforcement or is there any relationship between hydrogen content and strain energy?

3. Is the total hydrogen of black important or only the more reactive part of it?

The answer to the first question was provided by demonstrating the occurrence of hydrogen exchange between a tritium-labeled SBR and carbon black. Part of the surface hydrogen was, indeed, shown to transfer from the black to the polymer, and conversely, without SBR reinforcement being necessarily affected. In order to check this point and examine whether the presence of hydrogen on the surface of carbon black is related to its reinforcing ability and whether the latter is influenced by the total hydrogen content, tritium-labeled carbon black was heated to remove gradually its surface hydrogens. The reinforcing ability of the black, as derived from stress-strain curves, was then compared with the content of residual surface hydrogens. Figure 8 shows that a correlation is obtained between the latter and the strain energy retained by the vulcanizate.

However, heating carbon black to temperatures higher than 1400 °C not only removes the surface hydrogen atoms but can also be associated with structural modifications (even if graphitization is not yet important at such temperature). As a consequence, the decrease of the reinforcing potential of the black can result either from hydrogen removal or from a rearrangement, followed by the possible elimination of structural defects. These results nevertheless suggest the possibility of a chemisorption mechanism involving hydrogen abstraction from either carbon black or rubber (or both) resulting in polymer retention on the surface of the black.

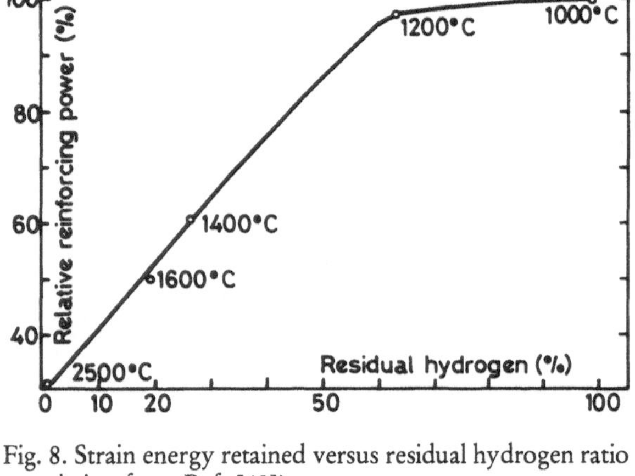

Fig. 8. Strain energy retained versus residual hydrogen ratio (with permission from Ref. [40])

In order to answer the third question, relating to the characterization of hydrogen atoms exhibiting different reactivities, the kinetics of the tritium-labeling reaction were investigated [33]. Assuming the existence of surface hydrogens of different reactivity and that each variety exhibits first order substitution kinetics, the data reported in Table 4 were recorded for a series of carbon blacks exhibiting different reinforcing abilities. k' represents the apparent rate constants of the substitution reactions and x, the hydrogen relative proportion. It thus appears that the blacks exhibit four different varieties of surface hydrogen. It appears moreover, that the labeling rate constants for a given type of hydrogen are of the same order of magnitude for all blacks, but differ significantly from one type of hydrogen to another. The relative proportions of each type vary depending on the nature of the black, the evolution of the most reactive hydrogens correlating quite well with the reinforcing potential of the filler. They could thus play a very significant part in the reinforcement of elastomers.

In order to confirm the great reactivity of the surface hydrogens, their substitution by chlorine atoms was studied. This reaction, performed using tritium-labeled carbons, was investigated at various temperatures and traced by loss of tritium. Figure 9 shows that tritium elimination starts at room temperature and by 750 °C all the surface hydrogen has reacted. In addition, it is noteworthy that the elimination of hydrogen occurs in four steps, each probably corresponding to a peculiar type of reactive hydrogen. The first and second steps correspond to about 10 % and 18 % hydrogen loss, respectively. These proportions are in agreement with the hydrogen types 1 and 2 yielded by the kinetic experiments. These studies reveal, therefore, that the reinforcing ability of carbon black towards elastomers is directly related only to the most reactive hydrogen atoms.

It appears beyond all doubt that filler-elastomer interactions result in the formation of chemical bonds between the polymer and the solid surface, which are due to a reaction of the macromolecule, either with the surface chemical groups or with the surface hydrogen atoms. The question arises, however, as to whether the formation of this bond is an essential condition of rein-

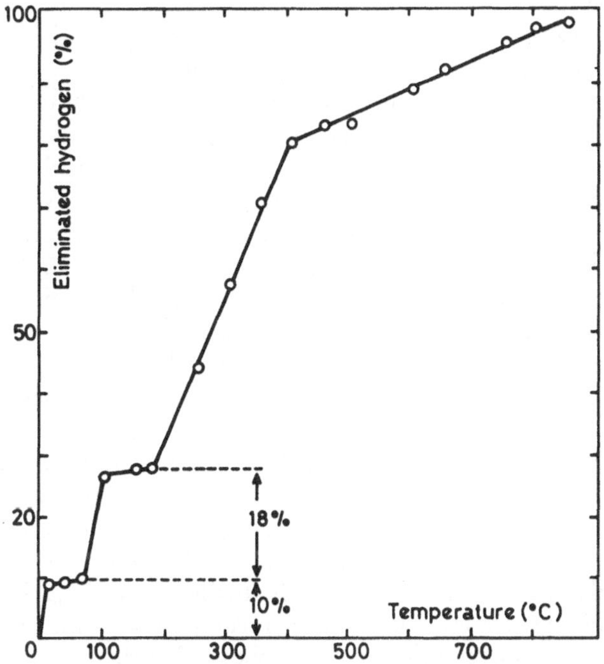

Fig. 9. Effect of chlorine treatment temperature on hydrogen elimination (from Ref. [41])

Table 4. Hydrogen exchange reaction

	Aro 1LS	Aro 3LS	Aromex	Aro 100	Aro 150	Aro 175
k'_1 (h^{-1})	0.090	0.097	0.080	0.106	0.101	0.105
x_1 (%)	2.0	2.7	5.3	7.9	9.3	10.2
k'_2 (h^{-1})	0.055	0.051	0.040	0.065	0.062	0.064
x_2 (%)	13.4	24.9	9.9	8.8	22.9	17.5
k'_3 (h^{-1})	0.028	0.025	0.020	0.030	0.026	0.026
x_3 (h^{-1})	40.6	31.7	25.3	29.5	24.6	37.9
k'_4 (h^{-1})	0.021	0.021	0.018	0.026	0.025	0.025
x_4 (h^{-1})	44.0	40.7	59.5	53.8	43.8	32.6

From Reference [33]

forcement, and what the consequences would be of a selective modification of the surface activity of the filler.

For this purpose, the effect of model compounds was studied, the surface energy of which was controlled by grafting silica with well characterized alkyl chains [34]. This selective modification resulted from the esterification of the surface silanol groups of silica with two alcohols: methanol and hexadecanol. After characterization of the modified surfaces (surface free energy, surface polarity, etc.), the effect of this surface treatment on filler-elastomer interactions was examined.

Two silicas, either fumed (*A*) or precipitated (*P*), were used (specfic surface areas, about 130 m^2/g). Table 5 provides the grafting ratios for the different modified silicas which are identified as AC_1 and PC_1 (methanol esterification), AC_{16} and PC_{16} (hexadecanol esterification), respectively.

Properties of the modified silicas were investigated, on the one hand by inverse gas-solid chromatography at zero surface coverage, which provided an estimation of their surface free energy, and on the other, by the study of their reinforcing potential.

Initial or modified silicas used as stationary phases were first characterized using, at zero surface coverage, probes of well known characteristics (non polar, i. e. alkanes, or polar). By performing adsorptions at zero covering ratio and using a proper selection of reference state, the thermodynamic parameters of the adsorption of n-alkanes at infinite dilution (ΔG^o, ΔH^o and ΔS^o) could be directly related to V_N, the net volume of retention of the probes [35, 36].

Figure 10 shows for silica *P* at 110 °C, the evolution of the free energies of adsorption of n-alkanes vs. the number of carbon atoms of the probe. Upon grafting of alkyl chains, a strong decrease of the free energy of adsorption is observed due to a drastic reduction in the surface reactivity of the filler; results which were confirmed by measurement of the surface free energy of the different samples.

Table 5. Grafting ratios of the modified silicas

Sample	AC_1	PC_1	AC_{16}	PC_{16}
Grafting ratio (chains/nm^2)	3.4	10.5	1.5	2.2

From Reference [34]

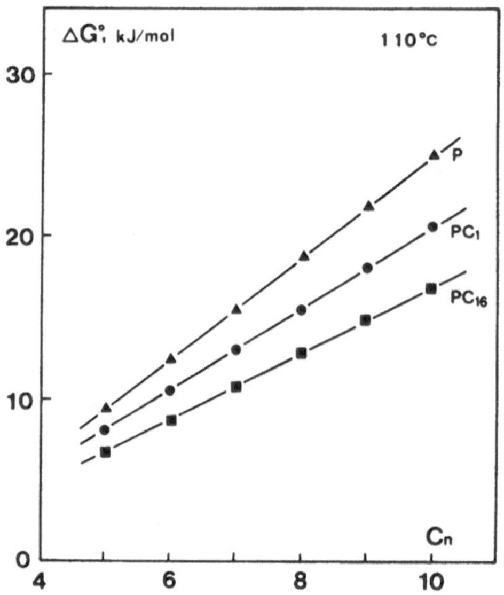

Fig. 10. Free energy of adsorption of n-alkanes versus number of carbon atoms of the probe (from Ref. [34])

It is known that the surface free energy of an adsorbent can be expressed by

$$\gamma_s = \gamma_s^d + \gamma_s^{spe}$$

where γ_s^d and γ_s^{spe} are the dispersive and specific component of the surface free energy, respectively. γ_s^d was deduced from the evolution of ΔG^o versus the number of carbon atoms of the probe, whose slope can be related to the surface tension of a surface made of CH_2 groups only [35]. The results obtained for the different samples at 20 °C are reported in Table 6. It appears that the grafting of alkyl chains onto the surface of the solid is associated with the deactivation of the latter. This effect is enhanced with grafted chains of increasing length. PC_{16} and AC_{16} indeed exhibit γ_s^d, in close

Table 6. γ_s^d of silicas at 20 °C (mJ/m^2)

Sample	Chromatography	Contact angle
P	98.2	–
PC_1	68.2	87
PC_{16}	35.6	47
A	75.3	–
AC_1	70.2	76
AC_{16}	38.7	50
Polyethylene	–	24–42

From Reference [34]

agreement with that of polyethylene, which is representative of a surface of small energy.

The specific component of the surface free energy of the different samples was provided by a comparison of the adsorption free energy of probes of different polarity. The generally accepted way of separating London dispersion effects from specific effects, in the case of infinite dilution chromatography, is to compare the chosen solute probe with a molecule able to exchange with the surface identical dispersive interactions [37–39]. Thus, from the comparison of the adsorption free energy of a polar probe with that of a real or hypothetical n-alkane of the same saturated vapour pressure, it is possible to estimate the magnitude of specific interactions. Comparison of Figures 11 and 12 thus points to a drastic reduction in the ability of the surface to exchange any specific interaction, once modified by grafting of hexadecyl chains (all probes, whatever their polarity, are indeed very close to the alkane straight line). The C_{16} grafts, then, seem to be able to form a layer, coating the silica particles and protecting it almost completely, or decreasing the accessibility of the surface high energy sites still available.

Inverse gas-solid chromatography at zero surface coverage thus points to a strong decrease in the surface free energy of silicas upon surface alkylation and to a shielding by grafted chains of the important silica specific forces. To move a step further towards relating the elastomer reinforcing ability of a filler to its surface activity, the consequences of the surface deactivation of silicas for rubber-filler interactions were studied.

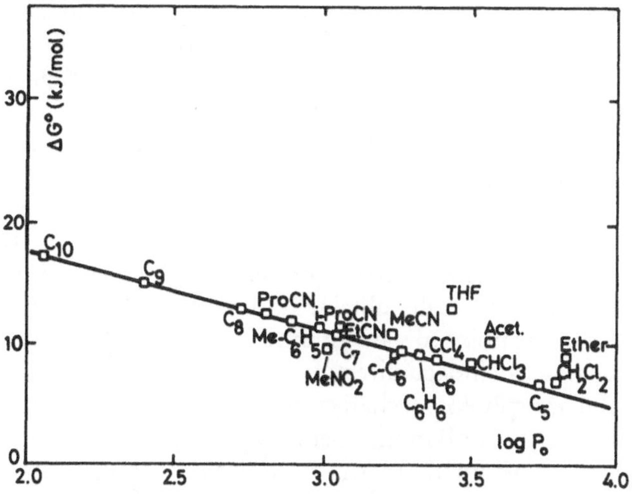

Fig. 12. Evolution of adsorption free energy vs. saturated vapour pressure of the probe (silica AC_{16}, 110 °C, from Ref. [42])

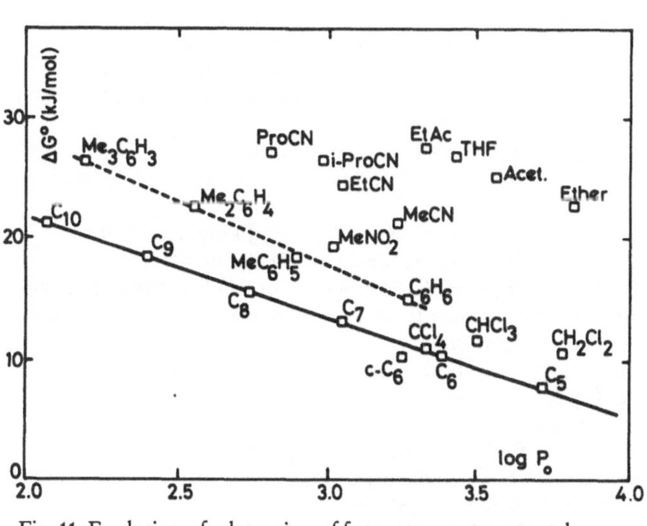

Fig. 11. Evolution of adsorption of free energy vs. saturated vapour pressure of the probe (silica A, 110 °C, from Ref. [42])

The organophilic character of the corresponding silicas should be of interest due to their improved compatibility towards elastomers. For this purpose, three commercial rubbers were used, a natural rubber, a styrene-butadiene and an acrylonitrile-butadiene copolymer. The corresponding filled vulcanizates were obtained by sulfur curing. The resulting materials were characterized particularly by measurement of their bound rubber and study of the rheological properties of the reinforced materials.

The evaluation of bound rubber can provide an estimation of the interactions an elastomer is able to exchange with a filler. It is measured by submitting, for an extended period, an uncured filler-elastomer blend to solvent extraction; bound rubber being the quantity of unextractable elastomer adhering to the filler. In Table 7 are reported the amounts of bound rubber

Table 7. Bound rubber

Elastomer	Silica	Bound rubber (%)
SBR	–	0
	P	24.8 ∓ 1.2
	PC_{16}	0
NBR	–	0
	P	31.2 ∓ 0.7
	PC_{16}	0

From Reference [34]

corresponding to samples P and PC_{16}. In contrast to the initial silicas, the modified fillers did not yield any bound rubber with NBR or SBR. Such results suggest that interactions between modified silica and elastomer are very weak and not associated with any chemisorption process.

During vulcanization the evolution of the rheological properties of formulations prepared from different fillers was used to assess the reinforcing ability of the solids. The results obtained are presented in Table 8 from the point of view of the α_F parameter [9]. From α_F one can indeed directly estimate the part played by the morphologic characteristics of the filler in the dispersion polymeric medium. It appears that for a given elastomeric matrix, α_F decreases strongly upon surface treatment of the solids, which lose from 70 % to 85 % of their reinforcing ability.

These results thus point to the important role played by filler-elastomer interactions in reinforcement processes.

Table 8. Reinforcement parameter

Sample	α_F NR	SBR	NBR
A	5.21	4.24	5.31
AC_1	3.90	1.41	2.63
AC_{16}	1.65	0.66	1.70

From Reference [34]

Conclusion

The improvement of properties of elastomeric materials associated with their blending with fillers, such as carbon black or silica, is called reinforcement. After an examination of the characterization of fillers and of some of the parameters which can be used for assessing their reinforcing ability, the nature of filler-elastomer interactions was scrutinized.

This led to the notions of occluded and bound rubber. The latter, which involves the occurrence of physical and chemical interactions between the surface of the filler and the elastomer, is thus strongly dependent on the filler surface activity.

In this respect, two examples were discussed. In the first, the kinds of interactions the surface active sites of the filler can exchange with an elastomer were considered and it was shown that the reinforcing ability of carbon blacks is directly related to their most reactive surface hydrogen atoms. In the second example, the consequences of selective deactivation of the filler surface on its reinforcing ability were investigated. It was shown that the surface deactivation, resulting from the grafting of alkyl chains, was associated with a decrease in the reinforcing potential of the filler.

References

1. Kraus G (1978) Rubber Chem Tech 51(2):297
2. Payne AR (1965) Dynamic properties of filler-loaded rubbers, In: Kraus G (ed) Reinforcement of elastomers, Interscience, New York, pp 69–123
3. Meinecke EA, Maksin S (1980) Coll & Polym Sci 258:556
4. Lammond TG, Price CR (1970) Rubber Chem Tech 43:941
5. Janzen J, Kraus G (1971) Rubber Chem Tech 44:1287
6. Medalia AI (ed) (1975) Filler aggregates and their effect on reinforcement, In: Le renforcement des élastomères, CNRS, Paris, pp 63–79
7. Kraus G, Janzen J (1975) Kautsch Gummi Kunstst 28:253
8. Donnet JB, Voet A (eds) (1976) Carbon Black, Dekker, New York
9. Wolff S (1970) Kautsch Gummi Kunstst 23:7
10. Soos I (1982) Ph D Thesis, Budapest
11. Medalia AI (ed) (1975) Filler aggregates and their effect on reinforcement, In: Les interactions entre les élastomères et les surfaces solides ayant une action renforçante, CNRS, Paris, p 63
12. Kraus G (1971) Rubber Chem Tech 44:199
13. Kraus G (1971) J Appl Polym Sci 15:1679
14. Sircar AK, Voet A (1970) Rubber Chem Tech 43:973
15. Kaufman S, Slichter WP, Davis DD (1971) J Polym Sci A2 9:829
16. Rivin D, Aron J, Medalia AI (1968) Rubber Chem Tech 41:330
17. Gessler AM (1969) Rubber Chem Tech 42:850
18. Voet A (1973) Kautsch Gummi Kunstst 26:254
19. Dannenberg EM (1975) Rubber Chem Tech 48(3):411
20. Harris JO, Wise RW (1965) In: Kraus G (ed) Reinforcement of elastomers, Interscience Publ, New York, Chap 9
21. Rivin D (1971) Rubber Chem Tech 44:307
22. Watson M (1965) Chemical Interaction of fillers and rubbers during cold milling, In: Kraus G (ed) Reinforcement of elastomers, Interscience, New York, pp 247–260
23. Waldrup MA, Kraus G (1969) Rubber Chem Tech 42:1155
24. Jamroz M, Kozlowski K, Sieniakowski H, Jachym B (1977) J Polym Sci 15:1359
25. Cashell EM, McBrierty VJ (1977) J Mater Sci 12:2011
26. Donnet JB, Geldreich L, Henrich G, Riess G (1964) Rev Gen Caout Plast 41:519
27. Donnet JB, Peter G, Riess G (1969) J Polym Sci 22:645
28. Donnet JB, Vidal A, Riess G (1971) J Chim Phys 68:1642
29. Donnet JB, Riess G, Majowski G (1971) Eur Polym J 7:1065
30. Papirer E, Donnet JB, Riess G, Van Tas N (1971) Angew Makromol Chem 19:65
31. Drappel S, Gauthier JM, Franta E (1983) Carbon 21(3):311
32. Papirer E, Voet A, Given PH (1969) Rubber Chem Tech 42(4):1200
33. Papirer E, Donnet JB, Heinkele J (1971) J Chem Phys 68:581
34. Donnet JB, Wang Meng Jiao, Papirer E, Vidal A (1986) Kautsch Gummi Kunstst 39(6):510
35. Dorris GM, Gray DG (1980) J Coll Interf Sci 77:353
36. Meyer EF (1980) J Chem Ed 57:121

37. Kiselev AV, Koteknikova TA, Nitkin Yu, Tsilipotkinia MV (1978) Koll Zh 40(5):865
38. Schmid R, Sapunov AV (eds) (1982) Non-formal kinetics, Verlag Chemie, Weinheim
39. Saint-Flour C, Papirer E (1983) J Coll Interf Sci 91:69
40. Voet A, Papirer E, Aboytes P, Schultz J (1971) Rev Gen Caout Plast 48(9):935
41. Donnet JB (1973) Br Polym J 5:213
42. Wang Meng Jiao (1984) Ph D Thesis, Mulhouse, France

Received December 11, 1986; accepted March 27, 1987

Authors' address:

A. Vidal
Centre de Recherches sur la
Physico-Chimie des Surfaces Solides – CNRS
24, avenue Président Kennedy
F-68200 Mulhouse, France

Discussion

PECHHOLD:

Could you say some more about the influence of the particle size and the influence of the aggregation of these carbon black particles?

VIDAL:

On the point of view of size, only small enough particles (< 10 µm) exhibit a reinforcing ability. If large sized particles can participate in an increase of modulus through a hydrodynamik effect, they nevertheless essentially contribute to an increase in the stress applied to the elastomer and thus to its weakening. As for the degree of aggregation of the particles, their so-called structure contributes to reinforcement through the occlusion of rubber, which is associated with an increase of the effective volume of the solid due to the filling of interparticulate voids by the elastomer. This can be interpreted, according to Kraus [12, 13], by a structure-concentration equivalence principle, stating that a low structure carbon black used at high concentration exerts the same effects as a high structure carbon black used at low concentration.

KILIAN:

What is your understanding of the Mullins softening? In unloading the sample, there is not very much of the "reinforcement" seen in the first cycle left. What is the role of the occluded rubber?

VIDAL:

The results I presented were related to small elongations. Occluded rubber, then, does not sustain the deformation. Going to much higher elongations may imply the rupture of filler-elastomer bonds and, as a consequence, a dewetting of the solid particles. At this point, occluded rubber will also be submitted to the deformation. Speaking now of the Mullins effect, it results not only from the destruction, upon deformation, of a tridimensional filler network but also from the slippage of some polymer chains weakly retained on the surface of the solid, as proposed by Dannenberg, for example [19].

KILIAN:

My problem is the occluded rubber effects. In the "bound rubber" (that means in the rubber layers around each of the filler particles), entropy of deformation will be stored during the first stretch. According to stretching calorimeter measurements and their interpretation, it is very likely that only tiny dissipative heats arise. To our current understanding, the properties in the layers around each of the filler particles should pass from a "glassy shell" at the surface of the filler somehow continuously into that of the rubbery matrix. But this hypothesis must be proven by further experiments.

VIDAL:

Occluded rubber is associated with a hydrodynamic effect. Occluded rubber chains will only be submitted to the deformation once the sample is strained to high enough elongations.

K. P. RICHTER:

Is it a weak crossing-over of the dimensions from a small to a greater particle?

VIDAL:

Small elementary particles have a higher tendency towards aggregation and yield aggregates with a large number of internal voids, hence of higher structure. But the selection of a reinforcing filler cannot be made only on a large versus small particle diameter basis. Nevertheless, only aggregates the size of which is smaller than 10 µm are classified as reinforcing.

OBRECHT:

In one figure you gave a comparison of the interaction between Decane unmodified and modified silica grafted with C_{16} alcohols. I personally was a bit surprised that you found a smaller interaction for the modified surface than for the unmodified surface, because, as a chemist, one has these prejudices that solvents like similar solvents and they dislike a sort of dissimilar solvents. I am really a bit surprised that the interaction is decreased for the modified surface. I really don't understand this.

VIDAL:

First it is an experimental observation. The modification of a surface with inactive molecules (alkanes) is associated with its deactivation (decrease of surface free energy). The particles are then coated with a layer of grafted chains which make the solid surface inaccessible. This results in the solutes only being able to interact with the grafted layer. Such a behaviour is confirmed by the fact that the heat of adsorption of the solute is then equal to its heat of liquefaction, thus suggesting gas-liquid rather than gas-solid interactions.

RIGBI:

In certain experiments we have carried out on mixtures of two polymers – one with a very low molecular weight and one with a very high molecular weight – and the interaction of carbon black, we found that the low molecular weight was absorbed preferentially right at the beginning.

After a very long time, the very high molecular weight polymer replaced the low molecular weight polymer. So, it is not only a question of immediate results, but also a question of time, as kinetics are involved in it.

HÄRTEL:

You know the famous work of Payne, who did some work on filler-filler interaction. We have strong evidence for filler-filler interaction, especially at very low amplitudes in dynamic measurements.

VIDAL:

The study of the rheological properties of suspensions made up from hydrocarbons (e. g. squalene used as molecular weight analog of elastomers) and silicas (modified or not) provides information related not only to the morphology of the dispersed solid phase but also to particle-particle as well as dispersion medium-particle interactions. Modification of the solid surface by grafting of C_{16} chains appears to be associated with a strong decrease of aggregate-aggregate interactions (disappearance of thixotropic processes and of pseudo-plastic behaviour, effective volume of filler strongly reduced). The particle-particle interactions which are so important in reinforcement are completely eliminated upon modification.

Progress in Colloid & Polymer Science Progr Colloid & Polymer Sci 75:213–230 (1987)

Filled van der Waals networks

H.-G. Kilian

Abteilung Experimentelle Physik, Universität Ulm, Ulm, F.R.G.

Abstract: New possibilities of characterizing filler-loaded networks with the aid of a generalized van der Waals formulation are presented. By arriving at a full description of stress-strain cycles, different means of cooperation within the composites are identified in the investigation of model systems. The role of filler-to-matrix contacts is discussed, including the Einstein-Smallwood boundary value problem. An interpretation of the Mullins softening is achieved, based on the description of a set of cyclic quasi-static experiments.

Key words: Networks, filler-loaded vulcanisates, Mullins softening, van der Waals theory of filled networks, Einstein-Smallwood effect.

Symbols

a	van der Waals parameter of global interactions between the chains
a_r	interaction parameter within the rubber network
a_f	interaction parameter within the filler network constituted by crosslinking filler particles only
C, C_λ	Einstein-Smallwood parameter
$D = \lambda - \lambda^{-2}$	deformation function of the „Gaussian continuum"
$D_m = \lambda_m - \lambda_m^{-2}$	maximum value of the deformation function
f	nominal force
f_h	enthalpy component of the force
f_s	entropy component of the force
G_o	modulus
$G_{oo} = \varrho RT/M_o$	
G_f	modulus accounting for the Einstein-Smallwood effect
G_{fv}	modulus with prefactor correction
H	enthalpy
$H^{(i)}$	enthalpy of the oscillatory freedom (i)
k_B	Boltzmann constant
$\varkappa$	fraction of filler to matrix bond
λ	macroscopic strain
$\lambda_{\max}$	maximum macroscopic strain achieved with the first stretch
λ_r	strain in the rubber matrix
λ_{rr}	matrix strain of a filled rubber
λ_{rf}	matrix strain of a filler network
λ_i	matrix strain in the filler-loaded network of the type (i)
λ_f	average total strain of the filler
λ_p	plastic strain of the filler-particle's ensemble
λ_{fe}	elastic strain of the filler
λ_m	maximum macroscopic strain
λ_{mr}	maximum matrix strain of the filled rubber
λ_{mf}	maximum matrix strain of the filler network
λ_{mrf}	maximum matrix strain of the filler network rubber
λ_{mrfv}	maximum matrix strain of the filler network rubber plus Einstein Smallwood
$\lambda_{m\ \max}$	maximum strain parameter after the first stretch
λ_b	remnant strain after the first stress-strain cycle
λ_{bb}	ad hoc assumption of how the remnant strain is developed in dependence on the macroscopic strain λ
M_c	molecular weight of the network chains
M_s	molecular weight of the Kuhn segment
M_o	molecular weight of the "stretching-invariant unit"
N	number of chains
n	volume density of chains
N_L	Loschmidt number
p	pressure
Q	heat exchanged during deformation
ϱ	mass density
R_g	gas constant
S	entropy
$S^{(i)}$	entropy of the i-th internal freedom
T	absolute temperature
U	internal energy
u_r	volume correction in the filled rubber
u_f	volume correction in the filler network
u_{rf}	volume correction in the filler network rubber
$u_{r\ \max}$	the minimum value of u_r after the first stretch up to $\lambda_{\max}$
$u_{f\ \max}$	the maximum value of u_f after the first stretch
y	number of stretching invariant units per chain
y_s	number of Kuhn segments per chain
ν	relative number of crosslinks
ν_r	relative number of matrix crosslinks
ν_f	relative number of filler-to-matrix crosslinks
ξ	prefactor correction
V	volume

1. Introduction

It is an interesting experience that the stress-strain pattern of rubbers yields information about "global" deformation phenomena only [1-5]. Being in the state of internal equilibrium, the "liquid-like short-range structure" is modified, on isothermal deformation, so as not to change the free enthalpy at all. Hence, rubbers can be brought into largely deformed states without any "free-enthalpy relevant" defect production. If the rubbery matrix within filled networks behaves in the same manner, an exceptional chance would become available of studying the global deformation mechanism running off. It should be possible to gain valuable information about the role of the filler-to-matrix contacts.

How the van der Waals model must be extended to achieve a description of the quasi-static deformation in filled rubbers, is reported in this paper. The quantitative treatment of the Mullins softening is the very first step in the direction towards a full description of irreversible processes running off in heterogeneously deforming rubberlike composites.

2. The phantom network

It was the ingenious idea of Kuhn [6] to describe the conformational behavior of macromolecular chains by defining the "phantom chain" (Gaussian chain). The mathematical representation is given by an "infinitely thin and infinitely long" fiber without any eigenvolume, being perfectly flexible and showing no interactions at all. The strain energy of the phantom-chain network (Gaussian network) is herewith straightforwardly derived under the assumption of having non-fluctuating crosslinks submitted to an affine transformation [1, 2, 6].

What is of crucial importance is that energy can only be stored if mass is continuously distributed across each of the phantom chains of infinite "contour-length". It is not necessary to know exactly the "molecular weight of the chains". This indifference is the consequence of having the kinetic energy equiparted, to make the chains "energy-equivalent subsystems of deformation" [4, 5, 8, 9]. The situation is, in principle, as in gases where different molecules have the same kinetic energy.

Under the given circumstances, it is not too surprising a finding that the internal energy of the Gaussian network is identical with that of an ideal gas [4, 8]

$$U = N\{3(k_B T/2)\} \tag{1}$$

k_B is Boltzmann's constant, T the absolute temperature. What cannot easily be rationalized is that the Gaussian chain has only three freedoms of kinetic energy, like a mass point in the mechanics. Since the internal energy depends on the temperature only, it suggests understanding the purely entropy-elastic Gaussian network as the limited model of the "ideal conformational gas network" [4-5, 7-9].

Clearly, Kuhn's fascinating idea that this phantom chain model should apply for describing the deformation in rubbers, relies upon very high degrees of abstraction which widely surpass those used for enabling the mass point to be a reasonable model in classical mechanics.

In consideration of the above circumstances, we are led to the mechanical equation of state [1, 2, 6]

$$f = nk_B TD \tag{2}$$

with

$$D = \lambda - \lambda^{-2} \tag{3}$$

n is the density of the chains per volume unit defined by

$$n = N/V\,. \tag{4}$$

Having energy-equivalent subsystems of deformation is the symmetry which determines the simple analytical form of Equation (2): the nominal force is found to be proportional to the deformation function of an incompressible, a priori isotropic continuum, as defined in Equation (3), multiplied by the density of the energy-equivalent permanent subsystems of deformation [1, 10]. The density of the subsystems of deformation of the Gaussian network is set to be identical with the density of the chains themselves.

A convenient description is then achieved by rewriting the volume density of the chains in the following manner [1]:

$$n = N/V = \varrho N_L/M_c \tag{5}$$

ϱ the the mass-density and N_L, Loschmidt's number. The molecular weight of the networks meshes, M_c,

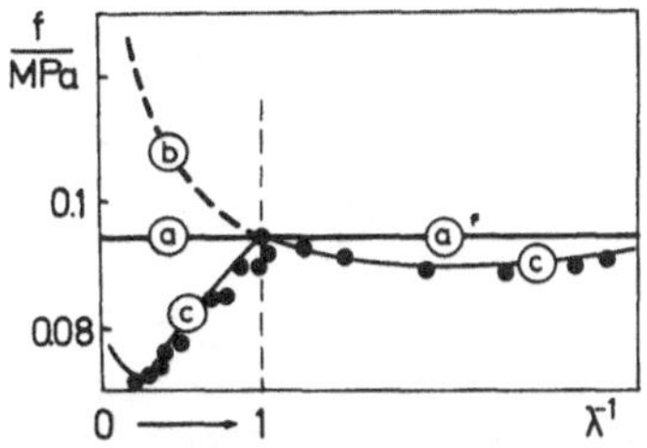

Fig. 1. Mooney plot of (a) the Gaussian phantom network, (b) Langevin and (c) experiments according to [16]. The solid line (c) is theoretical, according to [18], using the parameters: $G_o = 0.111$ MPa ($M_o = 68$ g mol^{-1}); $\lambda_m = 18$; $a = 0.2$

enters into the description to directly illuminate the single-chain characteristics of the Gaussian network model.

The modulus is then written as

$$G_o = \xi(nk_BT) = \xi(\varrho R_g T/M_c) \tag{6}$$

ξ may be also taken to account for unknown prefactor effects [1, 2, 11–15]. It is thus implied that the mean energy of deformation per phantom chain may become intrinsically modified, while the chains nevertheless continue to operate as energy-equivalent subsystems of deformation.

The high degrees of symmetry in the deformation behavior of the Gaussian network are best demonstrated by plotting the reduced force

$$f^* \equiv f/D = G_o \tag{7}$$

against λ^{-1} (Mooney plot). According to Figure 1, f^* is constant in the deformation modes of simple extension and of uniaxial compression which are equivalent to the modes of uniaxial and equi-biaxial deformation [16–18].

That experimental moduli of rubbers lie in the order of magnitude as predicted, must be seen as proof of the fundamental ideas behind the Gaussian network model:

1. First of all, it is evidenced that the deformation energy is practically of "global origins" so that it is possible to come near the truth by characterizing the conformational abilities of real chains with the aid of the phantom-chain model.

2. To have the strain energy equiparted is evidenced as long as the modulus is correctly calculated with the use of Equation (6) (including prefactor effects, as discussed in literature).

Yet calculations cannot reproduce the whole experimental stress-strain pattern at all as it is shown in Figure 1. It is an interesting question whether modifications of the Gaussian model can improve the reliability of this model by preserving most of its universal features.

To have the end-to-end distance distribution of the network chains always related to the Gaussian distribution is the finer characterization obtained by statistical treatment [1, 2]. However, these details do not enter into the formulation of the mechanical equation of state.

3. The van der Waals model

Two effects will be discussed:

1. Finite chain extensibility characterized by the maximum strain λ_m.

2. Global interactions between the energy-equivalent subsystems of deformation, described with the aid of the phenomenological parameter "a".

The two-parameter van der Waals equation of state can then be written as [4, 9]

$$f = G_o D\{1/(1 - D/D_m) - aD\}\,. \tag{8}$$

We would like to stress the fact that the corrections alone modify the global properties of the Gaussian network. From the maximum deformation function, written as

$$D_m = \lambda_m - \lambda_m^{-2} \tag{9}$$

we find that an infinitely large force would be necessary to bring the network into the state of maximum elongation: strains beyond these limits ($\lambda > \lambda_m$) are totally excluded. The constitutive singularity, defined by the van der Waals model at $\lambda = \lambda_m$, is a necessary energetic pole for realizing finite chain extensibility for physical causes.

3.1 The maximum strain parameter

The maximum strain parameter λ_m can be simply related to the chain length on the basis of the Gaussian coil model. Calling the molecular weight of the statistical Kuhn segment y_s, the molecular weight of the chains can be written as [1–2, 19].

$$M_c = y_s M_s \tag{10}$$

so that the maximum strain is straightforwardly defined by

$$\lambda_m = y_s M_s / (y_s^{1/2} M_s) = y_s^{1/2}. \tag{11}$$

3.2 *The modulus*

It is reasonable to suspect that the molecular conformations of the statistical segments should depend on the degree of elongation: the statistical segment is not at all stretching-invariant. It is relevant that different physical quantities, like entropy-changes or optical anisotropy, may be influenced therewith in a different manner. Describing the stress-strain behavior, it is evident that we have to know the total change of the strain-energy-relevant entropy of deformation, including, of course, every modification of the "diffusive entropy components per statistical segment". Since it is possible to fit the experimental stress-strain curves by using a constant modulus, it is possible to define the molecular-weight of the "stretching invariant unit", M_o, by writing

$$\xi / M_c = 1/M_o \tag{12}$$

so that the modulus defined in Equation (6) can be cast into the form

$$G_o = \xi(\varrho RT / y_s M_c) = \varrho RT / \lambda_m^2 / M_o. \tag{13}$$

The parameter ξ gives a number for characterizing the size of the stretching invariant unit which is likely to be different for each experimental method [20]. In the present case, of describing the stress-strain pattern, M_o defines the smallest unit within the chains that operates on deformation, so as not to change its conformational entropy. Anticipating the later finding that ξ is mostly found to be larger than 1, it is evident that:

The average conformational entropy per statistical segment is, in general, found to be diminished ($M_o < M_s$) on deformation, in unique dependence on the macroscopic strain parameter λ.

When drawing from the fit of calculations to the relative course of a stress-strain curve the size of λ_m, it is then interesting to make sure that the modulus is correctly calculated, with the aid of Equation (13) (provided that the molecular weight of the stretching invariant unit is known). Within the limits of experimental confidence, this was shown to hold true for a set of rubbers with different degrees of crosslinking (Fig. 2) [19].

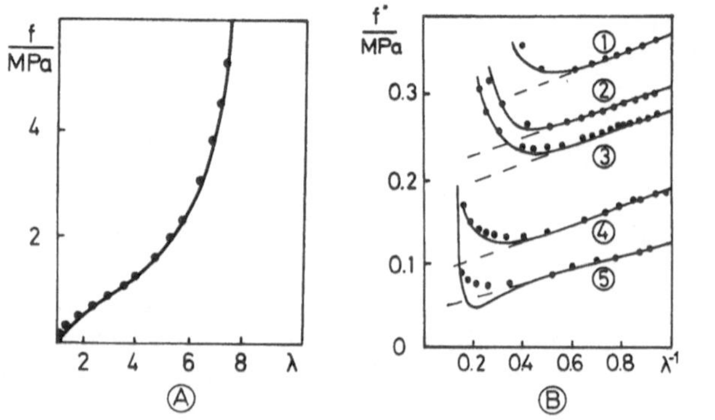

Fig. 2. (A) Stress-strain data of natural rubber in the mode of simple extension at room temperature, according to Treloar [1]. The solid line was computed with the aid of Equations (8) and Equation (13) using the parameters $\varrho = 9 \times 10^2$ kg m^{-3}, $T = 295$ K, $M_o = 68$ kg mol^{-1}, $\lambda_m = 9.7$, $a = 0.19$; (B) Mooney plot of experimental stress-strain data at room temperature for DCP-crosslinked natural rubber [52]. The solid lines are theoretical, according to [19], computed with the aid of the parameters $G_{oo} = \varrho RT/M_o = 36$ MPa ($M_o = 68$ g mol^{-1}); $a = 0.3$: (1) $\lambda_m = 11.5$, (2) $\lambda_m = 9.5$, (3) $\lambda_m = 7.8$, (4) $\lambda_m = 7.1$, (5) $\lambda_m = 6.5$

3.3 *The global interaction*

The quantitative fit to the experimental data can only be achieved by adjusting the van der Waals interaction paramter a to a definite, stretching invariant value. The van der Waals interaction term was recently shown likely to originate with the crosslink's fluctuations [15, 21–22]. To assign this parameter to positive values is necessary for bringing f^* below the constant value, as predicted by the Gaussian network theory (see Figs. 1 and 2).

3.4 *Multiaxial deformation modes*

Multi-axial deformation modes can be described with the aid of an extended van der Waals-theory [4, 9, 21] by only using the "elementary set of network parameters"

λ_m, a: global parameters
M_o: molecular parameter
ϱ: phenomenological continuums parameter.

It is evident here that real networks can uniquely be treated as "weakly interacting van der Waals conformational gas networks" [4, 9]. The maximum chain extensibility is, in all circumstances, related to the chain length parameter, y_s, in the most general case, according to [21]

$$\sum_{\nu=1}^{3} (\lambda_{m\nu})^2 = y_s \tag{14}$$

where the $\lambda_{m\nu}$ are the orthogonal components of the maximum strain vector.

It is interesting to remark on the finding that the interaction parameter a has been proven to be mode-independent. This strongly supports the idea that the chain's interactions originated with the crosslinks' fluctuations, uniquely bound to the global state of deformation as seems to be well described within the frame-work of the extended van der Waals theory [21].

4. The liquid-like properties

Having the isothermal stress-strain pattern as a nearly perfect mirror of global phenomena, is deeply related to finding liquid-like properties on the "local molecular level" of rubbers. Due to the Platzwechsel of chain segments, internal equilibrium is always established. Intrinsic properties on the local level of real networks should thus be transformed, in order to always satisfy the thermoelastic equations of state under constant temperature and pressure conditions [1, 2, 5, 8, 25, 26]

$$f = f_h + f_s$$

$$f_s = -T(dS/dL)_{T,p} = (dQ/dL)_{T,p}$$

$$f_s = f - T(df/dT)_{L,p}$$

$$f_h = (dH/dL)_{T,p} = T(df/dT)_{L,p}\,. \tag{15}$$

Clearly, the total force f is not predicted to depend on the "thermo-elastic term" $T(df/dT)_{L,p}$ according to which the elastic and thermal properties are uniquely correlated. In our experience, it is in fact sufficient for a full understanding of the thermo-elastic properties of rubbers when, according to the thermoelastic van der Waals equations of state, the thermal coefficients of the isotropic state of matter are used. This is demonstrated by documentary evidence from the plots drawn in Figure 3.

The generation of holes and anharmonic effects in the vibrational motions are the reasons behind the thermoelastic inversion in $Q(\lambda)$. Due to these effects,

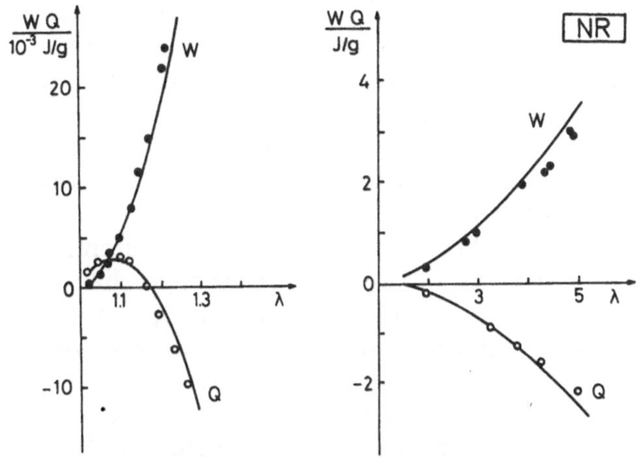

Fig. 3. The strain-energy $W(\lambda)$ and the heat exchanged $Q(\lambda)$ for natural rubber in the mode of simple extension under quasi-isothermal conditions, according to Godovsky [25]. The solid lines were computed [5] with the aid of Equation (15), using the parameters: $G_{oo} = 36$ MPa ($M_o = 68$ gmol^{-1}); $\lambda_m = 10$; $a = 0.2$; the linear thermal expansion coefficient in the isotropic state of matter: $\beta = 2.2\ 10^{-4}$ K^{-1}; the temperature coefficient of the rotational isomers, $\mu = 9.6\ 10^{-4}$ K^{-1}

entropy is produced, always satisfying the symmetry condition of internal freedoms or of "saturated" components $(\Delta\mu^{(i)}) = 0)$ [5, 46]

$$(dH^{(i)}/dL)_{T,p} = T(dS^{(i)}/dL)_{T,p} \tag{15a}$$

where $H^{(i)}$ and $S^{(i)}$ are the enthalpy and the entropy of the i-th of the orthogonal internal freedoms.

In the small-strain range, these effects are correctly predicted to overpower strongly the conformational

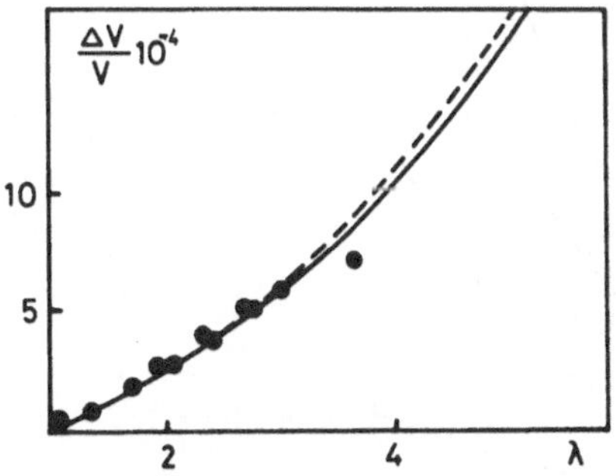

Fig. 4. The relative volume increase in simple extension measured by Göritz [27]. The solid line was computed with the aid of the van der Waals theory on the use of the equations as derived in [5], employing the parameters: $G_{oo} = 36$ MPa ($M_o = 68$ g mol^{-1}); $\lambda_m = 10$; $a = 0.18$; the isothermal compressibility: $\varkappa = 2.5\ 10^{-8}$ m^2 kg^{-1}. The dotted line was computed with the aid of the Gaussian network model

entropy in the network. At the largest extensions, on the other hand, the loss of "global conformational entropy" widely determines the heat exchanged during quasi-isothermal and quasi-static extension. It is satisfactory that the relative strain-induced volume increase, as measured by Göritz [27] can also be calculated fairly well (see Fig. 4).

Based on these results, we are able to summarize:

A full description of deformation is possible if the rubber is treated as an elastic liquid; the global properties of which are characterized as a weakly interacting "conformational gas network". This is why the characterization on the global level can be completed by defining two parameters only: the maximum strain parameter λ_m and the interaction parameter a.

In contrast to the behaviour of the Gaussian network, the global deformation phenomena of van der Waals networks are found to be related to structural properties of the network, like the mean chain length or functionality and mass of the crosslinks. Finite chain extensibility has the consequence that the elastic response depends, in fact, on the shape into which the network is enforced. For this reason, every real network displays, a sophisticated non-linear stress-strain behaviour when the shape is altered during deformation. It is indeed a substantial proof of the concepts behind the van der Waals model of networks that these phenomena can be rationalized within the limits of its generalized formulation [21, 47].

5. Filled rubbers

Now turning to rubbers filled with solid colloid particles we have to account for the heterogeneous stress-strain behaviour. Under quasi-static conditions, every description is simplified as long as there is a unique dependence on the macroscopic strain parameter λ [23, 28–33]. To postulate the existence of constrained states of mechanical equilibrium, we are led straight forwardly to the condition [23, 37].

$$\langle f_{\text{filler}}(\lambda_{fe})\rangle = \langle f(\lambda_r)\rangle; \; \lambda_{fe} \ll \lambda_r \tag{16}$$

where λ_{fe} or λ_r is the elastic strain within that phase which is indicated by the index. It should be noted that the mean strain, $\lambda_f = \lambda_{fe}\lambda_p$, that the solid filler particle's ensemble is submitted to, is mainly represented by larger "plastic components" $(\lambda_p - 1) \gg (\lambda_{fe} - 1)$ brought about by squeezing the colloid-particle ensemble into anisometric shapes. In general, one makes the soft matrix "overdrawn", so as to let its intrinsic strain exceed the macroscopic degree of extension $(\lambda_r > \lambda)$.

Since network elasticity is a global phenomenon, one immmediately realizes that global ways of cooperation should become active, by means of which the polymer matrix is elastically coupled with the filler particle's ensemble (Fig. 5). Every "reinforcement" observed in the first stretch should inform us in detail about the ways in which this cooperation actually runs and how the set of constraints is manipulated. Hope is engendered at the arrival at an understanding of the

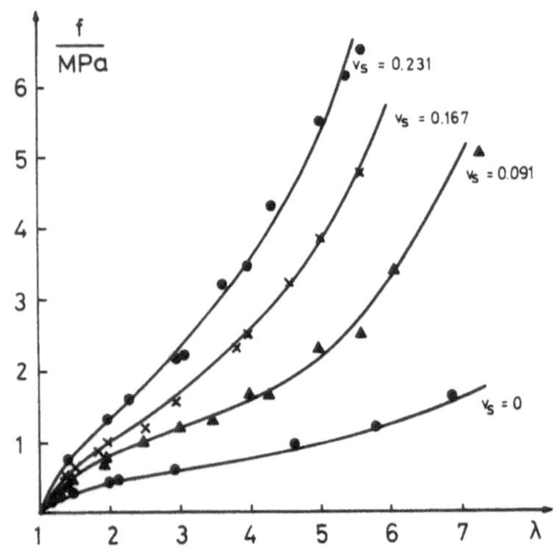

Fig. 5. The first quasi-static stress-strain curves for TiO_2-filled natural rubber at room temperature, according to Becker and Rademacher [43]. The filler volume fraction is indicated at each of these curves

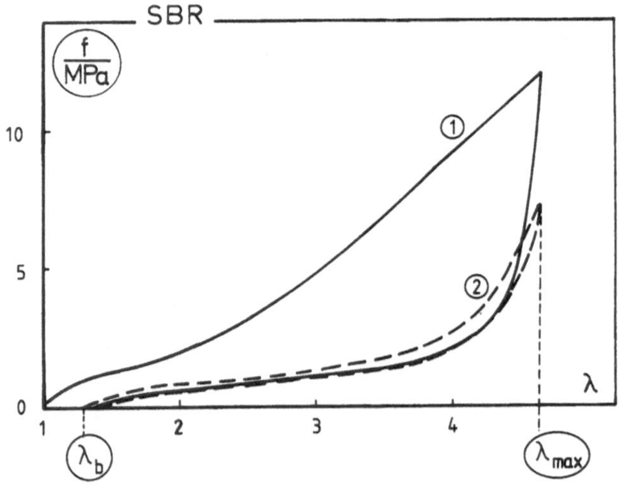

Fig. 6. Quais-static stress-strain cycles of SBR filled with carbon black (v = 0.217) performed under a constant strain rate ($d\lambda/dt$ = 3 10^{-3} s^{-1}) at room temperature. λ_{max} is the maximum macroscopic strain achieved while λ_b characterizes the remnant strain

quasi-static irreversible phenomena which happen to occur in stress-strain cycles (Fig. 6).

Under the given circumstances, the stress strain pattern of the composites is fully understandable when the mechanical equation of state of the rubber is known, as well as the manner in which the filler particle's ensemble is transformed, so that it can be determined how λ_r depends exactly on the macroscopic strain.

Also, to identify the role of the filler-to-matrix contacts (adhesion or chemical bonds) we may study the two limiting models:

Filled rubber wherein the filler-to-matrix contacts were made by adhesion

and

The filler-network that is constituted by permanent filler-to-matix contacts alone, thus being characterized by the existence of a set of large and heavy multifunctional cross-link "bunches"

A two-dimensional sketch of these limiting models is drawn in Figure 7.

What makes every description much more difficult is the effect of the so-called Mullins softening [34–36], according to which we find the first stress-strain cycle marked in a typical way by a very pronounced hysteresis. A representative experiment is shown in Figure 6.

We will now describe the deformation phenomena within these different composites, in turn, starting with the analytic formulation of the intrinsic strain within the rubber matrix.

5.1 The intrinsic deformation

On stretching, the configuration of the filler particle's ensemble will be brought into anisometric shapes, as is schematically illustrated in Figure 8. It is therefore exactly the way in which the filler particle's ensemble is transformed, which determines how the intrinsic rubber matrix strain depends on the macroscopic extension parameter λ.

A molecular statistical modelling of how these processes may work was given by Bueche [31, 36]. This transformation law, the "Bueche mode", turns out to be realized under certain circumstances only (see later). A more general treatment of the geometric contribution to the macroscopic strain, originating with a transformation of the shape of the fillller particle's ensemble into anisometric forms is, to our knowledge, not explicitly discussed in the literature.

As a result of not knowing the mechanism of deformation in rubbery composites, let us firstly formulate in more general terms how the intrinsic strain within the rubbery matrix is related to the macroscopic strain. We are thus led to the relationship [23, 28–29]

$$\lambda_r = (\lambda - v^{1/3}\lambda_f)/(1 - v^{1/3}) \quad (17)$$

where λ_f is the "strain parameter" which de facto only originates with the plastic components that are necessary for squeezing the filler-particle's ensemble into anisometric shape. v is the volume fraction of the filler.

Two limiting cases can be straightforwardly formulated:

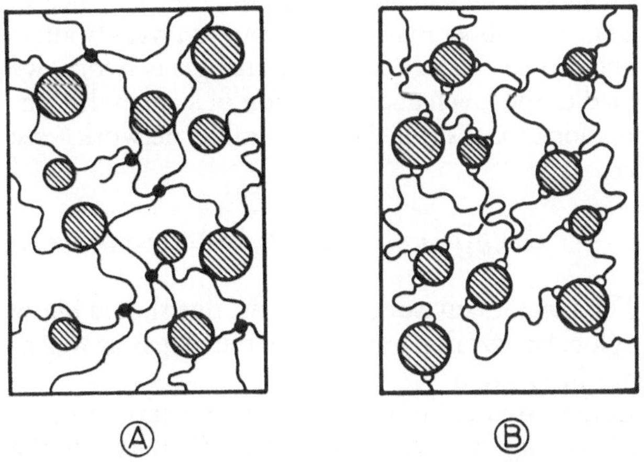

Fig. 7. Schematic representations of (A) a filled rubber and (B) a filler network

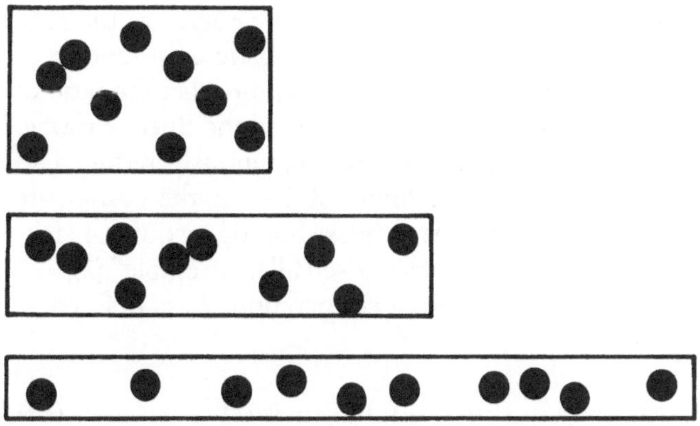

Fig. 8. Two-dimensional sketch of the strain-induced transformation of the filler-particle's ensemble shape in the mode of simple extension

The "elastic Bueche mode"

With $\lambda_f = 1$, we are in the totally elastic heterogeneous "Bueche mode" of deformation

$$\lambda_r = (\lambda - v^{1/3})/(1 - v^{1/3}) \tag{18}$$

wherein the transformation of the geometric shape of the filler particle's ensemble has to occur such as to always keep the sum of the projections of colloid particles into the three coordinate directions unchanged, whereby every overlapping must be carefully accounted for. The "overdrawing" of the matrix is maximum.

The "plastic filler mode"

The other limited way of transforming the composite is to postulate that the transformation of the whole composite should obey the law of an affine transformation. Introducing $\lambda_f = \lambda_r$ one derives, from Equation (16) that both the colloid phases should then be, on average, strained to the same degree, so that we are led to $\lambda_f = \lambda_r = \lambda$: this corresponds to the minimum strain-energy mode due to the "affine plastic transformation" of the whole filler particle's ensemble.

It is now interesting that both the above deformation modes do not deliver a satisfactory interpretation of experimental data obtained in the first stretch.

For filled rubber, the following model was found to give reliable results [23, 28–29]:

The "filled rubber mode"

The deformation of the filler particle's ensemble is formally described by defining two fractions, the first of which is assumed to be in the elastic Bueche mode, the other one strained to the same degree as the rubbery matrix λ_r, thus including plastic components of deformation. It is now a crucial point that the ratio of these fractions changes. Starting in the Bueche mode, its fraction should become continuously reduced so that in the heuristic limits of $\lambda \to \infty$ the composite should be submitted to a quasi-homogeneous transformation with a "fully plastizised filler ensemble" ($\lambda_{rr} = \lambda = \lambda_{fe}$). In this mode of cooperation, λ_{rr} is defined as given by

$$\lambda_{rr} = (\lambda - u_r)/(1 - u_r)\colon\ u_r = (v/\lambda)^{1/3}\,. \tag{19}$$

When stretching the sample for the first time, the reinforcement is predicted to become diminished. During the first stretch a set of not explicitly known constraints are thus assumed to become constantly modified.

The "filler network mode"

Filler networks are found to display an inverse strain dependence in the "first stretch reinforcement", known to be fairly well interpreted by the relation [37]

$$\lambda_{rf} = (\lambda - u_f)/(1 - u_f)\colon\ u_f = \{(\lambda - 1)/(\lambda_m - 1)v\}^{1/3}\,. \tag{20}$$

Starting at the smallest strains with an affine overall-transformation, reinforcement should continuously come into existence when the system is brought to larger strains. This phenomenon is believed to originate with constraints by which diffusive displacements of the filler particles become more and more restricted. This is, of course, the consequence of having the filler particles tightly fixed to the polymer matrix.

To discuss the deformation in composites in the manner as presented, implies an interesting view that at large deformations the global mechanism should be regulated by a λ-dependent modification of a system of constraints, which operates so as to have its topological features determined by the kind of filler-to-matrix contacts.

5.2 The two-phase model

The two-phase model is bound by the simple assumption that the rubbery matrix behaves as "normal" even in the presence of the filler, everywhere showing a homogeneous uniaxial force field. Hence, in treating the rubber matrix behaviour as a weakly interacting van der Waals conformational gas network, we are led to the mechanical equation of state in the "cooperation mode" i (filled rubber or filler network – see later)

$$f = G_o D(\lambda_{ri})\{1/(1 - D(\lambda_{ri})/D_{mi}) - a_i D(\lambda_{ri})\}\,. \tag{21}$$

For a quantitative understanding, one should know the global network parameters of the matrix as well as its intrinsic strain.

In the case of filler networks, the maximum strain parameter $\lambda_{mi} \equiv \lambda_{mf}$ is straightforwardly related to the mean distance between the colloid particles [23], $a(R)$, defined as in reciprocal dependence on the surface density of permanent filler-to-matrix contacts [15]; the maximum strain parameter depends on the filler's volume fraction and the radius of the filler particles R, according to

$$\lambda_{mf}^2 = 2R\alpha(R)(1 - v^{1/3})/v^{1/3} \,. \qquad (22)$$

In the case of the filled rubbers λ_{mr} is obtained from the degree of crosslinking within the polymer matrix [15].

It has been discussed elsewhere [15, 22, 23, 37] that the interaction parameter a should disappear for non-fluctuating crosslinks. The filler network should therefore be characterized by

$$a_f = 0 \,. \qquad (23)$$

Hence, we are led to the formulation of the mechanical equations of state for both of the limited models:

Filled rubber

$$f = G_o D(\lambda_{rr}) \{1/(1 - D(\lambda_{rr})/D_{mr}) - a_r D(\lambda_{rr})\} \qquad (24a)$$

$D_{mr} = \lambda_{mr} - \lambda_{mr}^{-2}$;
Parameters: λ_{mr}; a_r; M_o

Filler network

$$f = G_o D(\lambda_{rf})/(1 - D(\lambda_{rf})/D_{mf}) \qquad (24b)$$

$D_{mf} = \lambda_{mf} - \lambda_{mf}^{-2}$
$\lambda_{mf} = \{2R\alpha(R)(1 - v^{1/3})/v^{1/3}\}^{1/2}$
Parameters: $\alpha(R)$; M_o .

It is thus taken for granted that different cooperation on the global level is made effective with different filler-to-matrix contacts (like adhesion or chemical bonds). The deformation behavior of both systems is strongly related to the possibilities of how the filler particle's ensemble is enforced into anisometric shapes.

The heavy filler particles operate as multi-functional non-fluctuating crosslinks. With $a_f = 0$, filler networks should exhibit a stress-strain pattern which differs, in principle, from what is known for filled rubbers: filled to the same degree with identical particles and in addition brought to the same density of permanent crosslinks, the first stretch reinforcement is predicted to become "inverse" for both the limiting model networks. In the Mooney plot, this manifests itself by a crossing over, as is seen in Figure 9.

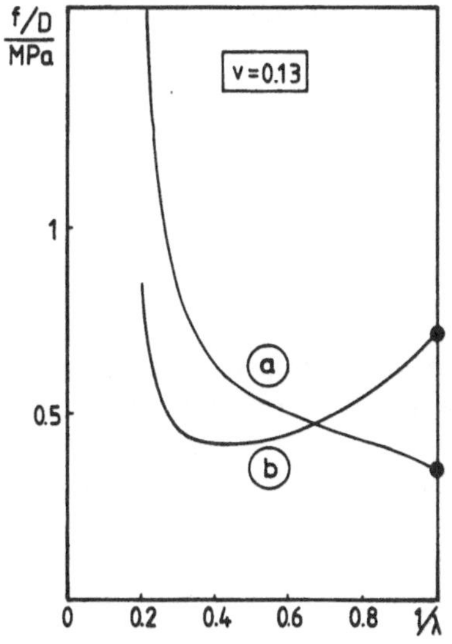

Fig. 9. The representation of (b) filled rubbers and (a) filler networks in the Mooney plot, each of these pairs showing the same density of permanent crosslinks [23]. The parameters: $G_{oo} = 35$ MPa ($M_o = 66$ g mol^{-1}); $2R\alpha(R) = 120$; $G_v = G_o\lambda_{mrf}^{-2}$: filled rubber: $G = G_v(1 - v^{1/3})^{-2}$; $a = 0.2$; filler network: $G = G_v$; $a_f = 0$

The apparent modulus derived from the stress-strain curve is very different in both cases. The difference arises from the fact that at the smallest strains filler networks display no reinforcement at all, while filled rubbers should operate in the Bueche mode.

5.3 The filler-network-rubber

A new type of network is obtained when additional filler-to-matrix bonds are made in a filled rubber. A sketch of such a "filler-network-rubber" is shown in Figure 10. The simplest way of describing the

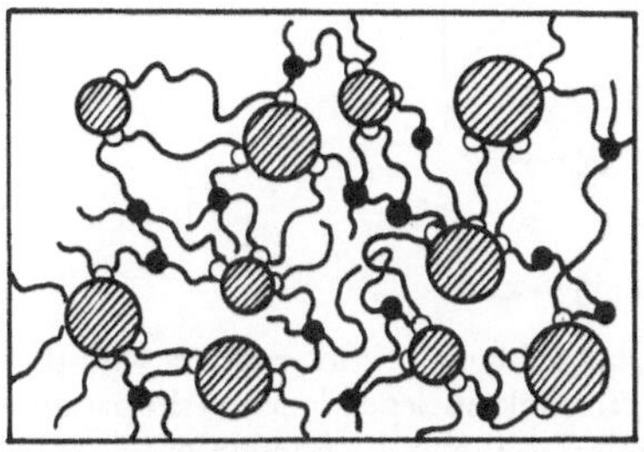

Fig. 10. The filler-network-rubber scheme

intrinsic strain in the rubbery matrix is given here by [37]

$$u_{rf} = \{xu_r^3 + (1-x)\, u_f^3\}^{1/3} \tag{25}$$

where

$$x = 1/\{1 + \varkappa(\lambda_m/\lambda_{mf})^2\} \tag{26}$$

is the relative fraction of the matrix bonds, simply characterized with the aid of the normalized parameter $\varkappa$ ($0 \leq \varkappa \leq 1$).

The density of the totality of crosslinks may be written as

$$\nu = \nu_r + \nu_f \tag{27}$$

and we are led to

$$\lambda_{mrf}^{\ -2} = \lambda_m^{-2} + \varkappa\lambda_{mf}^{\ -2}\,. \tag{28}$$

Bearing in mind that $a_f = 0$, the mean interaction parameter is given by

$$\langle a\rangle = xa_r\,. \tag{29}$$

It is significant to consider the physical idea behind defining the total degree of crosslinking as was done in Equation (27). "Tie-molecules" acting between filler particles are believed to come into existence with each of the filler-to-matrix bonds. These tie-molecules give a concrete picture of a possibly more complex cooperation, which leads to additional energy-equivalent subsystems of deformation. Each one of these new subsystems is assumed to comprise numbers of chains which themselves operate as energy-equivalent chains in the rubber matrix.

The apparent moduli of filler-network-rubbers with the same density of permanent crosslinks were predicted to vary strongly with the fraction $\varkappa$ of the tie-molecules, as is shown in Figure 11. When all of the composites compared have the same density of permanent crosslinks, the Mooney plots of filler-network-rubbers fall, in principle, between the limits fixed by the filled rubber and the filler network (Fig. 12).

5.4 *The Einstein-Smallwood effect*

The polymer matrix permanently adhered to the solid colloid particles is equivalent to, as expected, a defined distortion of the mean matrix field around each of the filler particles (Fig. 13). In the mean-field approach one is thus led to the "Einstein-Smallwood reinforcement", described in a very universal manner by [38–40]

$$G_f = G_o\,(1 + Cv) \tag{30}$$

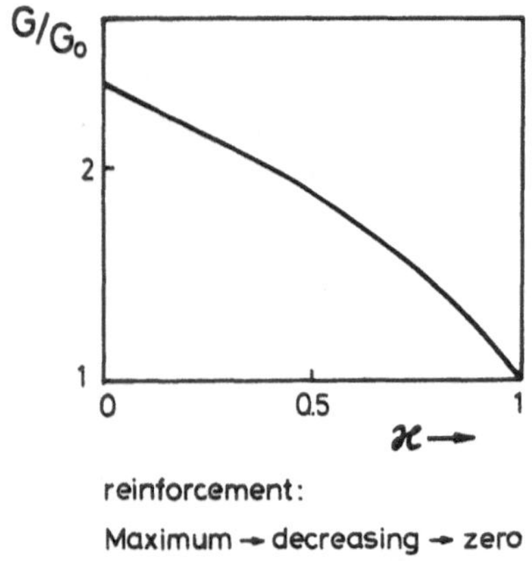

Fig. 11. The apparent modulus of filler-network-rubbers with the same number of permanent crosslinks in dependence on the parameter $\varkappa$ that is, de facto, identical to the molar fraction of filler-to-matrix bonds [37]

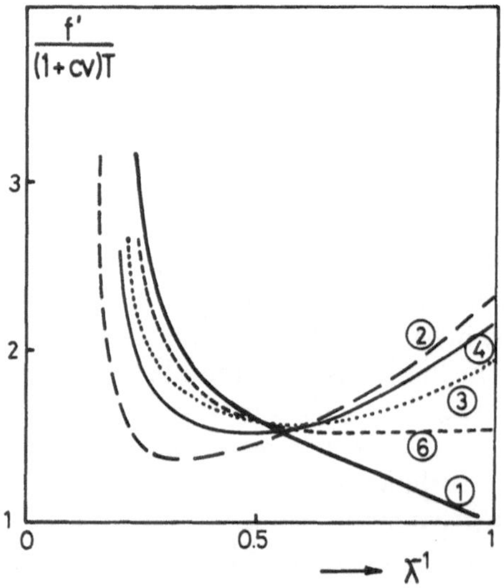

Fig. 12. Mooney plot of the different network models, all showing the same number of permanent crosslinks [37] ① $\nu = 0$, $\lambda_m = 10$, $\varkappa = 0$; ② $\nu = 0.2$, $\lambda_m = 12.3$, $\varkappa = 0$; ③ $\nu = 0.145$, $\lambda_m = 10.3$, $\varkappa = 1$; ④ $\nu = 0.115$, λ_m 10, $\varkappa = 1$; ⑥ $\nu = 0.20$ g, $\lambda_m = 20$, $\varkappa = 0.5$

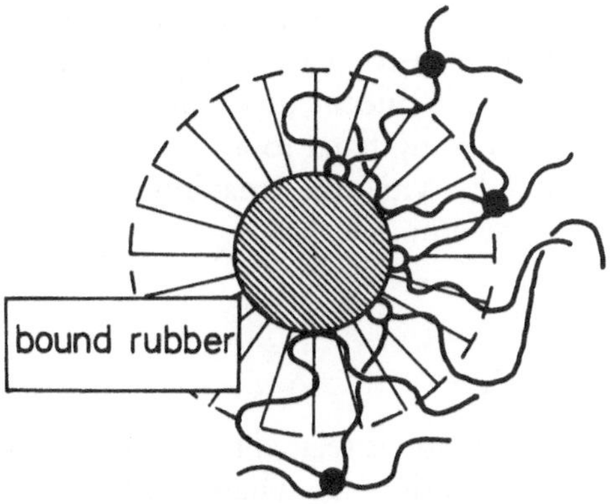

Fig. 13. Scheme of the interfacial layers with which the filler particles are assumed to be surrounded

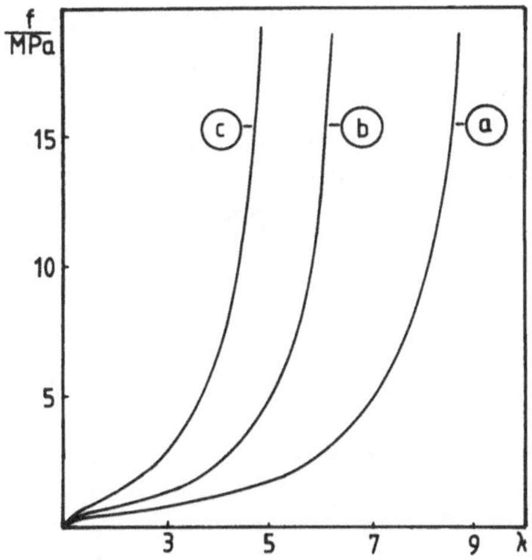

Fig. 14. Illustration of the Einstein-Smallwood-effect in the first quasi-static stretch of a filled rubber computed with the parameters [23]: $G_{oo} = 35$ MPa ($M_o = 66$ g mol^{-1}); $\lambda_m = 10$; $a = 0.2$: curve (a) $v = 0$; curve (b) $v = 0.2$, $C = 0$; curve (c) $v = 0.2$, $C = 2.5$

where for spherical filler particles the universal Einstein-coefficient C is equal to 2.5 independent of the size of the filler particles.

The storage properties of the network must become modified, according to Equation (30). With the aid of Equation (12), we are therefore led to

$$\lambda_{mrfv} = \lambda_{mrf}\,(1 + Cv)^{-1/2} \tag{31}$$

clearly indicating that the mean asymptotic global properties within the composite were altered in the presence of the solid colloid particles.
The mean maximum strain parameter characterizing the first stretch of filled rubber systems is predicted to be shifted to exceedingly small numbers when the volume fraction of the filler rises to higher values.

The order of magnitude of these Einstein-Smallwood reinforcement effects can be seen by evidence from the plot drawn in Figure 14. The reliability of this model can therefore be easily checked: the upturn in the stress-strain pattern of filled rubber systems should, in general, be shifted in the first stretch to increasingly lower macroscopical strains when the filler volume fraction rises to higher values.

5.4 *The mullins softening*

The Mullins softening should basically be related to global effects. This is strongly suggested by the observation that the stress-strain cycles after the first stretch display nearly the same relaxation as observed for unfilled rubbers [28, 42].

Bueche [31, 36] has presented a molecular interpretation by discussing the strain-induced modification of the polymer-matrix particle contacts, while Mullins himself discussed the phenomenon in terms of a phenomenological approach [32–33]. Rigbi [30] developed a somewhat different model.

(i) Our interpretation of the Mullins softening (see Fig. 6) should now be based on postulating two irreversible effets.

(ii) On shrinking, the configuration of the filler-particle's assembly cannot be recovered by the operation of the weak entropy-elastic forces within a rubbery matrix.

The "Einstein-Smallwood interfacial phenomena" should only become enforced during the first stretch.

After the first stretch, up to the macroscopic strain λ_{max}, the intrinsic strain on the rubbery matrix should consequenty be determined by [29, 37]

$$\lambda_{r\,max} = (\lambda - \varepsilon_{bb} - u_{max})/(1 - u_{max})$$

$$\varepsilon_{bb} = (\lambda_b - 1)(\lambda_{max} - \lambda)/(\lambda_{max} - (\lambda_b + 1)) \tag{32}$$

with

$$u_{max} = (xu_{r\,max} + (1 - x)u_{f\,max})^{1/3} \tag{33}$$

$$u_{r\,max} = v/\lambda_{max}\,;\; u_{f\,max} = (\lambda_{max} - 1)/(\lambda_{m\,max} - 1)_v\,. \tag{34}$$

The remaining strain is accounted for by the empirical parameter λ_b, the physics of which are not yet known exactly.

To have modifications of the constraints enforced only during the first stretch, is the reason for postulating that the Einstein-Smallwood phenomena should be irreversible too. This is phenomenologically described by

$$C_\lambda = C_{exp}\{-\varkappa_c(\lambda_{max} - \lambda)\} \tag{35}$$

with $\varkappa_c$ as an adjustable parameter. During the first steps of shrinking, descending and rapidly decreasing relaxations are thus allowed to run off within the interfacial layers, so that we are also led to

$$\lambda_{m\,max} = \lambda_{mrf}\,(1 + C_\lambda v)^{-1/2} \tag{36}$$

with

$$\lim_{\lambda \to 1} \lambda_{m\,max} = \lambda_{mrf}\,. \tag{37}$$

This formulation accounts for the state of matter within the interfacial layers becoming frozen in, so that substantial amounts of strain-energy may be stored within the composite.

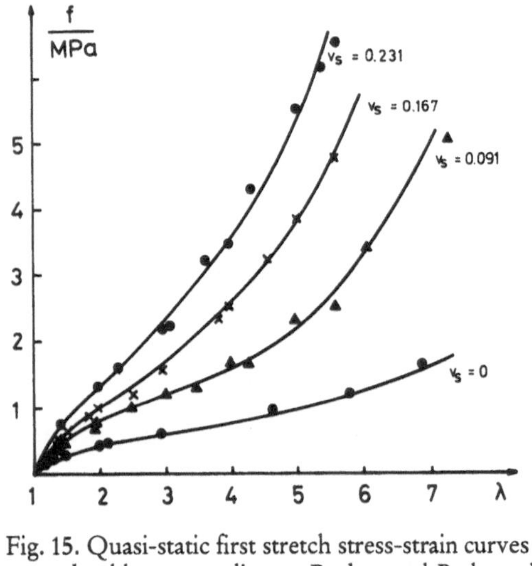

Fig. 15. Quasi-static first stretch stress-strain curves of TiO_2-filled natural rubber, according to Becker and Rademacher [43]: The solid lines were computed with the aid of the van der Waals equation of state (Eq. (24)), under the parameters: $G_o = 72$ MPa ($M_o = 34$ g mol^{-1}): this is an extraordinarily small value, exactly half of that which is usually determined (see other representations in this paper; the reason behind this discrepancy is unknown); $a = 0.16$; $\lambda_m = 15$; $C = 3$; the volume fractions of the filler as indicated at each of the curves

According to the model introduced, composites, once brought to the macroscopic strain λ_{max} should for $\lambda < \lambda_{max}$ run in the elastic Bueche mode, the apparent filler volume fraction being fixed to a value which is uniquely determined, with the maximum strain arrived at for the first time.

Results and discussion

The following discussion is related to a quantitative fit of calculations to experiments, whereby the reader is asked to draw the more detailed description from the original papers.

It can be seen from evidence that the first quasi-static stress-strain curves of natural rubber filled with TiO_2 (filled-rubber) can be fairly well computed with the set of parameters which is depicted in the caption of Figure 15 [43, 28, 29]. According to the plots drawn in Figure 16 [28, 29, 48], there is apparently no dependence of the maximum strain on the filler particle size, in excellent accord with what is demanded by the Einstein-Smallwood treatment. The filler particle size seems to influence the mean energy storage properties within the composite [28, 29, 52–54]. We found, for carbon black filler, the empirical relationship

$$G_{fv} = G_v\,(1/(1 - \alpha_v v) - \beta_v v^{1/3}) \tag{38}$$

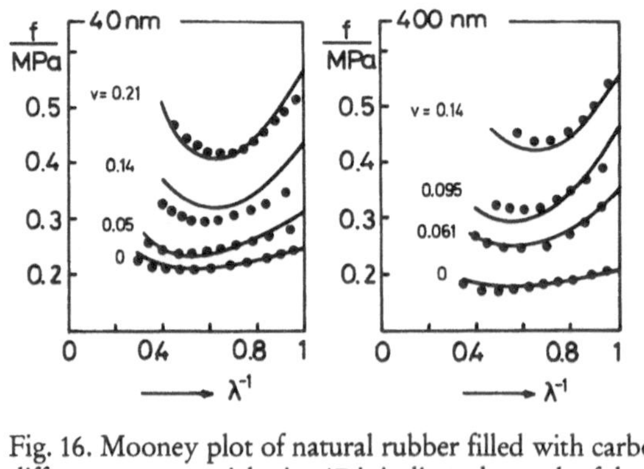

Fig. 16. Mooney plot of natural rubber filled with carbon black of different mean particle size (R is indicated at each of the drawings). The solid lines were computed with the following modified van der Waals equation of state [41]: $f = G_v D(1/(1-\eta) - a\,(\Phi)^{1/2})$; $\Phi = (\lambda^2 + 2/\lambda - 3)/2$; $\Phi_m = \varphi(\lambda_m)$; $\eta = (\Phi/\Phi_m)^{1/2}$ whereby the following set of parameters was used [29]: $G_{vv} = G_v(1/1 - \alpha_v v) - \beta_v v^{1/3})$ implying a unknown cooperative effect which only changes the energy storage properties within the composite: $\alpha_v = -1.36 \log (R/R_o) + 2.68$; $R_o = 1$ nm; $G_{oo} = 32$ MPa ($\varrho = 0.9$ g cm^{-3}; $Mpo = 68$ g mol^{-1}); $a = 0.25$; $C = 2.5$; $R = 40$ nm: $\lambda_m = 8.2$; $\alpha_v = 0.5$; $\beta_v = 0.8$; $R = 400$ nm: $\lambda_m = 9$; $\alpha_v = \beta_v = 0$

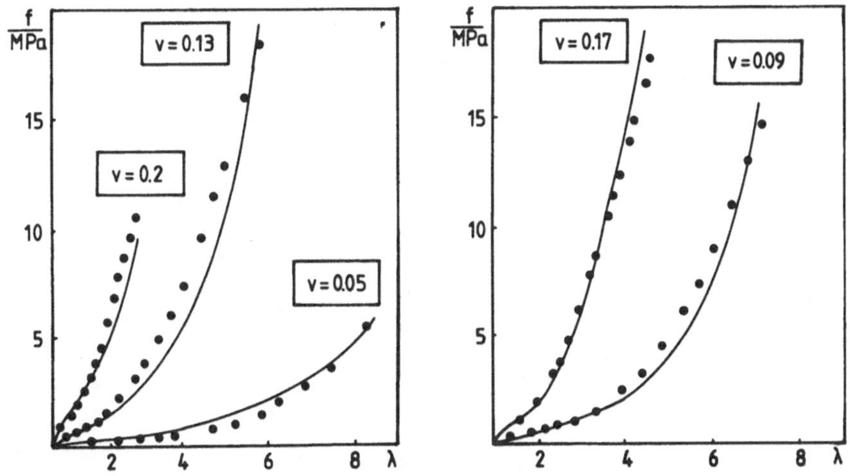

Fig. 17. Experimental stress-strain data of filler networks of natural rubber, filled with precipitated silica measured in the first stretch at room temperature under the constant strain rate of 300 %/min, according to [23]. The volume fraction of the filler is indicated at each of the curves. The solid lines were computed using one of the equations of state defined in Equation (24), employing the parameters: $G_{oo} = 35$ MPa ($M_o = 66$ g mol^{-1}); $2Ra(R) = 120$; $a_f = 0$

whereby the parameter α_v is related to the mean radius of the particles according to

$$\alpha_v = -1.36 \log (R/R_o) + 2.68\colon R_o = 1 \text{ nm}. \quad (39)$$

The physical reason behind this correction term is not yet known. For carbon black it seems to hold true that α_v and β_v are correlated (see Fig. 22) [23].

The filler networks' experimental data can be fitted when the global interaction parameter a_f is assigned the value of zero (Fig. 17) [23]. The elastic response within the rubbery maxtrix is thus found to behave in the first stretch strictly as predicted by the non-Gaussian Langevin theory [1, 23]. In spite of the very different network structure, we nevertheless succeeded in calculating the stress-strain curves of filled-rubbers that have been crosslinked to the same degree as the filler-networks previously discussed by only putting the global interaction parameter to the value of $a_r = 0.2$ (see Fig. 18).

Quasi-static experimental stress-strain cylces of filler-network-rubbers with varying amounts of additional filler-to-matrix contacts are fairly well fitted by our calculations (Fig. 19) [37].

In view of these interpretations, we are led to conclude that:

The conformational abilities of the chains of finite length within networks with non-fluctuating crosslinks, are correctly described by the single phantom-chain network-model. In quantitative respects, this holds true only if "extra entropy", originating with diffusive conformational modes within the statistical segments (y_s), is accounted for.

In having the conformational abilities of real chains in networks with non-fluctuating crosslinks well

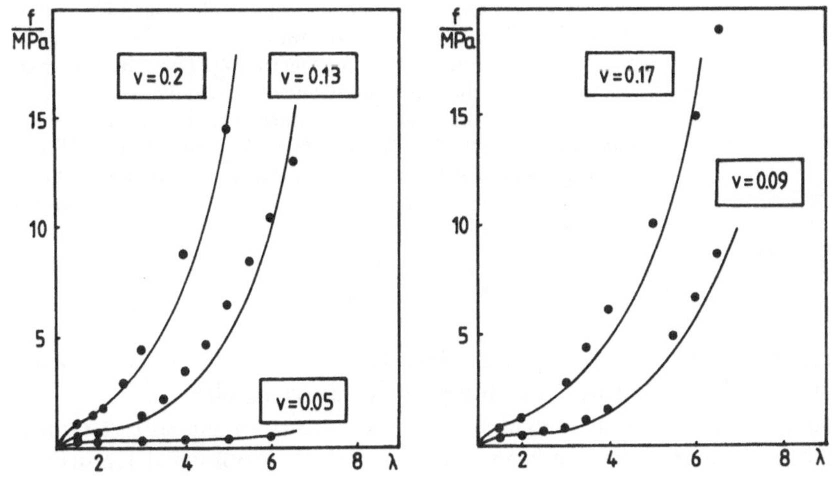

Fig. 18. First stress-strain curves of filled natural rubber filled with precipitated silica measured at room temperature under the constant strain rate of 300 %/min, according to [23]. The filler volume fraction is indicated at each of the plots. The solid lines are theoretical computed using parameters: $G_{oo} = 35$ MPa ($M_o = 66$ g mol^{-1}); the maximum strain was set to be equal to that one of the filler networks, thus computed with the parameter $2Ra(R) = 120$ and $a = 0.2$

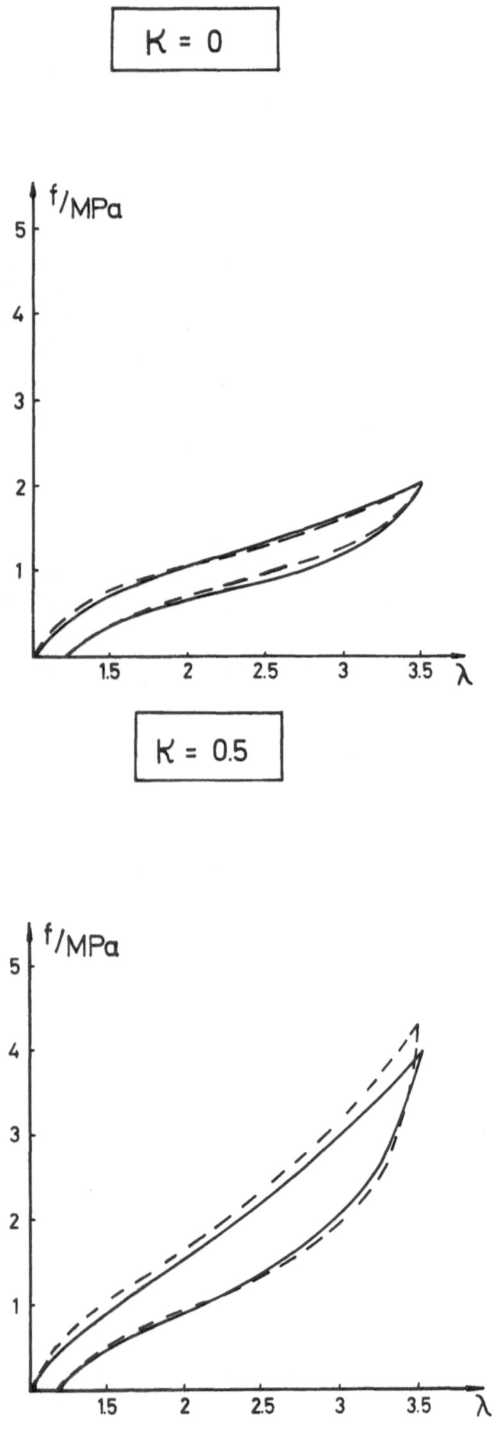

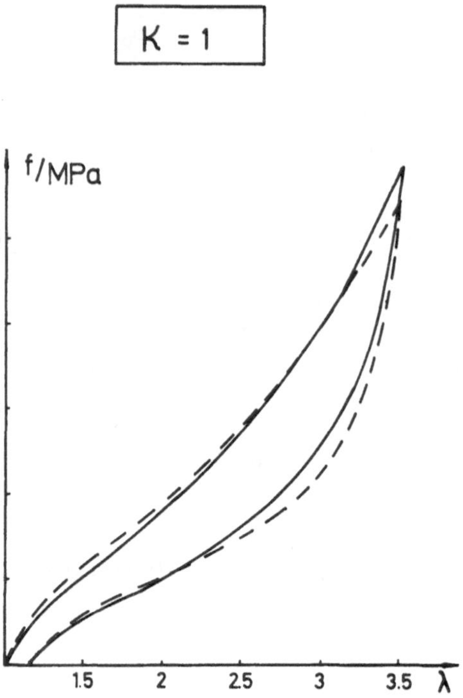

Fig. 19. Quasi-static stress strain cycles of silica-filled PNR-filler-network-rubbers deformed under the constant strain rate of 30 %/min at room temperature, according to [37]. The solid lines were computed with the aid of the adequate van der Waals equation of state employing the equations as defined in section 5.4. The parameters used were: $v = 0.2$, $G_{oo} = 36$ MPa ($M_o = 68$ g mol^{-1}); $2R\alpha(R) = 120$; $\lambda_m = 10$; $a_r = 0.2$; $\varkappa_c = 5$; the parameter $\varkappa$ is indicated at each of the curves; --- experimental; —— theoretical (K is identical with $\varkappa$)

represented with the aid of the phantom chain model, the question is provoked of how exclusion effects operate in densly packed polymer systems: which kind of sophisticated correlations on the molecular and on the global level are necessary for achieving a finer understanding of the above result?

It is evident that the density of the subsystems of deformation may depend on topological features of the network structure. While in filler networks and in filled rubbers the chains themselves act as autonomous subsystems of deformation, additional subsystems seem to come into play in filler-network rubbers (particle-to-particle tie-molecules).

To achieve, within the scope of the van der Waals model, an interpretation of the absolute modulus, there is a need to define the "stretching invariant unit". This unit is, in general, found to be smaller than the statistical segment, due to additional strain-induced entropy effects occuring within the Kuhn segments themselves. If that is shown to be true beyond doubt, the size of the stretching-invariant unit is an important molecular structure factor that affects the size of the modulus dependent on the real molecular situation within the statistical segments.

A cooperative elastic effect seems to exist that modifies the mean modulus for each kind of filler, bringing about, for example particle size effects which have been known to exist for a long time [28–29, 52–54].

Fundamental consequences of this kind can be recognized since the maximum strain parameter in the van der Waals equation of state is straightforwardly obtained from simply fitting calculations to the relative course of the stress-strain measurements. Absolute accord with the experimental data is then possible, within the scope of our approach, by assigning the size of the stretching invariant unit (M_o) and the prefactor parameters (α_v, β_v) to distinct and stretching invariant values. At same point in the future should be explained in molecular detail why the size of the stretching invariant unit is, in general, found to lie in the order of magnitude of about two chain-standing carbon units, or what the reasons really are behind the filler-individual volume-fraction and size-dependent modulus modification.

Cooperation between the rubbery matrix and the solid filler particle ensemble seems to be governed by a set universal phenomena:

The quasi-static response on deformation (obtained by defining a certain limitation in the time scale within which the deformation phenomena were discussed) is always found to be in a well-defined dependence on the macroscopic strain parameter λ.

The composite is heterrogeneously deformed, de facto storing the elastic strain energy within the soft entropy-elastic polymer matrix.

Plastic mechanisms of deformation are necessary for making the filler particle's ensemble into anisometric shapes.

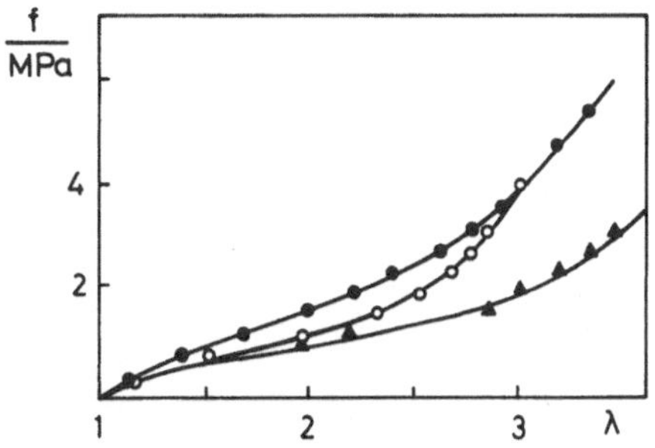

Fig. 20. Softening curves at 30 °C for styrene-butadiene rubber with $v = 0.14$ HAF carbon black as a filler, according to Bueche [36]. The solid lines were computed with the van der Waals-equation of state using the equations as defined in section 5.4: $G_{oo} = 31$ MPa ($M_o = 78$ g mol^{-1}); $C = 2.5$; $\lambda_m = 8.5$; $a = 0.2$; $\varkappa_c = 5$

For all of the composites investigated (filled rubbers, filler networks or filler-network-rubbers) the boundary value problem is well represented with the aid of Einstein-Smallwood's mean field approach. The force field distortions within the composites are analogous in the general sense of Einstein's considerations, not showing any effect with different filler-to-matrix contacts (chemical bonds or contacts by adhesion).

The stress-strain hysteresis should be related to a set of globally effective constraints, the plastic modification of which is made possible only during the first stretch. These effects imply the irreversible transformation of the filler-particle's ensemble, as well as of the "bound-rubber-interlayers" each of the filler particles is embeded.

These last statements are confirmed by the drawings shown in Figures 19 to 21, thus leading to the characterization of the Mullins effect:

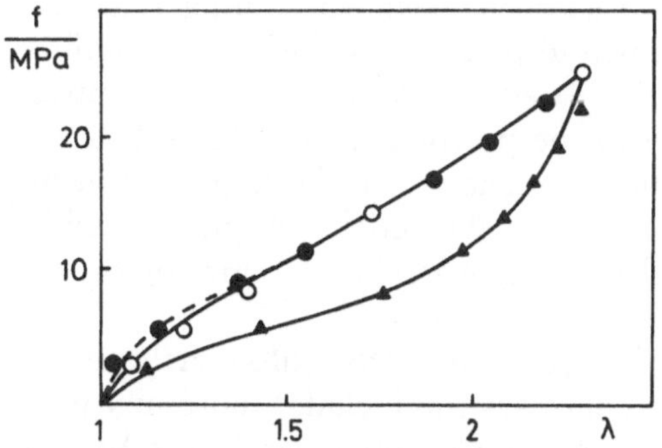

Fig. 21. Softening curves at 34 °C for styrene-butadiene rubber plus Hi Sil-233 ($v = 0.217$) according to Bueche [36]. ● first stretch; ○ data for material after it had recovered for 20 h at 115 °C. The solid lines were computed with the use of the parameters: $G_{oo} = 27$ MPa ($M_o = 90$ g mol^{-1}); $C = 2.5$; $\lambda_m = 8.5$; $a = 0.2$; $\varkappa_c = 5$

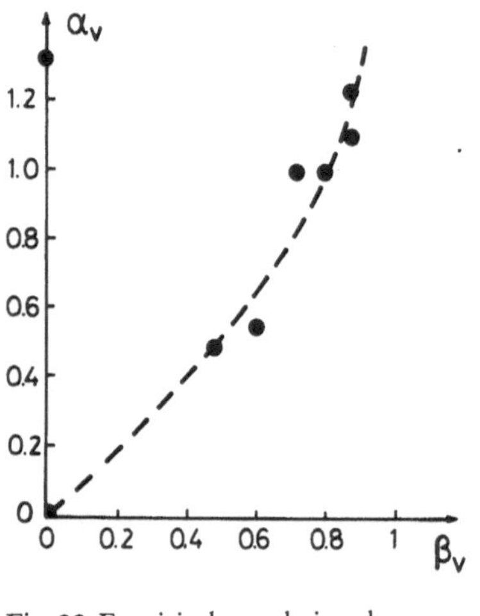

Fig. 22. Empirical correlations between α_v and β_v, according to [28–29]. TiO_2 in the origin, together with the 600 nm carbon black (see Fig. 16), silica filler at the top of the ordinate

The filler-particle's ensemble and the boundary-layers irreversibly reshuffled, is equivalent to postulating a frozen configuration of constraints after the first stretch, to then have an invariant "functional colloid structure" within the composite, invariant as long as subsequent straining is kept below the maximum extension $\lambda < \lambda_{max}$. The apparent fraction of solid fillers fixed to a defined value after the first stretch, these composites afterwards operate at strains $\lambda < \lambda_{max}$ in the mode with an invariant elastic Bueche component.

The energies lost after the first stress-strain cycle are likely to be stored within the filler-particle surrounding layers [42]. This is true only if the inner parts of these layers are in the glassy state. It is, then, the easiest way of coupling the solid Hookian-like filler particles to the "gas-like rubber matrix" (van der Waals gas network with weak interactions), when the interfacial layers are assumed to pass continuously in radial directions from the glassy state of matter into the rubber-elastic body of the polymer matrix [23, 37]. It thus becomes understandable that heating or swelling allows one to restore the initial isotropic state within the interfacial layers, wherein quantities or orientation-entropy were frozen.

It is thus deduced from these results that the universal features of filled systems include "shortrange properties" like those within the interfacial layers. We have to be aware that the strain-induced modification of constraints by which the global energy storage properties of the composite are regulated, may, of course, be coupled to the interlayer's irreversible deformation behavior. These considerations illuminate the interesting "colloid-physical properties" of filled rubber-elastic systems.

Individual global aspects come to light by comparing filled rubbers and filler networks: both of these showing the same density of permanent crosslinks, the energy storage properties found to be inverse, as is illustrated with the Mooney plots in Figure 12. Filler networks start without any small strain reinforcement approaching, in the heuristic limits of largest strains ($\lambda \to \lambda_{mf}$), the fully elastic Bueche mode. Filled rubbers, on the other hand, lose their initial reinforcement on elongation, ending up with the affine transformation at very large strains, thus being brought into the minimum strain-energy mode.

Therefore it is shown that the cooperation mechanism between the filler-particle's ensemble and polymer matrix was individually regulated by the kind of filler-to-matrix contacts as, for example, given by chemical bonds or adhesion.

The global properties of filler network rubbers with the same number of permanent crosslinks can apparently be modified only within the limits fixed by the filler network and by the filled rubber (see Fig. 20).

7. Final remarks

It is shown elsewhere that internal equilibrium can be asumed to be always established when the composites are deformed under quasi-static conditions [42]. An outstanding chance should therefore exist for achieving a finer understanding of intrinsic properties such as strain induced birefringence [1, 6, 20], heat-conductivity and its anisotropy [20] and similar processes behind the electrical conductivity [51]. Moreover, substantial hope is engendered for a successful treatment of relaxation phenomena in filled rubbers [28, 49, 50].

In this paper, empirical relations have often been used, for example, to define the λ-dependency of intrinsic deformation within the softer matrix. The last of these formulations was always held to be in accord with the evident structure-dependent asymptotic mechanical properties of the various systems under discussion. With an extended van der Waals approach and empirical prefactor modifications, a full description could be managed, illuminating through evidence surprisingly universal features of the heterogeneous quasi-static deformation behavior of filled rubbery systems.

Interesting new information was contributed on a very complicated scientific matter that might facilitate any new attempt at seeking a fuller understanding by a molecular-statistical theory.

Acknowledgements

We would like to thank the Deutsche Forschungsgemeinschaft, the Arbeitsgemeinschaft Industrieforschung (AIF) and the Deutsche Kautschuk-Gesellschaft, the companies Bayer AG and Degussa for generous support and promotion.

References

1. Treloar LRG (ed) (1975) The Physics of Rubber Elasticity, 3rd Ed, Clarendon Press, Oxford
2. Flory PJ (ed) (1953) Principles of Polymer Chemsitry, Cornell University Press, Ithaca
3. de Gennes PG (ed) (1979) Scaling Concepts in Polymer Physics, Cornell University Press, Ithaca
4. Kilian H-G (1981) Polymer 22:209
5. Kilian H-G (1982) Coll & Polym Sci 260:895
6. Kuhn W, Grün F (1942) Kolloid-Z 101:248
7. Kilian H-G (1981) Coll & Polym Sci 259:1084
8. Kilian H-G, Vilgis Th (1984) Makromol Chem 185:193
9. Kilian H-G (1979) Phys Bl 12:641
10. Green AE, Adkins JE (eds) (1970) Large Elastic Deformations, 2rd Ed, Clarendon Press Oxford
11. Graessley WW (1982) Adv Polym Sci 46:67
12. Dusek K (1984) Int Rubber Conf, Moscow
13. Oppermann W, Rehage G (1981) Coll & Polym Sci 259:117
14. Langley NR (1976) Macromolecules 1:348
15. Kilian H-G, Enderle HF, Unseld K (1986) Coll & Polym Sci 264:866
16. Pak H, Flory PJ (1979) J Polym Sci Phys Ed 17:1845
17. Rivlin RS, Saunders DW (1951) Philos Trans R Soc London A 243:251
18. Kilian H-G (1984) Int Rubber Conf, Moscow
19. Kilian H-G, Unseld K (1986) Coll & Polym Sci 264:9
20. Kilian H-G, Pietralla M (1988) in preparation
21. Enderle F, Kilian H-G (1987) Progr Coll & Polym Sci, in press
22. Vilgis Th, Kilian H-G (1986) Coll & Polym Sci 264:137
23. Kilian H-G, Schenk H, Wolff S (1987) Coll & Polym Sci, in press
24. Callen HB (ed) (1959) Thermodynamics, Wiley, Int Ed, New York
25. Godovsky YK (1977) Vysokomol Soed A 19:2359
26. Godovsky YK (1986) Adv Polym Sci 76:31
27. Göritz H (1982) Coll & Polym Sci 260:193
28. Kilian H-G (1986) Gummi, Faser-Kunststoffe 10:548
29. Kilian H-G (1986) Kautsch Gummi Kunstst 39:689
30. Rigbi Z (1980) Adv Polym Sci 36:21
31. Bueche F (1960) J Appl Polym Sci 4:107
32. Mullins L (1956) J Polym Sci 19:225
33. Mullins L (1956) J Polym Sci 19:237
34. Mullins L, Tobin NR (1965) J Appl Polym Sci 9:2993
35. Harwood JAC, Mullins L, Payne AH (1965) J Appl Polym Sci 9:3011
36. Bueche F (1961) J Appl Polym Sci 15:271
37. Kilian H-G, Schenk H, J Appl Polym Sci (1988) submitted publication, in press
38. Einstein A (1906) Ann Phys 19:289
39. Einstein A (1911) Ann Phys 34:581
40. Smallwood HM (1944) J Appl Polym Sci 15:758
41. Enderle HF, Kilian H-G, Vilgis Th (1984) Coll & Polym Sci 262:696
42. Ambacher H, Kilian H-G (1988) to be published
43. Becker GW, Rademacher AJ (1962) J Polym Sci 58:621
44. Mergenthaler, Kilian H-G, Pietralla M (1987) Progr & Coll Polym Sci
46. Kilian H-G, Höhne GWH, Trögele P, Ambacher H (1984) J Polym Sci Polym Symp 77:221
47. Kilian H-G (ed) (1986) German-Chinese Polymer Symposium, Pecking
48. Soos I (1982) Canditate Thesis: Characterization of the Rubber Filler Interaction, Budapest
49. Enderle HF, Kilian H-G, Vilgis Th (1984) Coll & Polym Sci 262:696
50. Enderle HF, Kilian H-G, Heise B, Mayer J, Hespe H (1986) Coll & Polym Sci 264:305
51. Blythe AR (ed) (1979) Electrical Porperties of Polymers, University Press, Cambridge
52. Westlinning H, Wolff S (1966) Kautsch Gummi Kunstst 19:470
53. Wolff S (1974) Kautsch Gummi Kunstst 27:511
54. Wolff S, Pöhnisch H, Hoffmann P (1975) Kautsch Gummi Kunstst 28:379

Received February 11, 1987;
accepted April 21, 1987

Author's address:

Prof. Dr. H.-G. Kilian
Abteilung für Experimentelle Physik
Universität Ulm
Oberer Eselsberg
D-7900 Ulm, F.R.G.

Discussion

GODOVSKY:

During your very interesting lecture, you mentioned a very important role of bounded rubber to the mechanical properties. Could you add something concerning the difference in mechanical properties, especially thermo-mechanical properties, of bounded rubber and the other part of the rubber?

KILIAN:

In the case of bounded rubber, it is primarily an entropy which is stored, having performed the first stretch. It cannot be understood under the assumption that it would be due to friction or dissipation, so I think you would agree that in this stretching calorimeter we are happy to have a device in which we have not to care so much about the dissipative heats developed. They are small by dynamic methods. The internal properties of the bound rubber, as Dr. Vidal has also found, should be proportional to the degree of crosslinking. This influences the properties that we found somewhat. Dr. Pietralla will make a remark, that the heat conductivity also reflects some special properties of the interfacial layers and can give some hints about, what the real structure is. There should be a continuous slope of properties, so that you always have a layer within some distance of the filler surface which is – with respect to the time you investigate it – in the glass transition temperature. But this must be proved much more convincingly: this is still a hypothesis now.

ILAVSKY:
You presented the stress-strain curve, and you can describe it very nicely by the van der Waals equation. I would expect, that for unfilled rubber you can have equilibrium, and some equilibrium stress-strain dependence. But for filled rubbers, I would expect that time dependencies are increasing as the filler content increases. So, I would like to know whether the stress-strain curve depends on the rate, and whether the constants you are using depend on time or not.

KILIAN:
For the relaxation spectrum you find very long relaxation times, probably due to the colloid structure of the composite. The fraction of extremely long relaxation times may not be considered to contribute to the phenomena. Making allowance for other relaxation processes, it is a very surprising finding that there are only tiny effects in the "quasi-static" limits. Even for different strain rates, there are no dramatic effects. Yet the time dependent measurements have not so far been done in a satisfactory manner.

ILAVSKY:
Can you comment on some dependencies on the temperature? If you use the same strain rate with 50 °C difference in temperature, are the parameters affected by the temperature?

KILIAN:
We expected larger effects than those as observed up to now. But there is a need to investigate this question very thoroughly.

RIGBI:
I would like to comment on this point. In my own work of 1967, I studied the Mullins-effect at very low temperatures, and there was a very definite effect of temperature and rate on the Mullins-effect. Let me just raise another point. Your picture of the reinforcement due to carbon black and due to silica as being so different, perhaps indicates why the rubber technologist today is already beginning to mix silica with carbon black to get very specific and favourable effects of reinforcement.

WEYMANS:
A good theory should not only explain existing experimental data but also predict something. Now supposing you know the stress-strain behaviour of an unfilled vulcanisate and you know the characteristic of a special filler, would you dare to say in your theory that you can predict the stress-strain behaviour of the filled vulcanisate?

KILIAN:
Only, if it is an ideal filler.

STADLER:
Your filled systems, in which you link to the filler, directly resemble the SBS block copolymers in some way. Can you treat these systems in a similar way, because there you have the advantage of knowing the size and number of chains starting from such styrene domains?

KILIAN:
It might be possible. But one might suspect that both components will be deformed. This gives additional problems. We have studied these effects in semicrystalline branched polyethylene where the crystal fraction is brought into anisometric shapes as well.

RIGBI:
You used a magnification factor for the strain between the particles. How similar is it to the Payne magnification factor?

KILIAN:
I think the first approach should only be correct if there are no particle-particle interactions. What is important is that the overdrawing of the rubber matrix increases the energy stored in the composite essentially at small strains. It might be, therefore, that the approach presented gives approximately reliable results.

Influence of the filler on the reinforcing mechanism of styrene-butadiene rubbers

G. Weymans and U. Eisele

Bayer-AG, Polymerphysik, Leverkusen, F.R.G.

Abstract: The ultimate mechanical properties of four series of filled SBR samples are analysed and discussed in detail. It is argued that two parameters (activity and concentration of filler particles) have a decisive influence on the homogeneity of the final network structure.

Key words: Styrene-butadiene rubber, filled networks, reinforcement, networks structure.

Only if they are reinforced by suitable fillers do rubbers normally satisfy the exacting mechanical property demands of industry. Although much progress has been made in interpreting the reinforcement mechanism, a thorough understanding of the quantitative influence of filler activity and concentration on the property pattern of vulcanized rubber is still lacking. Taking SBR, the most important tyre rubber, as an example, we will describe in this contribution the filler's significance to the high reinforcement of this polymer and its decisive influence on the network structure.

The stress-strain curves of four series of SBR specimens containing different active fillers (clay, Black N 660, N 330 and N 110) at different concentrations (0–120 parts by weight filler, on 100 parts by weight SBR – recipe otherwise identical) were measured (Fig. 1a–d) and the residual elongation of the rubber after cyclic stressing was determined (Fig. 4).

Figures 2 and 3 show how the elongation at break and tensile strength depend on the filler's activity and concentration: depending on the choice of these filler parameters it is possible to obtain mechanical property data that are several times higher than those of the unfilled vulcanizate.

In the case of the substantially inactive filler clay (cf. [1]) it is probable that even during compounding the higher shear gradient induced on the filler particles facilitates the dispersion of sulphur, thus enabling the rubber network to become increasingly homogeneous, until a saturation concentration (80 phr) has been reached. The slight residual activity of the clay simultaneously increases the crosslink density, with the result that the residual elongation of this SBR (Fig. 4) rises linearly with the clay concentration.

The specimens with carbon black show a marked stress and strain maximum at around 40 phr of filler. In addition, the stress maximum is shifted towards higher carbon black concentrations as the activity falls. Here we see not only the already discussed influence of the filler as a promoter of homogeneity, but also the decisive influence of the filler's activity on the tensile strength/elongation picture. Firstly, the crosslink density is increased and the mechanical property pattern thus improved (Fig. 2 and 3) by the binding of the filler to the polymer network as the additions of carbon black are raised. Within this concentration range, the residual elongation, like that of the clay-filled system, increases only slightly. At higher concentrations, homogeneous mixing of SBR and carbon black is hindered by the increasing tendency of the filler to agglomerate. The result is a more inhomogeneous network with impaired tensile strength and lower elongation at break values, which may even be inferior to those of the unfilled SBR. In cyclic stressing, the carbon black agglomerates are preferentially ruptured. The more

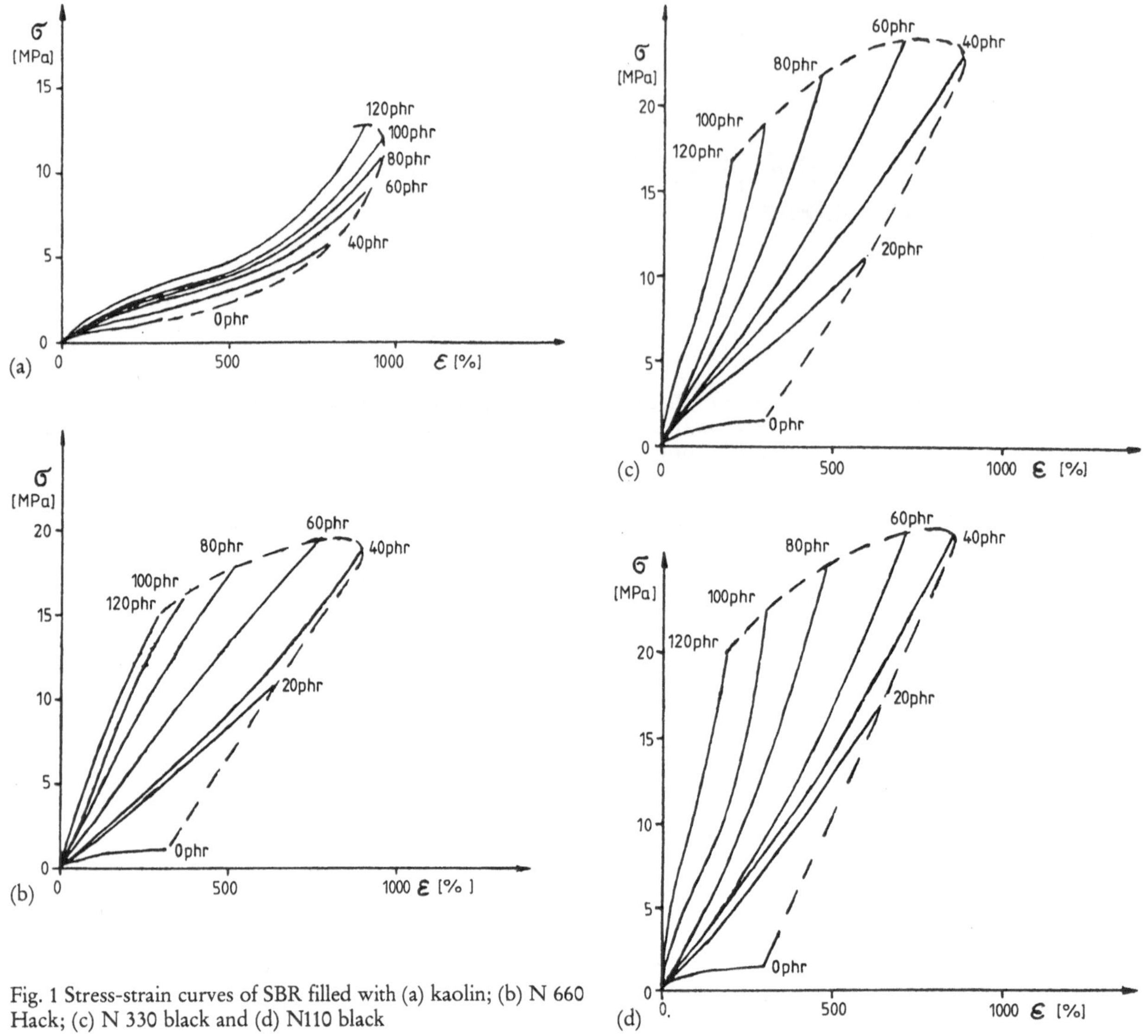

Fig. 1 Stress-strain curves of SBR filled with (a) kaolin; (b) N 660 Hack; (c) N 330 black and (d) N110 black

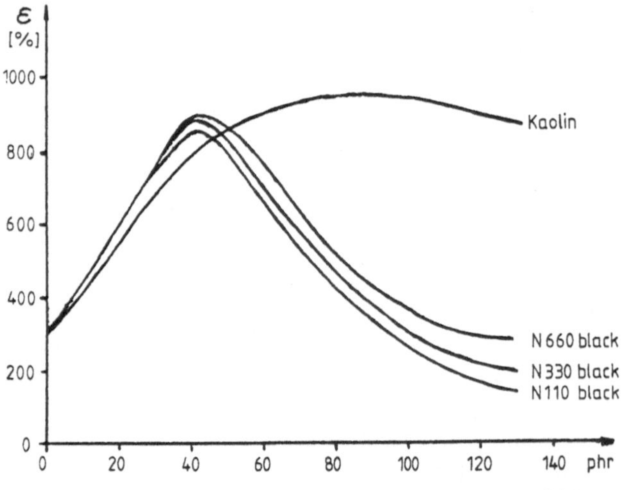

Fig. 2. Elongation at break for different concentrations of filler

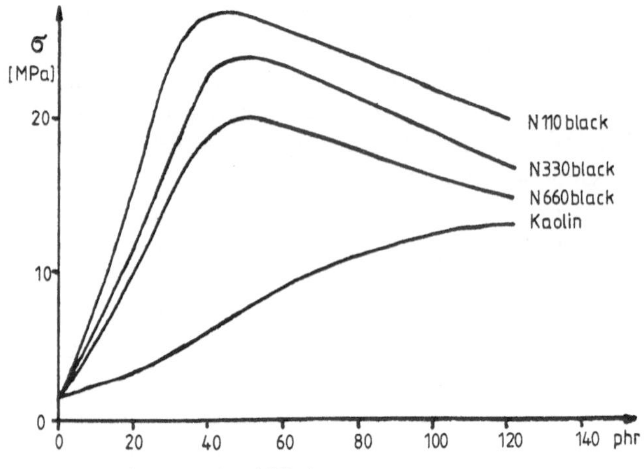

Fig. 3. Tensile strength of filled SBR

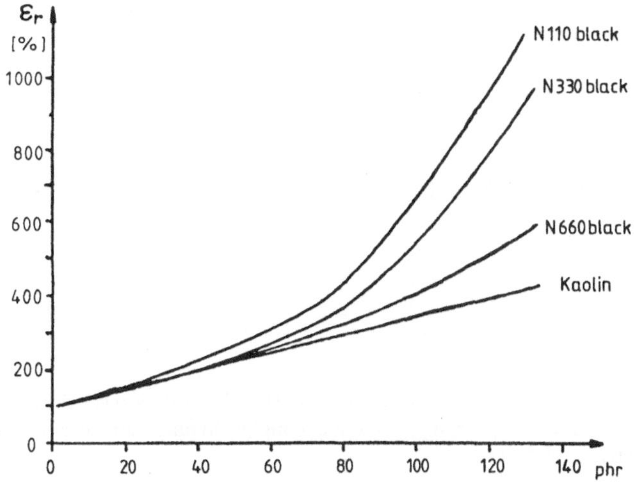

Fig. 4. Residual elongation after cyclic stressing

active the filler, the greater the extent to which the deformation processes are irreversible. Therefore the residual elongation increases greatly as the carbon black concentration is raised to the level that gives optimal mechanical rupture values.

In spite of these attempts to interpret (qualitatively) the influence of the filler on styrene-butadiene rubbers, it must be said that a general theory that would account quantitatively for the aforementioned filler parameters, together with the peculiarities of the raw polymer in question, has still to be developed. In view of the great importance of the filled networks, a quantitative understanding of the mechanical property profile (based on a knowledge of the structure of the unfilled rubber and of the filler parameters, for example) would be very helpful.

References

1. Müller P, Eisele U, Pampus G (1981) Angew Makromol Chemie 98:91–112

Received October 6, 1986;
accepted May 18, 1987

Authors' address:

G. Weymans
Bayer AG
FE Polymerphysik
Geb. B 406
D-5090 Leverkusen, F.R.G.

Progress in Colloid & Polymer Science Progr Colloid & Polymer Sci 75:234–238 (1987)

Filler-matrix coupling in rubbers as revealed by heat conduction experiments

D. Mergenthaler, M. Pietralla and H.-G. Kilian

Abteilung für Experimentelle Physik, Universität Ulm, Ulm, F.R.G.

Abstract: In this contribution we would like to demonstrate the ability of heat conduction experiments to yield information about the matrix filler coupling in filled rubbers and the orientation of the matrix molecules.

Key words: Rubber, filler, heat conduction, matrix-filler contact, orientation.

Introduction

When using heat conduction experiments as a tool for studying rubbers, it must be clear what kind of information can be obtained from these measurements on oriented samples. Heat conduction is a transport process of the diffusive type. Thus it depends strongly on the structure of the sample and on the orientation of the macromolecules. The latter induces an anisotropy of the energy transport. This has been discussed in detail in [1]. Usually, an anisotropy ratio is defined as

$$A = \frac{a^{\parallel}}{a^{\perp}} + = A(P_2, A_0) \quad (1)$$

which depends on the orientation parameter P_2 and an "intrinsic" anisotropy ratio A_0. For the latter the universal range of 5–25 can be estimated for non crystallizing polymers. The dependence on structure can be demonstrated by plotting the mean value

$$3a_{is} = a^{\parallel} + 2a^{\perp} \quad (2)$$

which should be an invariant under pure orientational effects. In principle, a may denote the thermal conductivity – or diffusivity – or the thermal resistance. What kind of averaging is appropriate cannot be decided a priori. If neither of the possible averages remain constant, then the changes of structure accompanying the deformation additionally influence the heat transport. For the discussion of these effects it is not sufficient to measure only the anisotropy ratio by the DeSénarmont method [1, 2] which is otherwise very convenient for studying oriented polymers. Hence we have improved this method in order to gain absolute values of the thermal diffusivity [3]. The latter is related to the thermal conductivity via

$$a = k / \varrho c_p \quad (3)$$

where a is thermal diffusivity; k, thermal conductivity; c_p, specific heat per mass; ϱ, density. The results presented are the first with this new apparatus.

State of affairs

Before discussing the possible influence of filler-matrix effects, how the heat conduction of a pure rubber changes upon orientation must be confirmed. There is a remarkable discrepancy in the literature about this question. The first person to try, to measure the heat conduction anisotropy of a stretched rubber was F. H. Müller [4] in 1962, who revived the De Sénarmont method. Within the limits of the accuracy obtained, no anisotropy was detected but an increase of the overall conductivity was seen. Two years before Tautz [5] published results of heat conduction measurements parallel to the stretching direction of rubber he found a strong increase up to a factor of five. Hennig [6] measured the conductivity perpendicular to the stretching direction of rubber sheets and found no variation for natural rubber and polyisobutylene. Hands [7] measured a strong decrease of the conductivity perpendicular to an equibiaxial stretched rubber, whereas Pietralla [8] found only a small increase of the directly measured anisotropy. These results do not fit together. If the conductivity perpendicular to the stretching direction remains constant or diminishes, the anisotropy must always be larger than the relative conductivity in the stretching direction. The only known measurements of this kind are those of Tautz [5]. It can well be argued that these measurements include a systematic error because the rubber bands used had been stretched using the thermocouples as "clamps". Hence, the increase of conductivity is mostly due to the increasing contact pressure at the thermocouples. The discrepancy is thus tempered but not removed.

Experimental

The experimental equipment used in our laboratory, as described in [2], has been improved by the use of a computer which collects the data. The data are then compared to the solution of the corresponding heat conduction equation. From the least squares fit the thermal diffusivity is determined. Five measurements in each direction (parallel and perpendicular to the stretching direction) are usually performed. The literature values of PMMA, PC, PS are reproduced with about 2 % accuracy. As a result we are able to measure the thermal diffusivity in every direction within the plan of a thin sample (thickness < 1.5 mm) at room temperature and above. As long as the specific heat of a stretched rubber does not change upon deformation – which holds true if no crystallization occurs – we can thus determine the thermal conductivity as well.

Results and Discussion

In Figure 1 the anisotropy ratio [1] of natural rubber is plotted versus the draw ratio. The increase of the anisotropy is of the same magnitude as published previously [8]. In order to remove effects of preparation history seen in the first stretching cycle, the sample had been swollen and deswollen prior to the first stretching. The solid line is calculated with the orientation parameter of a single finite chain (inverse Langevin function, see [9]). The chain length had been determined from the fitting of the van der Waals form to the force extension curve of the rubber [10]. The absolute values of the thermal diffusivity are shown in Figure 2. The striking feature is that in both directions the diffusivity increases upon stretching. The relative increase in the direction parallel to the stretching is thus higher than the anisotropy ratio, but nevertheless remarkably lower than that measured by Tautz [5]. The increase in the direction perpendicular to the stretching has not been observed before. In the measurements cited [6, 7] this value decreased or remained constant. But this observation may be due to contact problems of the stretched rubber which has an unusual thermal expansion behaviour. Our measurements are performed without any contact with the sample and without any surface coating, as has been used previously. They conclusively show a small increase in the anisotropy and a more pronounced increase of the

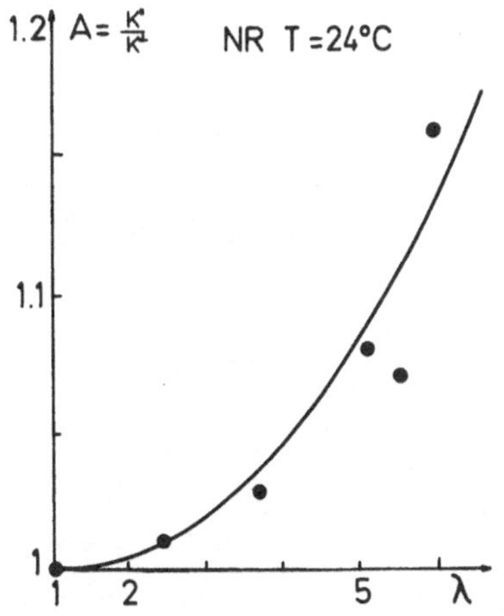

Fig. 1. Anisotropy ratio versus draw ratio of natural rubber cross-linked with 1.0 phr DCP. Prior to the first stretching, the sample had been swollen in toluene, and the measurements performed after deswelling. The solid curve has been calculated with an intrinsic anisotropy $A_0 = 7$ and the orientation parameter of a single chain with chain length $y = 272$

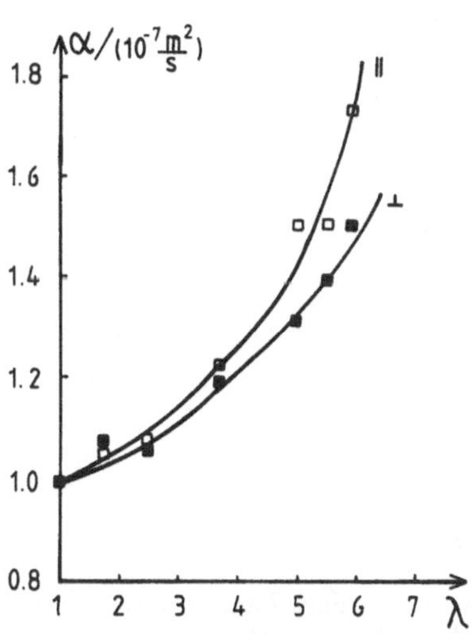

Fig. 2. The absolute values of the diffusivity of the same rubber parallel and perpendicular to the stretching direction. The overall diffusivity and so the conductivity run between these two curves and thus increase as well

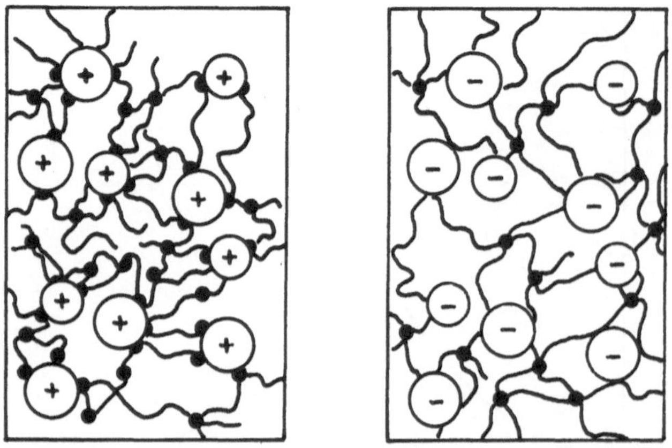

Fig. 3. Fillers bonded differently to the rubber matrix. Left (−) the filler, a deposited silica is only bonded by van der Waals forces. The junction points only exist in the matrix. Right (+) the same type of filler is bonded covalently to the matrix molecules by means of silan bridges. Hence the filler particles act as multifunctional junctions as well. The amount of filler in both rubbers is equal

overall conductivity, as has been already presumed by Müller [4]. The increase in the average conductivity is rather unexpected. It has never been observed in the case of amorphous glassy polymers. Thus there must be a distinct difference between oriented rubbers and glasses.

The behaviour reported is typical for rubber networks and has been confirmed by measurements from

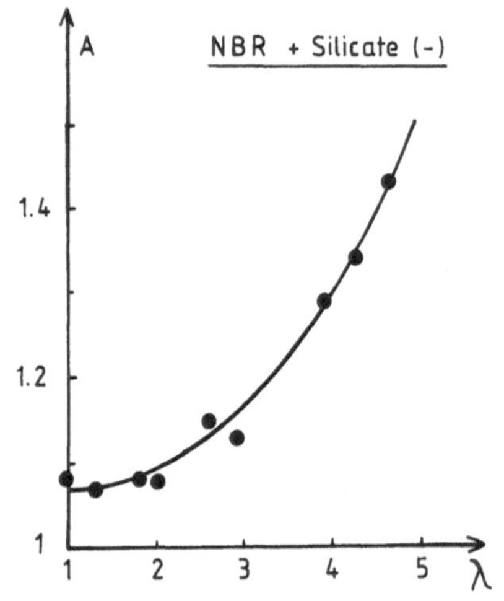

Fig. 4. The anisotropy ratio versus draw ratio of an acrylnitrile-butadiene rubber (NBR) filled with silica (−) bonded to the matrix. The small initial anisotropy is obviously due to the preparation of the rubber. Beyond that, the curve cannot be distinguished from that of an unfilled rubber. This has been observed with other fillers too

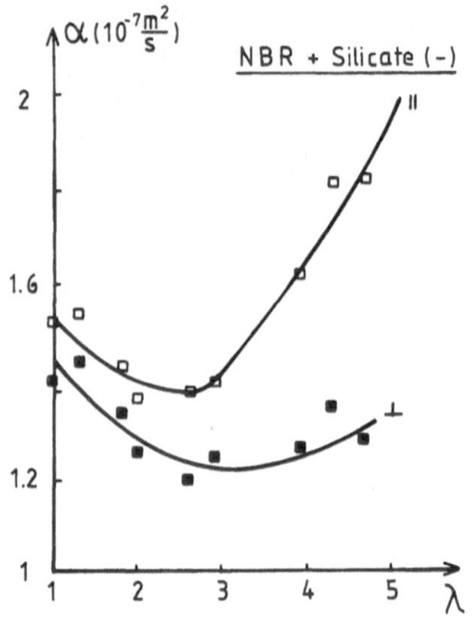

Fig. 5. The absolute values of the diffusivity parallel and perpendicular to the stretching direction of the filled NB-rubber (−) of Figure 4. Due to the filler, the values of the isotropic sample are higher than those of the pure matrix material which has a diffusivity of 1.22×10^{-7} m^2/s

non crystallizing rubbers like SBR. In the following, we would now like to present results from filled rubbers. The matrix was an Acrylnitril-butadiene rubber (NBR) and the filler, deposited silica. The type of filler matrix contact is depicted in Figure 3. A pure van der Waals contact of filler and matrix molecules is denoted by (−); a covalent contact where the matrix molecules have been bonded by silane bridges to the filler by (+). Within the matrix the density of junction points is equal. Thus, we have two rubbers with identical fillers but different matrix filler contact and can now study the influence of these contacts. In Figure 4 we see the anisotropy of the (−) contact. There is a small anisotropy in the unstretched state which is possibly due to the preparation of the rubber. The dependence upon stretching is nevertheless the same as for the natural rubber. The interesting feature follows from the absolute values of the diffusivity which firstly decrease in both directions before the upturn occurs (Fig. 5). Obviously, the heat transfer at the matrix filler boundary deteriorates. This becomes more evident when the stress is released. The values measured then coincide with those of the second and further stretching cycles (Fig. 6). They are above those of the first cycle, especially values in the parallel direction. One may tentatively interpret this as the formation of a superstructure fitting to the deformation mode. The anisotropy of the

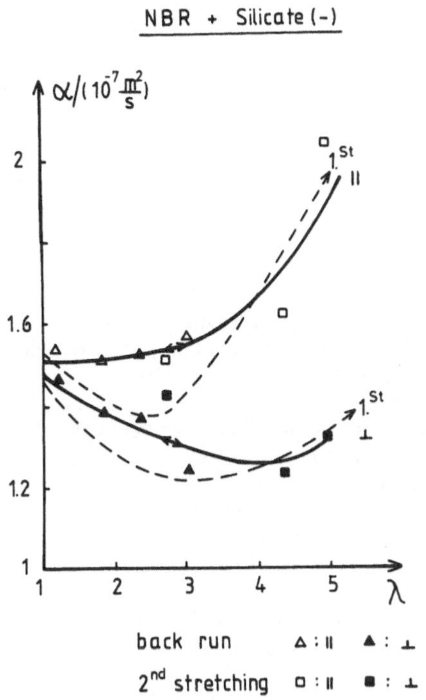

Fig. 6. The same as in Figure 5, but with the values of the back run and the second stretching included

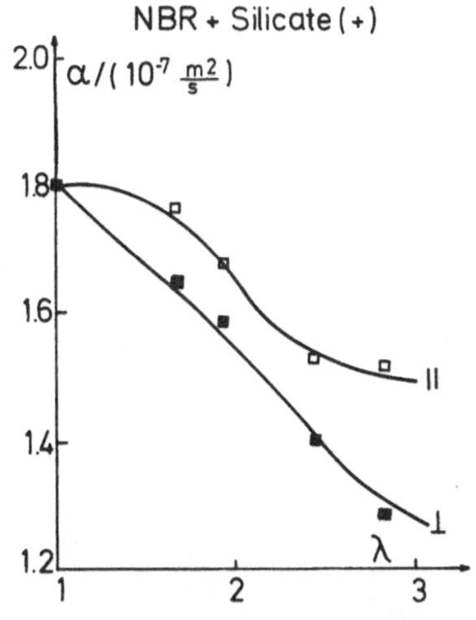

Fig. 8. The absolute values of the diffusivity parallel and perpendicular to the stretching direction of the (+)-filled sample. The covalent bonding of the silica filler enhances the diffusivity again, and is higher than that of the (—)-filled rubber. In contrast to the behaviour of this rubber, the values of the second stretching cycle are now smaller than those of the first. In both runs, the diffusivity decreases to about the values of the (—)-filled rubber. A possible upturn as has been observed with the latter (cf. Figs. 5, 6) cannot be measured because the sample ruptures at $\lambda = 3$

second cycle is that of the first within the margin of error. The behaviour of the anisotropy obviously does not change compared to the pure matrix. This shows

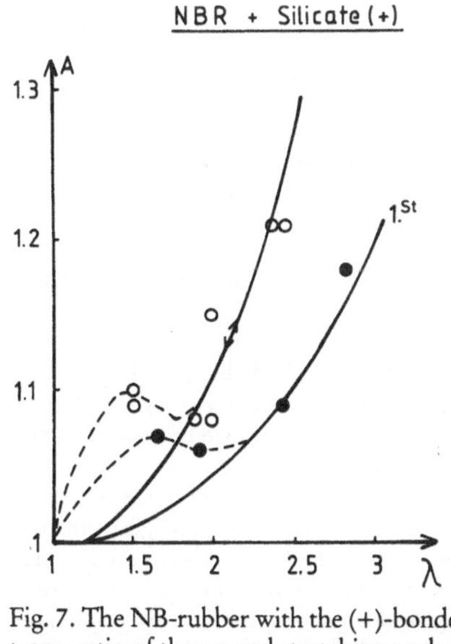

Fig. 7. The NB-rubber with the (+)-bonded silica filler. The anisotropy ratio of the second stretching cycle is higher than that of the first. The hump at small draw ratios can be attributed to some kind of orienting short range order regions

that the diffusivities in both the directions are altered by the same factor. If we now look at the results of the rubber with (+) contact things have changed (Figs. 7–9). The first and second stretching cycles reveal rather different anisotropies, the latter showing higher values throughout. Moreover, in both curves a hump at small draw ratios is seen. This type of curve is typical for the orientation of some kind of anisotropic short range order which, in this case, might be introduced by the bonding of the fillers. The interpretation of the absolute values shown in Figure 8 is more difficult. The diffusivity in both the directions decrease strongly but always remains higher than that of the rubber with the (—) contact filled with the same amount of silica. The improved contact thus enhances the overall conductivity. The second cycle (Fig. 9) now runs below the first in contrast to the (—) filled rubber. A more detailed interpretation of these findings would be purely speculative in the present state of our knowledge.

This was a short overwiew of our current investigations regarding the anisotropy and matrix-filler problems in rubbers. It is now obvious that the studies to be performed must embrace rubbers with an increasing number of covalent contacts per filler particle in order

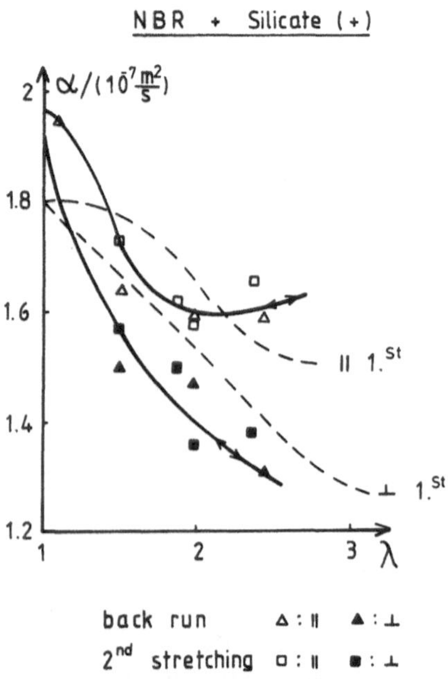

Fig. 9. Back run (□) and second stretching (△) of the (+)-filled rubber compared to the first rund (- - -) from Figure 8

to give a better insight into the role of these contacts. The interpretation suffers to some extent from the lack of a suitable theory of thermal conductivity in amorphous polymers. Nevertheless, the results are very promising and we hope that it has become clear that measurements of thermal diffusivity can give interesting results regarding the orientation and the structural effects of stretched rubbes.

Acknowledgement:

We would to thank Dr. V. Härtel from Metzeler Kautschuk GmbH for his kind co-operation and the preparation of the samples.

References

1. Pietralla M (1982) Habilitationsschrift Universität Ulm
2. Blum K (1981) Diplomarbeit Universität Ulm; Blum K, Kilian H-G, Pietralla M (1983) J Phys E, Sci Instr 16:807
3. Mergenthaler DB (1986) Diplomarbeit Universität Ulm
4. Hellmuth W, Müller FH (1962) Kolloid Z Z Polym 185:159
5. Tautz H (1960) Kolloid Z Z Polym 174:128
6. Hennig J (1964) Kolloid Z Z Polym 196:136
7. Hands D (1980) Rubber Chem Technol 53:80
8. Pietralla M (1981) Coll Polym Sci 259:111
9. Treloar LRG (1975) The physics of rubber elasticity, 3rd ed, Clarendon Press, Oxford
10. Eisele U, Heise B, Kilian H-G, Pietralla M (1981) Die Angewandte Makromolekulare Chemie 100:67

Received March 11, 1987;
accepted March 16, 1987

Authors' address:

D. Mergenthaler
Universität Ulm
Abt. Experimentelle Physik
Oberer Eselsberg
D-7900 Ulm, F.R.G.

Discussion

OBRECHT:

What was the amount of carbon black or of silica in your compounds? It is possible that things are quite different if you perform your experiments with a different amount of carbon black?

PIETRALLA:

The amount of silica in this case was about 14% by volume, and I cannot say what is happening with higher amounts of filler particles, because we haven't done all the experiments. In this case here, however, we believe we know what the difference is between the two kinds of fillers.

PECHHOLD:

Could you speculate about this strange appearance – that the parallel as well as the perpendicular heat conductivity goes up in the matrix material?

PIETRALLA:

Yes, one can speculate about the fact. It is interesting enough, because by simple analogy arguments, one would say that the only difference between a glass and a rubber is the exchange of atomic or molecular sites, and a different thermal expansivity. But it is well known that thermal conductivity is not affected by the exchange of sites, i. e. by the diffusive motion, contributing to only about 2%. Then, what follows for the difference between a rubber and a glass, seen by thermal conductivity, is only caused by the different volume expansion – a very small effect. Indeed, if one measures the overall conductivity of an isotropic sample, passing the glass temperature, one can account for the measured change in slope by the different thermal expansion. There is no real theory for thermal conductivity in oriented long chain melts. So we have only ideas at hand, not arguments.

From a heuristic point of view, one should check the difference in the dynamics between a low molecular weight glass-forming melt and a material of linear crosslinked molecules in the oriented state. In the isotropic state, their behaviour is identical. The possibility remains of travelling conformational modes carrying energy which may become effective upon orientation, on account of the decrease of scattering centres and scattering efficiency in oriented chain ensembles.

Progress in Colloid & Polymer Science Progr Colloid & Polymer Sci 75:239–242 (1987)

Interrelation between the orientation of the polymer chains and the mesogenic groups in crosslinked liquid crystalline polymers

R. Zentel

Institut für Organische Chemie, Mainz, F.R.G.

Abstract: In crosslinked liquid crystalline polymers, the polymer chains can be oriented by mechanical forces, and the mesogenic groups should orient independently in electric fields. Thus crosslinked liquid crystalline polymers should allow the investigation of the interrelation between the orientation of polymer chains and mesogenic groups. The possibility of orienting the mesogenic groups by mechanical forces (via the polymer chains) is of considerable interest, if liquid crystalline phases with special properties, e. g. ferroelectric properties, are considered. Synthetic routes to uncrosslinked and crosslinked polymers with these properties are discussed.

Key words: Liquid crystal polymers, elastomers, ferroelectric properties.

In order to understand the principal properties of crosslinked liquid crystalline polymers, one can imagine a network swollen with mesogenic groups [1] (see Fig. 1). In real liquid crystalline networks, these mesogenic groups are linked to or incorporated into the polymer chains in different ways [2] (see Fig. 2), but it is not necessary to specify these in order to obtain a basic understanding of the properties. Since all the mesogenic groups orient parallel to each other in the liquid crystalline phase, the polymer chains are in an anisotropic medium and consequently they adopt an anisotropic average chain conformation. That means that the average distance between two crosslinks in the network not only depends on the chain length between the crosslinks, but also on the orientation relative to the long axis of the mesogenic groups.

Special interest in these systems now arises from the fact that it should, in principle, be possible to orient each of the two components of the network by two independent methods. The polymer chains can be oriented, as is usual in networks, by the application of mechanical forces. The mesogenic groups should orient, as is usual for non-crosslinked liquid crystalline polymers in electric or magnetic fields. As the polymer chains and the mesogenic groups "feel" each other through space and since they are covalently linked to each other in addition, the orientation of one component must have an effect on the other. This implies that it is also possible to orient, e. g. the liquid crystalline phases, by mechanical forces. This is very important both for the phase assignment of these polymers [3] and to learn something about the orientation of polymer chains and mesogenic groups relative to each

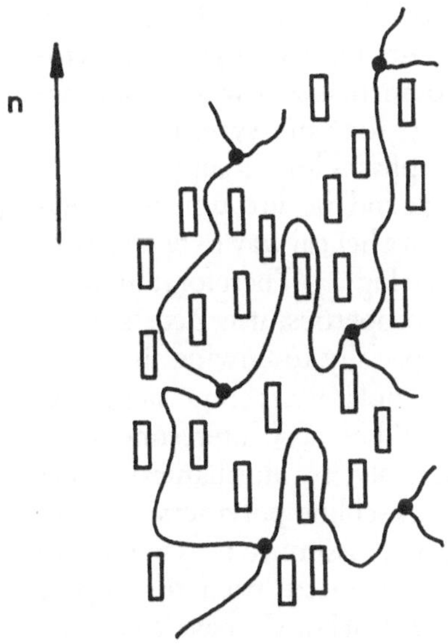

Fig. 1. Schematic representation of a network swollen with mesogenic groups

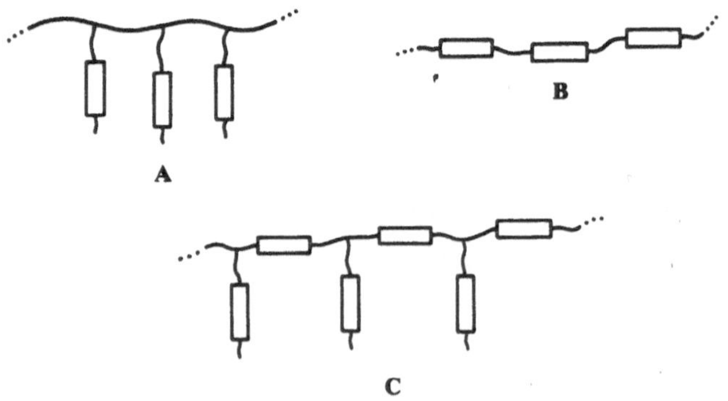

Fig. 2. Different types of liquid crystalline polymers that can be used to prepare crosslinked liquid crystalline polymers [2a]

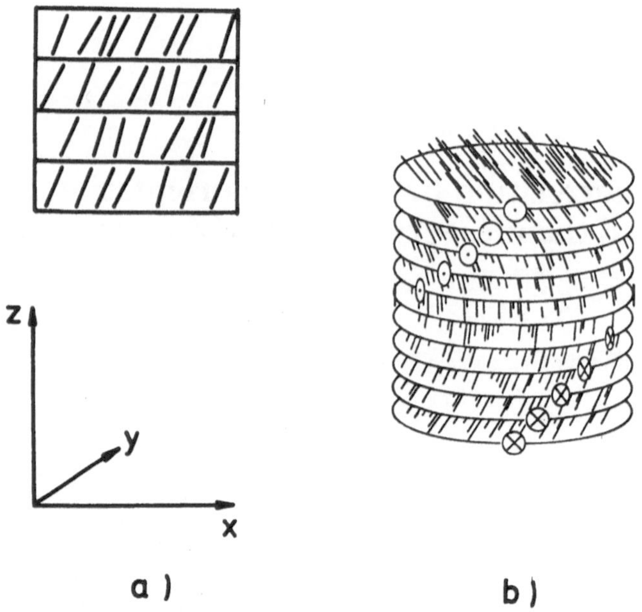

Fig. 3. Structure of the chiral smectic C phase; (a) arrangement of the mesogenic groups in layers (the structure within the layers is liquid-like), (b) helical superstructure over distances of some micrometres ($\odot$, $\otimes$): direction of the spontaneous polarization

other [4]. The possibility of orienting liquid crystalline phases by mechanical forces is of considerable importance, however, if liquid crystalline phases with very special properties, e. g. ferroelectric properties, are used. In such cases, it may be possible to orient the direction of the macroscopic electric dipole moment by mechanical forces. On the other hand by using electric fields it may be possible to orient the mesogenic groups and, at the same time, the polymer chains should rearrange and something like a shape variation in electric fields should result [5].

Liquid crystalline phases with ferroelectric properties (especially the chiral smectic C phase) have well been investigated for low molar mass liquid crystals [6]. They result in the smectic phases formed from chiral molecules, in which the long axes of the molecules are tilted with respect to the layer normal (see Fig. 3). Due to the presence of these chiral mesogens, the symmetry of the local environment is so low that a spontaneous polarization occurs for pure symmetry arguments. In addition, these phases form a superstructure in which both the tilt angle and the direction of spontaneous polarization twist in a helical way over distances of some micrometres (see Fig. 3). Therefore, in order to observe ferroelectric properties, it is necessary to orient the sample and especially to unwind the helix. This can be done for low molar mass chiral smectic C liquid crystals. But it would be very interesting to do this, perhaps even reversibly, by mechanical forces with crosslinked liquid crystalline polymers.

Accordingly, to prepare crosslinked polymers with ferroelectric properties, we firstly tried to prepare linear polymers with chiral smectic C phases. This can be done, e. g. with "combined" liquid crystalline polymers with chiral constituents [7] (polymers with the

mesogenic groups in the main chain as well as in the side groups, see Figure 2 (C) and Table 1).

Identification of the chiral smectic C phases was performed using a combination of texture observations with a polarizing microscope and X-ray measurements. X-ray measurements of oriented crystalline samples at room temperature prove a tilted arrange-

Table 1. "Combined" liquid crystalline polymers 1–3 with chiral phases and their characterization

$$\left[O \{CH_2\}_6 - O - \bigcirc - R^1 - \bigcirc - O \{CH_2\}_6 - OOC - CH - CO \right]_x$$

$$CH_3 - CH_2 - \overset{*}{C}H - CH_2 - O - \bigcirc - R^2 - \bigcirc - O \{CH_2\}_6$$
$$|$$
$$CH_3$$

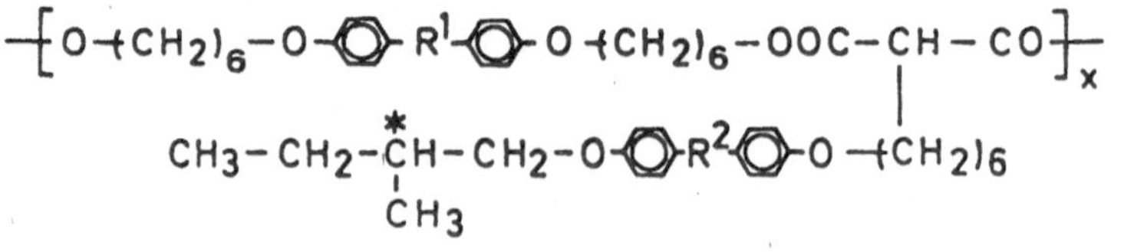

No.	R^1	R^2	Molecular weight (GPC)	Phase transition temperatures[a]) (°C)
1	$-N=N-$	$--$	3 200	k 108 s_C^* 115 n^* 128 i
2	$--$	$-N=N-$	30 000	k 130 s_C^* 136 i
3	$-N=N-$	$-N=N-$	36 000	k 107 s_C^* 111 n^* 137 i

[a]) k: crystalline phase, s_C^*: chiral smectic C, n^*: cholesteric, i: isotropic

Table 2. Uncrosslinked 4 and crosslinked 4′ liquid crystalline polymer and its characterization

$$\left\{\left(O-A-OOC-\underset{\underset{S_1}{|}}{CH}-CO\right)_{0.5}\left(O-A-OOC-\underset{\underset{S_2}{|}}{CH}-CO\right)_{0.5}\right\}_x$$

4

A : $(CH_2)_6-O-\bigcirc-N=N-\bigcirc-O(CH_2)_6$

S_1: $(CH_2)_6-O-\bigcirc\bigcirc-O(CH_2)_3-CH=CH_2$

S_2: $(CH_2)_6-O-\bigcirc-N=N-\bigcirc-O-CH_2-\overset{*}{\underset{\underset{CH_3}{|}}{CH}}-CH_2CH_3$

$$\downarrow \quad H\left(\underset{\underset{CH_3}{|}}{\overset{\overset{CH_3}{|}}{Si}}-O\right)_{6.5}\underset{\underset{CH_3}{|}}{\overset{\overset{CH_3}{|}}{Si}}-H$$

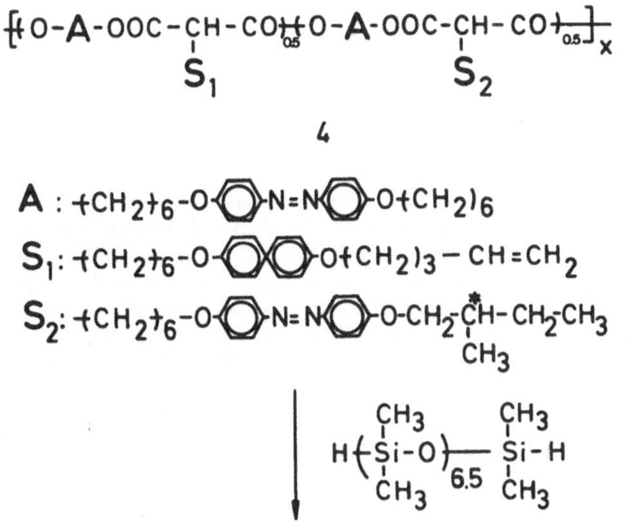

Crosslinked Polymer 4′

No.	Molecular weight (GPC)	Amount of crosslinking agent in mole %	Phase transition temperature[a] (°C)
4	35 000	—	k 115 s_A 133 n^* 142 i
4′	—	10	k 106 s_A 122 n^* 138 i

[a]) k: crystalline phase, s_A: smectic A, n^*: cholesteric, i: isotropic

ment of the long axes of the mesogenic groups. Temperature dependent X-ray measurements of unoriented samples show that the layer thickness of the tilted crystalline phase is retained in the smectic phase. This preservation of the layer spacing in the smectic phase is a strong indication of the preservation of the tilted structure and thus for the existence of a chiral smectic C phase.

The ferroelectric properties of these substances, which have not yet been measured, should, however, be very poor, since in addition to the structure, a strong dipole moment close to the chiral center is also needed, to obtain a strong polarization. These dipole moments have been absent in the polymers prepared so far, but work is currently in progress to synthesize polymers of similar structure which possess such dipole moments.

Crosslinked liquid crystalline polymers can be prepared, starting with these polymers, by a copolymerization of chiral repeating units and repeating units with olefinic double bonds [7] (see Table 2). Afterwards, some of the double bonds can be reacted with the Si-H bonds of a oligo (dimethyl-siloxane). Thus, crosslinked samples can be obtained, but for the first samples the chiral smectic C phase was lost. Only smectic A and cholesteric phases were found, which do not show ferroelectric properties. Work is now in progress to prepare crosslinkable polymers with chiral smectic C phases.

References

1. Brochard F (1979) J Phys 40:1049
2. a) Zentel R, Reckert G (1986) Makromol Chem 187:1915; b) Finkelmann H, Kock H-J, Rehage G (1981) Makromol Chem Rapid Commun 2:317; Finkelmann H, Kock H-J, Gleim W, Rehage G (1984) Makromol Chem, Rapid Commun 5:287; Finkelmann H (1984) Adv Polym Sci 60/61:99
3. Zentel R, Schmidt GF, Meyer J, Benalia M (1987) Liquid Crystals, in press
4. Zentel R, Benalia M (1987) Makromol Chem 188:665
5. Zentel R (1986) Liquid Crystals 1:589
6. Goodby JW, Leslie TM (1984) Mol Cryst Liq Cryst 110:175; Wahl JW, Leslie TM (1984) Ferroelectrics 59:161
7. Zentel R, Reckert G, Reck B (1987) Liquid Crystals 2:83

Received December 17, 1986;
accepted March 27, 1987

Author's address:
R. Zentel
Institut für Organische Chemie
Universität Mainz
J.-J.-Becher-Weg 18–20
D-6500 Mainz, F.R.G.

Discussion

KILIAN:

Can you estimate whether it is possible to enforce a change in shape by the weak entropy elastic forces, in the presence of the short-range interactions in the liquid crystalline state?

ZENTEL:

For this experiments, I started from some known facts. You can easily orient a liquid-crystalline polymer in electric fields. If you crosslink the same polymer and try to orient it in an electric field, you fail completely. The problem is that you have a two-phase system, and I do not know what happens at the interfaces between the low molar mass liquid crystal and the crosslinked polymer.

KILIAN:

My question was directed to a comparison of the competing forces, the liquid-crystalline ones, which are behind the phase transition into the liquid-crystalline state, and the entropy elastic network forces. They should be in the same order of magnitude for enforcing the shape transformation. You have not proved that?

ZENTEL:

No, I haven't. It is certain that the mesogenic groups cannot orient completely. If they oriented completely, the sample should have become dark. As this depends on the degree of swelling of the sample, if you work with very weakly crosslinked samples, where you have a high sol content, and swell the samples highly, it is possible to orient the mesogenic groups completely. If you work with higher crosslinked samples, you get some response.

WEYMANS:

Could you tell me the glass temperature and the molecular weight of the polymers you used?

ZENTEL:

The problem is, for this polymer, before crosslinking, the molecular weight was greater than 100,000, but after the sample was crosslinked there was no sol content remaining in the polymer. The glass transition temperature was 55 °C. But after the polymer was dispersed in this low molar mass liquid crystal, it was swollen a bit, so T_g was lower. The problem now is, as you cool down the sample, most of the low molar mass liquid crystal is repelled from the sample. So it is difficult to determine the glass transition temperature of these polymers, which were swollen at 80 °C for a longer time.

Progress in Colloid & Polymer Science Progr Colloid & Polymer Sci 75:243–247 (1987)

Comment: Networks and Theory

Some principal problems in statistical mechanics of networks and their relationship to other topics in physics and materials science

T. A. Vilgis

Max-Planck-Institut für Polymerforschung, Mainz, F.R.G.

Abstract: This paper discusses some ideas in the theory of rubber elasticity from an unusual but 'natural' view. The theory of rubber elasticity is embedded in the general physics of condensed matter. Rubbers are systems with quenched disorder.

Key words: Rubber elasticity, statistical mechanics, quenched disorder, materials science.

Introduction

During the meeting we discussed more or less the whole area of problems in thermomechanics of permanent networks – filled and unfilled – or temporary networks, i. e. polymer melts. Rubber elasticity is beside solution properties, the oldest subject in polymer physics and polymer chemistry and one might think there has been time enough for the solution of most problems. A lot of problems have indeed been solved, during the early 1940s and 1950s by the pioneering work of Kuhn [1], Treloar [2], Flory [3], James and Guth [4] to name only a few of the scientists in this field or research. It was soon recognized that the exceptional elastic property is to deform a solid with relatively weak forces (compared to non polymeric solids) to a large extent of deformation. Shortly after Kuhn's proof that macrolecules are long flexible molecules, rather than stiffer objects or aggregates (as proposed by Staudinger when he invented macromolecular chemistry) the idea of large deformation due to change of conformation was born. This makes rubber a very peculiar solid: the elasticity is almost due to a change of entropy rather than changes in energy.

The theory derived equations for the free energy are all of the type

$$F = \alpha N kT \sum_i \{\lambda_i^2 + \beta f(\lambda_i)\} \tag{1}$$

where α, β are constants depending on the theory, and $f(\lambda_i)$ is some function depending on the assumptions used as input for the theory. For example, α is $\frac{1}{2}$, allowing for fluctuation of the crosslinks (phantom limit) while $\alpha = 1$ for the affine limit, where the crosslinks are held fixed. Hence N is the number of crosslinks in a well-linked rubber. For β the situation is the same, i. e. $\beta = 0$ for the affine limit. The function $f(\lambda_i)$ is often used as the ideal gas term and $f(\lambda_i) = \log \lambda_i$. This form is still disputed, as we have also seen during this meeting. The origin of the log term is different according to various authors. For example, Flory's interpretation [3] comes from the condition that at least two chain ends must find themselves to form a crosslink. They must be in the same volume element. But all the classical theories are simple models far from reality.

In the next sections we will comment on the questions derived from real problems in theoretical physics and questions from the technological aspect. These are almost open questions and we are not going to solve them in this paper but keep them in mind for further discussions on networks.

Fundamental problems in calculating the free energy

In the early seventies it was recognized by Edwards that a rubber is a system with quenched disorder [5]. This means that there are "frozen" variables in the system. These frozen variables are given by the crosslinks. To see this, suppose first a polymer melt where the contacts between chains are mobile and sliding.

This is a typical "annealed" thermodynamic problem and in principal the free energy of a polymer melt can be calculated by the usual Gibbs formula

$$\beta F = - \log \int d\{x\}\, e^{-\beta \mathscr{H}(\{x\})} \tag{2}$$

where the $\{x\}$ are abstract variables characterizing the Hamilton function $\mathscr{H}$.

Imagine now that the contacts are suddenly fixed at same time, t, so that they are fixed forever by chemical bonds. Clearly the contacts cannot slide anymore and final answer will depend on the topology or the structure which has been picked up at the quench (≡ crosslinking) at the time t. But in this case, Equation (2) is no longer applicable and a new thermodynamic formulation has to be employed. The reason is very simple: One can imagine having two types of variables $\{x\}$ and $\{y\}$ where $\{x\}$ are some free degrees of freedom while $\{y\}$ are the quenched ones. Here $\{y\}$ are the crosslinks.

The free energy of a sample created at t is

$$\beta F_t(\{y\}) = - \log \int d\{x\}\, e^{-\beta \mathscr{H}(\{x\},\{y\})} \tag{3}$$

depending on the frozen variables $\{y\}$. Experimentally, the average over all posibilities of arranging $\{y\}$ will be relevant. These possibilities are described by the probability $P(y)$

$$P(\{y\}) = \int d\{x\}\, e^{-\beta \mathscr{H}(\{x\},\{y\})} \tag{3a}$$

and Equation (3) has to be averaged over the distribution $P(y)$ to give the observed free energy.

$$\beta F = \int dP\{y\} \log Z(\{y\}) = \langle \log Z(\{y\}) \rangle \tag{4}$$

where $Z(y) = \int d\{x\}\, e^{-\beta \mathscr{H}(x,y)}$. This formula suggests that the average of $Z(y)$ has to be done *after* deforming the rubber, since the answer of the system on a deformation depends on the configuration $\{y\}$. This is now a non Gibbsian way of performing the average. Mathematically, the average of the logarithm is very difficult to perform and a trick can be used [5]

$$\langle \log Z \rangle = \frac{\partial}{\partial n}\bigg|_{n=0} \langle Z^n \rangle . \tag{5}$$

Obviously it is much easier to perform the average of a power instead of the logarithm. Rewriting $Z^n(y)$in terms of multiple integrals

$$Z^n = \int \prod_{\alpha=1}^{n} dx^{(\alpha)}\, e^{-\beta \sum_{\alpha=1}^{n} \mathscr{H}(x^{(\alpha)},y)} \tag{6}$$

and defining

$$e^{-\beta F(n)} = \int dP(y) \int \prod_{\alpha=1}^{n} dx^{(\alpha)}\, e^{-\beta \sum_{\alpha=1}^{n} \mathscr{H}(x^{(\alpha)},y)} \tag{7}$$

the final answer is given by

$$F = \frac{\partial F}{\partial n}\bigg|_{n=0} . \tag{8}$$

This method is called the replica method because a system ($\mathscr{H}$) has been replicated n-times ($\prod_{\alpha=1}^{n} \mathscr{H}^{(\alpha)}$).

Using this method, the rubber problem has been solved by Edwards and Deam [6] and the reader is referred to that paper for details.

There are many interesting points in this reference. For example it allows, for the first time, an exact classification of the two basic models, the phantom network and the affine network [1, 4] (see Ref. [6] for details).

By rigorous application of the replica method it follows that the widely used form for the swollen network

$$F_{\text{gel}} = F_{\text{rubber}} + F_{\text{solution}} \tag{9}$$

is *basically* wrong. This is due to the appearance of the crosslink and the excluded volume in a similar but mathematically totally different way. The "free energy" $F(n)$ for a swollen rubber is

$$\begin{aligned} e^{-\beta F(n)} = \mathscr{N} \int \delta R^{(0)} \int \prod_{\alpha=1}^{n} \delta R^{(\alpha)} \frac{1}{2\pi i} \oint \frac{d\mu\, N!}{\mu^{N+1}} \\ \times \exp\Bigg\{ -\frac{3}{2l} \sum_{\alpha=0}^{n} \left(\frac{\partial R^{(\alpha)}}{\partial s}\right)^2 \\ - \sum_{\alpha=0}^{n} v^{(\alpha)} \int_0^L ds \int_0^L ds'\, \delta(R^{(\alpha)}(s) - R^{(\alpha)}(s')) \\ + \mu \prod_{\alpha=0}^{n} \int_0^L ds \int_0^L ds'\, \delta(R^{(\alpha)}(s) - R^{(\alpha)}(s')) \Bigg\} \end{aligned} \tag{10}$$

where $v^{(\alpha)}$ are the excluded volume parameters (binary cluster integrals) and μ the chemical potential of the N crosslinks.

Hence all coordinates $R^{(0)}(s)$ come from $P(y)$, the distribution of the crosslinks, while $R^{(\alpha)}(s)$ are the coordinates of the chain in the α-th replica (systems with $\alpha > 0$ are essentially the deformed systems). Equation (10) is, however, the mathematical form of a diagram (Fig. 1). The difference between the solution term (v) and the crosslink term (μ) is obvious: the crosslink is fixed (quenched) while the interaction, via v, can be everywhere along the chains. In mathematical terms this is expressed in Equation (10) as the sum of all replicas for the solution term and a product for the crosslink term. Thus, Equation (9) can never hold. The complete analysis of Equation (10) has been carried out in a paper by Ball et al. [7]. This analysis also provides some conclusions on the '$N \log \Pi\lambda_i$' term of Flory. No statistical mechanics of networks are able to find this term by a molecular model. However, the introduction of the ideal gas term by Flory seems to be correct, at first glance, but it can be used only for solutions, rather than gels. The reasoning behind this statement is that each crosslink takes away one degree of freedom [8] and the chains are in a volume and hence the free energy contains a term like '$N \log \Pi\lambda_i$'. But this is only correct if the system is not gelled – and therefore it is not a solid on larger length scales (larger than R_g). In this case, the total volume $V\lambda_x\lambda_y\lambda_z$ would be available for the system and the Flory term appears. Far away from the gelpoint, the monomers are localised in a volume with a linear dimension of the order of the radius of gyration of a strand between two crosslinks (this can be shown mathematically), but this volume is independent of deformation [6] and no Flory term is present.

It is interesting to note that the replica method was applied recently first to rubbers [5] and then to spin glasses [9] where it came as a real break-through. The

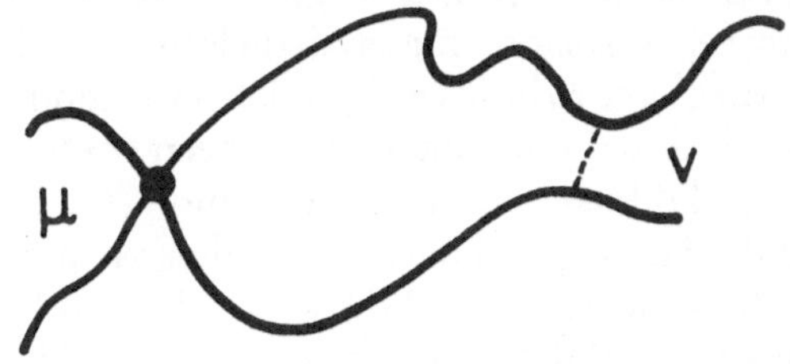

Fig. 1. The two fundamental interactions in a swollen rubber. μ defines a crosslink, which is fixed in space, while the excluded volume interaction v can be every where in space

spin glass is a system with quenched disorder, too. Quenched disorder in this case means that some atoms with a spin, for example Cu, are distributed randomly among others without a spin such as Mn (for an excellent review on spin glasses see [10]). The spins are then fixed on a lattice and the spin glass can be considered as a dilute magnet. The spin glass now shows a transition at some temperature where the spins freeze in random directions giving a finite magnetization. For calculating the magnetic properties of a spin glass the same procedure has to be applied as for rubbers. The spins are fixed at some lattice points (which means quenched disorder) then the magnetic properties have to be calculated. The free energy then has to be averaged over the disorder which is in principle the average over all possible topologies of arranging the spins. In order to establish a mean field theory, Sherrington and Kirkpatrick (SK) [11] used the Edwards Anderson Hamiltonian

$$H = - \sum_{\langle ij \rangle} J_{ij} S_i S_j \tag{11}$$

where J_{ij} are the exchange couplings of the spin at site i; the spin at sites j and S_i are Ising spins. The J_{ij} are random variables with a distribution

$$P(J_{ij}) = \left(\frac{z}{2\pi J_0^2} \right)^{d/2} e^{-\frac{z J_{ij}^2}{2 J_0^2}} \tag{12}$$

where z is the number of spins interacting. If z becomes large (of order of N, the total number of spins) the SK model is established. The free energy of a spin glass can now be calculated with the same formulation given by Equations (2)–(5) [9,11]. For the free energy, it is necessary to calculate $\langle Z^n \rangle_{J_{ij}}$ which can be written as

$$\langle Z^n \rangle_{J_{ij}} = Tr_{\{S_i^{(\alpha)}\}} \exp \left\{ \frac{J_0^2 \beta^2}{4z} \sum_{\alpha,\beta}^{n} \sum_{\langle ij \rangle} S_i^{(\alpha)} S_j^{(\alpha)} S_i^{(\beta)} S_j^{(\beta)} \right\} \tag{13}$$

where α, β are the replicas of the system.

Equation (13) makes the difference in comparison to the simple rubber model used above. By averaging over the disorder, interaction between two different replicas appears. This becomes more clear if a generalized order parameter is defined $q^{\alpha\beta} = \langle S_i^{(\alpha)} S_i^{(\beta)} \rangle$. The solution in [11] was the replica symmetric one, say $q^{\alpha\beta} = q$ and it has been shown [12] that this model is unstable if a magnetic field h is applied. There exists a critical line $h(T)$ below which the theory becomes unstable.

The SK model calculates a negative entropy at $T = 0$, too. A more accepted solution was given by Parisi, assuming replica symmetry breaking [13] by saying that a general order parameter function replaces $q = q^{\alpha\beta}$. This seems problematic since finally the limit $n \to 0$ has to be taken, but it has been shown that replica symmetry breaking induces non-ergodicity. The phase space is divided into small parts so that the phase space cannot be explored by the system in a finite time. This widely accepted model concludes with the result that the free energy has a multiple valley structure and the state of the system is represented by one of the valleys (not necessarily the lowest one) [14].

The ideas developed in spin glasses are applied nowadays to all sorts of complex systems and optimization problems like the Travelling Salesman problem [15] (which is the shortest closed self avoiding walk (!) between fixed points) neutral networks [16] or chip design.

Coming back to the rubber problem, we note that the solution given in References [6, 7] is a replica symmetric one. The reason is simple. We assumed that the crosslinks are fixed forever when a rubber is formed. That means, however, that no further conformation is available for the crosslinks and the free energy has only one minimum. This is different from the spin glass, where the spins are fixed at the lattice in space, but by flipping them one can create new conformations and cross from one valley of the free energy to another. Nevertheless this point needs further investigation [29].

To conclude this section we must remark that the use of the non-Gibbsian statistical mechanics becomes very important if one replaces the Gaussian statistics for the end-to-end distribution $P(R, L)$ and uses more complicated models. This has been shown in Reference [17] where the results obtained by the correct averaging

$$F \sim \int d^3R \log P(\underset{\sim}{\lambda} \cdot \underset{\sim}{R}, L)\, P(\underset{\sim}{R}, L) \tag{14}$$

are much better than experiments comparing an average like:

$$F \sim \log P(\underset{\sim}{\lambda} \cdot \langle \underset{\sim}{R} \rangle, L) \tag{15}$$

as is widely used. Note that Equation (14) is a simplified version of Equation (7).

Further remarks

Another subject of growing interest is the formation of rubbers [18, 19] and the relation of structure obtained by formation, and elasticity. The reader is referred to the contribution of Ilavsky [30] in this volume, where a lot of material is discussed in this context. The structure-elasticity relationship has been addressed by Boué, too (see also this volume), who showed that all classical models cannot explain the neutron data. He proposed another link to a very young field is physics – the physics of self similar structures or fractals, by assuming that the network could be self similar on some length scales. He used the Sierpinski gasket as a fractal model [20] concluding that larger fluctuations can be present in self similar structures compared to the classical phantom limit of James and Guth [4]. This is of relevance if one considers only ghost networks by saying that the chains can pass through each other, but I have the feeling that these approaches are less valid in an entanglement-dominated network, where much shorter length scales matter than the chemical distance between two crosslinks [21].

On the other hand, fractal structures can appear in the formation process. Winter's contribution in this conference describes this as well as work in connection with percolation ideas [19] and growth phenomena [22]. Using the idea of polymeric fractals, the experiments of Winter can be interpreted in this direction [23]. The term polymeric fractal accounts for a self similar structure build up from polymer chains which are themselves a random fractal [24] with a fractal dimension $d_f = 2$, if they are ideal Gaussian chains.

Our last point in this section now makes the link to materials science and the technological aspects of rubbers, a subject which has not been touched very much by theory . The general idea is based mainly on reinforcement of rubbers and their use in technical products, for example car tyres, where special properties are demanded. These properties are, for example, high temperature stability, large energy dissipation without large deformations, good adhesion properties and so on. But not only mechanical properties may play a role, but electrical properties, too, using rubbers, for example, as isolating material, dielectric break down properties have to be investigated. Very recently dielectric break down phenomena have been studied in the context of diffusion limited aggregation (see R. C. Ball in [25]). This approach uses the scaling properties of growth of the electric field in space. The problem becomes more difficult if, for some reason, the rubber has to be filled with metallic-like particles. Electrical properties of filled rubbers are interesting in connection with percolation but it is still unclear which role the matrix plays. Experiments in this context are in progress [26].

During the conference, investigations on mechanical properties were reproted. The simplest forms of rubber reinforcing are bimodal networks. These networks are made of long chains and short chains (see Picot's paper within this proceedings). It is assumed that the short chains are built-in randomly among the long ones, but Mark has reported cases in which the short chains aggregate before endlinking [27].

The result is an inhomogeneous rubber where regions of high and low degrees of crosslinks are present [27]. It is clear that the high crosslink regions act as "soft" fillers, too. In Kilian's paper a phenomenological theory for filled "model" networks has been presented, where fair agreement with experiments was reported. Nevertheless, these types of materials are cause for further research in materials science.

Final statement

This paper raises some questions about points discussed during the meeting. Clearly there are many more problems than presented here, and some of the remarks seem too short and not sufficiently quantitative. We did not discuss, for example, stress-induced crystallization [28] in this paper. The problems occurring in polymer melts have not been touched on here and we refer the reader to the reviews of Meissner and Laun within this proceedings. We hope to have at least shown that rubber elasticity rubber theory is not a subject on its own and connections to other aspects of physics are to be expected.

Acknowledgement

Helpful discussions with F. Boué, P. Goldbart and H.-G. Kilian are gratefully acknowledged. The author thanks the editors of this volume for the opportunity to present this paper, despite the fact that it was not presented at the meeting – the author felt that there were some points to be discussed further.

References

1. Kuhn W, Grün F (1942) Kolloid Z 101:248
2. Treloar LRG (ed) (1975) The Physics of Rubber Elasticity, Clarendon Press, Oxford
3. Flory PI (ed) (1959) Principles in Polymer Chemistry, Cornell University Press, Ithaca
4. James HM, Guth E (1943) J Chem Phys 11:455
5. Edwards SF (1971) In: Chrompft A, Newman S (eds) Polymer Networks, Plenum Press, New York
6. Deam RT, Edwards SF (1976) Phil Trans R Soc 11:317
7. Ball RC, Edwards SF (1980) Macromolecules 13:748
8. Edwards SF, Goodyear AJ (1972) J Phys A5:965
9. Edwards SF, Anderson PW (1975) J Phys F7:965
10. Binder K, Young AP, Rev Mod Phys
11. Sherrington D, Kirkpatrick S (1975) Phys Rev Lett 35:1972
12. de Almeida JRO, Thouless D (1978) J Phys A11:983
13. Parisi G (1983) Phys Rev Lett 50:1946
14. van Hemmen L, Morgenstern I (eds) (1983) Heidelberg Colloquium on Spin Glasses, Springer Verlag, Heidelberg
15. Mézard M, Parisi G (1986) J Phys, Paris 47:1285
16. Hopfield JJ (1982) Proc Natl Acad Sci USA 79:2554
17. Menduiña C, Freire JJ, Llorente MA, Vilgis TA (1986) Macromolecules 19:1212
18. Winter HH, this issue [23]
19. Stauffer D, Adam M, Delsanti M (1982) Adv Polym Sci 78:103
20. Mandelbrot BB (ed) (1982) The Fractal Geometry of Nature, Freeman, San Francisco
21. Vilgis TA (1987) this volume
22. Herrmann HJ (1986) Phys Rep 136:153
23. Vilgis TA, Winter HH (1987) to be published
24. de Gennes PG (ed) (1979) Scaling Concepts in Polymer Physics, Cornell University Press, Ithaca
25. Stanley HE, Ostrovsky N (eds) (1986) Growth and Form, D Reidel Publ, Amsterdam
26. Ezguerra TA, Mohammadi M, Kremer F, Vilgis TA, Wegner G (1987) J Phys C, Solid State, accepted
27. Mark J (1985) Acc Chem Res 18:202
28. Holl B, this issue
29. Goldbart P, Goldenfeld N (1987) Phys Rev Lett 58:2676
30. Ilavsky (1987) this issue
31. Boué F (1987) this issue

Received December 15, 1986;
accepted May 15, 1987

Author's address:

T. A. Vilgis
Max-Planck-Institut für Polymerforschung
Postfach 3148
D-6500 Mainz, F.R.G.

Author Index

Bastide, J. 152
Böhm, H. 23, 62
Boué, F. 152
Buzier, M. 152

Collette, C. 152

Deloche, B. 45
Demarmels, A. 146
Donnet, J. B. 201
Dubault, A. 45
Dušek, K. 11

Eisele, U. 231
Enderle, H. F. 55

Godovsky, Yu. K. 70
Grassl, O. 62

Havránek, A. 21
Herz, J. 45, 152

Ilavský, M. 11

Keller, A. 179
Kilian, H.-G. 55, 213, 234

Lapp, A. 152
Laun, H. M. 111, 136

Meissner, J. 146
Mergenthaler, D. 234
Müller, A. J. 179

Odell, J. A. 179
Oppermann, W. 49

Pakula, T. 171
Pechhold, W. 23, 62
Picot, C. 83
Pietralla, M. 234

Rennar, N. 49
Rigbi, Z. 1, 149

v. Soden, W. 23, 62
Stadler, R. 140

Vidal, A. 201
Vilgis, T. A. 4, 243

Weymans, G. 231
Winter, H. H. 104

Zentel, R. 239

Subject Index

birefringence 179
block copolymers 70
branching 111
– process 21

calorimetry, deformation 70
chain, dangling 21
computer simulation 171
cooperative motion 171
coupling 149
creep 149
critical phenomenon 104
crosslinking 152
crystal polymers, liquid 239

deformation, biaxial 55
– modes of 55
dense polymer system, model of 171
deuterium NMR 45
disorder, quenched 243
dynamics 152

Einstein-Smallwood effect 213
elastic properties of polymer melts 111
elastically active networks chains, concentration of 11
elasticity, rubber 49
elastomers 201, 239
elongation, multiaxial 146
elongational flow 179
– viscosity 179
energy contribution 70
entanglements 4, 49, 179
epoxy networks 11
equilibrium modulus 11
extensibility, finite 4

ferroelectric properties 239

filled networks 231
–, van der Waals theory of 213
filler 201, 234
– loaded vulcanisates 213
finite extensibility 55
fractal 104

gel point 104
glass transition 171

heat conduction 234
hydrogen bonding 140

mass distribution, molar 111
materials science 243
matrix-filler contact 234
meander model 23, 62
mechanics, statistical 243
melts, polymer 23, 111
model network 21
Mooney-Rivlin plot 49
Mullins softening 213

networks 4, 70, 213
–, epoxy 11
–, formation of 11
–, model 49
–, polymer 23, 104
–, polyurethane 11
– structure 231
–, swollen 62
–, transient 140
– theory 146
neutron scattering 4

orientation 234
–, molecular 111
orientational order 45

polydimethylsiloxane 45

polyisobutylene 146
polymer 21
– modification 140
– networks 23, 152
– melts 23, 111, 146
polystyrene 152
polyurethane 104
– networks 11

reinforcement 201, 231
relaxation 149, 152
–, mechanical 23
–, times spectrum 136
retardation spectrum 11
rheological properties of transient networks 140
rubber 45, 234
– deformation 152
– elasticity 49, 55, 243

self diffusion 171
slip link 4
sol fraction 11
shear compliance 23, 62
small angle neutron scattering 152
solution, semi-dilute 179
strain energy function 55
styrene-butadiene rubber 231
surface activity 201
surface free energy 201

thermoelasticity 4
thermomechanics 70
topological constraints 4
trapped entanglements 45

van der Waals theory 55
– of filled networks 213
viscoelasticity 21, 149
–, linear 146
vulcanisates, filler-loaded 213